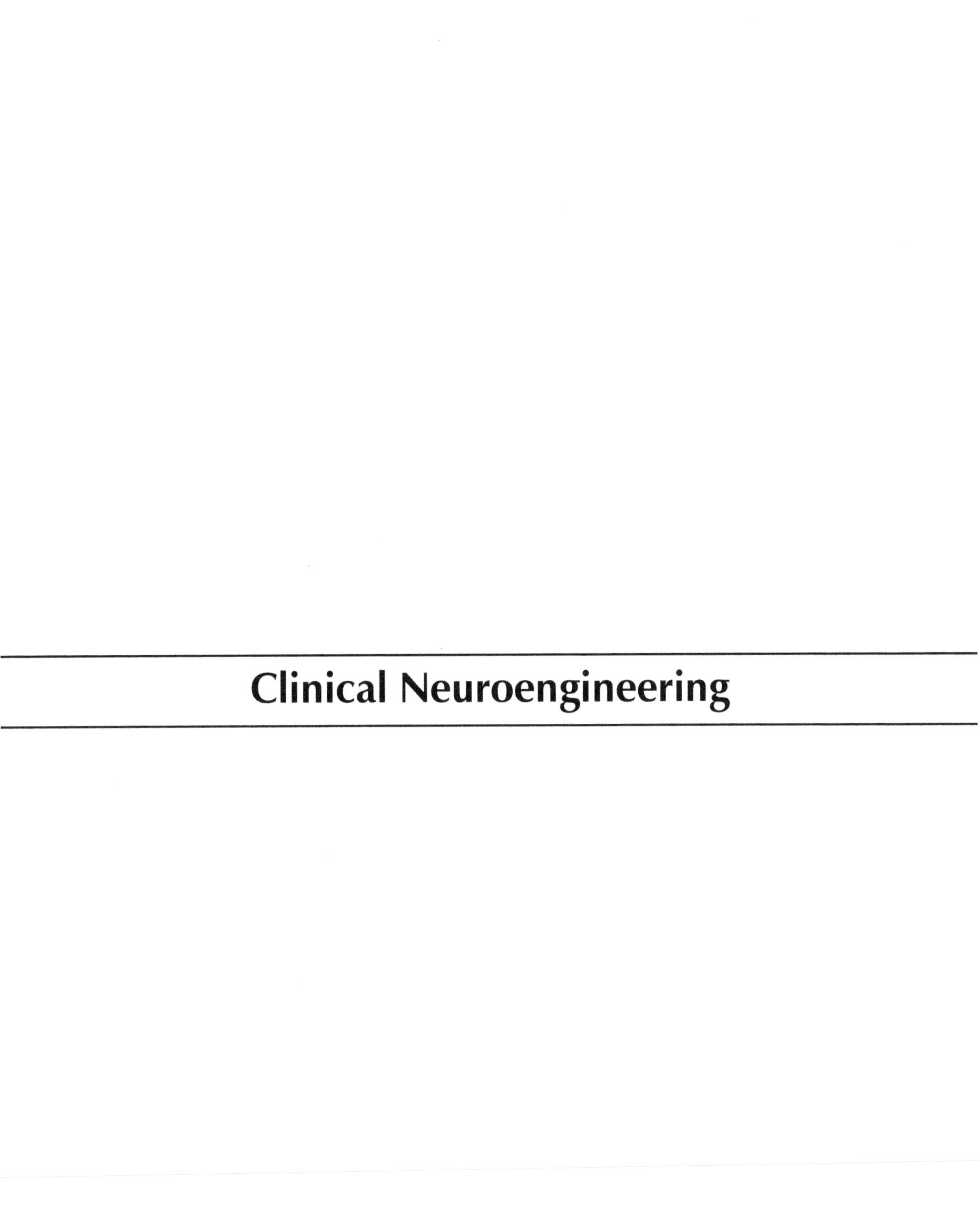

# Clinical Neuroengineering

# Clinical Neuroengineering

Editor: Evelyn Page

www.fosteracademics.com

www.fosteracademics.com

**Cataloging-in-Publication Data**

Clinical neuroengineering / edited by Evelyn Page.
p. cm.
Includes bibliographical references and index.
ISBN 978-1-63242-941-4
1. Biomedical engineering. 2. Neural networks (Neurobiology). 3. Neurology. I. Page, Evelyn.
R856 .C55 2020
610.28--dc23

© Foster Academics, 2020

Foster Academics,
118-35 Queens Blvd., Suite 400,
Forest Hills, NY 11375, USA

ISBN 978-1-63242-941-4 (Hardback)

This book contains information obtained from authentic and highly regarded sources. Copyright for all individual chapters remain with the respective authors as indicated. All chapters are published with permission under the Creative Commons Attribution License or equivalent. A wide variety of references are listed. Permission and sources are indicated; for detailed attributions, please refer to the permissions page and list of contributors. Reasonable efforts have been made to publish reliable data and information, but the authors, editors and publisher cannot assume any responsibility for the validity of all materials or the consequences of their use.

**Trademark Notice:** Registered trademark of products or corporate names are used only for explanation and identification without intent to infringe.

# Contents

**Preface** ........ VII

Chapter 1 **Exoskeleton assistance symmetry matters: unilateral assistance reduces metabolic cost, but relatively less than bilateral assistance** ........ 1
Philippe Malcolm, Samuel Galle, Pieter Van den Berghe and Dirk De Clercq

Chapter 2 **Analysis of sensory system aspects of postural stability during quiet standing in adolescent idiopathic scoliosis patients** ........ 12
Taeyong Sim, Hakje Yoo, Dongjun Lee, Seung-Woo Suh, Jae Hyuk Yang, Hyunggun Kim and Joung Hwan Mun

Chapter 3 **Decoding the grasping intention from electromyography during reaching motions** ........ 23
Iason Batzianoulis, Nili E. Krausz, Ann M. Simon, Levi Hargrove and Aude Billard

Chapter 4 **Recovery of kinematic arm function in well-performing people with subacute stroke** ........ 36
Gyrd Thrane, Margit Alt Murphy and Katharina Stibrant Sunnerhagen

Chapter 5 **Predictive value of the pendulum test for assessing knee extensor spasticity** ........ 46
Alyssa Whelan, Andrew Sexton, Melony Jones, Colleen O'Connell and Chris A. McGibbon

Chapter 6 **A robot-based gait training therapy for pediatric population with cerebral palsy: goal setting, proposal and preliminary clinical implementation** ........ 58
Cristina Bayón, Teresa Martín-Lorenzo, Beatriz Moral-Saiz, Óscar Ramírez, Álvaro Pérez-Somarriba, Sergio Lerma-Lara, Ignacio Martínez and Eduardo Rocon

Chapter 7 **Using principal component analysis to reduce complex datasets produced by robotic technology in healthy participants** ........ 73
Michael D. Wood, Leif E. R. Simmatis, J. Gordon Boyd, Stephen H. Scott and Jill A. Jacobson

Chapter 8 **Validation of wearable visual feedback for retraining foot progression angle using inertial sensors and an augmented reality headset** ........ 85
Angelos Karatsidis, Rosie E. Richards, Jason M. Konrath, Josien C. van den Noort, H. Martin Schepers, Giovanni Bellusci, Jaap Harlaar and Peter H. Veltink

Chapter 9 **Home-based transcranial direct current stimulation plus tracking training therapy in people with stroke** ........ 97
Ann Van de Winckel, James R. Carey, Teresa A. Bisson, Elsa C. Hauschildt, Christopher D. Streib and William K. Durfee

Chapter 10 **Amputee perception of prosthetic ankle stiffness during locomotion** ........ 107
Max K. Shepherd, Alejandro F. Azocar, Matthew J. Major and Elliott J. Rouse

Chapter 11 **Impact of antagonistic muscle co-contraction on in vivo knee contact forces**......117
Adam Trepczynski, Ines Kutzner, Verena Schwachmeyer, Markus O. Heller, Tilman Pfitzner and Georg N. Duda

Chapter 12 **Overground walking patterns after chronic incomplete spinal cord injury show distinct response patterns to unloading**......126
Christopher Schmidt Easthope, Luca Renato Traini, Lea Awai, Martina Franz, Georg Rauter, Armin Curt and Marc Bolliger

Chapter 13 **Identification of forearm skin zones with similar muscle activation patterns during activities of daily living**......138
Néstor J. Jarque-Bou, Margarita Vergara, Joaquín L. Sancho-Bru, Alba Roda-Sales and Verónica Gracia-Ibáñez

Chapter 14 **Assessment of a markerless motion analysis system for manual wheelchair application**......149
Jacob Rammer, Brooke Slavens, Joseph Krzak, Jack Winters, Susan Riedel and Gerald Harris

Chapter 15 **Speed-adaptive control of functional electrical stimulation for dropfoot correction**......161
Guangtao Chen, Le Ma, Rong Song, Le Li, Xiaoyun Wang and Kaiyu Tong

Chapter 16 **Manual wheelchair downhill stability: an analysis of factors affecting tip probability**......172
Louise Thomas, Jaimie Borisoff and Carolyn J. Sparrey

Chapter 17 **Kinect-based assessment of proximal arm non-use after a stroke**......184
K. K. A. Bakhti, I. Laffont, M. Muthalib, J. Froger and D. Mottet

Chapter 18 **Effects of a powered ankle-foot orthosis on perturbed standing balance**......196
Amber R. Emmens, Edwin H. F. van Asseldonk and Herman van der Kooij

Chapter 19 **Mechanisms of electrical vasoconstriction**......209
Mark Brinton, Yossi Mandel, Ira Schachar and Daniel Palanker

**Permissions**

**List of Contributors**

**Index**

# Preface

Every book is initially just a concept; it takes months of research and hard work to give it the final shape in which the readers receive it. In its early stages, this book also went through rigorous reviewing. The notable contributions made by experts from across the globe were first molded into patterned chapters and then arranged in a sensibly sequential manner to bring out the best results.

Neuroengineering or neural engineering is a field within biomedical engineering. It uses engineering techniques to repair, enhance, replace, understand or exploit the properties of neural systems. An important objective of this field is the augmentation and restoration of human function through direct interactions between artificial devices and the nervous system. Research is being conducted to develop an understanding of the coding and processing of information in the motor and sensory systems, how such processing is altered in the pathological state and how this can be manipulated using interactions with neuroprosthetics and brain-computer interfaces. The scope of neuroengineering is wide with applications in repair and rehabilitation, and neuromodulation. Neuroengineering and rehabilitation applies the principles of engineering and neuroscience for investigating the peripheral and central nervous system function. This can help in developing solutions for problems related to brain damage and brain malfunction. This book contains some path-breaking studies in neuroengineering and rehabilitation. It will also provide interesting topics for research which interested readers can take up. With state-of-the-art inputs by acclaimed experts of this field, this book targets students and professionals.

It has been my immense pleasure to be a part of this project and to contribute my years of learning in such a meaningful form. I would like to take this opportunity to thank all the people who have been associated with the completion of this book at any step.

**Editor**

# 1

# Exoskeleton assistance symmetry matters: unilateral assistance reduces metabolic cost, but relatively less than bilateral assistance

Philippe Malcolm[1,2*], Samuel Galle[2], Pieter Van den Berghe[2] and Dirk De Clercq[2]

**Abstract**

**Background:** Many gait impairments are characterized by asymmetry and result in reduced mobility. Exoskeletons could be useful for restoring gait symmetry by assisting only one leg. However, we still have limited understanding of the effects of unilateral exoskeleton assistance. Our aim was to compare the effects of unilateral and bilateral assistance using a within-subject study design.

**Methods:** Eleven participants walked in different exoskeleton conditions. In the *Unilateral* conditions, only one leg was assisted. In *Bilateral Matched Total Work*, half of the assistance from the *Unilateral* conditions was applied to both legs such that the bilateral sum was equal to that of the *Unilateral* conditions. In *Bilateral Matched Work Per Leg*, the same assistance as in the *Unilateral* conditions was provided to both legs such that the bilateral sum was the double of that of the *Unilateral* conditions. In the *Powered-Off* condition, no assistance was provided. We measured metabolic energy consumption, exoskeleton mechanics and kinematics.

**Results:** On average, the *Unilateral*, *Bilateral Matched Total Work* and *Bilateral Matched Work Per Leg* conditions reduced the metabolic rate by 7, 11 and 15%, respectively, compared with the *Powered-Off* condition. A possible explanation for why the *Unilateral* conditions effectively reduced the metabolic rate could be that they caused only very little asymmetry in gait biomechanics, except at the ankle and in the horizontal center-of-mass velocity. We found the highest ratio of metabolic rate reduction versus positive work assistance with bilateral assistance and low work per leg (*Bilateral Matched Total Work*). Statistical analysis indicated that assistance symmetry and assistance per leg are more important than the bilateral summed assistance for reducing the metabolic rate of walking.

**Conclusions:** These data bridge the gap between conclusions from studies with unilateral and bilateral exoskeletons and inform how unilateral assistance can be used to influence gait parameters, such as center-of-mass velocity.

**Keywords:** Exoskeleton, Symmetry, Asymmetry, Unilateral, Bilateral, Metabolic, Magnitude, Work, Ankle, Walking

## Background

In healthy people, walking is generally a symmetrical movement. Studies often assume full gait symmetry for simplifying data processing by analyzing only one side of the body [1]. While some studies report asymmetry even in healthy walking [2, 3], gait asymmetry is mainly a concern in various gait impairments. Unilateral amputees have slower forward speeds and longer stance durations during an intact leg stance compared with a prosthetic leg stance [4]. Hemiparesis patients often display asymmetry in step length [5–7] and gait mechanics [8, 9], and the elderly often show asymmetry in trunk acceleration [10]. Restricted mobility due to wearing an orthosis can also lead to gait asymmetry [11]. In many rehabilitation programs, assessing gait symmetry is an important aspect of evaluating rehabilitation effectiveness [12].

* Correspondence: pmalcolm@unomaha.edu
[1]Department of Biomechanics and Center for Research in Human Movement Variability, University of Nebraska Omaha, Omaha, NE 68182, USA
[2]Department of Movement and Sports Sciences, Ghent University, B-9000 Ghent, Belgium

Many devices have been proposed to reduce gait asymmetry. There are indications that ankle-foot orthoses can improve gait symmetry in stroke patients [13]. Powered prostheses have been developed and have been shown to renormalize the gait of unilateral amputees [14]. Training methods such as split-belt treadmill walking [15, 16] or walking with ankle weights [17] have been developed to help stroke survivors. Another technology that is rapidly evolving is the use of robotic exoskeletons [18]. In hemiparetic stroke patients, the ankle of the paretic side is thought to be primarily responsible for compensatory movements [19], which is why recent attempts to assist the gait of stroke patients have focused on assisting that side [20, 21]. In contrast with prostheses, most applications for exoskeletons focus on objectives other than gait symmetry (e.g., metabolic rate reduction [22–29], gait retraining [30, 31] and performance improvement [32]). Compared with such methods as split-belt treadmills or ankle weights, exoskeletons are an interesting new approach for restoring symmetry because they allow assistance with a specific timing [23–26] and magnitude [24, 28, 33] at specific joints.

However, there is still an incomplete understanding of how the effects of unilateral assistance at the ankle propagate through the rest of the body. Ankle exoskeletons are known to also indirectly affect the rest of the body, e.g., by providing hip assistance [24, 34–36] or center-of-mass rebound assistance [24, 33]. There have been multiple studies with unilateral [20, 21, 33, 37–40] and bilateral ankle exoskeletons [22, 24–26, 28, 29, 34, 36], but no studies have reported a within-subject comparison of unilateral and bilateral assistance. Thus, there is a lack of knowledge of whether different results between experiments are due to differences in exoskeleton hardware or controls. Understanding the differences between studies with unilateral and bilateral exoskeletons could improve this overall understanding and benefit applications of unilateral and bilateral exoskeletons.

Therefore, our aim was to compare the effects of unilateral and bilateral ankle exoskeleton assistance. Other experimental paradigms, such as a unilateral plantarflexion restriction [11] and an imposed asymmetric step frequency [41], have shown that gait asymmetry leads to increases in both the joint work and the metabolic rate. Based on these prior findings, we expected that unilateral exoskeleton assistance will still reduce the metabolic rate. However, we also expected that unilateral assistance will lead to asymmetry, mostly at the ankle, and will in turn lead to relatively smaller reductions in metabolic cost than bilateral assistance.

## Methods

### Participants

Thirteen healthy male volunteers participated. We retained data from eleven participants (21.3 ± 0.1 y, 74 ± 3 kg, 182 ± 2 cm, mean ± standard error). Data from two participants were excluded due to exoskeleton malfunctions. All participants provided informed consent. The protocol was approved by the ethical committee of the Ghent University Hospital.

### Exoskeleton hardware

We used bilateral exoskeletons similar to those we used previously ([26], Fig. 1). The exoskeletons consisted of a thermoplastic shell with a hinge joint at the ankle and pneumatic muscles that were 27 cm long and 3 cm in diameter. We adjusted the attachments of the pneumatic muscles such that they allowed 15° dorsiflexion at a low passive stretch force between 35 to 40 N. We mounted load cells (210 Series, Richmond Industries Ltd., Reading, UK) in series with the pneumatic muscles and a linear displacement sensor (SLS130, Penny&Giles, Christchurch, UK) between the foot and shank to monitor pneumatic muscle force, ankle angular velocity and moment arm, similar to [24]. The exoskeletons were worn with running shoes with foot switches mounted underneath the heels (Multimec 5E/5G, MEC, Ballerup, Denmark) to detect foot contact. The exoskeleton and sensors weighed 0.92 kg per side. The pneumatic muscles were connected via hoses to a valve station (CPE24, Festo, Esslingen am Neckar, Germany).

### Conditions

Participants walked on a treadmill at 1.25 $ms^{-1}$ under different conditions. To isolate the effects of assistance asymmetry, we tested four assistance conditions with and without asymmetric assistance in which either the amount of assistance per leg was matched or the bilateral sum of the assistance was matched. In two *Unilateral* conditions, the participants wore both exoskeletons but only one leg was assisted. In one of these, the dominant leg was assisted, while in the other, the non-dominant leg was assisted; these conditions were called the *Unilateral Dominant* and *Unilateral Non-Dominant* conditions, respectively. Leg dominance was determined by asking for the preferred take-off leg for jumping. In the *Bilateral Matched Total Work* condition, half of the rate of work from the *Unilateral* conditions was applied to each leg such that the bilateral sum was equal to that of the *Unilateral* conditions. In the *Bilateral Matched Work Per Leg* condition, the same rate of work per leg as in the assisted leg of the *Unilateral* condition was provided to both legs such that the bilateral sum was double of that of the *Unilateral* conditions (Fig. 2a). In the *Powered-Off* condition, participants walked with exoskeletons on both legs but with no assistance.

### Exoskeleton control

A control program (LabVIEW, National Instruments, Austin, TX, USA) permitted the specification of the timing

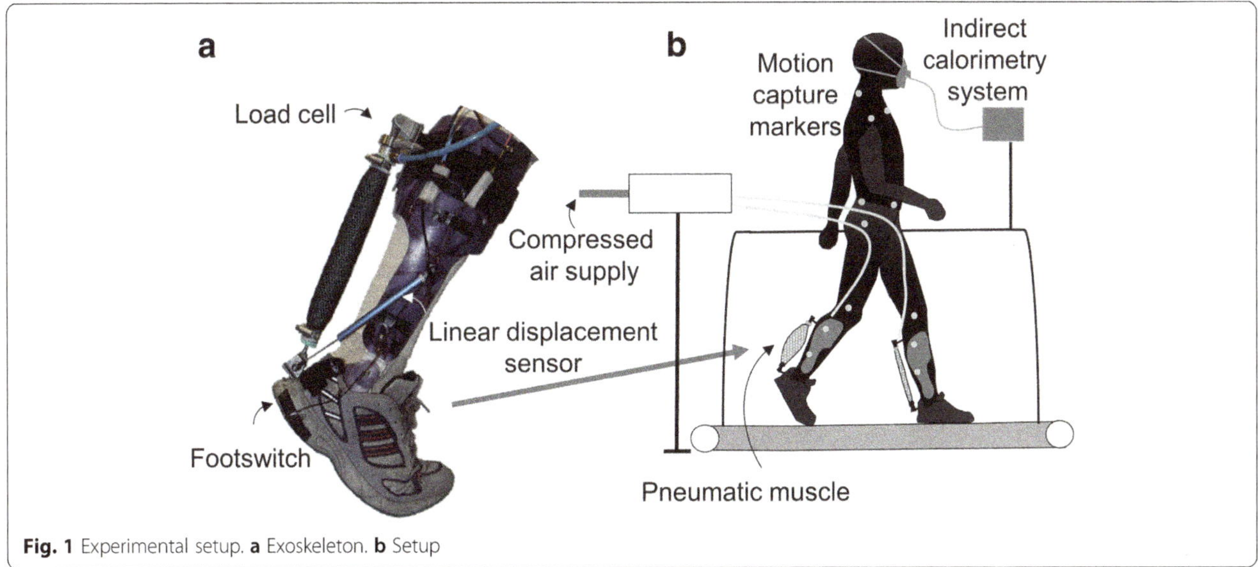


**Fig. 1** Experimental setup. **a** Exoskeleton. **b** Setup

and the rate of positive work. For real-time control, exoskeleton power was calculated by multiplying the exoskeleton moment by the angular velocity of the ankle. The exoskeleton moment and angular velocity of the ankle were determined based on the displacement sensors and load cells. The rate of positive work assistance was calculated by averaging the positive exoskeleton power over a moving window of 10 strides. A learning algorithm similar to that used by [24, 42, 43] adjusted the supplied air pressure such that the desired rate of positive work assistance was maintained. In the *Unilateral* conditions and the *Bilateral Matched Work Per Leg* condition, we set the controller to provide approximately 0.13 W $kg^{-1}$ per assisted leg. In the *Bilateral Matched Total Work* condition, we set the controller to provide approximately 0.13 W $kg^{-1}$ for the sum of both legs. These chosen assistance levels were below the level where we found that metabolic rate stops decreasing further with additional assistance in a study with the same device [24]. The control program also triggered the start and end of the actuation at desired percentages of the stride based on the previous stride. The pneumatic muscles were triggered to assist from 44 to 62%. In two previous studies with the same exoskeleton we used an end timing of 62-63% and determined that an onset timing of 42-43% is optimal in combination with different assistance magnitudes [24, 26], therefore the used settings were close to the optimal actuation pattern for the current device.

### Protocol

Before the data collection the participants walked for about 5 min in each exoskeleton condition. The order of the conditions was randomized and the entire habituation lasted a total duration of 20 min based on an earlier study with the same exoskeleton that indicates that participants need on average about 20 min to adapt to exoskeleton walking [44]. During the actual experiment, participants walked for 4 min in each condition alternating with 2 min of rest. In a supplementary analysis in a study with the same exoskeleton we found that participants adapt quickly enough to changes in conditions such that metabolic rate reaches steady state after about 2 min [24]. The *Powered--Off* condition was conducted twice, and we calculated the average to improve our estimate of the metabolic baseline. The conditions were randomized. Due to hardware malfunctions and scheduling constraints, we did not complete 4 out of the 66 trials and treated these as missing values.

### Data collection

We measured breath-by-breath $O_2$ consumption and $CO_2$ production via indirect calorimetry (COSMED, K4b2, Rome, Italy). We estimated the metabolic rate of the last 2 min of each condition based on the protocol described in [45]. We calculated reductions in the metabolic rate by subtracting the metabolic rate from *Powered-Off.* We recorded kinematics at a rate of 200 frames per s for 10 s per condition with motion capture (Pro Reflex, Qualisys AB, Gothenburg, Sweden). Marker positions and pneumatic muscle force data were filtered with a Butterworth low-pass filter with a 12 Hz cut-off. We calculated sagittal joint angles using Visual3D (C-Motion, Germantown MD, USA). We also calculated exoskeleton moment by multiplying exoskeleton force from the load cell by the moment arm calculated with motion capture from markers on the pneumatic muscle attachments and the ankle joint. Next, we calculated exoskeleton power by multiplying the ankle angular velocity by the exoskeleton moment, and we calculated the rate of positive work assistance by integrating the positive power over time and dividing by the stride cycle duration. We estimated the

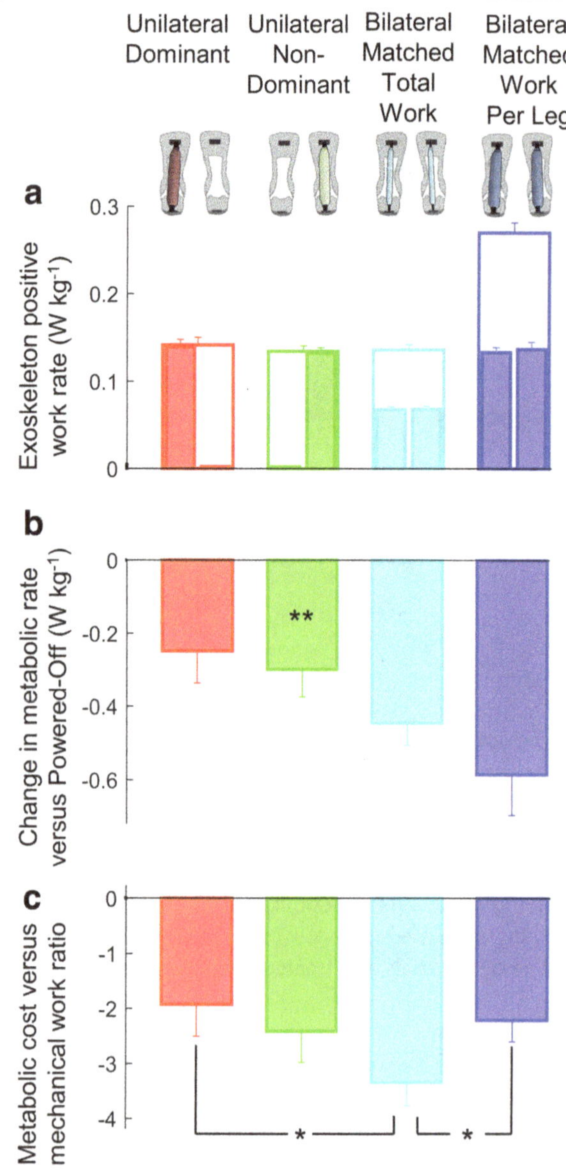


**Fig. 2** Exoskeleton and human energetics. **a** Exoskeleton positive work rate. Narrow filled bar plots represent the work rate per leg from the leg that is aligned on top of the figure. Empty wide bar plots represent the bilateral sum. **b** Change in metabolic rate versus the *Powered-Off* condition. **c** Metabolic cost versus mechanical work ratio. Colors represent the *Unilateral*, *Bilateral* and *Powered-Off* conditions indicated by figures on top of the chart columns. Error bars represent the standard error. Asterisks on top of the bar plots in panel b represent significant differences of *Unilateral* conditions versus the *Powered-Off* condition. Brackets indicate significant differences between powered conditions. Only the pairwise comparisons that are relevant for research questions 1 to 4 listed in the methods are analyzed (e.g., we did not compare the *Bilateral* conditions versus the *Powered-Off* condition since that was not relevant for the scope of this study). * is $p \leq 0.05$, ** is $p \leq 0.01$

center-of-mass position and velocity based on the pelvis markers [46]. We used a kinematic gait event detection algorithm to determine the timing of heel strike and toe off [47]. We calculated step length based on the gait event detection and treadmill speed. All timeseries data were stride-normalized based on the detected heel strike timings.

### Statistics

We used paired t-tests for specific comparisons of interest regarding mechanical, energetic and kinematic parameters. We compared (1) (the assisted leg in) the *Unilateral* conditions versus (the same leg in) the *Bilateral Matched Work Per Leg* condition to evaluate the direct effects of unilateral assistance; (2) (the unassisted leg in) the *Unilateral* conditions versus (the same leg in) the *Powered-Off* condition to evaluate the indirect effects on the unassisted leg; (3) the total effect on both legs in the *Unilateral* conditions versus the *Bilateral Matched Total Work* condition to evaluate the difference between focusing assistance on one leg versus distributing assistance over both legs; and (4) the total effect of the *Bilateral Matched Total Work* condition versus the *Bilateral Matched Work Per Leg* condition to assess the effects of assistance magnitude versus assistance symmetry. To compare the effects of assistance work rate asymmetry, the assistance work rate per leg and the total assistance work rate for both legs on the metabolic rate, we used a mixed-model ANOVA. We reported all descriptive results as the mean ± standard error.

## Results

### Exoskeleton mechanics

On average, in the *Unilateral* conditions, the exoskeleton provided $0.136 \pm 0.008$ W kg$^{-1}$ for the assisted leg. In the *Bilateral Matched Work Per Leg* condition, the exoskeletons provided $0.133 \pm 0.006$ W kg$^{-1}$ per leg (Figs. 2a and 3I-L). There was no significant difference in the exoskeleton work rate per assisted leg between the *Unilateral* and *Bilateral Matched Work Per Leg* conditions (*p*-value *Unilateral Dominant* versus *Bilateral Matched Work Per Leg* 0.403, *Unilateral Non-Dominant* versus *Bilateral Matched Work Per Leg* 0.276), indicating that we successfully matched the work rate per leg. In the *Bilateral Matched Total Work* condition, the exoskeletons provided $0.136 \pm 0.007$ W kg$^{-1}$ for the sum of both legs. There was no significant difference in the bilateral sum of the exoskeleton work rate for both legs between the *Unilateral* and *Bilateral Matched Total Work* conditions (*p*-value *Unilateral Dominant* versus *Bilateral Matched Total Work* 0.385, *Unilateral Non-Dominant* versus *Bilateral Matched Total Work* 0.682), indicating that we successfully matched the total work rate. The stride-average of exoskeleton moment for the assisted leg in the *Unilateral* conditions was 0.055 $\pm 0.004$ Nm kg$^{-1}$ (Fig. 3a, b). The stride-average of exoskeleton moment in the *Bilateral Matched Work Per Leg* condition was $0.056 \pm 0.004$ Nm kg$^{-1}$. There was no significant difference in the stride-average of exoskeleton moment per assisted leg between the *Unilateral* and *Bilateral*

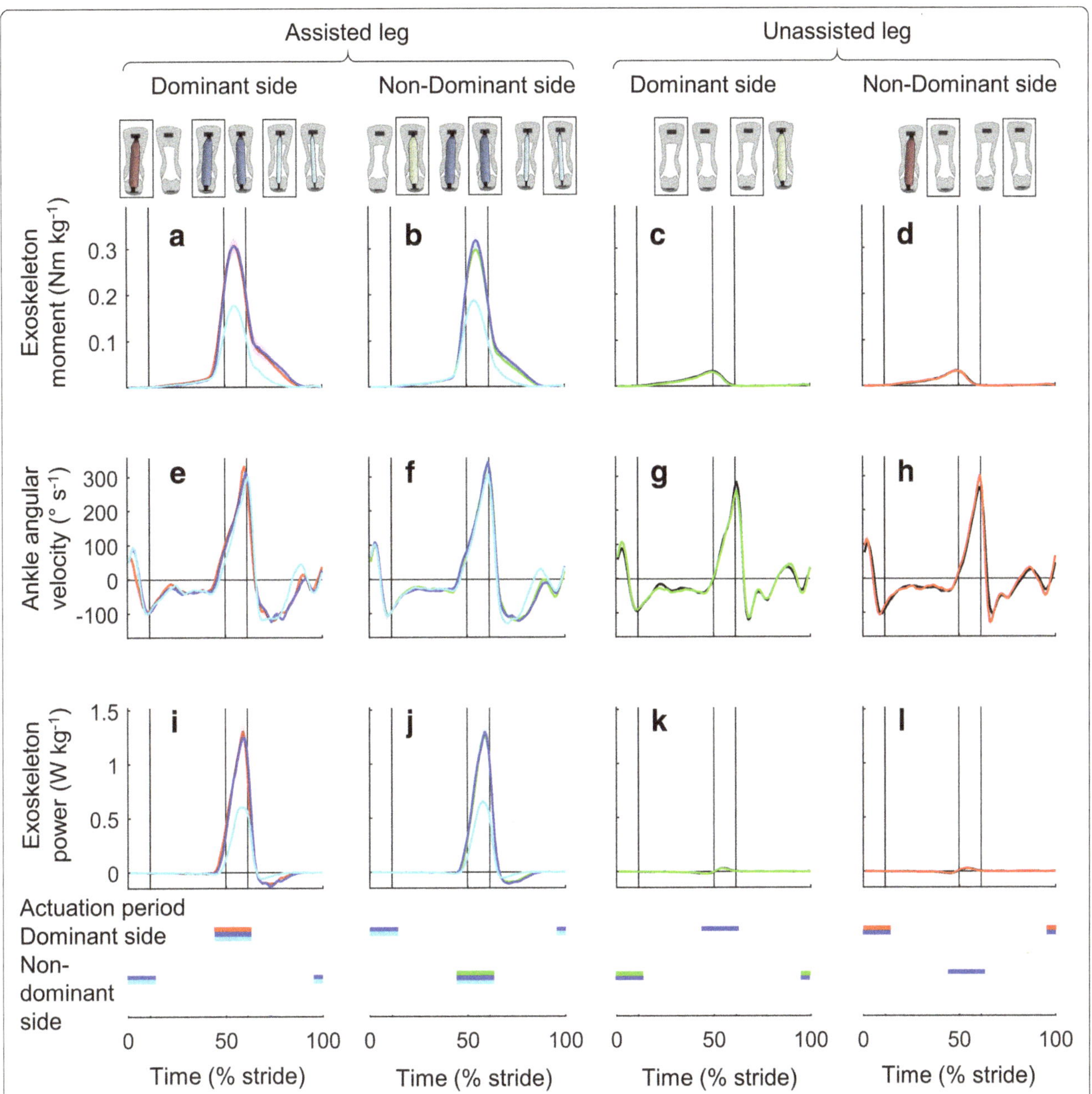


**Fig. 3** Actuation conditions. **a-d** Exoskeleton moment. **e-h** Ankle angular velocity. **i-l** Exoskeleton power. Left two columns show assisted leg. Right two columns show unassisted leg. Colored lines represent the *Unilateral*, *Bilateral* and *Powered-Off* conditions marked by rectangles in figures on top of the chart columns. Shaded area represents the standard error. Vertical lines show beginning and ending of single and double stance phases. Horizontal colored bars indicate the actuation period in conditions with corresponding colors

*Matched Work Per Leg* conditions (*p*-value *Unilateral Dominant* versus *Bilateral Matched Work Per Leg* 0.106, *Unilateral Non-Dominant* versus *Bilateral Matched Work Per Leg* 0.125). The stride-average of exoskeleton moment in the *Bilateral Matched Total Work* condition was $0.032 \pm 0.002$ Nm kg$^{-1}$.

### Metabolic rate

Exoskeleton assistance reduced the metabolic rate in all conditions compared with that of the *Powered-Off* condition. Changes in metabolic rate compared with that of the *Powered-Off* condition were $-6.4 \pm 3.3\%$ for *Unilateral Dominant*, $-7.9 \pm 2.2\%$ for *Unilateral Non-Dominant*, $-11.3 \pm 1.5\%$ for *Bilateral Matched Total Work*, and $-14.9 \pm 2.5\%$ for *Bilateral Matched Work Per Leg* (Fig. 2b). The *Unilateral Non-Dominant* condition significantly reduced metabolic rate versus *Powered-Off* (*p*-value 0.003). The *Unilateral Dominant* condition trended towards reducing metabolic cost but this was not significant (*p*-value 0.066). We found the highest reduction in the *Bilateral Matched Work Per Leg* condition. The highest ratio of metabolic rate reduction versus positive work rate assistance was found in the *Bilateral Matched Total Work* condition (− 3.340 ± 0.416 W per W, Fig. 2c). This ratio was higher than that of the *Unilateral Dominant* condition (− 1.927 ± 0.722, *p*-value 0.048) and that of the *Bilateral Matched Work Per Leg* condition (*p*-value 0.020). Using a mixed-model ANOVA, we found that the change in metabolic rate is related to the work rate per assisted leg and the work rate difference between both legs according to the following formula:

$$\begin{aligned} &\textit{Change in metabolic rate}\left(W\ kg^{-1}\right) = \\ &-4.26 \cdot \textit{Work rate per assisted leg}\left(W\ kg^{-1}\right) \\ &+2.75 \cdot \textit{Absolute work rate difference}\left(W\ kg^{-1}\right) \end{aligned}$$

(*p*-value Work rate per assisted leg is $3 \cdot 10^{-7}$, *p*-value work rate difference is $3 \cdot 10^{-4}$, Additional file 1).

### Spatiotemporal results

In the *Unilateral Non-Dominant* condition, the heel contact of the unassisted leg occurred 0.5 ± 0.2% later than in the *Powered-Off* condition (*p*-value 0.013) and 0.6 ± 0.2% later than in the *Bilateral Matched Work Per Leg* condition (*p*-value 0.014, Fig. 4). On average, in the *Unilateral Dominant* condition, the heel contact of the unassisted leg also occurred 0.4 ± 0.3% later than in the *Powered-Off* condition, but this difference was not significant (*p*-value 0.250). No significant differences were found in the step frequency or duty factor among the *Unilateral*, *Bilateral* and *Powered-Off* conditions.

### Joint kinematics

There were no significant differences in peak plantarflexion between the individual legs when comparing the same leg between the *Unilateral* and *Bilateral Matched Work Per Leg* conditions. However, the change in peak plantarflexion in the *Bilateral Matched Total Work* condition trended to be larger than the average of the assisted and unassisted legs in the *Unilateral* conditions (*p*-value *Unilateral Non-Dominant* versus *Bilateral Matched Total Work* $10^{-4}$, *Unilateral Dominant* versus *Bilateral Matched Total Work* $0.063^{-4}$, Fig. 5m). There were no other relevant significant differences in the evaluated peak joint angles of the ankle, knee or hip among the *Unilateral*, *Bilateral* and *Powered-Off* conditions (Fig. 5a-l).

### Center-of-mass kinematics

The minimum horizontal center-of-mass velocity before the heel strike of the assisted leg was lower in the *Unilateral* conditions than in the *Powered-Off* condition (*p*-value *Unilateral Dominant* versus *Powered-Off* 0.002, *Unilateral Non-Dominant* versus *Powered-Off* 0.022, Fig. 6). In the *Unilateral Non-Dominant* condition, the minimum horizontal velocity after the heel strike of the assisted leg was also lower than that in the *Bilateral Matched Work Per Leg* condition (*p*-value 0.017).

## Discussion

The aim of our study was to compare the effects of unilateral and bilateral exoskeleton assistance in a within-subject design. On average the *Unilateral* conditions reduced the

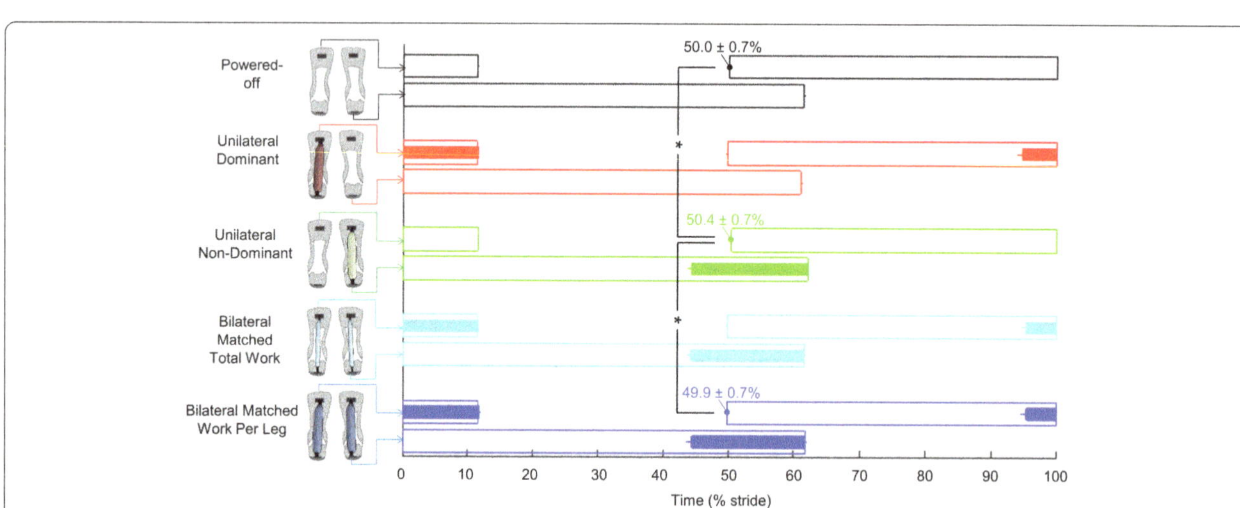


**Fig. 4** Spatiotemporal results. Horizontal axis indicates right-side stride cycle percentage. Empty bars represent the stance phase duration. Filled bars represent the assistance phase. Numbers indicate timing of opposite leg heel contact of conditions that show significant differences. Colors represent the *Unilateral*, *Bilateral* and *Powered-Off* conditions, and each horizontal line corresponds to the leg that is shown in the figures on the left. Error bars represent the standard error. Brackets represent significant differences between conditions. Only the pairwise comparisons that are relevant for the research questions listed in the methods are analyzed. * is $p \leq 0.05$

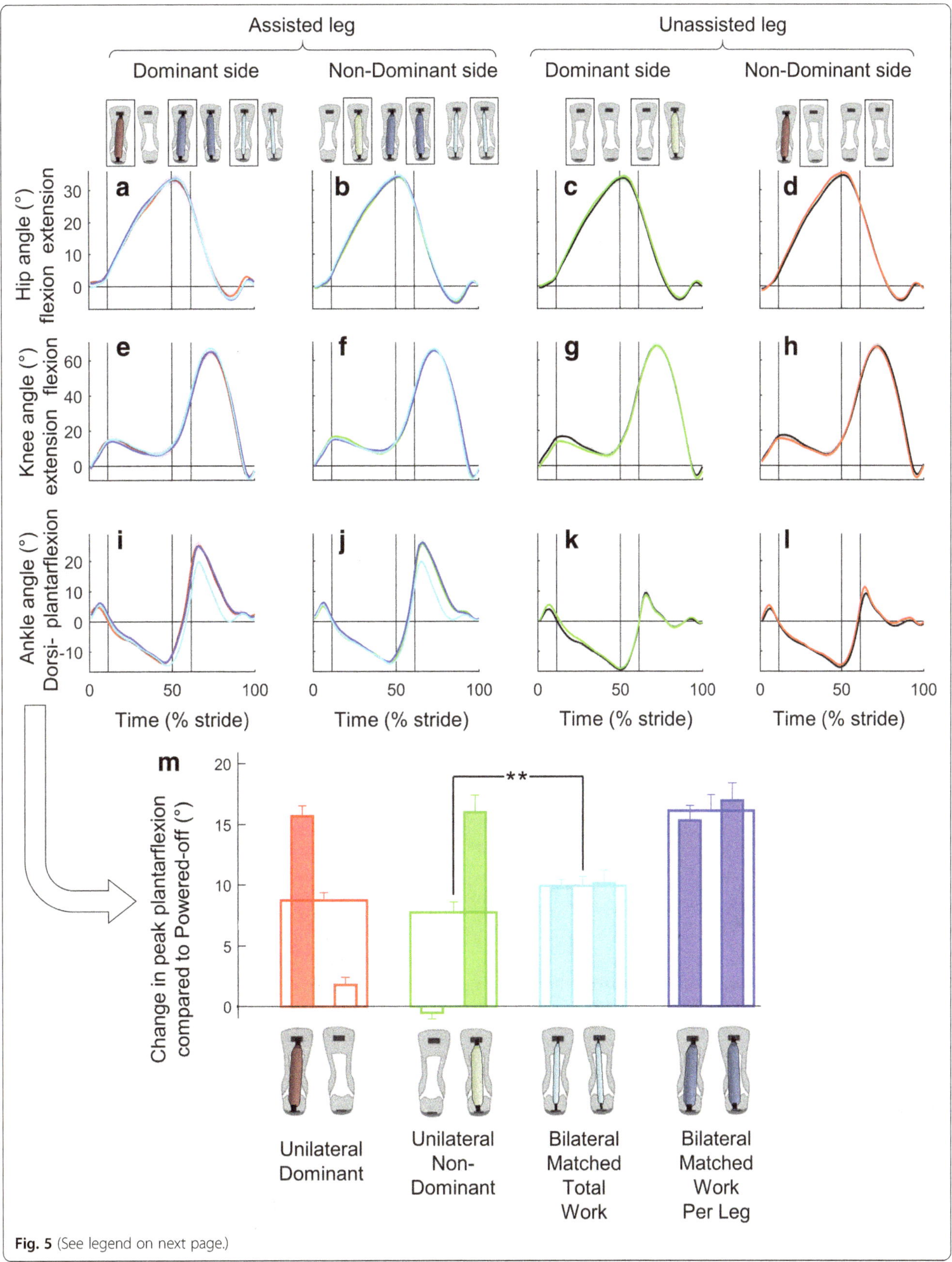


**Fig. 5** (See legend on next page.)

(See figure on previous page.)

**Fig. 5** Joint kinematics. **a-d** Hip. **e-h** Knee. **i-l** Ankle. Left two columns show assisted leg. Right two columns show unassisted leg. Colored lines are from the *Unilateral*, *Bilateral* and *Powered-Off* conditions marked by rectangles in figures on top of the chart columns. Shaded area represents the standard error. Vertical lines show beginning and ending of single and double stance phases. **m** Change in peak plantarflexion compared with *Powered-Off*. Narrow filled bar plots represent peak plantarflexion per leg from the leg that is aligned in the figure below. Empty wide bar plots represent the bilateral mean. Error bars represent the standard error. Brackets represent significant differences between powered conditions. Only the pairwise comparisons that are relevant for the research questions listed in the methods are analyzed. ** is $p \leq 0.01$

metabolic rate by 7% compared with the *Powered-Off* condition (Fig. 2). As expected, the highest absolute reduction was found in the bilateral condition with the highest assistance (*Bilateral Matched Work Per Leg*). However, we found the highest ratio of metabolic cost reduction versus positive work in the bilateral condition with low assistance per leg (*Bilateral Matched Total Work*), which indicates that a given amount of exoskeleton assistance is the most efficiently used when it is evenly distributed over both legs.

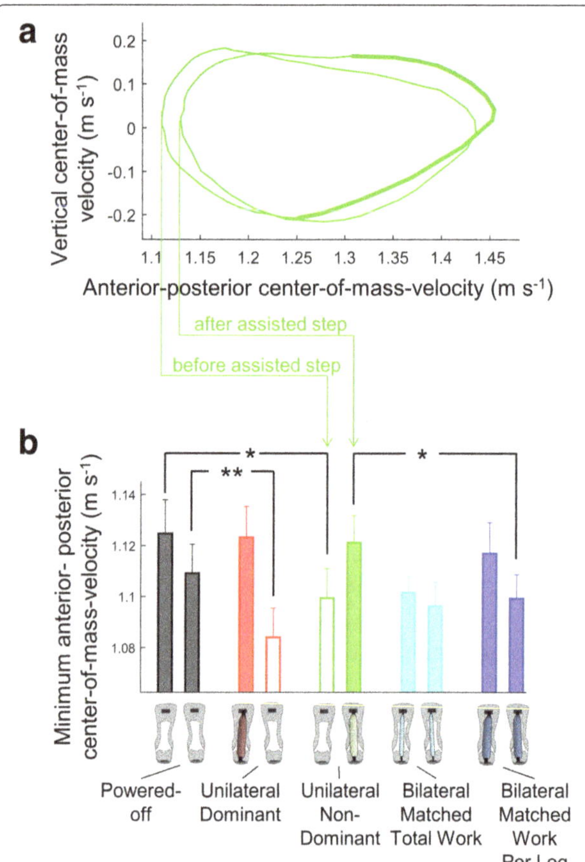


**Fig. 6** Center-of-mass kinematics. **a** Vertical center-of-mass velocity plotted versus horizontal center-of-mass velocity from the *Unilateral Non-Dominant* condition. Wider part of the line represents the assisted portion of the center-of-mass trajectory. **b** Minimum anterior-posterior center-of-mass velocity. Narrow empty bar plots represent the minimum velocity during the unassisted leg stance. Narrow filled bar plots represent the minimum velocity during the assisted leg stance. Empty wide bar plots represent the bilateral mean. Error bars represent the standard error. Brackets represent significant differences between powered conditions. Only the pairwise comparisons that are relevant for the research questions listed in the methods are analyzed. * is $p \leq 0.05$, ** is $p \leq 0.01$

Based on the literature, it was uncertain whether unilateral assistance would reduce the metabolic rate. The trends from simulations [4], measurements in populations with push-off asymmetry [48, 49] and experiments that evoke gait asymmetry in healthy participants [11, 41] indicate that push-off asymmetry can lead to energy losses and increased metabolic rates. An exoskeleton study with unilateral plantarflexion assistance had conditions that reduced the metabolic rate but also conditions that increased the metabolic rate by up to 22% compared with the *Powered-Off* condition [50]. In our experiment, the *Unilateral* conditions reduced the metabolic rate with a ratio of 2.2 W per W positive work rate. This ratio is close to the efficiency ratio of 2.5 found in the other unilateral exoskeleton study [50] and falls within the reported range for bilateral exoskeletons (1.6 [51] to 4.7 [26]).

A possible explanation for why the *Unilateral* conditions did not increase the metabolic rate compared with the *Powered-Off* condition could be that they caused only slight gait asymmetry (Additional file 2). The *Unilateral* conditions did not cause significant asymmetry in step frequency (Fig. 4), which could otherwise have led to an increased metabolic rate [41]. We only observed effects of assistance asymmetry in joint kinematics at the ankle (Fig. 5). Similarly, Wutzke et al. [11] found no significant kinematic asymmetry in any joint except the ankle during walking with unilateral ankle restriction. It seems that healthy participants can retain normal kinematics despite strong perturbations. We also did not find indications of increased spatiotemporal variability or increased imbalance in the *Unilateral* conditions (Additional file 3). This echoes findings from split-belt walking studies showing that healthy participants can walk comfortably even with large differences in belt speed [52, 53].

We observed asymmetry in the timing of opposite heel contact, but this asymmetry was less than 1%. The slower forward center-of-mass velocity during the assisted leg stance and faster velocity during the unassisted leg stance (Fig. 6) correspond to findings from simulations and experiments performed by Adamczyk and Kuo [4] with unilateral amputees. It is possible that this adaptation allows participants to save energy by allowing the exoskeleton to accelerate the center off mass during the assisted step and by saving effort during the unassisted step. A follow-up analysis of joint kinetics and EMG data would allow to interpret where the metabolic savings could come from and more specifically if some of the metabolic savings come from the unassisted leg in addition to the assisted leg.

We found the highest metabolic rate versus mechanical work ratio in the *Bilateral Matched Total Work* condition. The metabolic cost reduction resulting from assisting both legs with low assistance was higher than the average metabolic cost reduction resulting from the *Unilateral* conditions. We found a similar "bilateral surplus" effect (i.e., the opposite of "bilateral deficit" as in [54]) in peak plantarflexion. More specifically, we found that the increase in peak plantarflexion in the *Bilateral Matched Total Work* condition was higher than the mean of the assisted and unassisted leg in the *Unilateral* conditions.

The question remains why the assistance was the most efficient in the *Bilateral Matched Total Work* condition. Is it because of the low assistance asymmetry, or is it because of the assistance work rate per leg or the total assistance work rate for both legs? Using a mixed-model ANOVA with stepwise elimination, we found that the assistance work rate per assisted leg and the assistance work rate difference between both legs are more determining than the total assistance work for both legs (Additional file 1). This information could be used to bridge the gap between studies with unilateral and bilateral exoskeletons. For example, Jackson and Collins [50] found metabolic rate reductions of up to 0.16 W kg$^{-1}$, or 5%, compared with the *Powered-Off* condition with a unilateral exoskeleton that provided a rate of 0.20 W kg$^{-1}$ positive work. This value is similar to our reduction of 7% found for the *Unilateral* conditions. Based on the assistance work asymmetry coefficient from the mixed-model ANOVA, we can estimate that the optimal condition in the study by Jackson and Collins could have resulted in a reduction of 0.86 W kg$^{-1}$, or 28%, if they provided an additional assistance of 0.20 W kg$^{-1}$ to the other leg. This estimation is slightly higher than the best results recently obtained from bilateral exoskeletons (– 21% in [24], – 23% in [28]); however, this estimation assumes perfect assistance symmetry, which is difficult to achieve. Furthermore, this estimation does not account for potential interaction effects with timing and magnitude parameters of the actuation pattern and potential different effects depending on exoskeleton designs or control methods that are used. Therefore, we believe that the coefficient for the effect of assistance asymmetry that we found can only be used for rough estimates when applying it on other datasets than the dataset of the present study. The assistance asymmetry coefficient could also be used to estimate the losses in metabolic cost reduction due to assistance asymmetry in studies with bilateral exoskeletons.

A limitation of the interpretation of the coefficients for the effects of assistance work rate per leg and assistance work rate asymmetry that we found is that these coefficients were obtained from a dataset with only three work rate levels (*Powered-Off*, 0.068 and 0.135 W kg$^{-1}$) and only two assistance work rate asymmetry levels (no asymmetry and 100% asymmetry). It is likely that the effect of assistance asymmetry is smaller around low asymmetry levels. Another limitation is that the timing that was used was optimized for bilateral assistance. It could be that the smaller metabolic cost benefits from the *Unilateral* conditions are due to sub-optimal timing. To facilitate switching between conditions and to isolate the effect of assistance asymmetry from exoskeleton weight, we let the participants wear exoskeletons on both legs in the *Unilateral* conditions. It is uncertain if the effect of assistance asymmetry would have been smaller or larger if participants had worn only one exoskeleton in the *Unilateral* conditions. We did not calculate if any of the exoskeleton conditions had a net metabolic cost benefit compared to walking without exoskeleton. In previous studies we found that wearing the same type of bilateral exoskeleton while powered-off caused a metabolic penalty of 11% [24, 26]. Therefore, we can estimate that unilateral assistance while wearing both exoskeletons would increase metabolic cost by 4% compared to not wearing the exoskeleton and unilateral assistance while wearing only one exoskeleton would reduce metabolic cost by about 1.5% (assuming that the metabolic cost penalty for wearing a unilateral exoskeleton powered-off is half of the metabolic cost penalty for wearing a bilateral exoskeleton powered-off).

As expected, our participants had a symmetric walking pattern in the *Powered-off* condition. Therefore, it is logical that we found a positive coefficient for the effect of assistance work rate asymmetry, indicating that increasing the amount of assistance asymmetry is detrimental in healthy participants. Based on our data we do not know what would happen in patients who start off with an initial asymmetry such as hemiparetic stroke patients [5, 7–9]. Studies with a unilateral exoskeleton [21] and unilateral exosuits [20] indicate that unilateral assistance to the impaired leg can reduce the metabolic cost in stroke patients. However, we do not know exactly what level of assistance asymmetry is optimal. Awad et al. [20], found a correlation between improvement in propulsion symmetry and reduction in metabolic cost from an exosuit which suggests that tuning the assistance to reduce gait asymmetry as much as possible could be optimal. A follow-up analysis from the same group showed that reduction in metabolic cost was mostly correlated to center-of-mass power from the non-paretic limb [55]. Maybe, assisting stroke patients at the non-paretic side in addition to the paretic side could result in even greater reductions in metabolic cost? Similar protocols as our present study that compare different types of unilateral and bilateral assistance in patient populations with asymmetric gait could answer such questions.

## Conclusion

We found that *Unilateral* exoskeleton assistance can reduce the metabolic rate by 7 ± 3% compared with wearing a *Powered-Off* exoskeleton. Similar to results from walking with a unilateral prosthesis, we found slower forward center-of-mass velocity during the assisted leg stance and faster velocity during the unassisted leg stance, which is possibly an adaptation to reduce effort. We found the highest ratio of metabolic cost reduction versus positive work with *Bilateral Matched Total Work*. The results indicate that assistance work rate symmetry and the assistance work rate per leg are more important for metabolic cost reduction than the bilateral work rate sum. These results can help explain differences between studies with unilateral and bilateral exoskeletons and inform how unilateral assistance can potentially be used for practical applications, such as reducing gait asymmetry.

## Additional files

**Additional file 1:** Analysis of effect of total work, work per assisted leg and work asymmetry on change in metabolic rate. (PDF 43 kb)

**Additional file 2: Movie 1.** *Unilateral* condition. Participant walking on treadmill with assistance to the right leg and no assistance to the left leg. It can be qualitatively observed that there is little asymmetry in the gait pattern except in the ankle as shown quantitively in Figs. 4 and 5. (MP4 1264 kb)

**Additional file 3:** Spatiotemporal stability and balance results. A) Step width was calculated based on the average distance in the medio-lateral direction between the markers on the left and right foot during consecutive stance phases. B) Step width variability was calculated by taking the standard deviation of step width across different strides. C) Step length variability was calculated by taking the standard deviation of step time multiplied by treadmill speed across different strides. Bars represent population averages. Error bars represent the standard error. Colors represent the Unilateral, Bilateral and Powered-Off conditions shown in the figures on the bottom. Brackets represent significant differences between conditions. Only the pairwise comparisons that are relevant for the research questions listed in the methods are analyzed. ** is $p \leq 0.01$. (TIF 2283 kb)


#### Acknowledgments

The authors would like to thank Prof. Wim Derave for his advice in initiating the manuscript, Ing. Davy Spiessens for the technical support, Jozefien Speeckaert for her help with data collection, and Aqtor! (Ghent, Belgium) for their assistance in manufacturing the exoskeleton.

#### Funding

SG was supported by the Special Research Fund of Ghent University (BOF10/DOC/288). PM received partial support from the Center for Research in Human Movement Variability of the University of Nebraska Omaha and the NIH (P20GM109090).


#### Authors' contributions

PM, SG, PVDB and DDC designed the study; PM, SG and PVDB performed the data collection and data processing; PM, SG and PVDB performed the data analysis; PM, SG, PVDB and DDC interpreted the results of the study; PM and PVDB drafted the manuscript; and PM, SG, PVDB and DDC edited and revised the manuscript. All authors gave final approval for publication.

#### Competing interests

The authors declare that they have no competing interests.

#### References


1. Hsiao-Wecksler ET, Polk JD, Rosengren KS, Sosnoff JJ, Hong S. A review of new analytic techniques for quantifying symmetry in locomotion. Symmetry (Basel). 2010;2:1135–55.
2. Sadeghi H, Allard P, Prince F, Labelle H. Symmetry and limb dominance in able-bodied gait: a review. Gait Posture. 2000;12:34–45.
3. Gregg RD, Dhaher YY, Degani A, Lynch KM. On the mechanics of functional asymmetry in bipedal walking. IEEE Trans Biomed Eng. 2012;59:1310–8.
4. Adamczyk PG, Kuo AD. Mechanisms of gait asymmetry due to push-off deficiency in unilateral amputees. IEEE Trans Neural Syst Rehabil Eng. 2015;23:776–85.
5. Allen JL, Kautz SA, Neptune RR. Step length asymmetry is representative of compensatory mechanisms used in post-stroke hemiparetic walking. Gait Posture. 2011;33:538–43.
6. Balasubramanian CK, Bowden MG, Neptune RR, Kautz SA. Relationship between step length asymmetry and walking performance in subjects with chronic hemiparesis. Arch Phys Med Rehabil. 2007;88:43–9.
7. Roerdink M, Beek PJ. Understanding inconsistent step-length asymmetries across hemiplegic stroke patients: impairments and compensatory gait. Neurorehabil Neural Repair. 2011;25:253–8.
8. Farris DJ, Hampton A, Lewek MD, Sawicki GS. Revisiting the mechanics and energetics of walking in individuals with chronic hemiparesis following stroke : from individual limbs to lower limb joints. J. Neuroeng. Rehabil. 2015;12:1–12.
9. Mahon CE, Farris DJ, Sawicki GS, Lewek MD. Individual limb mechanical analysis of gait following stroke. J Biomech. 2015;48:984–9.
10. Kobayashi H, Kakihana W, Kimura T. Combined effects of age and gender on gait symmetry and regularity assessed by autocorrelation of trunk acceleration. J. Neuroeng. Rehabil. 2014;11:109.
11. Wutzke CJ, Sawicki GS, Lewek MD. The influence of a unilateral fixed ankle on metabolic and mechanical demands during walking in unimpaired young adults. J Biomech. 2012;45:2405–10.
12. Hesse S, Werner C, Bardeleben A, Barbeau H. Body weight-supported treadmill training after stroke. Curr Atheroscler Rep. 2001;3:287–94.
13. Tyson SF, Kent RM. Effects of an ankle-foot orthosis on balance and walking after stroke: a systematic review and pooled meta-analysis. Arch Phys Med Rehabil. 2013;94:1377–85.
14. Herr HM, Grabowski AM. Bionic ankle-foot prosthesis normalizes walking gait for persons with leg amputation. Proc Biol Sci. 2012;279:457–64.
15. Reisman DS, Wityk R, Silver K, Bastian AJ. Locomotor adaptation on a split-belt treadmill can improve walking symmetry post-stroke. Brain. 2007;130:1861–72.
16. Finley JM, Long A, Bastian AJ, Torres-Oviedo G. Spatial and temporal control contribute to step length asymmetry during Split-Belt adaptation and Hemiparetic gait. Neurorehabil Neural Repair. 2015.
17. Regnaux JP, Pradon D, Roche N, Robertson J, Bussel B, Dobkin B. Effects of loading the unaffected limb for one session of locomotor training on laboratory measures of gait in stroke. Clin Biomech. 2008;23:762–8.
18. Veneman JF. Emerging directions in lower limb externally wearable robots. Front Inf Technol Electron Eng. 2016;9184:1–8.

19. Lin PY, Yang YR, Cheng SJ, Wang RY. The relation between ankle impairments and gait velocity and symmetry in people with stroke. Arch Phys Med Rehabil. 2006;87:562–8.
20. Awad LN, Bae J, O'Donnell K, De Rossi SMM, Hendron K, Sloot LH, et al. A soft robotic exosuit improves walking in patients after stroke. Sci Transl Med. 2017;9.
21. Takahashi KZ, Lewek MD, Sawicki GS. A neuromechanics-based powered ankle exoskeleton to assist walking post-stroke: a feasibility study. J. Neuroeng. Rehabil. 2015;12:23.
22. Collins SH, Wiggin MB, Sawicki GS. Reducing the energy cost of human walking using an unpowered exoskeleton. Nature. 2015;522:212–5.
23. Ding Y, Panizzolo FA, Siviy CJ, Malcolm P, Galiana I, Holt KG, et al. Effect of timing of hip extension assistance during loaded walking with a soft exosuit. J Neuroeng Rehabil. 2016;13:87.
24. Galle S, Malcolm P, Collins SH, De Clercq D. Reducing the metabolic cost of walking with an ankle exoskeleton: interaction between actuation timing and power. J. Neuroeng. Rehabil. 2017;14:35.
25. Koller JR, Gates DH, DP Ferris, CD Remy. "Body-in-the-loop" optimization of assistive robotic devices: a validation study. In: Proceedings of Robotics: Science and Systems. AnnArbor: 2016. http://www.roboticsproceedings.org/rss12/p07.pdf.
26. Malcolm P, Derave W, Galle S, De Clercq D. A simple exoskeleton that assists plantarflexion can reduce the metabolic cost of human walking. PLoS One. 2013;8:e56137.
27. Mooney LM, Rouse EJ, Herr HM. Autonomous exoskeleton reduces metabolic cost of human walking during load carriage. J. Neuroeng. Rehabil. 2014;11:80.
28. Quinlivan BT, Lee S, Malcolm P, Rossi DM, Grimmer M, Siviy C, et al. Assistance magnitude versus metabolic cost reductions for a tethered multiarticular soft exosuit. Sci Robot. 2017;2:eaah4416.
29. Panizzolo FA, Galiana I, Asbeck AT, Siviy C, Schmidt K, Holt KG, et al. A biologically-inspired multi-joint soft exosuit that can reduce the energy cost of loaded walking. J Neuroeng Rehabil. 2016;13:43.
30. Hidler J, Nichols D, Pelliccio M, Brady K, Campbell DD, Kahn JH, et al. Multicenter randomized clinical trial evaluating the effectiveness of the Lokomat in subacute stroke. Neurorehabil Neural Repair. 2009;23:5–13.
31. Collins SH, Jackson RW. Inducing self-selected human engagement in robotic locomotion training. Seattle: ICORR; 2013.
32. Galle S, Malcolm P, Derave W, De Clercq D. Enhancing performance during inclined loaded walking with a powered ankle-foot exoskeleton. Eur J Appl Physiol. 2014.
33. Jackson RW, Collins SH. An experimental comparison of the relative benefits of work and torque assistance in ankle exoskeletons. J Appl Physiol. 2015;119:541–57.
34. Galle S, Malcolm P, Derave W, De Clercq D. Uphill walking with a simple exoskeleton: plantarflexion assistance leads to proximal adaptations. Gait Posture. 2015;41:246–51.
35. Koller JR, Jacobs DA, Ferris DP, Remy CD. Learning to walk with an adaptive gain proportional myoelectric controller for a robotic ankle exoskeleton. J. Neuroeng. Rehabil. 2015;12:97.
36. Mooney LM, Herr HM. Biomechanical walking mechanisms underlying the metabolic reduction caused by an autonomous exoskeleton. J. Neuroeng. Rehabil. J NeuroEng Rehabil. 2016;13:4.
37. Cain SM, Gordon KE, Ferris DP. Locomotor adaptation to a powered ankle-foot orthosis depends on control method. J. Neuroeng. Rehabil. 2007;4:48.
38. Ferris DP, Czerniecki JM, Hannaford B. An ankle-foot orthosis powered by artificial pneumatic muscles. J Appl Biomech. 2005;21:189–97.
39. Gordon K, Ferris D. Learning to walk with a robotic ankle exoskeleton. J Biomech. 2007;40:2636–44.
40. Kao P-C, Lewis CL, Ferris DP. Invariant ankle moment patterns when walking with and without a robotic ankle exoskeleton. J Biomech. 2010;43:203–9.
41. Ellis RG, Howard KC, Kram R. The metabolic and mechanical costs of step time asymmetry in walking. Proc Biol Sci. 2013;280:20122784.
42. Lee S, Crea S, Malcolm P, Galiana I, Asbeck A, Walsh C. Controlling negative and positive power at the ankle with a soft exosuit. Proc - IEEE Int Conf Robot Autom. Stockholm: IEEE; 2016. p. 3509–15. https://ieeexplore.ieee.org/abstract/document/7487531/.
43. Malcolm P, Quesada RE, Caputo JM, Collins SH. The influence of push-off timing in a robotic ankle-foot prosthesis on the energetics and mechanics of walking. J Neuroeng Rehabil. 2015;12:14.
44. Galle S, Malcolm P, Derave W, De Clercq D. Adaptation to walking with an exoskeleton that assists ankle extension. Gait Posture. 2013;38:495–9.
45. Brockway JM. Derivation of formulae used to calculate energy expenditure in man. Hum Nutr Clin Nutr. 1987;41:463–71.
46. Gard SA, Miff SC, Kuo AD. Comparison of kinematic and kinetic methods for computing the vertical motion of the body center of mass during walking. Hum Mov Sci. 2004;22:597–610.
47. Stanhope SJ, Kepple TM, McGuire DA, Roman NL. Kinematic-based technique for event time determination during gait. Med Biol Eng Comput. 1990;28:355–60.
48. Waters RL, Perry J, Antonelli D, Hislop H. Energy cost of walking of amputees: the influence of level of amputation. J Bone Joint Surg Am. 1976;58:42–6.
49. Platts MM, Rafferty D, Paul L. Metabolic cost of overground gait in younger stroke patients and healthy controls. Med Sci Sports Exerc. 2006;38:1041–6.
50. Jackson RW, Collins SH. An experimental comparison of the relative benefits of work and torque assistance in ankle exoskeletons: supplementary data [internet]. J Appl Physiol. 2015:541–57. Available from: http://biomechatronics.cit.cmu.edu/publications/Jackson_2015_JAP-%2D-Data.zip
51. Sawicki GS, Ferris DP. Mechanics and energetics of level walking with powered ankle exoskeletons. J Exp Biol Company of Biologists. 2008;211:1402–13.
52. Long AW, Roemmich RT, Bastian AJ. Blocking trial-by-trial error correction does not interfere with motor learning in human walking. J Neurophysiol. 2016;115:2341–8.
53. Hoogkamer W, Bruijn SM, Duysens J. Stride length asymmetry in split-belt locomotion. Gait Posture. 2014;39:652–4.
54. Bobbert MF, de GWW, Jonk JN, LJR C. Explanation of the bilateral deficit in human vertical squat jumping. J Appl Physiol. 2006;100:493–9.
55. Ding Y, Galiana I, Asbeck A, De Rossi S, Bae J, Santos T, et al. Biomechanical and physiological evaluation of multi-joint assistance with soft exosuits. IEEE Trans Neural Syst Rehabil Eng. 2016;25(2):1. https://ieeexplore.ieee.org/abstract/document/7394183/.

# 2

# Analysis of sensory system aspects of postural stability during quiet standing in adolescent idiopathic scoliosis patients

Taeyong Sim[1†], Hakje Yoo[1†], Dongjun Lee[1], Seung-Woo Suh[2], Jae Hyuk Yang[2], Hyunggun Kim[1*] and Joung Hwan Mun[1*]

**Abstract**

**Background:** The aim of this study was to quantitatively analyze quite standing postural stability of adolescent idiopathic scoliosis (AIS) patients in respect to three sensory systems (visual, vestibular, and somatosensory).

**Method:** In this study, we analyzed the anterior-posterior center of pressure (CoP) signal using discrete wavelet transform (DWT) between AIS patients ($n = 32$) and normal controls ($n = 25$) during quiet standing.

**Result:** The energy rate ($\Delta E_{EYE}\%$) of the CoP signal was significantly higher in the AIS group than that in the control group at levels corresponding to vestibular and somatosensory systems ($p < 0.01$).

**Conclusions:** This implies that AIS patients use strategies to compensate for possible head position changes and spinal asymmetry caused by morphological deformations of the spine through vestibular and somatosensory systems. This could be interpreted that such compensation could help them maintain postural stability during quiet standing. The interpretation of CoP signal during quiet standing in AIS patients will improve our understanding of changes in physical exercise ability due to morphological deformity of the spine. This result is useful for evaluating postural stability before and after treatments (spinal fusion, bracing, rehabilitation, and so on).

**Keywords:** Adolescent idiopathic scoliosis (AIS), Discrete wavelet transform (DWT), Postural stability, Balance, Sensory system, Visual, Vestibular, Somatosensory

## Background

Adolescent idiopathic scoliosis (AIS) is defined as a three-dimensional morphological deformity of the spine [1]. The severity of AIS is generally diagnosed by measuring Cobb's angle, the golden standard with 10 degrees diagnostic criteria on radiograph based X-ray images [1, 2]. The morphological deformity of the spine is intimately interconnected with the central-nervous system [3, 4], altering the sensory process [5–7] and causes balance dysfunction of the static stability [8–10]. The morphological deformity such as concave (the inside of scoliosis curve direction) and convex (the outside of scoliosis curve direction) of the spine can cause alternation in head position [11, 12] and asymmetric activity of the muscles surrounding the spine [13], which can result in trunk center of mass (CoM) change [14]. People with spinal deformity maintain their postural stability by rearranging these alterations through various interactions of sensory systems [11, 15, 16]. The processing for controlling postural instability is determined by sensory organization of visual, vestibular, and somatosensory systems [17]. Therefore, sensory processing of postural stability during quiet standing in AIS patients can be an important analytical factor [18–20].

The human body is designed to maintain a vertical orientation to the eyes and ground parallel in horizontal plane [21]. To maintain postural stability, the visual system adjusts the posture by providing information about the surrounding environment [17] while vestibular system adjusts the posture by managing the movement and position of the head [22]. In other words, both visual and vestibular systems help adjust the spatial orientation for maintaining

* Correspondence: hkim.bme@skku.edu; jmun@skku.edu
†Taeyong Sim and Hakje Yoo contributed equally to this work.
[1]Department of Bio-Mechatronic Engineering, College of Biotechnology and Bioengineering, Sungkyunkwan University, Natural Sciences Campus, 2066 Seobu-ro, Jangan-gu, Suwon-si, Gyeonggi-do 16419, South Korea
Full list of author information is available at the end of the article

postural stability [23, 24]. The somatosensory system helps support the body to maintain postural stability; additionally, provides information on the improvement of joint stability to maintain the static state through controlling muscle activation by the nervous system [25–27]. This kinematic control ability is organized in close association with the input from the visual and vestibular systems [23, 28]. The analysis of the center of pressure (CoP) can be applied to interpret a postural instability with respect to sensory processing during quiet standing in AIS patients [18, 29, 30]. The CoP signal is distinguished by visual (< 0.1 Hz), vestibular (0.1–0.5 Hz), and somatosensory (0.5–1.0 Hz) systems according to the frequency [31–34]. In fact, different power spectrum intervals have been reported in various studies, including 0–0.5, 0.5–2, and > 2 Hz [35]; 0–0.1, 0.1–0.5, and 0.5–1 Hz [31], and 0–0.3, 0.3–1, and 1–3 Hz [36]. However, according to recent studies, criteria of 0–0.1, 0.1–0.5, and 0.5–1 Hz are frequently used to analyze the effect a certain medicine treatment on the postural stability of patients with unilateral vestibular disorder [34], evaluate postural sway dynamics in quiet standing [32], and investigate balance functions in migraineur with and without vertigo [33]. These criteria correspond to each sensory system were calculated by measuring oscillations made on a stabilometer in a second [31].

A sensory organization test (SOT) has been used to evaluate sensory information while maintaining postural stability in AIS patients [8–10]. The SOT method is used to observe posture control during standing according to environmental changes of the three sensory systems (visual, vestibular, and somatosensory) [8–10]. The SOT method evaluates the postural control ability in six states. Individual sensory information is identified through comparative analysis of two of these six states. For example, vestibular sensory information is evaluated by dividing the result of the protocol that is responsible for the vestibular system into results of the protocol of the three sensory systems [9]. The SOT has been used to analyze the effect of a spinal brace on postural control for AIS [8], the difference in sensory information according to the degree of spinal deformation [9], and the relationship between somatosensory function changes and postural control in AIS patients [10]. However, using SOT to quantify the functions of visual, vestibular, and somatosensory systems during quiet standing has some limitations. First, a spatial constraint arises due to the size of the equipment used to analyze the sensory system, for example, Smart Equitest (NeuroCom International Inc., Clackamas OR, USA): size = 1350 × 1550 × 2390 mm. Second, to identify the complex sensory integration, sensory information should be acquired simultaneously. Consequently, the individual results can be obtained from two separate SOT experiments [8–10]. This method also has limitations in multiple analysis of the sensory integration process for maintaining the stability of standing posture [8–10].

In this study, we applied discrete wavelet transform (DWT) method to analyze the sensory integration process, using only CoP signal without additional equipment, in order to evaluate postural instability during quiet standing. DWT method is used to decompose a signal composed of high and low frequencies. It is appropriate to interpret effects of three sensory systems (visual, vestibular and somatosensory systems) on maintaining postural stability through decomposing the CoP signal [37]. Additionally, we evaluated sensory information characteristic in postural instability depending on AIS severity according to Cobb's angle.

## Methods

### Participants

Fifty-seven subjects participated in this study to analyze the effect of morphologic deformation on postural instability, including 32 AIS patients (all females, age = 14.3 ± 2.2 yr., height = 155.8 ± 9.2 cm, weight = 44.0 ± 8.3 kg, cobb angle = 31.1 ± 15.8°) without history of surgical treatment or rehabilitation and 25 adolescents (all females, age = 14.8 ± 4.2 yr., height = 156.8 ± 6.3 cm, weight = 44.2 ± 7.6 kg) without musculoskeletal diseases of similar age. In order to analyze posture control according to AIS severity, we divided the subjects into three cases of severity based on the largest Cobb's angle (10° < mild ≤20° ($n$ = 12); 20° < moderate ≤40° ($n$ = 10); and severe > 40° (n = 10) [38, 39]. All experimental protocols were approved by the Institutional Review Boards of Korea University Guro Hospital. All experiments were performed in accordance with relevant approved guidelines and regulations of Sungkyunkwan University. Before the experiment, all participants provided informed consent to participate.

### Experimental setup and data collection

We used anterior-posterior CoP during quiet standing to analyze the effect of spinal morphologic deformity on postural instability. The CoP defined as the point location of the ground reaction force vector has been used as a parameter for evaluating balance control ability of CoM [40]. Subjects stood upright in a comfortable stance, and they were asked to hold their arms along the side of their body while standing quietly on a force plate. The standing condition was divided into two conditions during quiet standing (with eyes open and with eyes closed) in order to analyze the difference in the sensory system according to change of visual information [37]. Differing eyes open and eyes closed conditions, the presence and absence of visual information, an important design parameter, allows observers to measure activities of vestibular and somatosensory systems in postural control [37, 41]. A CoP signal in eyes open condition was used to analyze the effect of normal sensory information from visual, vestibular, and somatosensory systems on posture control. On the contrary, that from eyes closed condition facilitated the analysis of the effect of vestibular and somatosensory information on posture control with limited visual information.

In addition, difference of energy contents between eyes open and eyes closed conditions could allow quantitative analysis about the effect of differences among three sensory systems on posture control. Subjects were asked to stare at a dot marked at a horizontal distance from their eyes. In the eyes open experiment, subjects kept staring at the dot. In the eyes closed experiment, subjects instantly closed their eyes at the start of the experiment [37]. Each subject continued quiet standing with eyes open or with eyes closed for a period of 70s [42]. This trial was repeated six times. Average values of each eyes open and eyes closed condition were used. CoP data were obtained using force plates (OR6–2000, AMTI Inc., Newton, Massachusetts, USA) and sampled at 120 Hz [43].

### Data processing

To compare the difference in sensory system between AIS and control groups, CoP signal was analyzed using a DWT to compare between conditions of eyes open and eyes closed during quiet standing [37]. DWT can be decomposed into a high frequency and a low frequency signal through the scaling function ($\phi_{(j,k)}(t)$) and the wavelet function ($\psi_{(j,k)}(t)$), respectively (Eq. (1)).

$$f(t) = \sum_k \sum_j^{\infty} c_{j,k}\phi_{j,k}(t) + \sum_k \sum_j^{\infty} d_{j,k}\psi_{j,k}(t) \qquad (1)$$

In the above Equation, $c_{j,\,k}$ and $d_{j,\,k}$ denote approximated coefficients and detailed coefficients, respectively. $j$ and $k$ denote decomposition level ($j = 1, 2, \ldots, J$) and discrete location of the CoP signal ($f(t)$). The detail signal ($W(j, k)$) and approximated signal ($S(j, k)$) of the CoP signal are obtained using the following Eqs. (2) and (3).

$$W(j,k) = \int_{-\infty}^{\infty} f(t)\psi_{j,k}(t)dt \qquad (2)$$

$$S(j,k) = \int_{-\infty}^{\infty} f(t)\phi_{(j,k)}(t)dt \qquad (3)$$

From the Eqs. (4) and (5), energy content ($E(j)$) and total energy content ($E_T$) of the corresponding decomposition level are calculated using detail signal and approximated signal obtained above, respectively. Where, $K$ ($j$) is discrete location in level $j$ [37].

$$E(j) = \sum_{k=0}^{K(j)} (W(j,k))^2 \qquad (4)$$

$$E_T = \sum_{k=0}^{K(j)} (S(j,k))^2 + \sum_{j=1}^{J} \sum_{k=0}^{K(j)} (W(j,k))^2 \qquad (5)$$

The energy content in eyes closed case ($E_{EC}\%(j)$) is calculated as a percentage of the ratio between total energy content ($E_T$) and energy content corresponding to decomposition levels ($E_{EC}(j)$). The energy content in eyes open case ($E_{EO}\%(j)$) is calculated using the same method (Eqs. (6) and (7)).

$$E_{EC}\%(j) = \left(\frac{(E_{EC}(j))}{E_T}\right) * 100\% \qquad (6)$$

$$E_{EO}\%(j) = \left(\frac{(E_{EO}(j))}{E_T}\right) * 100\% \qquad (7)$$

In order to analyze energy content differences between eyes closed and eyes open conditions, energy rate ($\Delta E_{EYE}(j)$) was quantified by the following Eq. (8).

$$\Delta E_{EYE}(j) = \left(\frac{E_{EC}\%(j) - E_{EO}\%(j)}{E_{EO}\%(j)}\right) * 100 \qquad (8)$$

Figure 1 presents the decomposition process of CoP signal using DWT. Decomposition level $j$ is calculated considering the characteristics of both CoP signal and wavelet function. For each level, the CoP signal is decomposed by applying a high pass filter and a low pass filter. Detail coefficients ($d_1$~$d_j$) are obtained through level *1* to $j$ high-pass filter while the approximation coefficient ($c_j$) at level $j$ is obtained through the low-pass filter.

### Statistical analysis

Statistical analysis was performed using SPSS 15.0 software (SPSS Inc., Chicago, IL, USA) for analyzing changes in the sensory system between conditions of eyes open and eyes closed during quiet standing in AIS and control groups. Energy contents using DWT were analyzed by calculating the average and standard deviations. To analyze the difference in energy contents between eyes open and eyes closed conditions in the same group and the difference in the energy rate between AIS and control groups, independent $t$-tests were performed

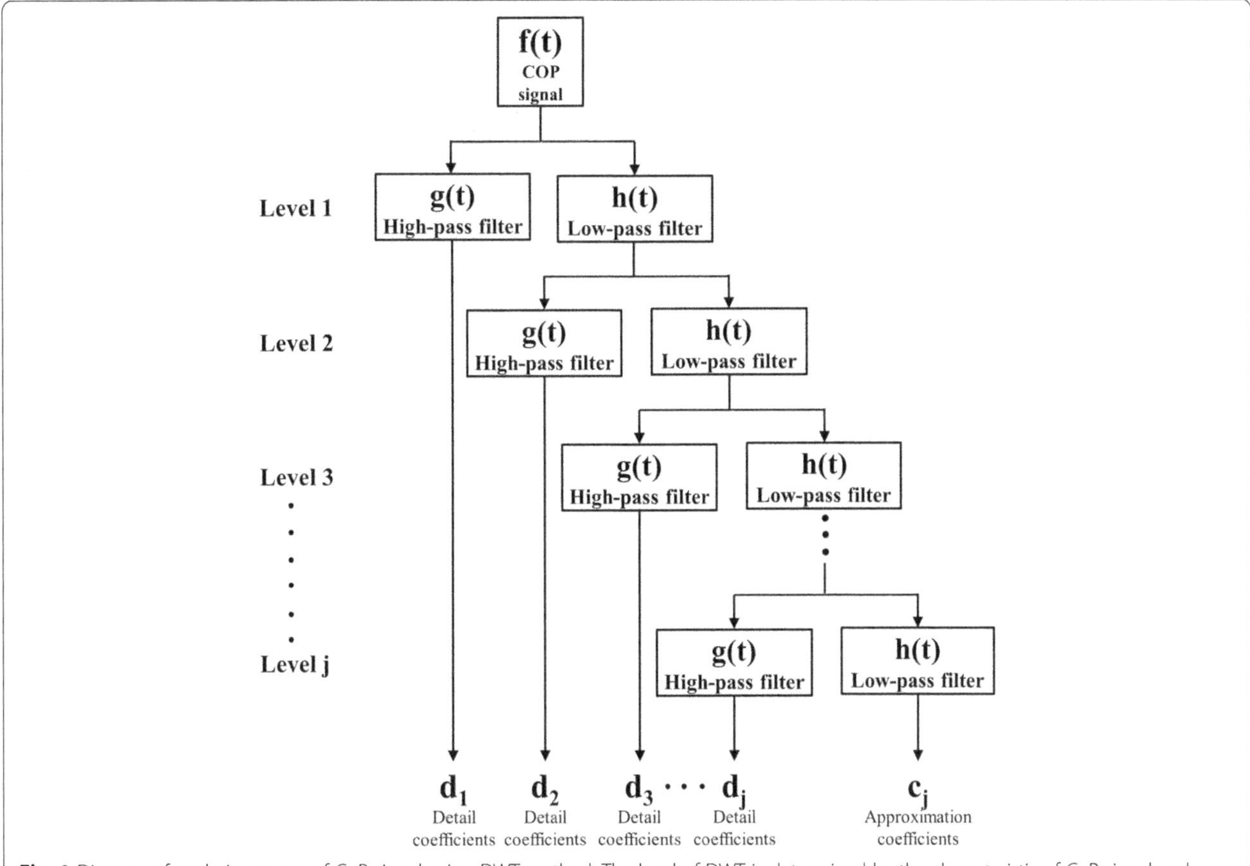


**Fig. 1** Diagram of analysis process of CoP signal using DWT method. The level of DWT is determined by the characteristic of CoP signal and wavelet function

for all variables. Analysis of variance (ANOVA) was performed on variables to analyze difference in energy rate between AIS patients with three different severity levels based on Cobb's angle. Differences among the three groups of AIS were analyzed using Bonferroni post-*hoc* test. Statistical significance level was set at $p$ value of less than 0.05.

## Results

### Discrete decomposition of CoP signal using DWT and analysis of sensory system in both control and AIS groups

To quantitatively analyze the sensory system known to control postural stability in AIS and control groups during quiet standing, we acquired ten detail coefficients for low frequency components and one approximate coefficient for high frequency components of the CoP signal using DWT. Figure 2 shows an example of reconstructing the results into low frequency signal and high frequency signal of CoP. The downward direction denotes an increasing discrete level. The signal is divided into an approximation signal (high frequency signal) and detail signal (low frequency signal). The open-loop and closed-loop signals are separated by 1 Hz [44]. In addition, the closed-loop signal is distinguished by three sensory systems (Somatosensory: 0.5–1.0 Hz; Vestibular: 0.1–0.5 Hz; Visual: < 0.1 Hz) according to the frequency range [31–34].

Figures 3a and b show results of transforming low frequency signals of CoP signals into energy content (%) during quiet standing with eyes open and eyes closed in the control group. The vertical axis shows the magnitude of energy content (%) in the corresponding frequency range. The magnitude of energy content (%) for each level section shows a similar value in eyes open and eyes closed subjects. The major part of the energy content (%) was represented in the closed-loop signal. The highest magnitude of energy content (%) in the closed-loop signal was shown at the vestibular section (eyes open = 29.97 ± 4.72%; eyes closed = 43.77 ± 4.92%), followed by that at the somatosensory section (eyes open = 16.97 ± 4.06%; eyes closed = 25.25 ± 3.58%) and the visual section (eyes open = 14.91 ± 2.18%; eyes closed = 12.97 ± 2.24%). Figure 3c shows difference in energy content (%) between eyes open and eyes closed subjects for each level section.

Figures 4a and b show results in the AIS group. The highest magnitude of energy content (%) was

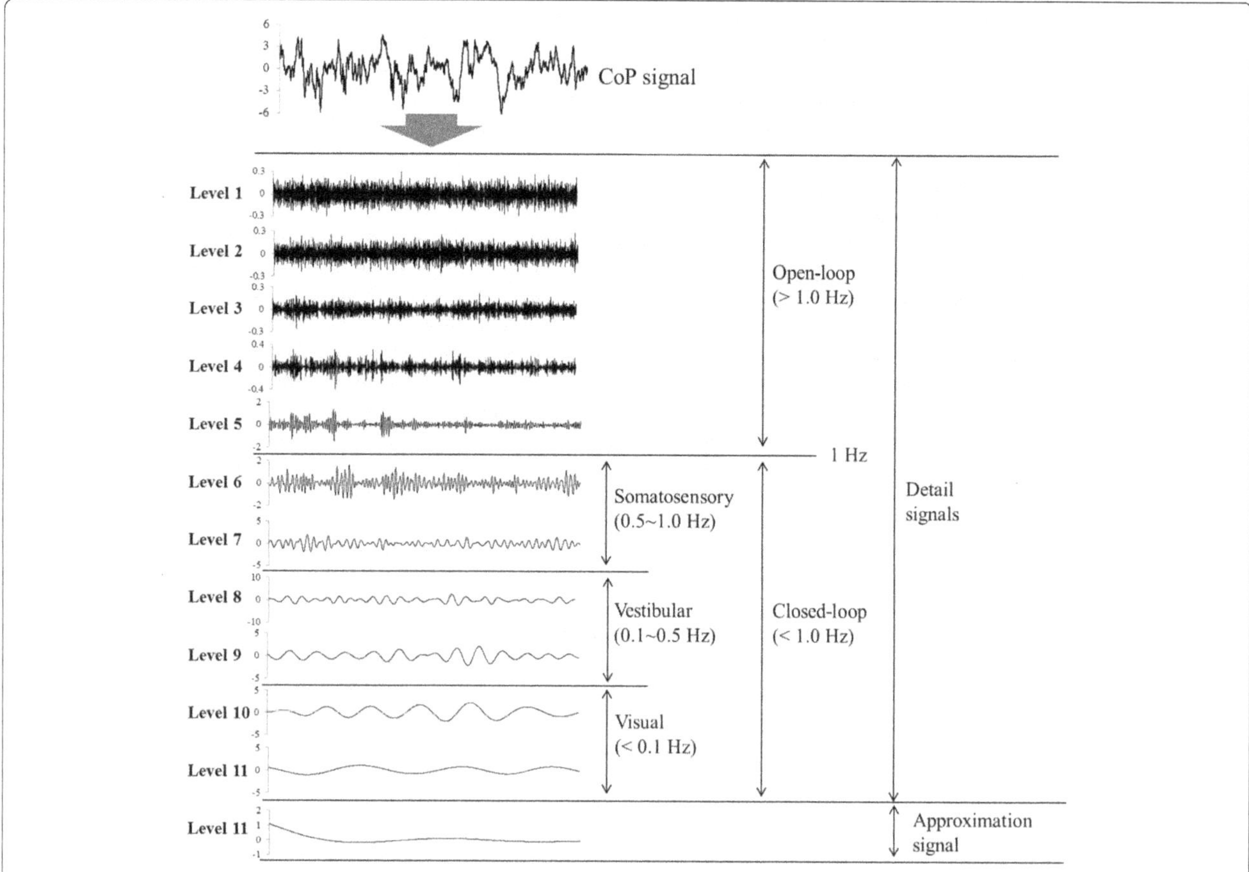


**Fig. 2** Results of the discrete decomposition of the CoP signal into 11 level low-frequency signals. This decomposition is the result of using 11 level discrete wavelet transform during quiet standing with eyes open in people without AIS. Based on level 5, open-loop and closed-loop were divided. The closed-loop (levels 6~ 10) group is divided into three sections (levels 6, 7: somatosensory (0.5~ 1.0 Hz); levels 8, 9: vestibular (0.1~ 0.5 Hz); levels 10, 11: visual (below 0.1 Hz))

found at the vestibular section (eyes open = 30.93 ± 5.09%; eyes closed = 53.39 ± 7.09%), followed by that at somatosensory section (eyes open = 18.29 ± 3.97%; eyes closed = 31.94 ± 6.33%) and the visual section (eyes open = 15.55 ± 3.61%; eyes closed = 14.52 ± 3.41%). Figure 4c shows difference of energy content (%) between eyes open and eyes closed subjects for each level section. $E_{EC}$% represents a significantly higher value than $E_{EO}$% at levels 4~ 9 while $E_{EO}$% shows a higher value than $E_{EC}$% at level 10 ($p < 0.05$ or $p < 0.01$).

### Difference of energy content (%) between control and AIS in eyes open and eyes closed subjects

Table 1 and Fig. 5 show results comparing the energy rate between control and AIS groups for each level section. Table 1 shows quantitative results (mean ± standard deviation) of the difference in energy rate (%) between control and AIS groups. Figure 5 shows statistical results. It graphically illustrates the result of energy rate. The vertical axis shows the percentage of energy rate ($\Delta E_{EYE}$%). A positive energy rate is shown when the magnitude of $E_{EC}$% is larger than that of $E_{EO}$% while a negative energy rate is shown in the opposite case. As a result, the energy rate of the AIS group was significantly higher than that of the control group at levels 4 to 9 ($p < 0.05$ or $p < 0.01$).

### Analysis of sensory system between eyes open and eyes closed according to severity of AIS

Table 2 shows energy contents between eyes open and eyes closed subjects according to the severity of AIS for each level section. $E_{EC}$% showed a significantly higher tendency than $E_{EO}$% at levels from 4 to 9 (all $p < 0.01$). On the contrary, $E_{EO}$% was relatively higher at level 10 ($p < 0.05$). In many frequency sections, the difference in energy contents between eyes open and eyes closed subjects was more significant in the closed-loop signal than that in the open-loop signal. Significant differences were shown in energy content (%) between eyes open and eyes closed subjects in the frequency sections of somatosensory

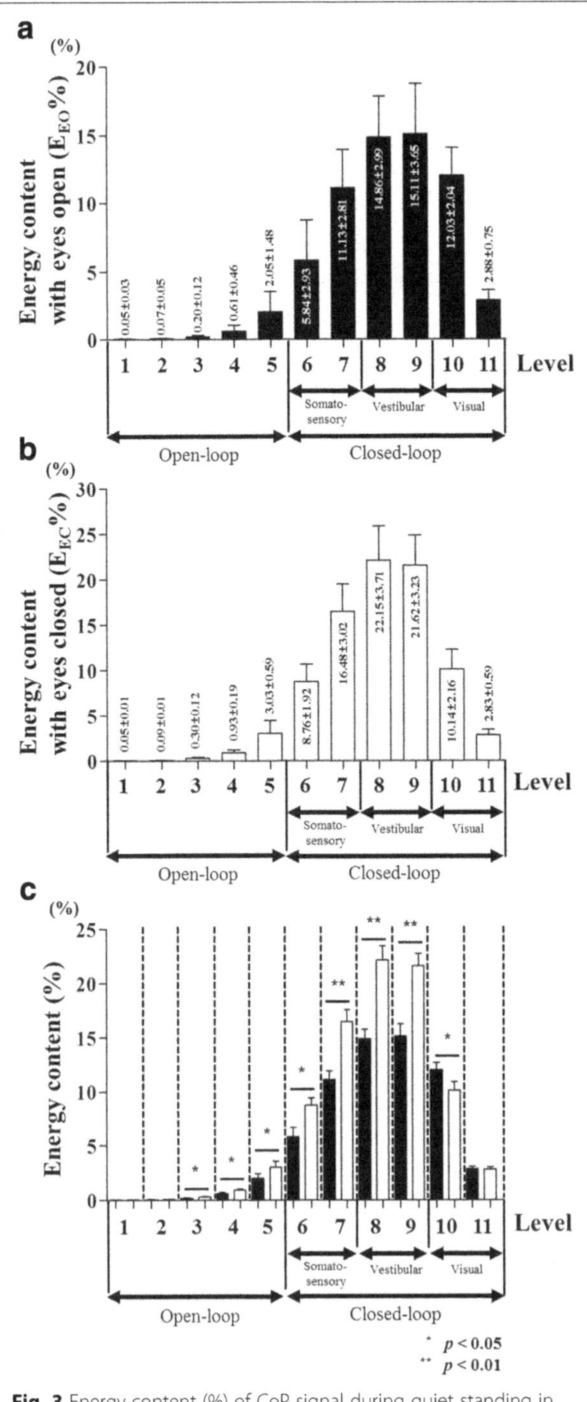


**Fig. 3** Energy content (%) of CoP signal during quiet standing in control group. **a** Energy content ($E_{EO}$%) for each level section with eyes open condition (**b**) energy content ($E_{EC}$%) for each level section with eyes closed condition (**c**) statistical comparison of difference between $E_{EO}$% and $E_{EC}$% for each level section

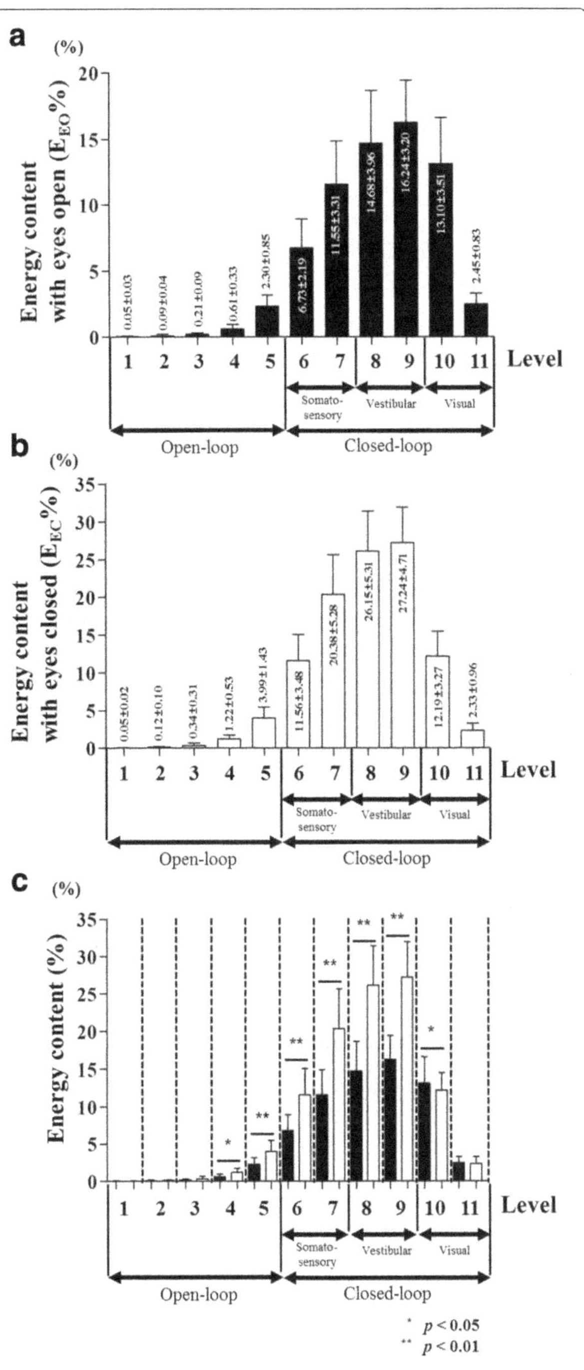


**Fig. 4** Energy content (%) of CoP signal during quiet standing in AIS group. **a** Energy content ($E_{EO}$%) for each level section with eyes open condition (**b**) energy content ($E_{EC}$%) for each level section with eyes closed condition (**c**) statistical comparison of difference of $E_{EO}$% and $E_{EC}$% for each level section

(levels 6 and 7) and vestibular (levels 8 and 9) in all severity groups.

Table 3 and Fig. 6 show difference of energy rate ($\Delta E_{EYE}$%) according to the severity of AIS for each level section. The result shows significant differences among severity groups in open-loop (level 2) and closed-loop (levels 7~ 10) signals. At levels 7 to 10, the energy rate was increased significantly depending on the severity ($p < 0.05$ or $p < 0.01$). At level 6, there was a significant difference in energy rate between mild and severe groups ($p < 0.05$). In addition, the energy rate of the severe group was increased

**Table 1** Quantification result (mean ± standard deviation) of $\Delta E_{EYE}\%$ depending on level in control and AIS groups

| | | Open-loop | | | | | Closed-loop | | | | | |
|---|---|---|---|---|---|---|---|---|---|---|---|---|
| | | | | | | | Somatosensory | | Vestibular | | Visual | |
| | Level | 1 | 2 | 3 | 4 | 5 | 6 | 7 | 8 | 9 | 10 | 11 |
| $\Delta E_{EYE}\%$ (%) | Control group (n = 25) | 2.22 ±9.46 | 32.86 ±16.50 | 50.92 ±21.05 | 53.13 ±16.59 | 48.05 ±21.95 | 49.87 ±23.17 | 47.04 ±27.71 | 49.06 ±26.31 | 42.82 ±20.96 | 12.19 ±19.96 | −1.91 ±9.05 |
| | AIS group (n = 32) | − 1.49 ±11.51 | 26.91 ±17.47 | 58.47 ±21.38 | 63.86 ±23.48 | 72.78 ±25.46 | 72.66 ±25.56 | 81.33 ±24.39 | 79.32 ±23.39 | 69.55 ±21.24 | −7.81 ±16.12 | −17.74 ±12.34 |

more significantly than the moderate group at levels 1 and 11 ($p < 0.05$).

## Discussion

In order to analyze postural stability with respect to three sensory systems during quiet standing in AIS patients, the DWT method was applied in this study. The DWT method for CoP signal decomposition has been applied in various fields. It has been used to analyze difference of postural stability based upon gender [45], determine effect of vestibular dysfunction on postural control [46], and evaluate difference of postural control ability after cochlear implant surgery [47]. In this study, we analyzed the difference of postural instability between AIS group ($n = 32$) and control group ($n = 25$) of the similar age, and interpreted differences in sensory information in terms of visual, vestibular, and somatosensory systems. For analysis of the three sensory systems used for postural stability, we divided 11 frequency sections of the CoP signal using the DWT method [42]. The decomposed CoP signal was divided into open-loop and closed-loop signals based on 1 Hz [44]. In Figs. 3 and 4, the variation in energy content (%) showed similar tendencies in both the control and AIS groups. The energy content (%) was higher in the closed-loop than that in the open-loop signal because the feedback mechanism through neurological information played an important role in postural control [48]. In the open-loop signal, sensory feedback information is not used because the posture is directly controlled by generated motor control in the central nervous system [48, 49]. However, using only open-loop does not provide enough information to maintain posture stability. Therefore, posture control using closed-loop via sensory feedback information is needed [48, 49]. The lowest energy content was observed at level 1, which was gradually increased up to vestibular sensory system level (levels 8 and 9), and then decreased. The highest energy content appeared in levels according to vestibular input. The next highest energy content was shown in somatosensory system levels (Figs. 3 and 4). Morphological deformity of the spine can alter the head position [11, 12] which changes the CoM of trunk segment [14]. The vestibular system has a role of controlling variation of head position at the quite standing posture [11, 15]. The altered trunk CoM is rearranged by ankle and hip

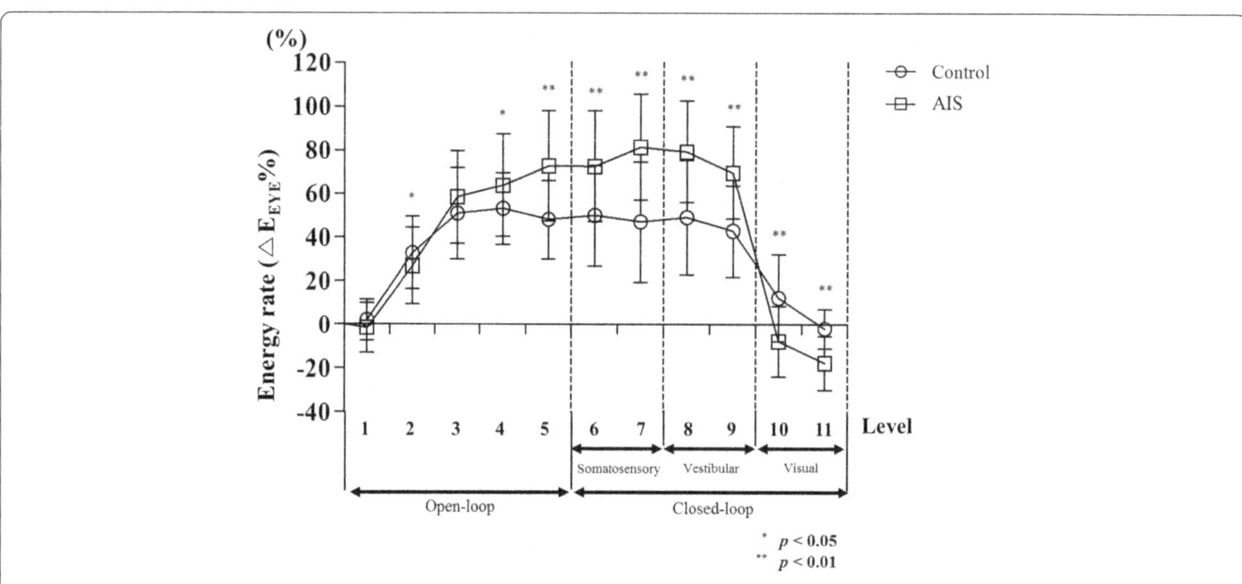


**Fig. 5** Difference in energy rate ($\Delta E_{EYE}\%$) of energy contents between control and AIS groups. The energy rate of each group was obtained using energy content during quiet standing with eyes open and eyes closed

**Table 2** Quantification result (mean ± standard deviation) of energy contents depending on level in AIS groups by severity

| | | | Open-loop | | | | | Closed-loop | | | | | |
|---|---|---|---|---|---|---|---|---|---|---|---|---|---|
| | | | | | | | | Somatosensory | | Vestibular | | Visual | |
| | Level | | 1 | 2 | 3 | 4 | 5 | 6 | 7 | 8 | 9 | 10 | 11 |
| Severity (n = 32) | Mild (n = 12) | $E_{EO}$% (%) | 0.03 ±0.01 | 0.10 ±0.06 | 0.22 ±0.11 | 0.73 ±0.38 | 2.56 ±0.86 | 6.01 ±1.71 | 12.44 ±2.74 | 15.66 ±3.83 | 16.51 ±3.32 | 14.13 ±2.31 | 2.54 ±0.38 |
| | | $E_{EC}$% (%) | 0.03 ±0.01 | 0.12 ±0.03 | 0.34 ±0.11 | 1.23 ±0.36 | 4.09 ±0.72 | 8.81 ±1.02 | 18.34 ±4.99 | 25.23 ±6.01 | 25.54 ±2.67 | 12.85 ±1.47 | 2.11 ±0.76 |
| | | *p*-value | NS | NS | NS | $p < 0.01$ | $p < 0.01$ | $p < 0.01$ | $p < 0.01$ | $p < 0.01$ | $p < 0.01$ | $p < 0.05$ | NS |
| | Moderate (n = 10) | $E_{EO}$% (%) | 0.04 ±0.03 | 0.07 ±0.03 | 0.15 ±0.09 | 0.51 ±0.33 | 2.03 ±0.99 | 7.71 ±2.91 | 12.52 ±3.33 | 14.69 ±4.58 | 16.92 ±3.57 | 13.92 ±3.07 | 2.02 ±0.83 |
| | | $E_{EC}$% (%) | 0.05 ±0.01 | 0.10 ±0.03 | 0.25 ±0.06 | 0.85 ±0.06 | 3.45 ±1.72 | 14.23 ±4.06 | 24.26 ±5.12 | 26.64 ±6.52 | 29.18 ±6.61 | 10.81 ±2.64 | 1.71 ±0.58 |
| | | *p*-value | NS | NS | NS | $p < 0.01$ | $p < 0.01$ | $p < 0.01$ | $p < 0.01$ | $p < 0.01$ | $p < 0.01$ | $p < 0.05$ | NS |
| | Severe (n = 10) | $E_{EO}$% (%) | 0.06 ±0.05 | 0.10 ±0.04 | 0.26 ±0.02 | 0.61 ±0.26 | 2.29 ±0.65 | 6.57 ±1.66 | 9.50 ±3.15 | 13.69 ±3.42 | 15.23 ±2.66 | 12.88 ±1.67 | 2.87 ±0.99 |
| | | $E_{EC}$% (%) | 0.06 ±0.04 | 0.15 ±0.17 | 0.41 ±0.16 | 1.54 ±0.72 | 4.31 ±1.78 | 12.21 ±2.59 | 19.20 ±4.25 | 26.73 ±3.47 | 27.40 ±4.28 | 10.12 ±2.99 | 3.11 ±0.97 |
| | | *p*-value | NS | NS | NS | $p < 0.01$ | $p < 0.01$ | $p < 0.01$ | $p < 0.01$ | $p < 0.01$ | $p < 0.01$ | $p < 0.05$ | NS |

*NS* Not significant

joints control strategy of the somatosensory system [16]. The energy content (%) showed relatively low values at visual levels. However, visual information also plays an important role in posture control as it provides continuous information to maintain a horizontal relationship between the body and the surrounding environment [17]. Figure 3c shows statistical result of the difference in energy content (%) acquired in the control group under eyes open and eyes closed conditions. Figure 4(c) shows the result for the AIS group. For control and AIS groups, significant differences in energy contents between eyes open and eyes closed conditions were shown at levels 4~ 10 ($p < 0.05$ or $p < 0.01$). At levels 4~ 9, the energy content (%) of eyes closed condition was higher than that of eyes open condition. Both groups showed higher energy content (%) in the eyes closed condition at levels 4 and 5 which corresponded to the open-loop signal. This is because the human body uses energy content (%) in the open-loop signal to compensate for body stiffness arising from the lack of visual information in eyes closed condition. The joint velocity increment for the posture rectification in eyes closed condition might increase joint stiffness [50]. Effective stochastic activity fluctuation of the open-loop control mechanisms is closely related to stiffness variations in their musculoskeletal system [51, 52]. In addition, the high reliance on eyes closed condition at somatosensory (levels 6 and 7) and vestibular (levels 8 and 9) might be compensation for the lack of visual information [37, 53]. On the other hand, the lower energy content (%) in eyes closed condition rather than eyes open was found at level 10 because visual information was not provided in the eyes closed condition [37].

The difference in energy rate ($\Delta E_{EYE}$%) for control and AIS groups between eyes open and eyes closed conditions was analyzed (Table 1 and Fig. 5). The purpose of this analysis was to determine the effect of morphological deformation of the spine on the sensory system. Results revealed a significant difference in energy rate ($\Delta E_{EYE}$%) between control and AIS groups at levels corresponding to vestibular and somatosensory systems (AIS > control; all $p < 0.01$). This study was mainly focused on the results that corresponded to vestibular and

**Table 3** Quantification result (mean ± standard deviation) of $\Delta E_{EYE}$% depending on level in AIS groups by severity

| | | Open-loop | | | | | Closed-loop | | | | | |
|---|---|---|---|---|---|---|---|---|---|---|---|---|
| | | | | | | | Somatosensory | | Vestibular | | Visual | |
| | Level | 1 | 2 | 3 | 4 | 5 | 6 | 7 | 8 | 9 | 10 | 11 |
| $\Delta E_{EYE}$% (%) | Mild | −1.11 ± 10.79 | 37.81 ± 18.59 | 58.77 ± 25.55 | 68.95 ± 26.10 | 60.08 ± 27.53 | 59.89 ± 26.02 | 60.36 ± 29.43 | 66.42 ± 27.98 | 58.54 ± 21.25 | −20.59 ± 12.26 | −22.93 ± 11.62 |
| | Moderate | −0.12 ± 10.51 | 27.49 ± 16.98 | 57.15 ± 23.99 | 64.84 ± 28.61 | 70.05 ± 25.96 | 72.22 ± 31.69 | 81.59 ± 27.65 | 76.29 ± 20.79 | 64.13 ± 23.25 | −10.95 ± 14.63 | −20.12 ± 10.10 |
| | Severe | −3.27 ± 14.25 | 19.43 ± 18.47 | 56.48 ± 16.72 | 57.79 ± 19.96 | 88.21 ± 25.27 | 85.87 ± 21.20 | 102.04 ± 18.51 | 95.24 ± 23.82 | 85.99 ± 21.09 | −0.82 ± 10.32 | −12.17 ± 12.30 |

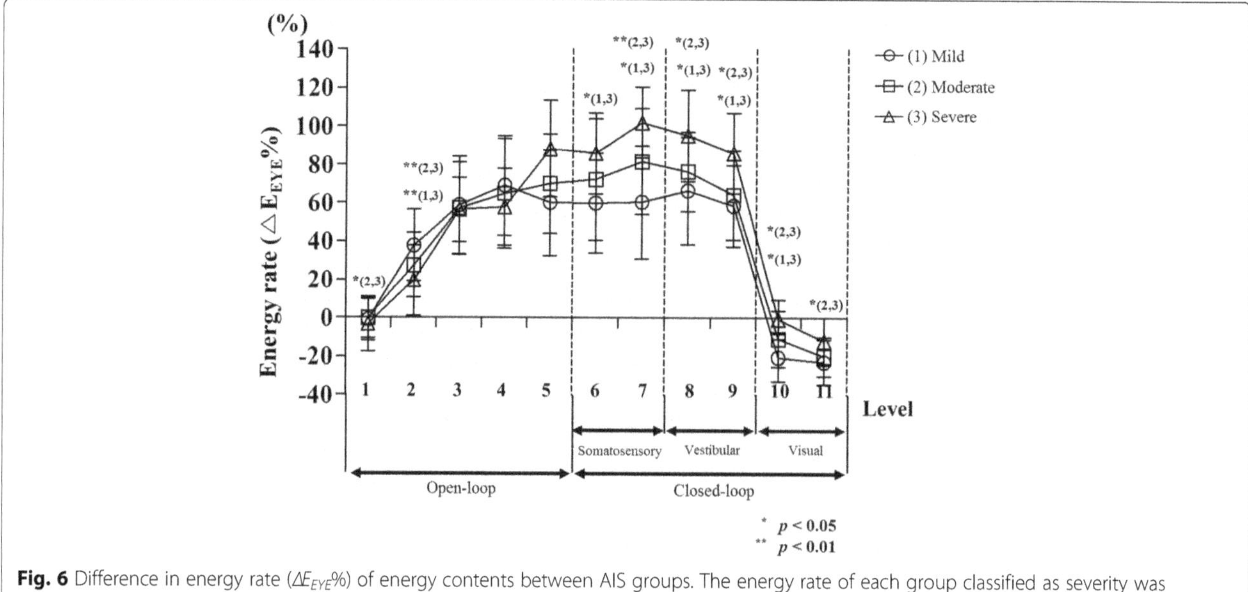


**Fig. 6** Difference in energy rate ($\Delta E_{EYE}$%) of energy contents between AIS groups. The energy rate of each group classified as severity was obtained using energy content during quiet standing with eyes open and eyes closed

somatosensory system because the open-loop signal (levels 1~ 5) was not required for feedback information of the sensory system and the energy rate differences were very low in visual sensory levels. The vestibular system works to control the balance of the head [54]. The somatosensory system maintains postural stability through kinematic information which is generated from muscles and joints [17]. The $\Delta E_{EYE}$% of the AIS group in the vestibular levels was significantly higher than that of the control group because the AIS group showed greater energy content (%) in eyes closed condition than that in eyes open condition. The human body is composed of multiple segments that are continuously moved by joints to align the body over the base of support to maintain postural stability [55–57]. However, the morphological deformity of the spine may alter the orientation of the head in 3-dimensional planes [11, 12]. The change in head position is not only the main cause of a change in trunk CoM [58]. It also changes the ability to provide information about the vestibular system [59, 60]. The human body adopts a strategy that utilizes more vestibular information in maintaining postural stability in case of visual information paucity while in eyes closed condition [37, 61]. For this reason, the $\Delta E_{EYE}$% of the AIS group was higher than that of the control group in levels corresponding to the vestibular system. The $\Delta E_{EYE}$% of the AIS group was also higher than that of the control group in somatosensory levels (levels 6 and 7). The morphological deformation of the spine causes an asymmetric activity of muscles between concave and convex side of spine [62]. Musculoskeletal impairments can change the head position [11, 12], alter trunk CoM position [14], and cause postural instability [62]. The somatosensory system can rearrange the altered CoM of trunk with ankle and hip joint control strategy to maintain postural stability [8, 17]. Therefore, AIS patients relied on the somatosensory system which plays a role in controlling the equilibrium of the posture through information from joints and muscles [10, 17].

In this study, we also analyzed the effect of morphological deformation of the spine on the sensory system depending on AIS severity (Tables 2, 3, and Fig. 6). Results revealed that the $\Delta E_{EYE}$% had high values as morphological deformity worsened at levels 8 and 9 which corresponded to the vestibular system ($p < 0.05$). In addition, a high energy rate ($\Delta E_{EYE}$%) was observed as morphological deformity worsened at somatosensory level (level 7) and visual sensory level (level 10) ($p < 0.05$ or $p < 0.01$). At level 6 (severe vs. mild) and level 11 (severe vs. moderate), energy rate difference was only observed between two severities ($p < 0.05$). The energy rate was gradually increased according to the degree of AIS severity. However, it is difficult to propose a generalized conclusion about whether a clear difference in the sensory system can maintain postural stability depending on AIS severity. This is because the standard deviation showed a large variation according to each severity level, as shown in the graph of Fig. 6. However, the energy rate ($\Delta E_{EYE}$%) tended to be higher as overall severity increased. This implies that the energy content (%) of the CoP signal through the DWT method is one of important kinematic variables to distinguish AIS severity.

## Conclusion

AIS patients showed high dependency on vestibular and somatosensory systems in maintaining their postural stability compared to the control group when CoP signal was analyzed using the DWT method to determine differences in sensory systems. These results can be interpreted as compensation due to instable posture caused by asymmetry of concave and convex curves in the spine and altered head position from morphological deformity of the spine. In the present study, posture control ability was also analyzed according to the degree of AIS severity. Although it is difficult to conclude any clear difference among severities due to high standard deviation of each severity, the tendency of the energy rate according to the degree of severity differed. Evaluating CoP signals from quiet standing posture of AIS patients, regarding sensory system using DWT method can give helpful reference to compare postural stability before and after various treatments such as spinal fusion, bracing and rehabilitation. Future study is needed to deduce more generalized and concrete results by acquiring more data and interpreting with SOT. This will allow more clear interpretation of changes in physical exercise ability caused by morphological deformation of the spine.

**Acknowledgements**
This work was supported by the National Research Foundation of Korea (NRF) grant funded by the Korea government (MSIP; No. NRF-2016R1A2B3009013).

**Authors' contributions**
TS and HY wrote the paper and performed experiments and contribute to data analysis; DL assisted our research, and helped in the revision phase of our manuscript; SWS and JHY performed experiments and examined all subjects; HK and JHM designed the experiments and led the manuscript. All authors discussed and commented on the manuscript. All authors read and approved the final manuscript.

**Competing interests**
The authors declare that they have no competing interests.

**Author details**
[1]Department of Bio-Mechatronic Engineering, College of Biotechnology and Bioengineering, Sungkyunkwan University, Natural Sciences Campus, 2066 Seobu-ro, Jangan-gu, Suwon-si, Gyeonggi-do 16419, South Korea.
[2]Department of Orthopedics, Scoliosis Research Institute, Korea University Medical College, Guro Hospital, 148 Gurodong-ro, Guro-gu, Seoul 08308, South Korea.

**References**

1. Cheung CWJ, Zheng Y. Development of 3-D ultrasound system for assessment of adolescent idiopathic scoliosis (AIS). In: Lim CT, JCH G, editors. WCB 2010, IFMBE Proceedings, vol. 31; 2010. p. 584–7.
2. Cheung CJ, Law S, Zheng Y. Development of 3-D ultrasound system for assessment of adolescent idiopathic scoliosis (AIS): and system validation. Conf Proc IEEE Eng Med Biol Soc. 2013:6474–7.
3. Herman R, Mixon J, Fisher A, Maulucci R, Stuyck J. Idiopathic scoliosis and the central nervous system: a motor control problem. The Harrington lecture, 1983. Scoliosis Research Society. Spine. 1985;10(1):1–14.
4. Dobosiewicz K. Neurophysiological mechanism of the unloading reflex as a prognostic factor in the early stages of idiopathic scoliosis. Eur Spine J. 1997;6(2):93–7.
5. Nault M, Allard P, Hinse S, Le Blanc R, Caron O, Labelle H, et al. Relations between standing stability and body posture parameters in adolescent idiopathic scoliosis. Spine. 2002;27(17):1911–7.
6. Assaiante C, Mallau S, Jouve JL, Bollini G, Vaugoyeau M. Do adolescent idiopathic scoliosis (AIS) neglect proprioceptive information in sensory integration of postural control? PLoS One. 2012;7(7):e40646.
7. Veldhuizen A, Weber D, Webb P. The aetiology of idiopathic scoliosis: biomechanical and neuromuscular factors. Eur Spine J. 2000;9(3):178–84.
8. Gur G, Dilek B, Ayhan C, Simsek E, Aras O, Aksoy S, et al. Effect of a spinal brace on postural control in different sensory conditions in adolescent idiopathic scoliosis: a preliminary analysis. Gait Posture. 2015;41(1):93–9.
9. Haumont T, Gauchard GC, Lascombes P, Perrin PP. Postural instability in early-stage idiopathic scoliosis in adolescent girls. Spine. 2011;36(13):E847–E85.
10. Guo X, Chau WW, Hui-Chan CW, Cheung CS, Tsang WW, Cheng JC. Balance control in adolescents with idiopathic scoliosis and disturbed somatosensory function. Spine. 2006;31(14):E437–40.
11. Chansirinukor W, Wilson D, Grimmer K, Dansie B. Effects of backpacks on students: measurement of cervical and shoulder posture. Aust J Physiother. 2001;47(2):110–6.
12. Chow DH, Kwok ML, Cheng JC, Lao ML, Holmes AD, Au-Yang A, et al. The effect of backpack weight on the standing posture and balance of schoolgirls with adolescent idiopathic scoliosis and normal controls. Gait Posture. 2006;24(2):173–81.
13. Cheung J, Halbertsma JP, Veldhuizen AG, Sluiter WJ, Maurits NM, Cool JC, et al. A preliminary study on electromyographic analysis of the paraspinal musculature in idiopathic scoliosis. Eur Spine J. 2005;14(2):130–7.
14. Maurer C, Mergner T, Bolha B, Hlavacka F. Vestibular, visual, and somatosensory contributions to human control of upright stance. Neurosci Lett. 2000;281(2):99–102.
15. Angelaki DE, Cullen KE. Vestibular system: the many facets of a multimodal sense. Annu Rev Neurosci. 2008;31:125–50.
16. Horak FB, Shupert CL, Dietz V, Horstmann G. Vestibular and somatosensory contributions to responses to head and body displacements in stance. Exp Brain Res. 1994;100(1):93–106.
17. Page P, Frank C, Lardner R. Assessment and treatment of muscle imbalance: the Janda approach. Chicago: Human Kinetics; 2010:19–25.
18. De Santiago HA, Reis JG, Gomes MM, Da Silva Herrero CF, Defino HL, De Abreu DC. The influence of vision and support base on balance during quiet standing in patients with adolescent idiopathic scoliosis before and after posterior spinal fusion. Spine J. 2013;13(11):1470–6.
19. Peterka RJ, Loughlin PJ. Dynamic regulation of sensorimotor integration in human postural control. J Neurophysiol. 2004;91(1):410–23.
20. Colnat-Coulbois S, Gauchard GC, Maillard L, Barroche G, Vespignani H, Auque J, et al. Management of postural sensory conflict and dynamic balance control in late-stage Parkinson's disease. Neuroscience. 2011; 193:363–9.
21. Micozzi MS. Fundamentals of complementary and alternative medicine. St. Louis: Saunders Elsevier; 2014:240–6.
22. Mitchell DE, Dai C, Rahman MA, Ahn JH, Della Santina CC, Cullen KE. Head movements evoked in alert rhesus monkey by vestibular prosthesis stimulation: implications for postural and gaze stabilization. PLoS One. 2013;8(10):e78767.
23. Horak FB. Postural orientation and equilibrium: what do we need to know about neural control of balance to prevent falls? Age Ageing. 2006;35(Suppl 2):ii7–ii11.
24. Horak FB. Postural compensation for vestibular loss. Ann N Y Acad Sci. 2009; 1164(1):76–81.
25. Riemann BL, Lephart SM. The sensorimotor system, part II: the role of proprioception in motor control and functional joint stability. J Athl Train. 2002;37(1):80–4.
26. Riemann B, Lephart S. The sensorimotor system, part I: the physiologic basis of functional joint stability. J Athl Train. 2002;37:71–9.
27. Blackburn T, Guskiewicz KM, Petschauer MA, Prentice WE. Balance and joint stability: the relative contributions of proprioception and muscular strength. J Sport Rehabil. 2000;9(4):315–28.
28. Ting LH, McKay JL. Neuromechanics of muscle synergies for posture and movement. Curr Opin Neurobiol. 2007;1(6):622–8.
29. Beaulieu M, Toulotte C, Gatto L, Rivard C, Teasdale N, Simoneau M, et al. Postural imbalance in non-treated adolescent idiopathic scoliosis at different periods of progression. Eur Spine J. 2009;18(1):38–44.

30. De Abreu DCC, Gomes MM, De Santiago HAR, Da Silva Herrero CFP, Porto MA, Defino HLA. What is the influence of surgical treatment of adolescent idiopathic scoliosis on postural control? Gait Posture. 2012;36(3):586–90.
31. Oppenheim U, Kohen-Raz R, Alex D, Kohen-Raz A, Azarya M. Postural characteristics of diabetic neuropathy. Diabetes Care. 1999;22(2):328–32.
32. Kirchner M, Schubert P, Schmidtbleicher D, Haas CT. Evaluation of the temporal structure of postural sway fluctuations based on a comprehensive set of analysis tools. Physica A Stat Mech Appl. 2012;391(20):4692–703.
33. Ongun N, Atalay NS, Degirmenci E, Sahin F, Bir LS. Tetra-ataxiometric posturography in patients with migrainous vertigo. Pain Physician. 2016;19:E87–96.
34. Karapolat H, Celebisoy N, Kirazli Y, Bilgen C, Eyigor S, Gode S. Does betahistine treatment have additional benefits to vestibular rehabilitation? Eur Arch Otorhinolaryngol. 2010;267(8):1207–12.
35. Bizid R, Jully JL, Gonzalez G, François Y, Dupui P, Paillard T. Effects of fatigue induced by neuromuscular electrical stimulation on postural control. J Sci Med Sport. 2009;12(1):60–6.
36. Nagy E, Toth K, Janositz G, Kovacs G, Feher-Kiss A, Angyan L, et al. Postural control in athletes participating in an ironman triathlon. Eur J Appl Physiol. 2004;92(4–5):407–13.
37. Chagdes JR, Rietdyk S, Haddad JM, Zelaznik HN, Raman A, Rhea CK, et al. Multiple timescales in postural dynamics associated with vision and a secondary task are revealed by wavelet analysis. Exp Brain Res. 2009;197(3):297–310.
38. Sperandio EF, Alexandre AS, Liu CY, Poletto PR, Gotfryd AO, Vidotto MC, et al. Functional aerobic exercise capacity limitation in adolescent idiopathic scoliosis. Spine J. 2014;14(10):2366–72.
39. Wang WJ, Hung VWY, Lam TP, Ng BKW, Qin L, Lee KM, et al. The association of disproportionate skeletal growth and abnormal radius dimension ratio with curve severity in adolescent idiopathic scoliosis. Eur. Spine J. 2010;19(5):726–31.
40. Lemay JF, Gagnon DH, Nadeau S, Grangeon M, Gauthier C, Duclos C. Center-of-pressure total trajectory length is a complementary measure to maximum excursion to better differentiate multidirectional standing limits of stability between individuals with incomplete spinal cord injury and able-bodied individuals. J Neuroeng Rehabil. 2014;11(1):8.
41. Guerraz M, Bronstein AM. Ocular versus extraocular control of posture and equilibrium. Neurophysiol Clin. 2008;38(6):391–8.
42. Kohen-Raz R, Volkman FR, Cohen DJ. Postural control in children with autism. J Autism Dev Disord. 1992;22(3):419–32.
43. Kim J, Kim C, Lee J, Kwon Y, Eom G, Tak G, et al. Human postural control against external force perturbation applied to the high-back. Int J Precis Eng Manuf. 2009;10(4):147–51.
44. Collins JJ, De Luca CJ. Open-loop and closed-loop control of posture: a random-walk analysis of center-of-pressure trajectories. Exp Brain Res. 1993; 95(2):308–18.
45. Maatar D, Fournier R, Lachiri Z, Nait-Ali A. Discrete wavelet and modified PCA decompositions for postural stability analysis in biometric applications. J Biomed Sci Eng. 2011;4(08):543–51.
46. Suarez H, Sotta G, San Roman C, Arocena S, Ferreira E, Geisinger D, et al. Postural response characterization in elderly patients with bilateral vestibular hypofunction. Acta Otolaryngol. 2013;133(4):361–7.
47. Suarez H, Ferreira E, Alonso R, Arocena S, San Roman C, Herrera T, et al. Postural responses applied in a control model in cochlear implant users with pre-lingual hearing loss. Acta Otolaryngol. 2016;136(4):344–50.
48. Human motor behavior: an introduction. New Jersey: Lawrence Erlbaum Associates; 2014. p. 93–116.
49. Martin ST, Kessler M. Neurologic interventions for physical therapy. St. Louis: Saunders Elsevier; 2015:33–55.
50. Albertsen IM, Ghédira M, Gracies JM, Hutin É. Postural stability in young healthy subjects–impact of reduced base of support, visual deprivation, dual tasking. J Electromyogr Kinesiol. 2017;33:27–33.
51. Lacour M, Bernard-Demanze L, Dumitrescu M. Posture control, aging, and attention resources: models and posture-analysis methods. Neurophysiol Clin. 2008;38(6):411–21.
52. Collins JJ, De Luca CJ. The effects of visual input on open-loop and closed-loop postural control mechanisms. Exp Brain Res. 1995;103(1):151–63.
53. Ray CT, Horvat M, Croce R, Mason RC, Wolf SL. The impact of vision loss on postural stability and balance strategies in individuals with profound vision loss. Gait Posture. 2008;28(1):58–61.
54. Treleaven J. Sensorimotor disturbances in neck disorders affecting postural stability, head and eye movement control. Man Ther. 2008;13(1):2–11.
55. Masso PD, Gorton GE III. Quantifying changes in standing body segment alignment following spinal instrumentation and fusion in idiopathic scoliosis using an optoelectronic measurement system. Spine. 2000;25(4):457–62.
56. Hsu WL, Scholz JP, Schöner G, Jeka JJ, Kiemel T. Control and estimation of posture during quiet stance depends on multijoint coordination. J Neurophysiol. 2007;97(4):3024–35.
57. Wu J, McKay S, Angulo-Barroso R. Center of mass control and multi-segment coordination in children during quiet stance. Exp Brain Res. 2009;196(3):329–39.
58. Lamontagne A, Paquet N, Fung J. Postural adjustments to voluntary head motions during standing are modified following stroke. Clin Biomech. 2003; 18(9):832–42.
59. Demer JL, Oas JG, Baloh RW. Visual-vestibular interaction in humans during active and passive, vertical head movement. J Vestib Res. 1993;3(2):101–14.
60. Demer JL, Viirre ES. Visual-vestibular interaction during standing, walking, and running. J Vestib Res. 1996;6(4):295–313.
61. Schwesig R, Goldich Y, Hahn A, Müller A, Kohen-Raz R, Kluttig A, et al. Postural control in subjects with visual impairment. Eur J Ophthalmol. 2011;21(3):303.
62. Zabjek KF, Coillard C, Rivard CH, Prince F. Estimation of the Centre of mass for the study of postural control in idiopathic scoliosis patients: a comparison of two techniques. Eur Spine J. 2008;17(3):355–60.

3

# Decoding the grasping intention from electromyography during reaching motions

Iason Batzianoulis[1*], Nili E. Krausz[2,3], Ann M. Simon[2,3], Levi Hargrove[2,3,4] and Aude Billard[1]

**Abstract**

**Background:** Active upper-limb prostheses are used to restore important hand functionalities, such as grasping. In conventional approaches, a pattern recognition system is trained over a number of static grasping gestures. However, training a classifier in a static position results in lower classification accuracy when performing dynamic motions, such as reach-to-grasp. We propose an electromyography-based learning approach that decodes the grasping intention during the reaching motion, leading to a faster and more natural response of the prosthesis.

**Methods and Results:** Eight able-bodied subjects and four individuals with transradial amputation gave informed consent and participated in our study. All the subjects performed reach-to-grasp motions for five grasp types, while the elecromyographic (EMG) activity and the extension of the arm were recorded. We separated the reach-to-grasp motion into three phases, with respect to the extension of the arm. A multivariate analysis of variance (MANOVA) on the muscular activity revealed significant differences among the motion phases. Additionally, we examined the classification performance on these phases. We compared the performance of three different pattern recognition methods; Linear Discriminant Analysis (LDA), Support Vector Machines (SVM) with linear and non-linear kernels, and an Echo State Network (ESN) approach. Our off-line analysis shows that it is possible to have high classification performance above 80% before the end of the motion when with three-grasp types. An on-line evaluation with an upper-limb prosthesis shows that the inclusion of the reaching motion in the training of the classifier importantly improves classification accuracy and enables the detection of grasp intention early in the reaching motion.

**Conclusions:** This method offers a more natural and intuitive control of prosthetic devices, as it will enable controlling grasp closure in synergy with the reaching motion. This work contributes to the decrease of delays between the user's intention and the device response and improves the coordination of the device with the motion of the arm.

**Keywords:** Myoelectric control, Upper limb prosthesis, Pattern recognition, Reach-to-grasp motion

## Background

The loss of a hand impacts all aspects of daily life, including work, recreation and communication [1]. Prosthetic devices can help restore motor abilities lost after amputation, and can improve the quality of life for amputees. Despite these potential benefits, many people with upper-limb amputations do not use a prosthesis [2]. Studies examining the causes of rejection [3–5] report that secondary prosthesis rejection, i.e. after a period of use, is primarily due to dissatisfaction with prosthetic comfort, function or control. Improving prosthesis control systems could thus presumably provide additional functionalities to users and could minimize prosthesis rejection.

Surface electromyography (EMG) has been widely studied as an intuitive human-machine interface for controlling intelligent external devices, such as prosthetic hands [6, 7]. As amputees generally have a limited number of independent EMG sites available for controlling a multi-degree of freedom (DOF) prosthesis, we cannot rely on a one-to-one EMG-to-DOF control. Surgical methods, such as targeted muscle reinnervation (TMR) [8, 9] and regenerative peripheral-nerve interfaces (RPNIs) [10], can enable the control of a larger number of DOFs.

*Correspondence: iason.batzianoulis@epfl.ch
[1] Learning Algorithms and Systems Laboratory (LASA), School of Engineering, École Polytechnique Fédérale de Lausanne (EPFL), Route Cantonale, CH-1015 Lausanne, Switzerland
Full list of author information is available at the end of the article

Advanced signal-processing approaches could also be used to control multi-DOF prostheses with fewer independent EMG sites. EMG-based pattern-recognition systems are proposed for estimation of, both independent and simultaneous [11, 12], hand and wrist movements [13–15]. Using extrinsic hand muscles, pattern recognition has effectively classified functional hand-grasp patterns [13] and even individual finger movements [16, 17].

In these previous studies, generally subjects performed muscle contractions while maintaining their arm in a fixed position. However, training a classifier in a static position, as mentioned above, results in lower classification accuracy when the limb is in different positions or performs dynamic motions [18].

Reach-to-grasp movements are important activities of daily living that require dynamic motions. A few studies have attempted to decode grasping intention from EMG during reach-to-grasp motions, but only with able-bodied subjects [19, 20]. When reaching to grasp an object, the opening and closing of the hand is in coordination with the motion of the arm [21, 22], see Fig. 1a. More specifically, the human hand opens rapidly in the early stages of the reaching cycle, while the fingers converge gradually to their final configuration [21, 23].

A self-paced reaching motion of an able-bodied hand could take approximately 1 s to complete [19, 24]. In contrast, the activation of prosthetic hands could occur more than one second after the onset of the motion [25, 26] (see Fig. 1a). This makes unnatural the actuation of a prosthetic hand, lacking the natural arm-hand coordination. It also slows down the reach-to-grasp motion. It is crucial for prosthetic devices to react promptly to human intentions to enable for natural and intuitive operations [27]. To

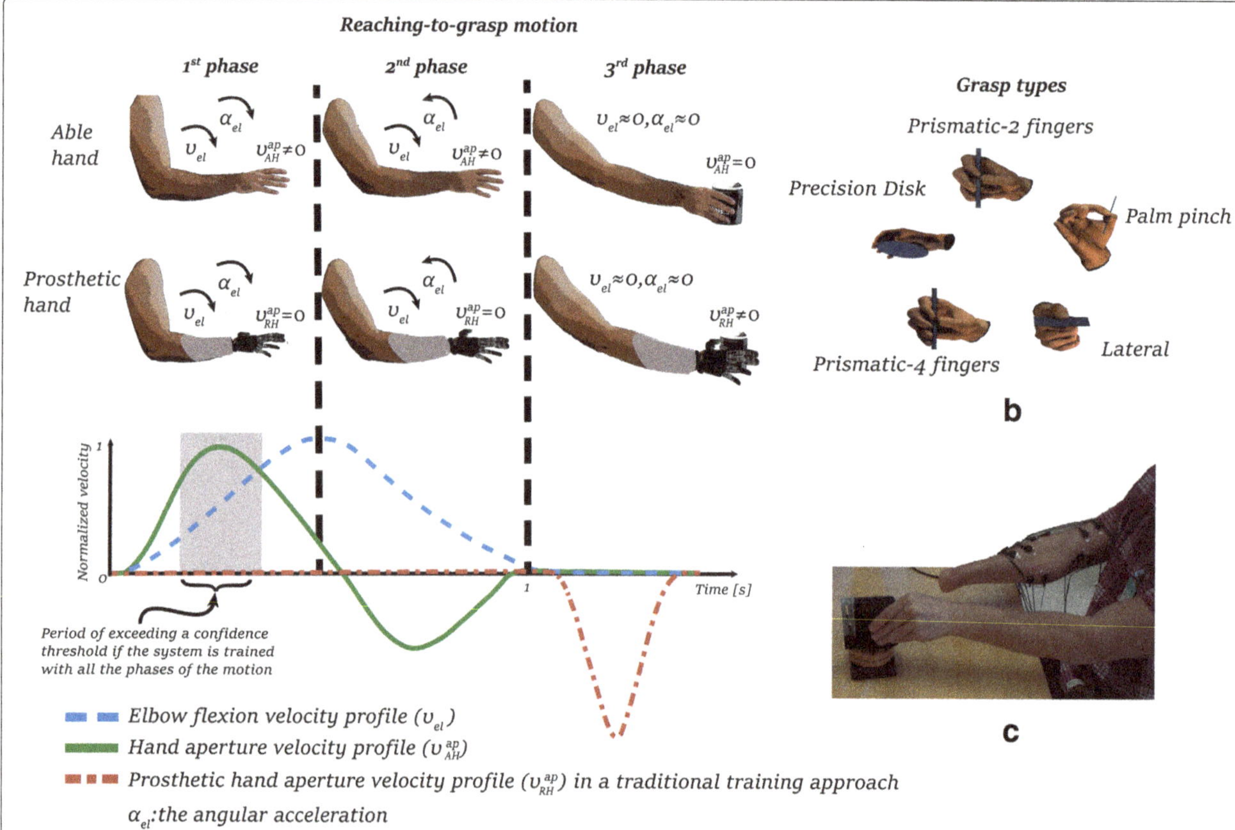


**Fig. 1 a** Typical profiles of velocities of the elbow and hand aperture in abled-bodied subject [21, 45, 46] compared to that generated with a traditional prosthetic device, as presented in [25, 26]. During reaching, the aperture of the human hand (solid green line) changes in coordination with the extension of the arm (dashed blue line). In contrast, the prosthetic hand (dash-dotted red line) begins its motion later in the reach-to-grasp cycle, once the elbow is fully extended. In our approach, we separate the reach-to-grasp motion into three phases (denoted by dashed vertical lines) according to the angular acceleration of the elbow joint $a_{el}$. We distinguish between acceleration, deceleration and rest phases. We present that a pattern recognition system, trained including the reaching motion, could gain efficient prediction confidence early in the reaching motion and, thus, activate faster a prosthetic device. **b** The selected five grasp types used in our classification, following the names and using figures from the taxonomy of [47]. **c** Experimental set-up for training the system with amputee subjects in data recordings. EMG-information from the amputated arm are recorded while the subject performs the reach and grasp motion with his/her intact arm

convey a seamless coordination between the device and the residual arm, it is important for the device to identify the grasping intention during the reaching phase.

In our previous work [19], we show that detection of the grasp type in synchrony with the reaching motion could enable smooth coordination of hand closure with the reaching motion, thus providing a more natural and seamless motion of the arm and a robotic hand. In our approach, the classification performance is related to the occurrence of hand pre-shape during reaching motions, following the natural pre-shape phase as documented in [23, 28]. Our prior study was limited to able-bodied subjects. Here, we extend this approach to decoding residual EMG in individuals with a below-elbow amputation. We compare the performance of four different classifiers: LDA, two SVMs, and an Echo State Network (ESN). Additionally, we explore a relationship between classification performance and the phases of the reach-to-grasp motion.

## Methods

### Experimental protocol

Eight able-bodied subjects (6 males and 2 females 25–32 years old) with no known neurological or physical deficits and four unilateral transradial amputees participated in the experiment. All able-bodied subjects were right-handed and performed the experiment with their dominant hand. All subjects were naive to pattern recognition control with the exception of three of the amputee subjects. Two of the amputee subjects had undergone a TMR surgery. Table 1 presents demographic information about the amputee subjects.

During the experiment, both the able-bodied subjects and the amputee subjects sat in front of a table, facing a computer screen, with their elbows at a 90°angle. The able-bodied subjects had their right hand closed on the table and they were asked to reach the object and grasp it with a predefined grasp type and to mentain the same wrist orientation for all grasp types. Custom computer software, called Control Algorithms for Prosthetics Systems (CAPS) [9], prompted users to initiate the reach-to-grasp motion and to go back to a resting position. Subjects performed the motion at their own pace. They were tasked to reach and grasp an object placed 30 cm away from the initial position of their hand. Once they had reached the objects, they were asked to remain in the same posture until they receiving a cue from CAPS to go back to rest position. The duration of each trial was 4 s with a 10 s rest between trials, to avoid fatigue. All subjects performed 30 trials for five grasp types, resulting in 150 trials in total.

In the experiments with the amputee subjects, the subjects were asked to reach the object and grasp it with their intact hand while trying to replicate the motion with their phantom limb, see Fig. 1c. These subjects started their self-paced motion when cued by the experimenter. Whenever a subject perceived an irregular or unexpected muscle contraction, the experiment was paused and the trial was repeated. Regular breaks were taken in order for the subjects to relax from the stress and effort of contracting their phantom limb. All the amputee subjects were able to complete the experiments.

**Table 1** Demographic information of the four transradial amputatee subjects

| Subject | TMR | User of myoprosthesis | Age | Years since amputation |
|---|---|---|---|---|
| TR1 | Yes | No | 25 | 7 |
| TR2 | No | myoelectric prosthesis | 53 | 38 |
| TR3 | Yes | myoelectric prosthesis | 51 | 2 |
| TR4 | No | myoelectric prosthesis | 68 | > 30 |

TR1 and TR3 underwent a TMR operation for neuroma pain, not for improving prosthetic control. TR1 is not a user of a myoprosthesis due to financial reasons

### Apparatus and pre-processing

Custom computer software [9] was used for signal acquisition, with EMG signals acquired at 1000 Hz with a 30–350 Hz band-pass filter using TI ADS1298 biosignal amplifiers. The EMG activity of 12 muscles was recorded: *Trapezius, Deltoid Anterior, Deltoid Medial, Deltoid Posterior, Biceps Brachii long head, Triceps Brachii long head, Brachialis, Flexor Digitorum Superficialis, Extensor Digitorum Communis, Flexor Carpi Ulnaris, Extensor Carpi Ulnaris, Flexor Carpi Radialis* (seven muscles of the upper arm and five muscles of the forearm). To construct a linear envelope, full-wave rectification was performed, followed by smoothing with a low-pass seventh-order Butterworth filter with cut-off frequency at 20Hz. At the end of this step, each channel was normalized by the maximum value recorded in the trials. A goniometer was placed on the subjects' elbow for measuring the onset and extension of the elbow. Please see the Additional file 1 for the exact location of the EMG sensors.

### Phases of the motion

As illustrated in Fig. 1a, during natural reach-to-grasp motion, the opening and closing of the hand is coordinated with the extension of the elbow [21, 22]. Typically for able-bodied subjects, the hand opens rapidly in the early stages of the reaching motion and decreases its velocity while approaching the object and reaching the final configuration [23]. The hand's velocity peak occurs before the peak velocity of the elbow extension [21]. Thus, the hand reaches its final grasp-posture after the peak velocity of the elbow extension. Because EMG recordings from upper and lower arm muscles encapsulate information on the hand motion, the EMG patterns will likely differ in the different phases, specifically before and after the elbow extension velocity peak. Taking inspiration from this behavior, we divided the reach-to-grasp motion into three phases with respect to the extension of the elbow joint. The first phase is defined as the interval from motion onset (i.e. when the angular velocity of the elbow joint exceeds a velocity threshold) until the angular velocity of the elbow reaches its maximum. The second phase is the interval between the aforementioned maximum angular velocity and the end of the reaching motion (i.e. when the angular velocity of the elbow drops below a velocity threshold). We define the third phase as the phase after the completion of the elbow extension. More specifically, we selected 25% of the duration of the reaching motion selected after the velocity drops below a threshold. The velocity threshold was set at 10% of the maximum angular velocity recorded for each subject.

We normalized the time of the duration of the reaching cycle. The reaching cycle corresponds to the time interval between the motion onset and the end of the extension of the elbow, i.e. when the angular velocity of the elbow drops below the velocity threshold. We performed a one-way multivariate analysis of variance (MANOVA) on the average values of the 12 EMG channels over the three phases for each grasp type. The Wilks Lambda test and the Pillai-Barlett Trace test were used to compare the results to a significance level of 5% ($a = 0.05$). We present the results in "Results" section.

To further investigate the three phases, we grouped the EMG signals of the classes together for each phase. The signals were divided into sliding time windows and the average activity of each channel was extracted, creating a vector of N elements (N corresponds to the number of EMG channels). A principal component analysis (PCA) was performed with the data from the third phase and the two remaining phases were projected into the new hyperplane. The distribution of the data on the first two principal components was fitted to Gaussian Mixture Models (GMMs) for each phase, and the number of Gaussian components was optimized by the Bayesian Information Criterion (BIC). We performed an analysis on the complete muscle set ($N = 12$) and an analysis using only the muscles of the forearm ($N = 5$).

### Classification methods for decoding the grasping intention

The preprocessed EMG signals were analyzed using a sliding time-window of 150 ms with an increment of 50 ms. The time window and increment lengths were chosen to be between the preferred values for an online implementation, as suggested in [29]. We used no dimensionality-reduction method (such as PCA) in this step. For each grasp type, 10 trials were randomly selected as the testing set. The remaining 20 trials of each grasp type constituted the training and validation sets. A four-fold cross-validation was performed to optimize the hyper-parameters for each classification method.

The classification accuracy of four classification methods was compared, namely for an LDA classifier, an SVM with linear kernel, an SVM with a Radial Basis Function (RBF) kernel and an ESN. For each classification method, one classifier was trained per subject. We did not attempt any inter-subject training. We inserted the classification outcome of each time window into a Majority Vote (MV) algorithm, that uses a buffer of 0.5 s history to predict the winning class.

In the cases of LDA and SVM, we extracted three features for each time window and introduced them into the classifier. Following previously described methods for EMG pattern recognition [30], we chose three features; the average activation of each time window, its waveform length, and the number of slope changes. In the case of ESN, we did not perform any feature extraction, treating the problem as a multidimensional time-series problem.

#### *Linear discriminant analysis*

LDA is one of the most commonly used classification algorithms for biomedical signals due to its performance and robustness. LDA finds a linearly optimal combination of the features in order to separate between classes. A fitting function estimates the parameters of a Gaussian distribution for each class and finds the probability of each point belonging to a class. Despite the linear nature of LDA, it has been shown to perform well in the classification of EMG signals [31].

#### *Support vector machine*

We tested two types of kernels, i.e. a linear kernel and an RBF kernel. In the case of the linear kernel, after a grid search, we optimized the penalty factor $C$. Likewise in the case of the RBF kernel, we optimized by a grid search both the penalty factor $C$ and the $\gamma$ parameter.

#### *Echo state networks*

ESN [32] is an effective recurrent neural network (RNN) that has attracted substantial interest due to its performance in time-series [33, 34]. The core of ESN is a large fixed reservoir. The reservoir contains a large number of randomly and sparsely connected neurons. The determination of the readout weights is the only trainable part; the weights can be obtained simply by linear regression. The necessary and sufficient condition to generate the echo state is based on information from the dynamic reservoir, such as the spectral radius of the internal weight-matrix. We optimized the three hyper-parameters; number of neurons, spatial radius and regularization parameter, by a grid search.

### Physical prosthesis control

For the purpose of the online implementation, we used the RIC hand [35], a prototype prosthesis with two degrees of actuation, that is able to perform two grasping postures: hand open, power grasp and prismatic-2 fingers. Due to its design, the RIC hand can offer access to a lower level control of the actuators. In this control scheme, we could control the actuators directly hence avoid any delays that arise from using control interfaces offered by commercial prosthesis. We mounted the prosthesis on a socket fitted to the user's residual forearm and a goniometer was placed on the elbow joint to record arm extension. We collected 12 EMG signals from the arm, preprocessed and classified them in real time and we inserted the classification output into a majority vote algorithm. The buffer of the majority vote was 0.5 s. Once the majority vote confidence exceeded a threshold of 0.5, i.e. more than half of the votes belonged to the same class, the corresponding command was sent to the prosthesis.

One subject, TR4, participated in a real-time control experiment. During the training phase, we cued the subject to perform 20 reach-to-grasp motions for each trained grasp type. The collected EMG signals were used to train two SVMs with RBF kernels: the first classifier used EMG from all the motion phases, whereas the second classifier was trained with EMG collected after arm extension (only the third phase). During the testing phase, the subject performed two sets of 20 reach-to-grasp trials for each classifier with the prosthesis turned on. Prior to the testing phase, the subject controlled the prosthesis for 10–15 min to familiarize themself with the device control. We used two metrics to compare the performance of the classifiers: the classification accuracy, and the time to reach a 0.5 majority-vote confidence level. We performed a two-sample t-test to validate the null hypothesis and to determine if there were significant differences between the results from the two classifiers.

## Results

### Phases of the motion

To examine the muscle-activation patterns during the reaching motion, we divided the recorded EMG signals in two groups: muscles of the forearm and muscles of the upper arm. Figure 2 presents representative examples of the average EMG activity of each muscle group in normalized time; the blue color corresponds to the muscles of the forearm and the red color corresponds to the

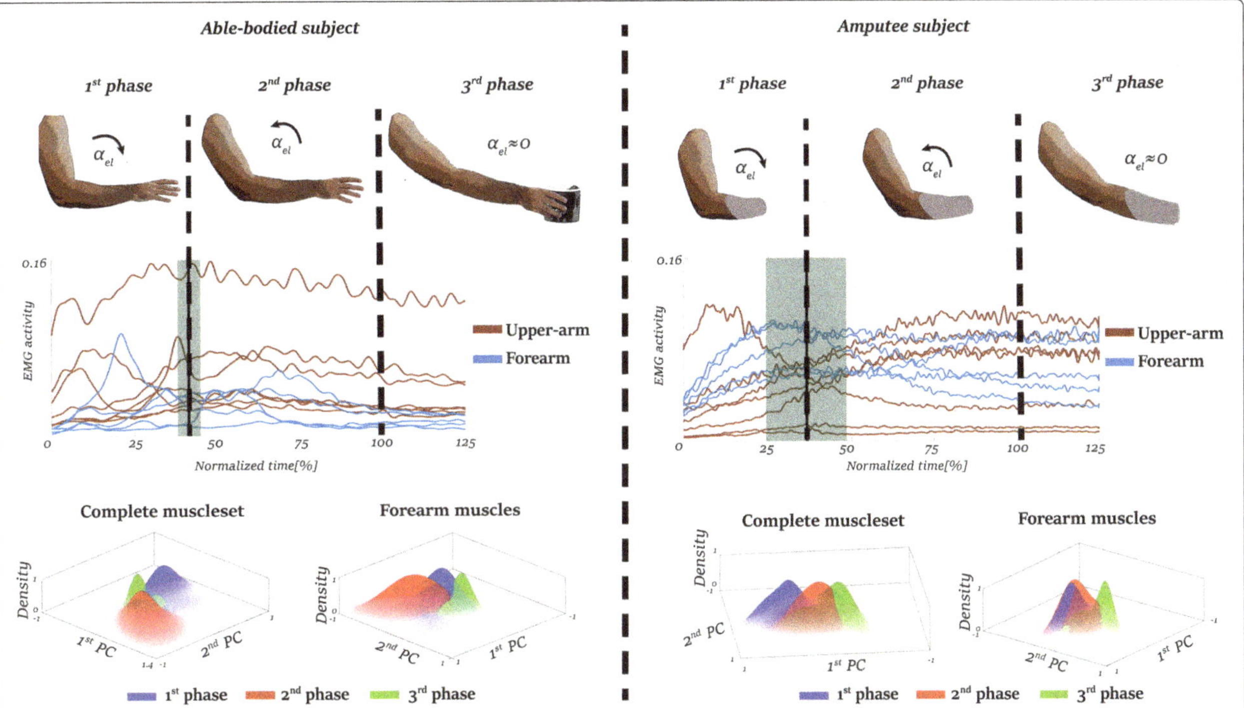


**Fig. 2** Representative examples of the EMG activity and the phases of the motion of the able-body subject 4 (left) and the amputee subject TR4 (right). The graphs in the middle correspond to the linear envelope of the EMG signals of the upper-arm (red lines) and the forearm (blue lines). The grey shadow areas correspond to the standard deviations of the timings where the shifts between the phases occurred. The graphs on the bottom of the figure show a representation with Gaussian Mixture Models (GMMs) of the EMG activity of the three phases projected on the first two components of third phase after performed Principal Component Analysis (PCA). The analysis was performed on the complete muscle set ($N = 12$) and when using only the muscles of the forearm ($N = 5$). The GMR representation shows limited overlap between the three phases, indicating differences on the EMG activity of the phases. Occasionally, an extended overlap occurred between the first and second phases as presented in the bottom-right graph. However, the third phase had rarely overlapped with any of the other two phases. Please see the Additional file 1 for more information regarding the EMG activity and its GMM representation for all the participants

muscles of the upper arm. The vertical dashed lines highlight the average time of the shift from the $1^{st}$ phase to the $2^{nd}$ phase, with the green shaded areas corresponding to the standard deviation of that shift. We calculated these events from the kinematic data recorded by the goniometer. The mean reaching time varied between $0.97 \pm 0.16s$ to $1.26 \pm 0.3s$ for able-bodied subjects and from $1.13 \pm 0.23s$ to $1.7 \pm 0.3s$ for amputee subjects

We identified the maximum elbow-joint angular velocity at 30–45% of the reaching motion for all the participants. We found no significant difference between able-bodied and amputee subjects regarding the timing of maximum elbow-joint angular velocity ($p = 0.55$, $t-value = 1.45$). The activation pattern of the distal muscles (muscles of the forearm) differed between amputee and able-bodied participants. In particular, the activation of the distal muscles in able-bodied subjects occurred earlier than in amputees. The muscular activity of the forearm muscles of able-bodied subjects reached a peak from 20–60% of the motion, decreasing as the motion is came closer to completion.

The EMG activity of forearm muscles of the amputee subjects increases gradually during the reaching. Whereas, the proximal muscles remain at a constant level of activation after the maximum angular velocity is reached. This difference in activation timing could have an effect on classification performance.

We compared the average activity across the three phases with a one-way MANOVA. As it rejected the null hypothesis, we found significant differences between the phases ($p < 0.01$, *Degrees of Freedom (DoF)*= 2) for all the subjects (able-bodied and amputees). Figure 2 presents the Gaussian models of the phases on the first two principal components for the complete muscle-set and the muscles of the forearm, respectively.

Although some models partially overlap, they have different mean values for all subjects, regardless of the muscle-set. In able-bodied subjects, the third-phase models are concentrated around the origin and have standard deviations smaller than the other phase models.

For amputees, the third-phase models are concentrated around the origin, similar to the able-bodied results. however, these models cover an area larger than the corresponding models for the able-bodied subjects. Also, a larger overlap was found between the models first and second phases for all amputee subjects, with larger distance from the models of third-phase.

We compared the average muscle activity during the three phases for each class (i.e. grasp type) by performing a one-way MANOVA. The one-way MANOVA failed to reject the null hypothesis ($p < 0.001, DoF = 2$), indicating significant differences between the means of the phase models for all the subjects. The significant differences between the three phases show that the data from all classes in the space change continuously, thus reducing the ability for a classifier to generalize across the three phases if it is trained on only one of them.

### Decoding the grasping intention

As stated previously, we compared the performance of four classifiers (LDA, SVM with linear kernel, SVM with RBF kernel and ESN). Figure 3 presents the average classification accuracy of each classifier over a time interval of $2s$. After performing an analysis of variance (ANOVA) for a significance level of 5%-$a = 0.05$, we did not notice any significant differences between the classifiers' performance for each group of classes ($\{p = 0.7, F - value = 0.43\}$, $\{p = 0.5, F - value = 0.97\}$ and $\{p = 0.8, F - value = 0.35\}$ for 5, 4 and 3 classes respectively). However, the SVM classifier with the RBF kernel performed better than the other classifiers with $60.45 \pm 8.2\%$, $65.82 \pm 8\%$ and $77.4 \pm 5.88\%$ classification accuracy for 5,4 and 3 classes, respectively, but this difference was not significant. This was followed by the SVM with the linear kernel, the ESN and the LDA. As the SVM-RBF classifier achieved slightly better performance, the rest of the results correspond to the performance of this classifier.

Figure 4 presents the classification performances of the five grasp types in each of the three motion phases. Poor classification performance occurred during the first phase in both amputee and able-bodied subjects. Accuracy improved in the subsequent two phases (see Fig. 4a and d). Grasp types; precision disk, palm pinch and lateral grasp, yielded the best performance in second and third phases for amputee subjects (see Fig. 4b and c). The lateral grasp improved from $58.1 \pm 6.2\%$ in the second phase to $60.45 \pm 8.2\%$ in the third phase. Accordingly, the precision disk and palm pinch increased from $60.2 \pm 10.2\%$ to

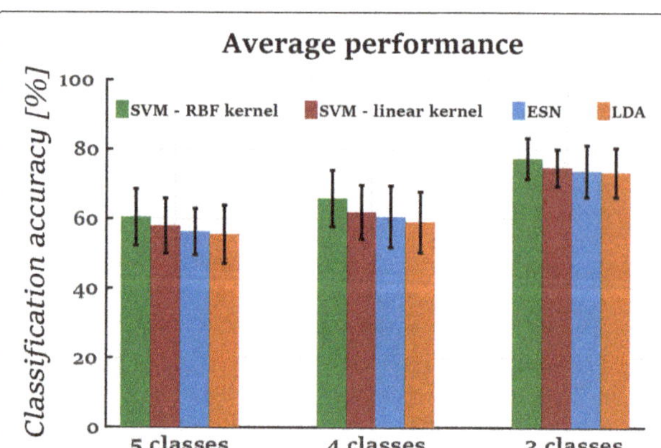


**Fig. 3** The average performance of all the classification models through out the whole trajectory of $2s$ among all the subjects.No significant differences were noticed between the performances of the classifiers for each group of classes($p = 0.7$, $p = 0.5$ and $p = 0.8$ for 5, 4 and 3 classes respectively

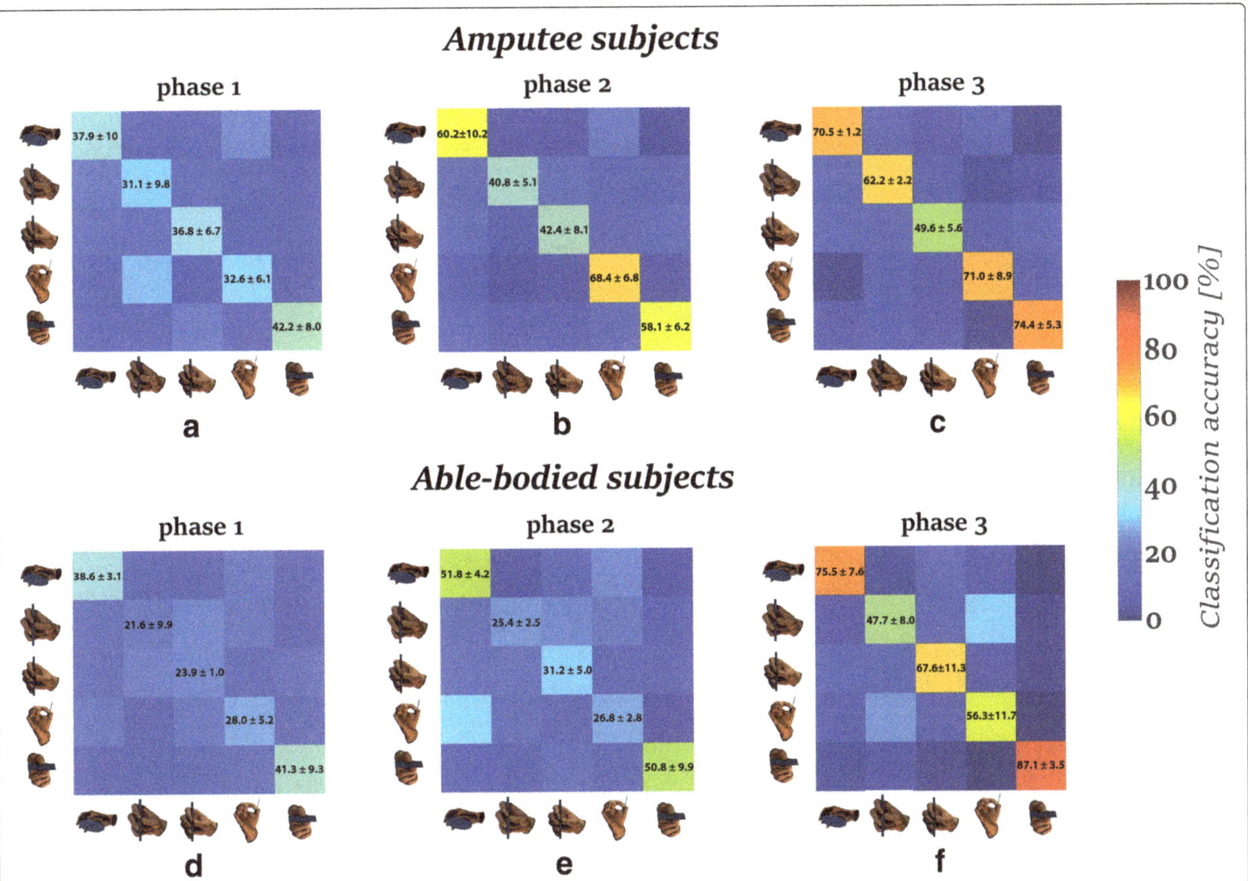


**Fig. 4** The confusion matrices for all the motion phases (phases 1, 2 and 3). The confusion matrices present the average classification accuracies and their standard deviations for the five grasp types. The matrices on the top correspond to the classification performances among the amputee subjects while the matrices on the bottom correspond to the classification performance of the able-bodied subjects. The horizontal axis of the confusion matrices correspond to the predictions while the vertical axis correspond to the ground-truth. The color of the tile was assigned according to the colormap of the classification accuracy on the right. The grasp types are represented by the corresponding figures as it is shown in Fig. 1

$70.5 \pm 1.2\%$ and $68.4 \pm 6.8\%$ to $71 \pm 8.9\%$, respectively. The precision disk and lateral grasp types had the best classification accuracy for the able-bodied subjects as well (see Fig. 4e and f). These grasp types' performances increased from $60.2 \pm 10.2\%$ and $50.8 \pm 9.9\%$ in the second phase to $75.5 \pm 7.6\%$ and $87.1 \pm 3.5\%$ in the third phase, respectively. We noticed the worst performance in the prismatic-4 fingers for the amputee subjects, with $49.6 \pm 5.6\%$ in the third phase (Fig. 4c), and in prismatic-2 fingers for the able-bodied subjects, with $47.7 \pm 8\%$ in the third phase (Fig. 4f). The prismatic-2 fingers and palm pinch were misclassified for one another in the third motion phase for the able-bodied subjects about 25−30% (see Fig. 4f). This indicates that the muscular activity during the preshaping of the fingers is similar for these grasps types. The reason for this could the similarity of the two grasp types as they differ mainly on the configuration of the middle finger.

Figure 5a-c presents the evolution of average classification performance of the control group, that consist of the eight able-bodied subjects, and the individual performance of all the amputee subjects, until 2 s after the motion onset. For the cases of 4−grasp and 5−grasp types, the classification performance of each classifier follows the same profile: poor classification performance in the first phase of the motion, and the performance increases as the hand approaches the object. Subject TR1 achieved the best performance of all amputee subjects, with performance comparable to that of the able bodied subjects, whereas TR2 had the lowest performance. TR3 and TR4 exceeded the level of 60–70% in the accuracy at the end of the first phase and the beginning of the second phase, and they stayed at this level until the end of the third phase.

In the case of 3−grasp types, TR3 and TR4 appeared better performance than the control group in the first phase, reaching 80% of accuracy during the shift to the second phase. In the second phase, the performance among the subjects TR1, TR3 and TR4 reached an accuracy of $90 \pm 10\%$, which is higher than the corresponding performance of the control group ($76 \pm 20\%$). The

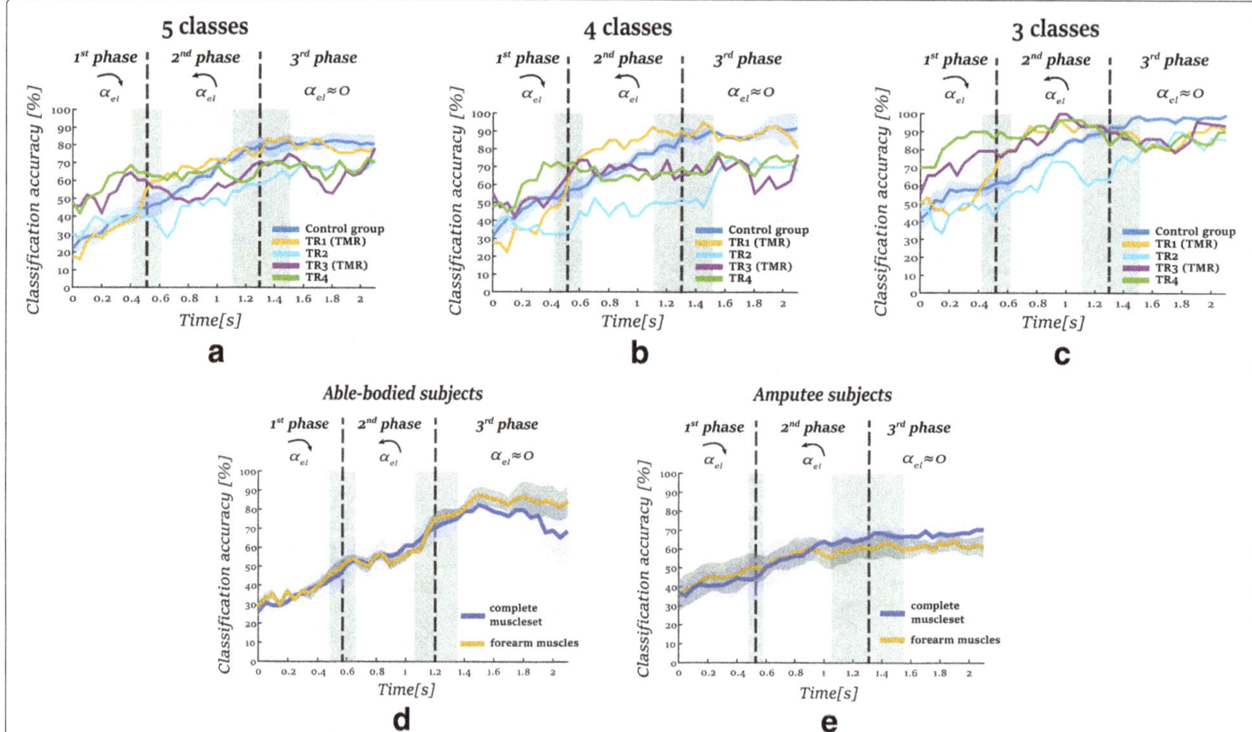


**Fig. 5** The evolution of classification performance and standard error through time for 2s after the motion onset. The vertical dashed lines correspond to the average moments of the shifts of between the phases while the shaded areas present the corresponding standard deviations accordingly. Figures **a**, **b** and **c** present the evolution of the classification performance of the control group and the amputee subjects for 5, 4 and 3 classes accordingly. Figures **d** and **e** compare the classification performance on 5 classes when using the forearm muscles and complete muscle-set as an input to the classifier a_e_l correspond to the angular acceleration of the elbow joint. The arrows above the angular acceleration represent its direction

classification accuracy in the third phase for the these amputee subjects stayed above 80%, though lower than the control group (95 ± 5%).

*Reducing the number of channels*

In this section, we compare classification performance for when using only the forearm EMG with that of using the complete muscle set. An SVM classifier with an RBF kernel was trained for each muscle set: the complete muscle set and the five muscles of the forearm. Figure 5d-e present the evolution of the classification accuracy for a duration of 2 s. As shown, using fewer EMG sites led to decreased performance for the amputee subjects. From the end of the second phase and afterwards, the average classification accuracy decreased significantly from 67.2 ± 8.4% to 60 ± 8.2% when predicting among 5−grasp types ($p < 0.01$, *t-value*=4.33). Although the reduction of the EMG sites available had a large impact on the performance of amputee subjects, when using only the muscles of the forearm the performance of able-bodied subjects was higher than when using the full muscle set.

*Comparing the performance of TMR versus non-TMR subjects*

In this section, we compare the classification performance between TMR subjects (TR1 and TR3) and non-TMR subjects (TR2 and TR4). In the case of 5 classes, the average classification accuracies and standard errors in the first, second and third phases of the TMR subjects are 44.1 ± 10.1%, 62.7 ± 86.5% and 74.9 ± 6.1%, respectively. The corresponding performances of the non-TMR subjects in the three motion phases are 47.6 ± 8.9%, 54.9 ± 7.5% and 67 ± 2.5%. The classification accuracies are at the same level also in the case of 4 classes. More specifically, the average classification accuracies and standard errors of the TMR subjects are 45.5 ± 8.9%, 76.45 ± 6.4% and 77.7 ± 8.4% in the first, second and third phase, respectively. The corresponding performances of the non-TMR subjects in the three motion phases are 48.33 ± 10.7%, 58.3 ± 6.9% and 77.72 ± 8.6%. In the case of 3−grasp types, both the TMR and non-TMR groups have improved accuracies with respect to the other two cases (5 and 4 classes). In the first and third phase, the average classification accuracies of the non-TMR group are at the same level as the TMR group; 68.6 ± 8.8% and 64 ± 14.4% for the first phase and 87.6 ± 3.4% and 83.6 ± 4% for the third phase, respectively. However, the accuracy of the TMR subjects exceeds the one of the non-TMR subjects in the second phase; 90.2 ± 4.6% and 77.8 ± 10.9%, respectively. As stated above, TR1 (a TMR-subject) has the better performance among the amputee subjects.

## On-line evaluation

To demonstrate the usability of our proposed approach for controlling a prosthetic hand, we present an on-line implementation of our approach. We followed the protocol described in "Physical prosthesis control" section. One amputee subject took part in this validation. The subject performed reach-to-grasp motions while commanding the device to close in one of two grasp types: a power grasp or a prismatic-2 fingers grasp. In total, the subject performed two sets of 20 trials for each training approach: training over all phases or training only over the third phase.

An SVM with an RBF kernel was trained off-line, while the testing was performed on-line, where the subject performed 20 reaching motions for each of the aforementioned training approaches. As soon as the classifier reached the confidence threshold of 0.5, the corresponding motor commands were sent to the prosthetic hand to drive the fingers to their desired final posture. We assessed the performance through two metrics: classification accuracy and time to generate a confident prediction on grasp type. Results are shown in Fig. 6b-c.

When trained with all three phases of the motion, the pattern recognition system showed a higher performance in terms of the overall classification accuracy compared to the one that used only the third phase. The overall classification accuracy, when using all three phases for training the system, was $80 \pm 5\%$ whereas the corresponding accuracy when the system was trained only with the third phase was $55 \pm 5\%$, see Fig. 6c. When using only the third the system identified poorly the power grasp type, despite having similar performance for the prismatic-2 fingers grasp type.

The system that trained on all 3 phases was also faster at delivering a robust prediction. On average, it offered a confident prediction 25–40% earlier than the system trained only on the 3rd phase. For the correctly classified trials, when the system trained over all the phases the time needed to exceed the confidence level was significantly lower($p = 0.0194$, *t-value*=2.49) than when it used only the third phase, $0.3 \pm 0.10$ and $0.42 \pm 0.12s$, respectively (Fig. 6b). The pattern recognition system trained with all the phases reached the confidence threshold at $0.26 \pm 0.04s$ for the prismatic-2 fingers grasp, which was significantly faster ($p = 0.003$, *t-value*=3.49) than the system trained only with the third phase. Regarding the power grasp, the system trained with all the phases reached the confidence threshold faster, but not significantly ($p = 0.3841$, $t - value = 0.93$), $0.37 \pm 0.13s$ than $0.48 \pm 0.18s$, respectively. A video of the on-line evaluation is presented in the Additional file 2.

## Discussion

We present an approach for decoding the grasping intention during reach-to-grasp motions. Although the classification results for our proposed approach are comparable with previous studies [31, 36, 37], it is different for two main reasons. Previous studies examine the classification performance of different hand gestures, including wrist motion, hand open/close, a small number of grasp types (2-4) and in some cases the resting condition [38–40]. Whereas, we focus only on grasping gestures with different finger configurations. Most importantly, previous studies examined static hand-gestures, whereas we investigate the classification of EMG activity during dynamic motions. The inclusion of the reaching motion in the training procedure increases classification performance and confidence, enabling faster activation of the prosthetic device while yielding a seamless and intuitive interaction with the user.

When a hand reaches for an object, the velocity and acceleration profile of the motion are coordinated with the motion of the fingers and the wrist, and the fingers function in a synergistic manner [28, 41]. It is shown that the reach-to-grasp motion consists of many components [42, 43]. Specifically, the motion can be separated into two phases; (1) the reaching phase, when the hand approaches the object while the fingers are pre-shaping [43],and (2) grasping phase, where the hand has traveled the distance to the object and the fingers have taken their final form. This gradual molding of the fingers is revealed through different patterns of muscle activation visible during the reaching motion, which we noticed this in our analysis on able-bodied subjects. Although we cannot observe preshaping in amputees, we assume that this pattern of muscle activation would be preserved partially and would be revealed through different patterns of muscle contractions, as we progress in reach-to-grasp movement.

Taking inspiration from human behavior, we examine the classification performance with respect to the velocity of elbow extension. In particular, we segment the reach-to-grasp motion into three phases: (1) the first phase- where the velocity of the motion increases, (2) the second phase- where the velocity of the motion decreases and (3) the third phase- when the reaching motion is complete. As the average activity of the EMG signals between the three phases is significantly different, training a classifier with only one phase could increase the difficulty of generalizing over the three phases. To highlight these differences, we model the first two principal components with Gaussian Mixture Models (GMMS) for each phase and show phase models occupy different spaces and they only partially overlap (Fig. 2). Hence, classification during different phases of the reaching motion could reduce the variability of the EMG signals, thus increasing the classification accuracy. The lack of motion after the contraction of the

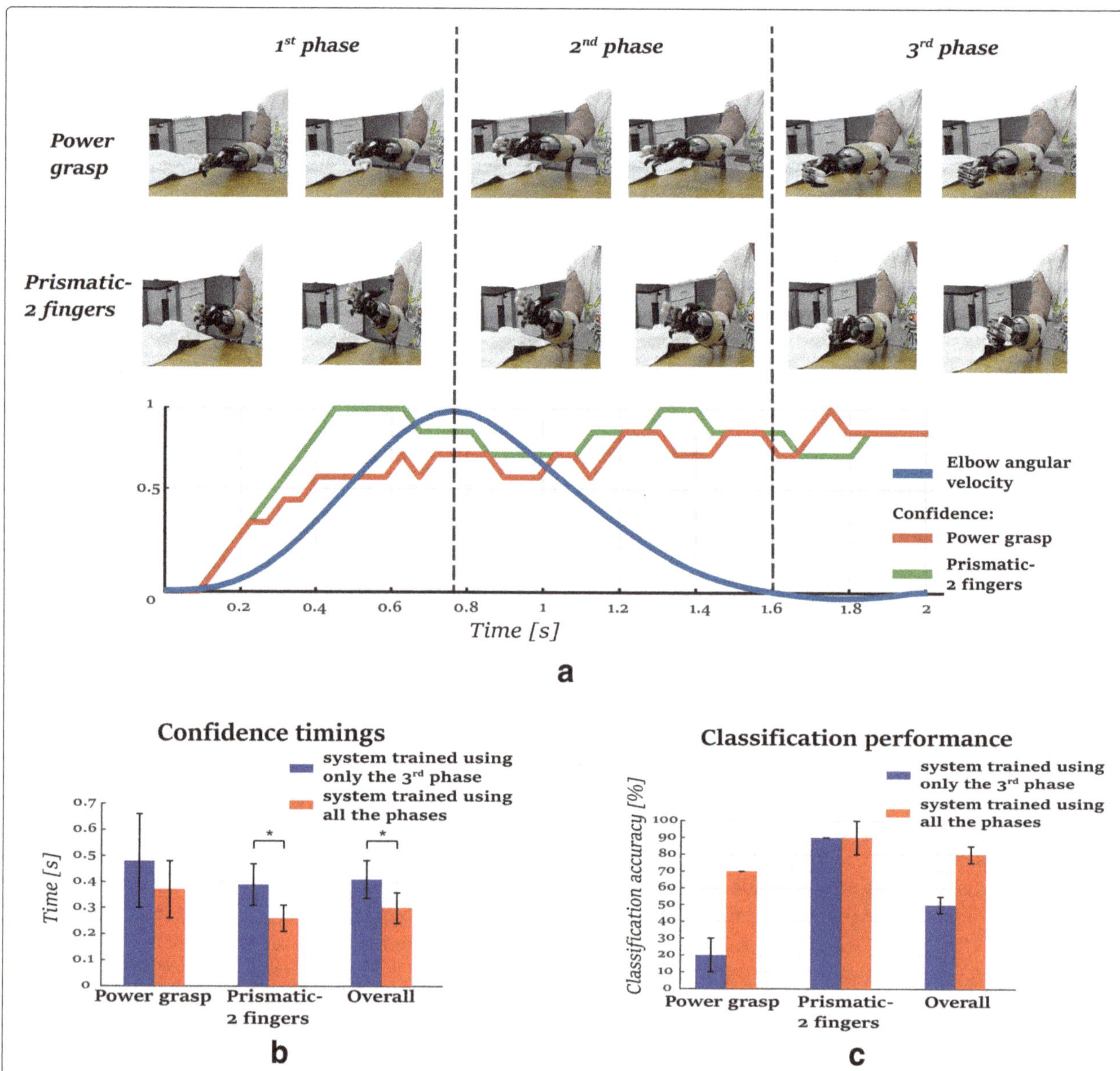


**Fig. 6** The results of the on-line evaluation. a) Screen-shots of two examples of the on-line implementation showing an activation of the prosthetic device during the second phase. The graph presents the confidence of the majority vote with the elbow's angular velocity profile. b) the average time and standard deviations until the confidence threshold was reached for the correctly classified trials. The pattern recognition system exceeds the confidence threshold of 0.5 significantly faster ($p = 0.019$) when trained including all the phases of the motion. c) the classification accuracy of the testing phase of the on-line evaluation. The pattern recognition system presents better accuracy when trained including all the phases of the motion

muscles could lead to different EMG patterns. As shown in Fig. 2, amputee subjects contract their forearm muscles even in the latter stages of the reaching motion, whereas the EMG of able-bodied subjects converges to lower levels in the final stages. Furthermore, in the case of able-bodied subjects, the fingers pre-shape during the early stages of the reaching motion [43] which results in earlier activation of the forearm muscles. As no pre-shape occurs in transradial amputees, they potentially contract the muscles but solely to close their phantom hand. This could lead to high accuracies in the prediction of the grasp type, even from the first phase with a smaller number of grasp types, as presented in Figs. 5c and 6c.

Therefore, to increase the efficiency of the classification approach, it is important to look into the patterns of the muscular activation. The authors in [39] point out that the muscle activation differs with respect to the arm position and that examining the EMG patterns is important. In our work, we elaborate on the EMG pattern during reach-to-grasp motions, both on able-bodied subjects and individuals with amputation.

In our real-time evaluation, we intend to highlight the negative impact that the lack of good classification over the entire duration of the reaching motion could have in the natural coordination of motion of the prosthesis with the arm. More specifically, we compare the performance of a classifier when it is trained only with one phase (i.e. the third motion phase) against our approach that takes the overall motion into account. Previous approaches [13–15] train a pattern recognition system while maintaining the arm in a fixed position while monitoring the contraction of the muscles. This arm configuration is similar to our third phase, where the extension is complete and the arm remains in the same position. Our results show that the muscle contractions when having the arm fixed are different from the contractions when the arm extends hence the pattern recognition fails to generalize. This leads to lower classification accuracy that results in a slower reaction of the prosthesis. This outcome is aligned with the findings of [38], where a classifier that takes into account different arm positions outperformed a single-position classifier. The benefits of a dynamic training protocol are also shown in [40]. Our work is complementary to these approaches in that it focuses on the timing of classification. As in [40], we address the problem of dynamically estimating the grasp type. To reduce the time needed for reaching a sufficient classification confidence, so as to provide faster reaction time, we focus on combining detection mechanisms.

Relating muscle activation of amputee subjects to the classification accuracy in Fig. 2, we noticed that as the activation level increases, performance also increases. The evolution of the classification accuracy follow the same trend on all the subjects: lower classification in the first phase an higher in the second and third phase (Fig. 5). Although it seems that forearm muscles are most important for classification performance, as they are responsible for finger motion, the muscles of the upper arm can help improve the classification performance. Our results show a decrease in accuracy for our amputee subjects when we remove upper arm EMG data by 10% on average, see Fig. 5e. This outcome is aligned with the findings of [20], where it is shown that the activation of the proximal muscles is statistically different when the arm reaches to grasp objects with different characteristics or orientations. Although our experimental protocol constrains subjects to a single-hand orientation, the decreased accuracy when removing the upper-arm EMG indicates that the proximal muscles are important for classification accuracy during reaching.

In addition to the importance of the proximal muscles, we notice an improvement on the performance of the individuals who undergo a TMR operation. More specifically, the classification accuracy on the TMR subjects becomes better than the non-TMR subject on the second motion phase, whereas it stays at the same level of performance in the third phase. These results are aligned with studies [8, 9] that indicated the potential benefits of TMR on the classification accuracy. However, considering the small sample size of the group in our study, these results should be taken with a grain of salt. The improvements that TMR operations could provide on the performance of a myoelectric pattern-recognition system should be further investigated.

We compare four different classification methods, but find no significant differences in classification performance. LDA perform well: delivering similar results with SVM with either a linear or RBF kernel. The performance of the Echo State Network was on a level similar to the other classification methods. It is worth mentioning that no feature extraction is performed on the EMG before being inserted in the ESN. In this case, we let the random reservoir select the features and then train a linear regressor for classification. This indicates that a random projection of the EMG signals to a very high dimensional space could be sufficient for achieving good classification results.

This work presents an approach to improve the reaction time of a hand prosthetic device through a systematic assessment of the accuracy of a myoelectric pattern recognition system over different phase periods during reach-to-grasp motions. The EMG signals are collected from seven muscles of the upper arm and five muscles of the forearm, as we focus on a potential application for individuals with transradial amputation. Our approach could potentially be implemented for proximal amputations in cases where the user has an enhanced ability to control a myoprosthesis, e.g. after undergoing a TMR operation. TMR has been shown to increase the accuracy of a myoelectric pattern-recognition system also for the case of transhumeral amputation [9, 44]. This improvement increases the number of degrees of freedom that individuals with transhumeral amputation can control. A potential extension of our approach to a proximal upper-limb amputation could be possible for individuals with TMR. This extension would however require further work to model the activation of the residual muscles of the upper arm during reach-to-grasp motions and to select a smaller number of grasp types, for increasing the classification confidence.

An important extension of our approach is the introduction of a robotic control scheme that derives from the natural motion of the human hand, and that imparts a human-like behavior to the prosthesis. As the EMG activation is significantly different among the three phases of the motion, a combination of different classifier for each phase, or a combination of those, could improve the classification performance. This approach could also be applied in conjunction with a synergistic closure of the hand [28], tackling the problem of the high dimensionality of the task. A direct extension of this work could be the coupling of the closure of the hand with the motion of the arm; this could provide a more natural coordination between the hand and the arm. The introduction of different hand orientations and an additional wrist control could be a further expansion of the approach.

Another interesting extension of the proposed approach could be the introduction of the kinematics of the arm, towards a multi-sensor pattern-recognition system. In this case, the angular position or velocity could correspond to a parameter of the system, providing information regarding the phase of the motion and potentially improving the accuracy of the system.

## Conclusion

In this work, we have presented an electromyography-based approach for decoding the grasping intention during reach-to-grasp motions. Four able-bodied subjects and four individuals with transradial amputation participated in our study. In order to examine the evolution of the classification accuracy over the reach-to-grasp motion, we separated the motion into three phases: (1) the first phase- where the velocity of the motion increases, (2) the second phase- where the velocity of the motion decreases and (3) the third phase- when the reaching motion is complete. Our results have shown that it is possible to decode the grasping intention before the end of the reaching motion, especially during the second motion phase. The inclusion of the muscular activity of the upper arm to the pattern recognition algorithm increases its accuracy by 10% on average. As a proof of concept, we have evaluated our approach with an individual with a transradial amputation controlling a myo-prosthesis in real-time. The real-time evaluation shows a significant improvement in classification accuracy as well as in the reaction time of the device when all the motion phases are included in the training data. Further extensions should involve the evaluation of our proposed approach in in real-life conditions as well as with different hand's orientations.

#### Acknowledgements
The authors would like to thank Kristi Turner, Dr. Laura A. Miller and Graci Finco for their assistance during the experiments and Denys Lamotte for his assistance in the analysis. Data not contained in the manuscript and supporting materials will be made available upon request.

#### Funding
This research was supported by the Swiss National Science Foundation through the National Center of Competence in Research Robotics (51NF40 – 160592) and United States' National Institutes of Health (R01HD081525).

#### Authors' contributions
IB LH and AB conceived and designed the experiments. IB conducted the experiments, analyzed the data, and wrote the paper; NEK assisted in the experiments; AMS contributed to the design of the experiment and assisted in the experiments. All the authors contributed to the revision of the manuscript. All authors read and approved the final manuscript.

#### Competing interests
The authors declare that they have no competing interests.

#### Author details
[1]Learning Algorithms and Systems Laboratory (LASA), School of Engineering, École Polytechnique Fédérale de Lausanne (EPFL), Route Cantonale, CH-1015 Lausanne, Switzerland. [2]Center for Bionic Medicine, Shirley Ryan AbilityLab, E Erie St., 60611 Chicago, IL, USA. [3]Dept. of Physical Medicine and Rehabilitation, Northwestern University, N Lake Shore, 60611 Chicago, IL, USA. [4]Dept. of Biomedical Engineering, Northwestern University, 60208 Evanston, IL, USA.

#### References

1. Freeland AE, Psonak R. Traumatic below-elbow amputations. Orthopedics. 2007.
2. Raichle KA, Hanley MA, Molton I, Kadel NJ, Campbell K, Phelps E, Ehde D, Smith DG. Prosthesis use in persons with lower- and upper-limb amputation. J Rehabil Res Dev. 2008.
3. Need-Directed Design of Prostheses and Enabling Resources in Amputation, Prosthesis Use, and Phantom Limb Pain: An Interdisciplinary Perspective. New York: Springer; 2010.
4. Oltstlie k, Lesjo IM, Franklin RJ, Garfelt B, Skjeldal OH, Magnus P. Prosthesis rejection in acquired major upper-limb amputees: a population-based survey. Disabil Rehabil. 2012.
5. Cordella F, Ciancio AL, Sacchetti R, Davalli A, Cutti AG, Guglielmelli E, Zollo L. Literature review on needs of upper limb prosthesis users. Front Neurosci. 2016.
6. Novak D, Riener R. A survey of sensor fusion methods in wearable robotics. Robot Auton Syst. 2014.
7. Earley EJ, Hargrove LJ, Kuiken TA. Dual window pattern recognition classifier for improved partial-hand prosthesis control. Front Neurosci. 2016.
8. Miller LA, Stubblefield KA, Lipschutz RD, Lock BA, Kuiken TA. Improved myoelectric prosthesis control using targeted reinnervation surgery: A case series. IEEE Trans Neural Syst Rehabil Eng. 2008.
9. Kuiken TA, Li BA, Guanglin amd Lock, Lipschutz RD, Miller LA, A SK, Englehart K. Targeted muscle reinnervation for real-time myoelectric control of multifunction artificial arms. JAMA. 2009.
10. Urbanchek MG, Wei B, Baghmanli Z, Sugg K, Cederna PS. Long-term stability of regenerative peripheral nerve interfaces. Plast Reocnstractive Surg. 2011.
11. Fougner AL, Stavdahl O, Kyberd PJ. System training and assessment in simultaneous proportional myoelectric prosthesis control. J Neuroengineering Rehabil. 2014.
12. Young AJ, Smith LH, Rouse EJ, Hargrove LJ. Classification of simultaneous movements using surface emg pattern recognition. IEEE Trans Biomed Eng. 2013.
13. Smith LH, Kuiken TA, Hargrove LJ. Evaluation of linear regression simultaneous myoelectric control using intramuscular emg. IEEE Trans Biomed Eng. 2016.
14. Gonzalez-Vargas J, Dosen S, Amsuess S, Yu W, Farina D. Human-machine interface for the control of multi-function systems based on electrocutaneous menu: Application to multi-grasp prosthetic hands. PLoS ONE. 2015.
15. Li G, Schultz AE, Kuiken TA. Quantifying pattern recognition-based myoelectric control of multifunctional transradial prostheses. IEEE Trans Neural Syst Rehabil Eng. 2010.

17. Khushaba RN, Kodagoda S, Takruri M, Dissanayake G. Toward improved control of prosthetic fingers using surface electromyogram (emg) signals. Expert Syst Appl. 2012.
18. Scheme E, Biron K, Englehart K. Improving myoelectric pattern recognition positional robustness using advanced training protocols. In: 33rd Annual International Conference of the IEEE EMBS. Boston; 2011.
19. Batzianoulis I, El-Khoury S, Pirondini E, Coscia M, Micera S, Billard A. Emg-based decoding of grasp gestures in reaching-to-grasping motions. Robot Auton Syst. 2017.
20. Martelloni C, Carpaneto J, Micera S. Characterization of emg patterns from proximal arm muscles during object- and orientation-specific grasps. IEEE Trans Biomed Eng. 2009.
21. Rand MK, Shimansky YP, Stelmach ABMI, Stelmach HE. Quantitative model of transport- aperture coordination during reach-to-grasp movements. Exp Brain Res. 2008.
22. Wang J, Stelmach G. E. Coordination among the body segments during reach-to-grasp action involving the trunk. Exp Brain Res. 1998.
23. Jeannerod M. The timing of natural prehension movements. J Motor Behav. 1984.
24. Haggard P, Wing A. Coordinated responses following mechanical perturbation of the arm during prehension. Exp Brain Res. 1995.
25. Ghazaei G, Alameer A, Degenaar P, Morgan G, Nazarpour K. Deep learning-based artificial vision for grasp classification in myoelectric hands. J Neural Eng. 2017.
26. Amsuess S, Vujaklija I, Goebel P, Roche AD, Graimann B, Aszmann OC, Dario F. Context-dependent upper limb prosthesis control for natural and robust use. IEEE Trans Neural Syst Rehabil Eng. 2016.
27. Farrell TR, Weir RF. The optimal controller delay for myoelectric prostheses. IEEE Trans Neural Syst Rehabil Eng. 2007.
28. Santello M, Soechting JF. Gradual molding of the hand to object contours. J Neurophys. 1998.
29. Smith LH, L J Hargrove TAK B A Lock. Determining the optimal window length for pattern recognition-based myoelectric control: Balancing the competing effects of classification error and controller delay. IEEE Trans Neural Syst Rehabil Eng. 2011.
30. Englehart K, Hudgins B. A robust, real-time control scheme for multifunction myoelectric control. IEEE Trans Biomed Eng. 2003.
31. Daley H, Englehart K, Hargrove L, Kuruganti U. High density electromyography data of normally limbed and transradial amputee subjects for multifunction prosthetic control. J Electromyogr Kinesiol. 2012.
32. Jaeger H. The "echo state" approach to analyzing and training recurrent neural networks. Technical report, German National Research Institute for Computer Science. 2001.
33. Li D, Han M, Wang J. Chaotic time series prediction based on a novel robust echo state network. IEEE Trans Neural Netw Learn Syst. 2012.
34. Xu M, Han M. Adaptive elastic echo state network for multivariate time series prediction. IEEE Trans Cybern. 46(10):2173–2183.
35. Lenzi T, Lipsey J, Sensinger JW. The ric arm- a small anthropomorphic transhumeral prosthesis. IEEE Trans Mechatron. 2016.
36. He J, Zhang D, Jiang N, Sheng X, Farina D, Zhu X. User adaptation in long-term, open-loop myoelectric training: implications for emg pattern recognition in prosthesis control. J Neural Eng. 2015.
37. Peerdeman B, Boere D, Witteveen H, Veld RH, Hermens H, Stramigioli S, Rietman H, Veltink P, Misra S. Myoelectric forearm prostheses: State of the art from a user-centered perspective. J Rehabil Res Dev. 2011.
38. Y Geng YW O W Samuel, Li G. Improving the robustness of real-time myoelectric pattern recognition against arm position changes in transradial amputees. BioMed Res Int. 2017.
39. J Liu XS, D Zhang, Zhu X. Quantification and solutions of arm movements effect on semg pattern recognition. Biomed Signal Process Control. 2014.
40. Yang D, Jiang L, Osborn L, Gu Y, Liu H. Dynamic training protocol improves the robustness of pr-based myoelectric control. Biomed Signal Process Control. 2017.
41. Wing AM, Turton A. Grasp size and accuracy of approach in reaching. J Mot Behav. 1986.
42. Jeannerod M, Paulignan PWY. Grasping an object: one movement, several components. In: Novartis Foundation Symposium. New Jersey: Wiley.
43. T Supuk TBTKodek. Estimation of hand preshaping during human grasping. Med Eng Phys. 2005.
44. Hargrove LJ, Turner K, Kuiken TA, Miller LA. Myoelectric pattern recognition outperforms direct control for transhumeral amputees with targeted muscle reinnervation: A randomized clinical trial. Sci Rep. 2017.
45. Gentilucci M, Benuzzi F, Bertolani L, Daprati E, Gangitano M. Language and motor control. Exp Brain Res. 2000.
46. Bongersa RM, Zaala FT, Jeannerod M. Hand aperture patterns in prehension. Hum Mov Sci. 2012.
47. Feix T, J Romero H-BS, Dollar AM, Kragic D. The grasp taxonomy of human grasp types. IEEE Trans Hum Mach Syst. 2015.

4

# Recovery of kinematic arm function in well-performing people with subacute stroke

Gyrd Thrane[1,2*], Margit Alt Murphy[2] and Katharina Stibrant Sunnerhagen[2,3]

**Abstract**

**Background:** Most motor function improvements in people who have experienced strokes occur within the first 3 months. However, individuals showing complete or nearly complete arm function recovery, as assessed using clinical scales, still show certain movement kinematic deficits at 3 months, post-stroke. This study evaluated the changes in upper extremity kinematics, in individuals demonstrating minor clinical motor impairments, 3–12 months post-stroke, and also examined the association between kinematics and the subjects's self-perceived hand abilities during the chronic stage, 12 months post-stroke.

**Methods:** Forty-two subjects recovering from strokes and having Fugl-Meyer upper extremity motor assessment scores ≥60 were included from the Stroke Arm Longitudinal Study at the University of Gothenburg (SALGOT). Kinematic analyses of a drinking task, performed 3, 6, and 12 months post-stroke, were compared with kinematic analyses performed in 35 healthy controls. The Stroke Impact Scale-Hand domain was evaluated at the 12-month follow-up.

**Results:** There were no significant changes in kinematic performance between 3 and 12 months, post-stroke. The patients recovering from stroke showed lower peak elbow extension velocities, and increased shoulder abduction and trunk displacement during drinking than did healthy controls, at all time points. At 12 months, post-stroke, better self-perceived arm functions correlated with improved trunk displacements, movement times, movement units, and time to peak velocity percentages.

**Conclusion:** Kinematic movement deficits, observed at 3 months post-stroke, remained unchanged at 12 months. Movement kinematics were associated with the patient's self-perceived ability to use their more affected hand.

**Keywords:** Stroke, Upper extremity, Kinematics, Movement analysis, Recovery, Motor impairment

## Background

Post-stroke upper extremity motor deficits are common, and initially range from total paralysis to near full upper extremity function [1, 2]. Many patients improve substantially during the first 4 post-stroke weeks, with the majority of improvements occurring within the first 3 months [3–5]. Upper extremity motor function improvements beyond the sub-acute stage might be induced by rehabilitative therapies [6–8], but the improvement observed using clinical scales is small during the sub-acute stage, compared with that occurring during the acute stage [6]. This finding may partly be attributed to the ceiling effect associated with clinical scales, such as the Fugl-Meyer Assessment upper extremity motor part (FMA-UE) [9]. More accurate motor control measurements might guide more precise treatment and more accurate monitoring of treatment results.

A kinematic analysis, using multiple optoelectronic high-speed cameras, provides objective and detailed measurement of movement quality and performance during reaching tasks [10–13]. The kinematic characteristics measured post stroke include movement time, smoothness, and velocity, as well as joint angles and movement strategies.

* Correspondence: gyrd.thrane@uit.no
[1]Department of Health and Care Sciences, UiT The Arctic University of Norway, Tromsø, Norway
[2]Institute of Neuroscience and Physiology, Rehabilitation Medicine, Sahlgrenska Academy, University of Gothenburg, Gothenburg, Sweden
Full list of author information is available at the end of the article

Post stroke motor and sensory impairments might lead to increased numbers of corrections and speed alterations during upper extremity movements, impairing movement smoothness [14] and resulting in longer completion times [15]. Adequate velocity control is essential for the quality of post-stroke arm movements, and requires intact feed-forward and feedback systems [10]. Individuals recovering from strokes take longer to complete reaching tasks. Additionally, the peak velocity appears later and they spend relatively more time in the deceleration phase of reaching than do healthy people [12, 16, 17]. These individuals also extend their elbows less during reach-to-grasp movements, and the elbow joint peak angular velocity discriminates well between people suffering from mild to moderate strokes and healthy controls [12]. Post-stroke motor impairments might also lead to compensatory shoulder or trunk movements during upper extremity tasks [12, 18, 19]. The increased trunk movement during reaching can be part of a strategy to improve the control, speed, and smoothness of arm and hand movements as well as to compensate for reduced elbow extension [19–21]. Further, compensatory shoulder movement patterns might be related to abnormal recruitment of the shoulder abductor muscles [13, 19, 22, 23].

Three months post-stroke, individuals with submaximal upper extremity FMA-UE scores (60–65 points) have lower tangential and angular peak velocities and use more trunk displacement than do healthy controls. Furthermore, individuals with full FMA-UE scores, 3 months post stroke, do not demonstrate the same elbow angular velocities as do healthy controls. They also use more trunk movements during reaching and demonstrate greater arm abduction during drinking [24]. However, whether the remaining movement deficits resolve over time remains unclear. Previous research have shown improvement that the Fugl Meyer upper extremity score might improve even after 3 months in subgroups of patients [25]. Because movement quality might improve due to targeted exercise during the chronic stage of stroke recovery [26, 27], we hypothesized that kinematic performance might improve, in well-performing individuals, later than 3 months post stroke.

Few studies have examined how kinematic measures relate to self-perceived arm and hand functioning during the performance of daily activities [28]. Therefore, the extent to which kinematic performance deficits are related to patient-perceived functional problems is also uncertain. Thus, this study examined kinematic movement deficits, between 3 and 12 months post-stroke, in individuals with complete or near-complete recovery, as assessed using the FMA-UE, and examined the degree to which the kinematic performance at 12 months was associated with a patient's self-perceived ability to use their more affected arm.

## Methods

### Study design

The Stroke Arm Longitudinal Study at the University of Gothenburg (SALGOT) consisted of a cohort of patients recovering from first-time strokes who were admitted to the stroke unit at Sahlgrenska University Hospital during an 18-month period in 2009–2010. The present study included only those study patients who had experienced unilateral ischemic or hemorrhagic strokes, were ≥ 18-years-old, demonstrated upper extremity impairment at 1 or 2 days post-stroke (Modified Motor Assessment Scale score < 15) [29], and had FMA-UE scores ≥60 at 3 months post-stroke. The included patients were evaluated at 3, 6, and 12 months post-stroke. In addition, 35 healthy controls who considered themselves in good health and did not have any medical conditions affecting their arm functions were also recruited for kinematic testing.

#### Kinematic analysis

A motion capture system (ProReflex MCU240 Hz, Qualisys, Gothenburg, Sweden), with five optoelectronic cameras, was used to obtain movement data during a standardized drinking task [12]. The task included five phases: reaching and grasping the glass, lifting the glass from the table to the mouth, taking one sip of water, placing the glass behind a line marked on the table, and returning the glass to its initial position. Participants were instructed to sit against the back of a chair during the whole task, but they were not restrained in that position. In the initial position, the participants sat close enough to reach the drinking glass using their less-affected arm, without trunk displacement. The pronated hand rested on the table, with the wrist line close to the edge of the table. The chair height was adjusted so that, in the sitting position, the participant demonstrated 90° of flexion in the hips and knees; the table height was adjusted so that the forearm was placed in the horizontal plane with the elbow at 90° of flexion. Retroreflective markers were placed on the third metacarpophalangeal joint of the hand, styloid process of the ulna on the wrist, lateral epicondyle of the elbow, middle part of the acromion on the right and left shoulders, upper part of the sternum, forehead, and the upper and lower edges of the glass. The task was performed five times at a comfortable speed, with the mean time of the 3 middle trials used for the analysis [30]. Data were captured using the more affected arm of the patients recovering from stroke and using the non-dominant hand of the healthy controls.

Endpoint variables were calculated from the hand marker and included the total movement time (s), number of movement units (n), peak velocity during reaching (cm/s), and time to peak velocity percentages (%). The total movement time included the time required to complete the whole drinking task and was measured from the point where the hand marker surpassed 2% of the peak velocity of the reaching phase through to the point where it returned back to 2% of the peak velocity of the returning phase. The number of movement units reflected the number of velocity peaks during the task, excluding the drinking phase [31]. A difference between the local minimum and the next maximum velocity value that exceeded the 20 mm/s amplitude limit signified a velocity peak. The time between two subsequent peaks had to be at least 150 ms to be defined as a movement unit [12]. The minimum number of movement units was 4, one for each movement phase. The peak hand velocity and the time to the peak velocity percentage during reaching, indicating the relative acceleration time and initial movement effort, were calculated.

Elbow angles were calculated between the vectors joining markers at the wrist and elbow, and at the elbow and shoulder. The joint angle at maximal extension and the peak angular velocity of the elbow joint during reaching were computed. Trunk displacement was defined as maximal forward displacement (mm) of the sternal marker from the initial position during the entire task. Shoulder abduction in the frontal plane, while drinking, was defined as the maximal angle between the vector joining the shoulder and elbow markers, and the vertical vector from the shoulder marker [12].

### Clinical scales

Motor impairment of the more affected arms of patients recovering from strokes was assessed using the FMA-UE. The FMA-UE includes 33 items that assess movement, coordination, and reflex actions of the shoulder, elbow, forearm, wrist, and hand. Each item is scored on a 3-point ordinal scale (0, cannot perform; 1, performs partially; and 2, performs fully); the test has a total possible score range of 0–66 [32]. The sensations for light touch and position were assessed using the FMA domain for sensation (FMA-Sensation), which has a total score range of 0–12; a score of 12 indicates normal sensation [32]. Covariates included age and body height. For the stroke group, we gathered information regarding the stroke type (hemorrhagic or ischemic) and the time elapsed since the stroke. The stroke severity was assessed using the National Institutes of Health Stroke Scale (NIHSS) score at the time of hospital admission. NIHSS scores range from 0 to 42, with a score $< 6$ indicating a mild stroke and a score $> 20$ indicating a severe stroke.

The Stroke Impact Scale (SIS) is a 64-item, self-reported questionnaire divided into eight domains [33]. The five hand function domain items were used to quantify the patient's self-perceived ability to use the more affected arm to carry heavy objects, turn a door knob, open a can or jar, tie a shoelace, or pick up a coin. Each item was rated on a 5-point Likert-scale, with 5 indicating no difficulties and 1 indicating that the activity could not be performed. Data from the 12-month assessment were used in the analysis, with the results being presented as percentages of the maximum score. The reliability of the SIS hand domain (SIS-Hand) is reported to be adequate, valid, and sensitive to changes in patients recovering from strokes [33].

### Statistical analyses

The statistical analyses were performed using R language for statistical computing, version 3.4.2 [34]. The distribution of the included variables was evaluated by visually inspecting histograms, qq-plots, and boxplots. Normally distributed variables are described using means and standard deviations (SD); non-normally distributed variables are described using medians and quartiles.

Longitudinal changes were assessed using linear mixed models in the Lme4 package [35]. Fixed effects included time and the cofactors of age, body height, and FMA-UE score. The participants were grouped according to their FMA-UE results into either the submaximal (FMA-UE, 60–65 points) or maximal (FMA-UE, 66 points) subgroup. Random effects included the inter-subject variation of intercept and inter-subject variation in change across the three time points. Eight different analyses were run for the eight dependent variables. A step-up strategy was used to evaluate which factors contributed significantly to the model [36]. Model selection started with an intercept-only model that included both random and fixed intercepts. Because development over time was the primary interest, a time model was constructed first. The random effect of time was tested first, followed by the fixed effect of time. An adjusted model was then built by adding the fixed effect of the FMA group, followed by age, body height, side of stroke lesion and the concordance between the dominant hand and the affected side. At each step of model building, the log-likelihood ratio test was used to determine the significance of each inserted variable. The restricted maximum likelihood criteria were used for testing random effects, whereas the maximum likelihood criteria were used to test the significance of the fixed factors [36]. Variables that contributed significantly ($p < 0.05$) to the model at each step were included [36]. The time and adjusted models were then run without outliers (defined as more than 3 SDs away from the mean) to assess their potential effect on the results. Because most of the kinematic variables had at least one outlier, the Wilcoxon rank-sum test was used to determine the differences between the kinematic variables in both the patients and healthy controls. The location difference (LD), which reflects the median difference between groups, was used to describe the effect size [37]. The *p*-values from the 24 comparisons (8 dependent variables × 3 time points) were adjusted for the false discovery rate using the Benjamini-Hochberg method [38]. The association between the SIS-Hand and kinematic variables, at 12 months post-stroke, was analyzed using Spearman's correlation.

The power of the linear mixed models was calculated using the Simr package [39]. Because the data were already available, we used the variance components from the actual data for the power calculations. The lowest detectable effect size was set to 0.75 SDs during the 9-month follow-up period. The power of detecting a true change, over time, of more than 0.75 SDs was > 95% in all models.

## Results

Forty-two subjects met the inclusion criteria for this study. Figure 1 shows the flow chart of the inclusion process. The included patients were younger ($p = 0.003$) and suffered less severe strokes, according to their admission NIHSS scores ($p < 0.001$), than were the SALGOT patients not meeting the inclusion criteria. Mild and moderate strokes were predominant, although the admission NIHSS scores ranged between 0 and 24. Table 1 shows the characteristics of the included participants.

Figures 2 and 3 show the longitudinal changes in the kinematic variables; the descriptive characteristics of the kinematic variables are shown in Table 2. The fixed effects of time on kinematic variables are shown in Table 3; the effect sizes were close to 0 for the 8 variables. Only the number of movement units showed a small, statistically significant increase over time ($p = 0.042$) in the time- and adjusted models. The random effects analysis showed that the variance of movement time (variance [V], 0.01; covariance [COV], – 0.05; $p = 0.006$) and arm abduction (V, 0.76; COV, – 5.99; $p = 0.005$) decreased over time. The trunk displacement variance showed a small increase over time (V, 0.58; COV, 7.69; $p = 0.041$). Table 3 also shows the effects of time, adjusted for covariates that were significant in the multivariate analysis. The multivariate analysis showed that participants with maximal FMA-UE scores demonstrated shorter movement times (– 1.40 s; standard error [SE], 0.31 s), fewer movement units (– 2.08; SE, 0.55), higher peak velocities (10.91 cm/s; SE, 2.20 cm/s), and higher peak elbow angular velocities (17.66°/s; SE, 5.89°/s) than those with submaximal scores (60–65). Increased body height resulted in lower elbow peak angular velocities (– 1.13°/s; SE, 0.32°/s) and larger elbow angles (less extension) at

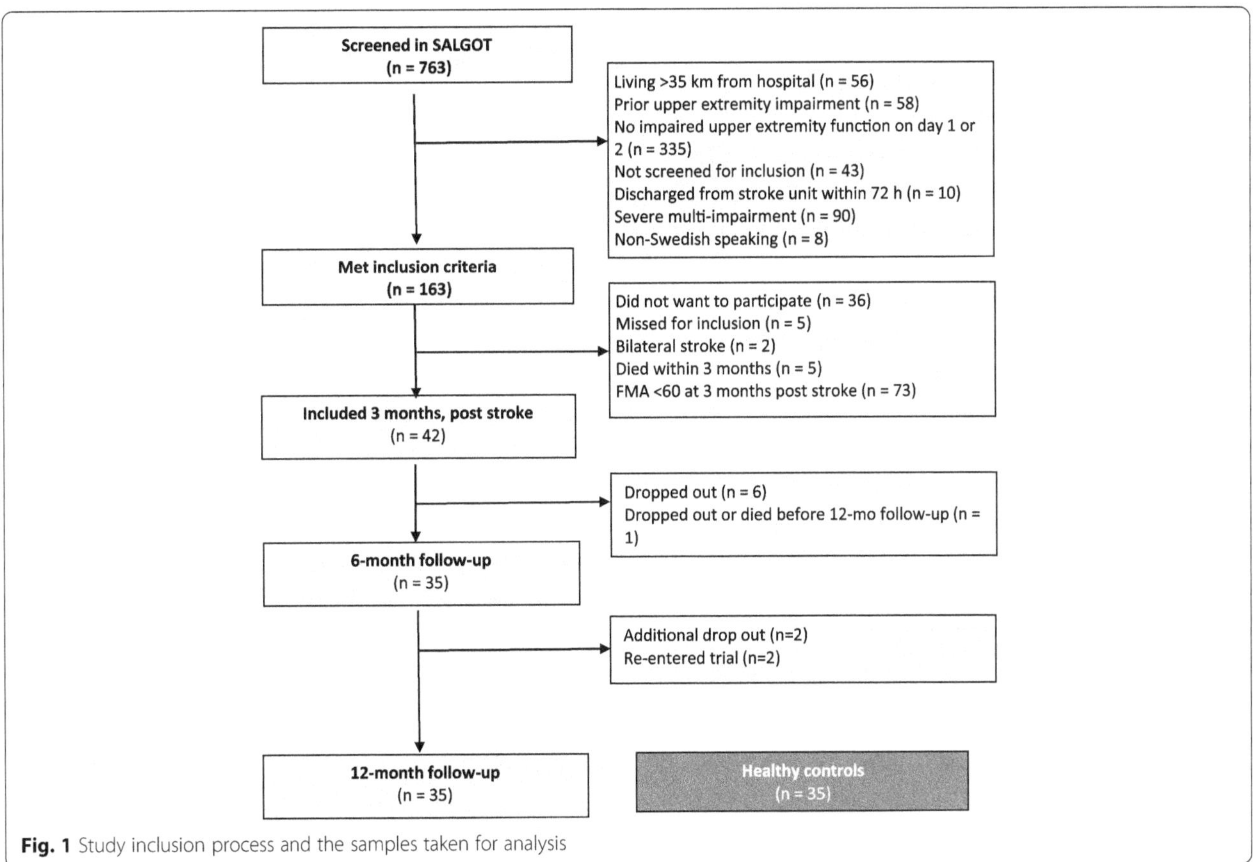


**Fig. 1** Study inclusion process and the samples taken for analysis

**Table 1** Demographic and clinical characteristics of the study population

| | Stroke Group | Healthy Group |
|---|---|---|
| Admission | ($n = 42$) | ($n = 35$) |
| Age, mean (SD) | 64.9 (12.2) | 60.1 (12.7) |
| Female, n (%) | 16 (38%) | 15 (42%) |
| Ischemic stroke, n (%) | 32 (87%) | |
| Stroke location, n (%) | | |
| * Left hemisphere | 19 (45%) | |
| * Right hemisphere | 23 (55%) | |
| Dominant arm affected | 21 (50%) | |
| NIHSS, median (QQ) | 4 (3–6) | |
| 3 days post-stroke | | |
| FMA-UE, median (QQ) | 56 (43–62) | |
| FMA-Sensation, $n > 12$ | 10 | |
| Physical or occupational therapy | 42 (100%) | |
| > 3 times/week 100% | 42 (100%) | |
| 3 months post-stroke | | |
| FMA-UE, median (QQ) | 65 (64–66) | |
| FMA-Sensation, n > 12 | 2 | |
| Physical or occupational therapy | 22 (52%) | |
| < 2 times/week | 10 (24%) | |
| 2–3 times/week | 11 (26%) | |
| > 3 times/week | 1 (2%) | |
| 6 months post-stroke ($n = 35$) | | |
| FMA-UE, median (QQ) | 66 (63–66) | |
| FMA-Sensation, n > 12 | 1 | |
| Physical or occupational therapy | 11 (31%) | |
| < 2 times/week | 5 (14%) | |
| 2–3 times/week | 5 (14%) | |
| > 3 times/week | 1 (3%) | |
| 12 months post-stroke (n = 35) | | |
| FMA-UE, median (QQ) | 66 (62.5–66) | |
| FMA-Sensation, n > 12 | 2 | |
| SIS-Arm ability, median (QQ) | 95 (78–100) | |
| Physical or occupational therapy | 6 (17%) | |
| < 2 times/week | 4 (11%) | |
| 2–3 times/week | 2 (6%) | |
| > 3 times/week | 0 | |

SD, Standard deviation; QQ, 1st Quartile–3rd Quartile; NIHSS, National Institutes of Health Stroke Scale; FMA–UE, Fugl-Meyer Upper Extremity Motor score; FMA-Sensation, the upper extremity part of the Fugl-Meyer–Sensation score; SIS, Stroke Impact Scale

the point of maximal extension. Subjects with a left side stroke lesion showed less arm abduction during drinking (– 7.3°, SE 3.2°), compared to right side stroke lesion. There were only minor influences on the effect of time.

Figures 2 and 3 also displays the differences between the kinematic variables in the recovering patients and the healthy controls. There were no cross-sectional differences between these groups of individuals with respect to the total movement time, number of movement units, or peak hand velocity at any of the measured time points. There tended to be a shorter relative time to the peak velocity among recovering patients than among healthy controls, but this was only statistically significant at 6 months, post-stroke (LD, – 3.137%; $p = 0.039$). The elbow peak angular velocity was significantly lower in the patients at 3 (LD, – 23.9°/s; $p = 0.001$), 6 (LD, – 18.2°/s; $p = 0.006$), and 12 months post-stroke (LD, – 18.9°/s; $p = 0.026$). The recovering patients used more arm abduction while drinking at 3 (LD, 8.7°; $p = 0.003$), 6 (LD, 7.6°; $p = 0.013$), and 12 months (LD, 5.7°; $p = 0.026$) post-stroke than did the healthy controls. Similarly, they also demonstrated more trunk displacement at 3 (LD, 14 mm; p = 0.001), 6 (LD, 10 mm; $p = 0.016$), and 12 months post-stroke (LD, 10 mm; $p = 0.019$) than did the healthy controls. At 12 months, the SIS-Hand score distribution for those recovering from strokes was skewed towards higher values, with the majority (20/35) reporting no difficulties in any hand item or only a little difficulty using the hand in a single item; their total scores were 95–100. In the whole stroke group, a higher SIS-Hand score was associated with shorter movement times ($\rho = -0.376, p < 0.001$), fewer movement units, ($\rho = -0.295$, $p = 0.002$), higher time to peak velocity percentages ($\rho = 0.38, p < 0.001$), and less trunk displacement ($\rho = -0.205, p = 0.035$).

## Discussion

This study evaluated the development of upper extremity kinematic performance between 3, 6, and 12 months post-stroke in patients exhibiting only minor clinical upper extremity impairment. Three months post stroke, the total movement time, the number of movement units, the peak velocity, the time to peak velocity, and the elbow angle at maximal extension had reached a level similar to healthy controls. A small increase in the number of movement units was observed, indicating a possible but clinically insignificant decline in movement quality between 3 and 12 months. The peak angular velocity of the elbow, arm abduction during drinking and forward trunk displacement was significantly different from healthy controls at 3 months post stroke. From 3 to 12 months post stroke there was no development and the recovering patients still showed lower elbow peak angular velocity and more forward trunk movement during reaching than did the healthy controls; they also used more arm abduction during drinking. Twelve months post-stroke, improved movement times,

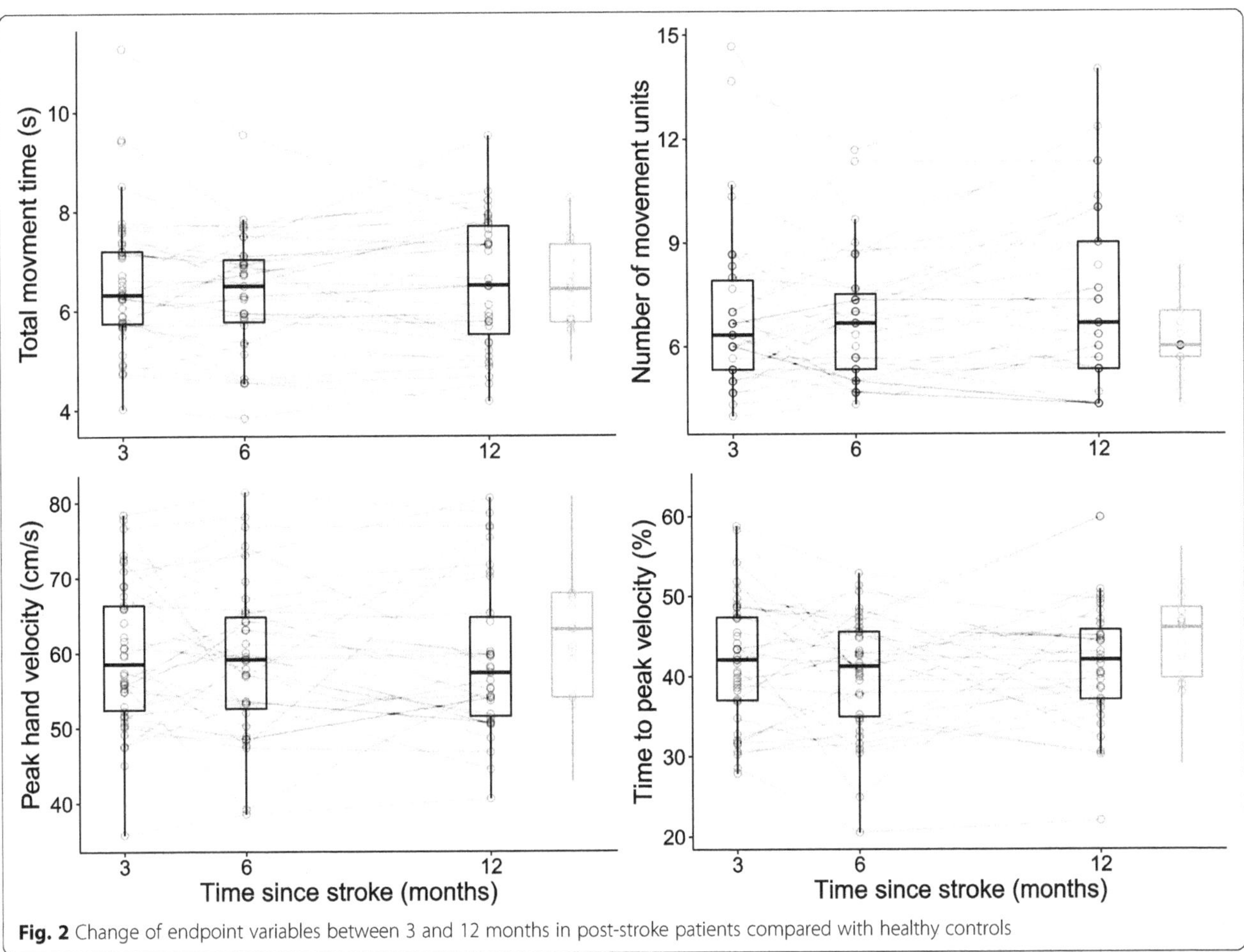


**Fig. 2** Change of endpoint variables between 3 and 12 months in post-stroke patients compared with healthy controls

movement units, and relative time to peak velocity values during reaching were associated with improved abilities to use the more affected arm.

To our knowledge, this is the first study to show upper extremity kinematic deficits that do not resolve within 12 months, post-stroke, in high-functioning individuals recovering from strokes. Previous cohort studies of arm kinematics have evaluated movement time and different measures smoothness up to 3 months after stroke [40–42]. However, the longitudinal course of trunk displacement, peak angular velocity of the elbow and arm abduction during drinking has not been evaluated. The deficits seen in trunk, shoulder, and elbow movement control may be part of a synergetic pattern commonly observed in more severely affected stroke survivors. Individuals with upper extremity impairments use fewer joint combinations during pointing movements than do healthy controls [43], and the contribution of elbow extension during reaching is often reduced [18]. A pathological coupling between the shoulder and elbow that reduces the degrees of freedom has been proposed [44–46], and there is evidence for this coupling in kinematic [13, 44] and electromyographic studies [23]. Another typical way of compensating for impaired elbow or shoulder function is to use the trunk to improve speed, available distance, movement quality, and precision [21, 47–49]. Whereas the upper extremity is predominantly activated through contralateral corticospinal pathways, the trunk muscles are activated bilaterally [50, 51]. Therefore, trunk movements might be more readily available and used during reaching to compensate for upper extremity impairments, such as limited elbow extension velocity. A novel finding of the present study is that high-functioning individuals recovering from strokes also use alternative strategies to compensate for minor movement deficits, and they rely on these strategies for at least 12 months, post-stroke.

In this study, patients recovering from strokes reached levels similar to those of the healthy controls for the endpoint kinematic measures of movement time, smoothness, and time to peak velocity at 3 months, and no further improvements were observed. This agrees with previous research that showed improvements in movement time and smoothness up to 6 or 8 weeks after stroke, and little or no improvement between 8 and 26 weeks [40, 41]. Although the recovering patients did not differ significantly from the controls with respect to movement times

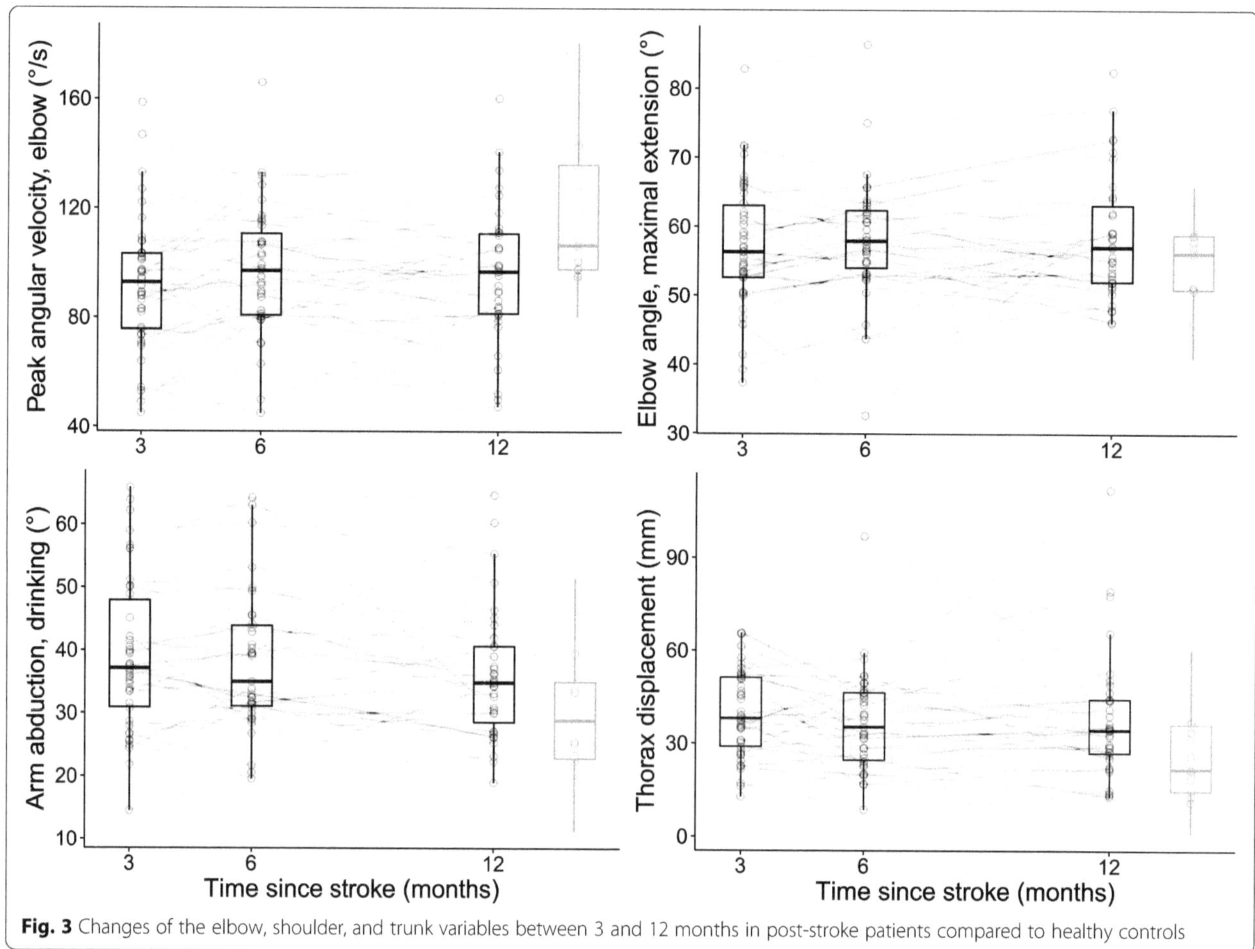


**Fig. 3** Changes of the elbow, shoulder, and trunk variables between 3 and 12 months in post-stroke patients compared to healthy controls

and numbers of movement units, these variables correlated with self-perceived arm ability (SIS-Hand) at 12 months, post-stroke. However, this association was weak, possibly due to the large variability in movement times and units for individuals in the stroke group, especially among those who scored ≥95% on the SIS-Hand evaluation. The correlation between the increased time to peak velocity percentages and self-perceived hand ability should be interpreted in light of the tendency towards lower percentages among individuals in the stroke group than among the healthy controls. Among the kinematic variables that were significantly different between the patients and controls, increased trunk displacement associated with the stroke group showed a weak but significant

**Table 2** Descriptive statistics for the kinematic analyses

| | Stroke Group | | | Healthy controls (mean (SD)) |
|---|---|---|---|---|
| | 3 months (mean (SD)) | 6 months (mean (SD)) | 12 months (mean (SD)) | |
| N | 42 | 35 | 35 | 35 |
| Movement time, s | 6.58 (1.38) | 6.37 (1.18) | 6.59 (1.32) | 6.55 (1.03) |
| Movement units, n | 6.86 (2.26) | 6.76 (1.84) | 7.33 (2.54) | 6.43 (1.43) |
| Peak hand velocity, cm/s | 59.74 (9.67) | 59.26 (10.28) | 59.15 (10.26) | 61.93 (9.32) |
| Time to peak velocity, % | 42 (8) | 40 (7) | 42 (8) | 45 (7) |
| Elbow peak angular velocity, °/s | 92.24 (24.26) | 97.14 (24.37) | 95.87 (25.12) | 118.58 (26.43) |
| Elbow angle, ° | 57.16 (9.19) | 58.23 (8.92) | 58.43 (9.04) | 54.78 (7.21) |
| Arm abduction, ° | 39.18 (12.31) | 37.97 (11.15) | 36.05 (10.44) | 29.97 (10.35) |
| Trunk displacement, mm | 39.81 (14.19) | 36.96 (16.53) | 38.23 (20.40) | 26.19 (1.03) |

*SD* Standard deviation. Velocity variables are measured in the reaching phase. The elbow angle is the minimum elbow angle in reaching indicating maximum extension of the elbow joint. Arm abduction during is measured during the drinking phase

**Table 3** Fixed effect of time on kinematic performance changes between 3 and 12 months, post-stroke

| Kinematic variable | Time β (SE) | Time Adjusted model β (SE) |
|---|---|---|
| Movement time (s) | 0 (0.02)[b] | 0 (0.02)[b,c] |
| Movement units (n) | 0.06 (0.03)[a] | 0.06 (0.03)[b, c, a] |
| Peak hand velocity (cm/s) | 0 (0.15) | 0 (0.15)[c] |
| Time to peak velocity (%) | 0 (0) | |
| Elbow peak angular velocity (°/s) | 0.31 (0.27) | 0.30 (0.28)[c,d] |
| Elbow angle, maximum extension (°) | 0.11 (0.12) | 0.13 (0.12)[d] |
| Arm abduction (°) | −0.32 (0.17)[b] | − 0.32 (0.17)[b,e] |
| Trunk displacement (mm) | −0.09 (0.30)[b] | |

[a] $p$-value > 0.05
[b] Adjusted for the random effect of time
[c] Adjusted for the maximal/submaximal Fugl-Meyer Upper Extremity Motor score
[d] Adjusted for body height
[e] Adjusted for stroke location (left/right). Velocity variables are measured in the reaching phase. The elbow angle is the minimum elbow angle in reaching indicating maximum extension of the elbow joint. Arm abduction during is measured during the drinking phase

correlation with difficulties using the more affected arm. The correlations between kinematic variables and self-perceived hand abilities, shown in this study, are similar in size to correlations previously shown using the ALBILHAND measure [28]. Even small to moderate correlations indicate that minor kinematic performance deficits might relate to the perceptions that these individuals have of their ability to use their affected hands during functional tasks.

Several reasons might explain why some of the movement deficits do not resolve within the first year, post-stroke. The post-stroke period between 3 and 12 months is outside the critical period for neurobiological improvement [6]. This and the lower level of rehabilitation provided by health care systems beyond the first 3 post-stroke months, especially for well-performing patients, might explain the plateau observed for some kinematic measures. On the other hand, since well-functioning patients are able to use their affected upper extremities during daily activities, movement performance might be expected to reach a level similar to that of healthy controls during the first year, post-stroke. In the present study, this was true for some endpoint kinematics, such as movement time and smoothness, but not for the elbow angular velocity and compensatory arm and trunk movements. An explanation for this finding might be that slowness and clumsiness are more easily perceived by the patient. Thus, the demands of daily life may constantly force the recovering patient to improve their functioning, whereas minor modifications in arm and trunk movements may become habitual and difficult to address without professional guidance. Because high-functioning stroke survivors succeed in their most important activities by using compensatory movements, they may not be motivated to improve minor deficits in their movement patterns [52, 53].

This study has some limitations that should be acknowledged. The results from the current study are most relevant for people suffering from mild stroke impairments. Further, the results of our kinematic analyses are specific to the drinking task [54]. However, the use of a common purposeful task, such as drinking, improves the ecological validity of the results. The functional task used in this study included drinking, making it inappropriate to repeat numerous trials. However, the use of the average from more than 3 attempts could possibly have improved the stability of the kinematic analysis. Since patients in the stroke group performed the kinematic testing several times, unlike the controls, a learning effect cannot be ruled out; however, it is probably negligible due to the number of repetitions needed to improve upper extremity movement kinematics [13]. The relatively high dropout rate at 12 months (16%) might impact the external validity of the results. In addition, the outliers might have impacted the results, although several measures were taken to control for this impact. A ceiling effect was present in the SIS-Hand score, indicating that most of the patients in the stroke group had reached adequate hand ability or that the SIS-Hand evaluation was insufficiently sensitive to changes at 12 months, post-stroke.

## Conclusions

This study evaluated changes in upper extremity movement kinematics in high-functioning individuals between 3 and 12 months, post-stroke, and examined the association of these changes with the patients' self-perceived ability to use their most affected hands during the chronic stage of stroke recovery. Our results show that little or no change in kinematic performance occurred between 3 and 12 months, post-stroke. The endpoint measures reached levels similar to those of the healthy controls within the first 3 months after the stroke. Further, the impairment of the elbow extension velocity and the altered movement patterns of the arm and trunk did not resolve during the 3–12-month post-stroke period. Compensatory movement patterns may help patients succeed in performing basic activities, but may also limit the possibilities for retraining the affected arm during daily activities and lead to more permanent movement changes [46]. The correlation between the patient's self-perceived ability to use their more affected hand and the patient's observed kinematic performance indicated that kinematic performance deficits might be clinically important. Previous research has demonstrated the possibility of retraining kinematic performance in people with mild stroke impairment [26]. Thus, depending on the complexity of a patient's goals, we recommend the evaluation of kinematic performance in these individuals to help develop individually targeted interventions.


#### Acknowledgements

We thank Eva-Lena Bustrén and Hanna Persson for help in collecting data and the Riks-Stroke Collaboration for help collecting clinical background data.

#### Funding

This study was funded, in part, by the Swedish Research Council (VR 2012-70X-22122-01-3), Swedish Heart and Lung Foundation, Swedish Brain Foundation, Norrbacka Eugenia Foundation, Foundation of the Swedish National Stroke Association, Willhelm and Martina Lundgren's Foundation, Hjalmar Svensson's Research Foundation, and the Promobilia Foundation.


#### Authors' contributions

GT analyzed the data, interpreted the data, and drafted the manuscript. MAM took part in the design of the study, collection of the data, interpretation of results and critical revision of the manuscript. KSS took part in design of the study, interpretation of results and critical revision of the manuscript. All authors read and approved the final manuscript.

#### Competing interests

The authors declare that they have no competing interests.


#### Author details

[1]Department of Health and Care Sciences, UiT The Arctic University of Norway, Tromsø, Norway. [2]Institute of Neuroscience and Physiology, Rehabilitation Medicine, Sahlgrenska Academy, University of Gothenburg, Gothenburg, Sweden. [3]Center for Person-Centered Care (GPCC), University of Gothenburg, Gothenburg, Sweden.


#### References


1. Lawrence ES, Coshall C, Dundas R, Stewart J, Rudd AG, Howard R, Wolfe CD. Estimates of the prevalence of acute stroke impairments and disability in a multiethnic population. Stroke. 2001;32(6):1279–84.
2. Persson HC, Parziali M, Danielsson A, Sunnerhagen KS. Outcome and upper extremity function within 72 hours after first occasion of stroke in an unselected population at a stroke unit. A part of the SALGOT study. BMC Neurol. 2012;12:162.
3. Nakayama H, Jorgensen HS, Raaschou HO, Olsen TS. Recovery of upper extremity function in stroke patients: the Copenhagen stroke study. Arch Phys Med Rehabil. 1994;75(4):394–8.
4. Hendricks HT, Van LJ, Geurts AC, Zwarts MJ. Motor recovery after stroke: a systematic review of the literature. Arch Phys Med Rehabil. 2002;83(11): 1629–37.
5. Verheyden G, Nieuwboer A, De WL, Thijs V, Dobbelaere J, Devos H, Severijns D, Vanbeveren S, De WW. Time course of trunk, arm, leg, and functional recovery after ischemic stroke. Neurorehabil Neural Repair. 2008;22(2):173–9.
6. Dobkin BH, Carmichael ST. The specific requirements of neural repair trials for stroke. Neurorehabil Neural Repair. 2016;30(5):470–8.
7. Lo AC, Guarino PD, Richards LG, Haselkorn JK, Wittenberg GF, Federman DG, Ringer RJ, Wagner TH, Krebs HI, Volpe BT, et al. Robot-assisted therapy for long-term upper-limb impairment after stroke. N Engl J Med. 2010;362(19):1772–83.
8. Wolf SL, Winstein CJ, Miller JP, Taub E, Uswatte G, Morris D, Giuliani C, Light KE, Nichols-Larsen D. Effect of constraint-induced movement therapy on upper extremity function 3 to 9 months after stroke: the EXCITE randomized clinical trial. JAMA. 2006;296(17):2095 104.
9. Gladstone DJ, Danells CJ, Black SE. The Fugl-Meyer assessment of motor recovery after stroke: a critical review of its measurement properties. Neurorehabil Neural Repair. 2002;16(3):232–40.
10. Stewart JC, Gordon J, Winstein CJ. Control of reach extent with the paretic and nonparetic arms after unilateral sensorimotor stroke: kinematic differences based on side of brain damage. Exp Brain Res. 2014;232(7):2407–19.
11. Beer RF, Dewald JP, Dawson ML, Rymer WZ. Target-dependent differences between free and constrained arm movements in chronic hemiparesis. Exp Brain Res. 2004;156(4):458–70.
12. Alt Murphy M, Willen C, Sunnerhagen KS. Kinematic variables quantifying upper-extremity performance after stroke during reaching and drinking from a glass. Neurorehabil Neural Repair. 2011;25(1):71–80.
13. Cirstea MC, Mitnitski AB, Feldman AG, Levin MF. Interjoint coordination dynamics during reaching in stroke. Exp Brain Res. 2003;151(3):289–300.
14. Kamper DG, McKenna-Cole AN, Kahn LE, Reinkensmeyer DJ. Alterations in reaching after stroke and their relation to movement direction and impairment severity. Arch Phys Med Rehabil. 2002;83(5):702–7.
15. Shumway-Cook A, Woollacott MH. PART IV reach, grasp, and manipulation. In: Motor control Translating research into clinical practice. 5th ed: Walter Kluwer, Philadelphia; 2016.
16. McCrea PH, Eng JJ, Hodgson AJ. Biomechanics of reaching: clinical implications for individuals with acquired brain injury. Disabil Rehabil. 2002; 24(10):534–41.
17. Thielman GT, Dean CM, Gentile AM. Rehabilitation of reaching after stroke: task-related training versus progressive resistive exercise. Arch Phys Med Rehabil. 2004;85(10):1613–8.
18. Roby-Brami A, Feydy A, Combeaud M, Biryukova EV, Bussel B, Levin MF. Motor compensation and recovery for reaching in stroke patients. Acta Neurol Scand. 2003;107(5):369–81.
19. Cirstea MC, Levin MF. Compensatory strategies for reaching in stroke. Brain. 2000;123(5):940–53.
20. Roby-Brami A, Fuchs S, Mokhtari M, Bussel B. Reaching and grasping strategies in hemiparetic patients. Mot Control. 1997;1:72):91.
21. Mandon L, Boudarham J, Robertson J, Bensmail D, Roche N, Roby-Brami A. Faster reaching in chronic spastic stroke patients comes at the expense of arm-trunk coordination. Neurorehabil Neural Repair. 2016; 30(3):209–20.
22. Dewald JP, Pope PS, Given JD, Buchanan TS, Rymer WZ. Abnormal muscle coactivation patterns during isometric torque generation at the elbow and shoulder in hemiparetic subjects. Brain. 1995;118(Pt 2):495–510.
23. Roh J, Rymer WZ, Perreault EJ, Yoo SB, Beer RF. Alterations in upper limb muscle synergy structure in chronic stroke survivors. J Neurophysiol. 2013; 109(3):768.
24. Thrane G, Sunnerhagen KS, Persson HC, Opheim A, Alt Murphy M. Kinematic upper extremity performance in people with near or fully recovered sensorimotor function after stroke. Physiotherapy theory and practice. 2018. https://doi.org/10.1080/09593985.2018.1458929.
25. Persson HC, Opheim A, Lundgren-Nilsson Å, Alt Murphy M, Danielsson A, Sunnerhagen KS. Upper extremity recovery after ischaemic and haemorrhagic stroke: part of the SALGOT study. Eur Stroke J. 2016;1(4):310–9.
26. Michaelsen SM, Dannenbaum R, Levin MF. Task-specific training with trunk restraint on arm recovery in stroke: randomized control trial. Stroke. 2006; 37(1):186–92.
27. Wu CY, Chen CL, Tang SF, Lin KC, Huang YY. Kinematic and clinical analyses of upper-extremity movements after constraint-induced movement therapy in patients with stroke: a randomized controlled trial. Arch Phys Med Rehabil. 2007;88(8):964–70.
28. Alt Murphy M, Willen C, Sunnerhagen KS. Movement kinematics during a drinking task are associated with the activity capacity level after stroke. Neurorehabil Neural Repair. 2012;26(9):1106–15.
29. Barkelius K, Johansson A, Korm K, Lindmark B. Reliabilitet-och validitetsprövning av Motor assessment Scale enligt Uppsala Akademiska sjukhus-95. Nord Fysioter. 1997;1:121.
30. Alt Murphy M, Persson HC, Danielsson A, Broeren J, Lundgren-Nilsson A, Sunnerhagen KS. SALGOT–stroke arm longitudinal study at the university of Gothenburg, prospective cohort study protocol. BMC neurology. 2011;11:56.
31. Alt Murphy M, Sunnerhagen KS, Johnels B, Willen C. Three-dimensional kinematic motion analysis of a daily activity drinking from a glass: a pilot study. Journal of NeuroEngineering and Rehabilitation. 2006;3:18.

32. Fugl-Meyer AR, Jaasko L, Leyman I, Olsson S, Steglind S. The post-stroke hemiplegic patient. 1. A method for evaluation of physical performance. Scand J Rehabil Med. 1975;7(1):13–31.
33. Duncan PW, Wallace D, Lai SM, Johnson D, Embretson S, Laster LJ. The stroke impact scale version 2.0. Evaluation of reliability, validity, and sensitivity to change. Stroke. 1999;30(10):2131–40.
34. R Core Team. A language and environment for statistical computing. In: In., R version 3.4.2 (2017-09-28) – "short summer" edn: R Fondation for statistical Computing; 2017.
35. Bates D, Mächler M, Bolker B, Walker S. Fitting linear mixed-effects models using lme4. 2015. 2015;67(1):48.
36. West BT. Linear mixed models. In: *Linear Mixed Models*. Edn: Chapman and Hall/CRC; 2006. p. 9–49.
37. Hollander M, Wolfe DA. Nonparametric statistical methods. New York: Wiley; 1973.
38. Benjamini Y, Hochberg Y. Controlling the false discovery rate: a practical and powerful approach to multiple testing. J R Stat Soc Ser B Methodol. 1995;57(1):289–300.
39. Green P, MacLeod CJ. SIMR: an R package for power analysis of generalized linear mixed models by simulation. Methods Ecol Evol. 2016;7(4):493–8.
40. van Kordelaar J, van Wegen E, Kwakkel G. Impact of time on quality of motor control of the paretic upper limb after stroke. Arch Phys Med Rehabil. 2014;95(2):338–44.
41. van Dokkum L, Hauret I, Mottet D, Froger J, Metrot J, Laffont I. The contribution of kinematics in the assessment of upper limb motor recovery early after stroke. Neurorehabil Neural Repair. 2014;28(1):4–12.
42. Metrot J, Froger J, Hauret I, Mottet D, van Dokkum L, Laffont I. Motor recovery of the ipsilesional upper limb in subacute stroke. Arch Phys Med Rehabil. 2013;94(11):2283–90.
43. Reisman DS, Scholz JP. Aspects of joint coordination are preserved during pointing in persons with post-stroke hemiparesis. Brain. 2003;126(Pt 11): 2510–27.
44. Sukal TM, Ellis MD, Dewald JPA. Shoulder abduction-induced reductions in reaching work area following hemiparetic stroke: neuroscientific implications. Exp Brain Res. 2007;183(2):215–23.
45. Levin MF. Interjoint coordination during pointing movements is disrupted in spastic hemiparesis. Brain. 1996;119(Pt 1):281–93.
46. Levin MF, Kleim JA, Wolf SL. What do motor "recovery" and "compensation" mean in patients following stroke? Neurorehabil Neural Repair. 2009;23(4): 313–9.
47. DeJong SL, Schaefer SY, Lang CE. Need for speed: better movement quality during faster task performance after stroke. Neurorehabil Neural Repair. 2012;26(4):362–73.
48. Massie CL, Malcolm MP. Instructions emphasizing speed improves hemiparetic arm kinematics during reaching in stroke. NeuroRehabilitation. 2012;30(4):341–50.
49. Robertson JV, Roby-Brami A. The trunk as a part of the kinematic chain for reaching movements in healthy subjects and hemiparetic patients. Brain Res. 2011;1382:137–46.
50. Palmer E, Ashby P. Corticospinal projections to upper limb motoneurones in humans. J Physiol. 1992;448:397–412.
51. Ferbert A, Caramia D, Priori A, Bertolasi L, Rothwell JC. Cortical projection to erector spinae muscles in man as assessed by focal transcranial magnetic stimulation. Electroencephalogr Clin Neurophysiol. 1992;85(6):382–7.
52. Taub E, Uswatte G, Mark VW, Morris DM. The learned nonuse phenomenon: implications for rehabilitation. Europa medicophysica. 2006;42(3):241–56.
53. Maslow AH. Toward a psychology of being: D. Van Norstrand Company, inc; 1968.
54. Wu CY, Wong MK, Lin KC, Chen HC. Effects of task goal and personal preference on seated reaching kinematics after stroke. Stroke. 2001;32(1):70–6.

5

# Predictive value of the pendulum test for assessing knee extensor spasticity

Alyssa Whelan[2], Andrew Sexton[1], Melony Jones[3], Colleen O'Connell[2,3] and Chris A. McGibbon[1,2*]

## Abstract

**Background:** The pendulum test is commonly used to quantify knee extensor spasticity, but it is currently unknown to what extent common pendulum test metrics can detect spasticity in patients with neurological injury or disease, and if the presence of flexor spasticity influences the test outcomes.

**Methods:** A retrospective analysis was conducted on 131 knees, from 93 patients, across four different patient cohorts. Clinical data included Modified Ashworth Scale (MAS) scores for knee extensors and flexors, and years since diagnosis. BioTone™ measures included extensor strength, passive and active range of motion, and pendulum tests of most affected or both knees. Pendulum test metrics included the relaxation index (RI), 1st flexion amplitude (F1amp) and plateau angle (Plat), where RI=F1amp/Plat. Two-way ANOVA tests were used to determine if pendulum test metrics were influenced by the degree of knee flexor spasticity graded by the MAS, and ANCOVA was used to test for confounding effects of age, years since injury, strength and range of motion (ROM). In order to identify the best pendulum test metrics, Receiver Operator Characteristic analysis and logistic regression (LR) analysis were used to classify knees by spasticity status (none or any) and severity (low/moderate or high/severe).

**Results:** Pendulum test metrics for knee extensors were not influenced by degree of flexor spasticity, age, years since injury, strength or ROM of the limb. RI, F1amp and Plat were > 70% accurate in classifying knees by presence of clinical spasticity (from the MAS), but were less accurate (< 70%) for grading spasticity level. The best classification accuracy was obtained using F1amp and Plat independently in the model rather than using RI alone.

**Conclusions:** We conclude that the pendulum test has good predictive value for detecting the presence of extensor spasticity, independent of the existence of flexor spasticity. However, the ability to grade spasticity level as measured by MAS using the RI and/or F1amp may be limited. Further study is warranted to explore if the pendulum test is suitable for quantifying more severe spasticity.

**Keywords:** Spasticity, Knee extensor, Knee flexor, Pendulum test, Relaxation index, Modified Ashworth scale, Classification, Logistic regression, Receiver operator characteristic

## Background

Muscle spasticity can be a painful and debilitating complication that negatively impacts function and quality of life in people with upper motor neuron injury from neurological disease or trauma [1], such as acquired brain injury (trauma, stroke), cerebral palsy, multiple sclerosis and spinal cord injury. Management of spasticity typically involves pharmacologic intervention and/or ongoing physical therapy [2, 3], but a significant barrier to effective treatment prescription is the inability to quantify spasticity in the clinic [4].

Spasticity is typically assessed by inducing a rapid stretch of the muscle, or administering a "stretch-reflex" test. Clinical tests such as the Modified Ashworth Scale [5] and Tardieu Scale [6] apply this method to quantify spasticity subjectively, but their inter-rater reliability [7–10] and validity [11] have been questioned. Several studies have examined objective approaches to quantifying spasticity in the clinic using wearable sensor technologies during passive muscle stretches [9, 12–15], but there is not yet a clear consensus on testing protocol and how to best translate the resulting electrophysiological and biomechanical signals

* Correspondence: cmcgibb@unb.ca
[1]Institute of Biomedical Engineering, University of New Brunswick, Fredericton, NB E3B 5A3, Canada
[2]Faculty of Kinesiology, University of New Brunswick, Fredericton, NB, Canada
Full list of author information is available at the end of the article

into clinically relevant measures of spasticity. As such, the Modified Ashworth Scale remains a commonly used method of quantifying spasticity in clinical settings.

For the knee joint, the Wartenberg pendulum test [16] offers a potential solution for translation to clinical assessment. The pendulum test offers a simple approach whereby gravity induces the stretch-reflex of knee muscles by dropping the lower-leg from a resting horizontal position, and observing its oscillatory behavior throughout the passive movement [17, 18]. The test's easy implementation and execution with commonly available sensors (e.g. electromyography with video [19], goniometry [20, 21], and other accessible devices such as the Wii remote [22]) has contributed to it emerging as an objective and reliable way to quantify spasticity in the knee extensors [17, 23–25]. Nevertheless, there is still a lack of consensus on what pendulum test metrics are most relevant to clinical spasticity assessment, and importantly, whether the pendulum test is sensitive to knee flexor spasticity. Thus the purpose of this study was two-fold:

1) To determine if the measurement of extensor spasticity is influenced by flexor spasticity during the pendulum test in patients with neurological injury or disease; and
2) To determine which pendulum test metrics are the closest indicators of clinical muscle spasticity, as represented by the Modified Ashworth Scale (MAS).

## Methods

This is a retrospective analysis of knee spasticity measurement data from a multi-site study to evaluate wearable sensor-based systems for acquiring objective measures of muscle tone in the clinic.

### Participants

#### *Ethics, consent and screening*

The study was approved by the University Research Ethics Board (REB) and by the REB or Institutional Review Board (IRB) for each of three participating clinical sites. The three clinical sites were medium to large rehabilitation hospitals in eastern Canada (Stan Cassidy Centre for Rehabilitation, Fredericton, New Brunswick, and Nova Scotia Rehabilitation Hospital, Halifax, Nova Scotia) and U.S. (Spaulding Rehabilitation Hospital, Boston, Massachusetts). Prior to data collection, all participants in the study provided informed, signed consent. Data were collected at the three sites between September 2011 and May 2014.

The site coordinator (non-therapist) approached patients meeting the inclusion criteria and asked if they were willing to participate in the study during a future regularly scheduled visit. Those who were willing were informed of the study purpose, risks, discomforts, potential benefits, and their rights to privacy and the use of data. Once a participant agreed to the outlined procedures and written consent was obtained, they were enrolled in study. Additional details of the larger study can be found elsewhere [13].

*Inclusion criteria were*: Male or female active inpatient or outpatient, sixteen years of age or older, and currently receiving services at study site for one or more of the following diagnoses: acquired brain injury (ABI: strokes, trauma, etc.), spinal cord injury (SCI: incomplete any level or complete C7 and below), multiple sclerosis (MS: meeting 2010 MacDonald criteria [26]), and cerebral palsy (CP: hemiplegic or diplegic); medically stable; and exhibits some degree of abnormal tone in either upper or lower limbs, specifically at the elbow and/or knee joint(s). *Exclusion criteria were*: Joint conditions such as osteoarthritis, rheumatoid arthritis, etc. that would confound measurement of spasticity; bariatric or with little measureable surface EMG signal; viral or bacterial infection; open skin lesions, and; not capable of autonomous consent.

### Measurements

Clinical assessment was performed on the patient as regularly scheduled. This included manual assessment of knee extensors and flexors, where the therapist used the Modified Ashworth Scale (MAS). The MAS is a ubiquitous instrument for clinical spasticity assessment, in which the therapist performs a manual stretch-reflex and rates the patient's spasticity on a 6-point scale with categories 0, 1, 1+, 2, 3 and 4, as defined elsewhere [5]. There were no participants with MAS = 4, which by definition is indicative of full rigidity [5] and cannot be assessed with a pendulum test.

The participant's age, height, weight, leg length (for the pendulum test), diagnosis (ABI/CVA, MS, CP, or SCI), affected extremity (right, left or both), and the month and year of onset or injury, were also recorded. The participant was then assessed using the BioTone™ system, as follows.

#### *Active and passive knee extension*

While seated, a fibre-optic goniometer (FOG, ShapeSensor™, Measurand, Fredericton NB) was positioned on the limb using the Neoprene/Velcro cuffs and secured in place as seen in Fig. 1 (top). For passive knee extension, the therapist moved the participant's limb slowly to the full extension. For active knee extension, the subject was instructed to move their lower leg against gravity from a flexed to fully extended position. For both active and passive tests, the minimum angle achieved (peak extension angle) was recorded and used to estimate the degree of contracture and paresis.

#### *Isometric muscle strength*

Knee extensor strength was measured with the limb strength measurement device (LSMD) shown in Fig. 1

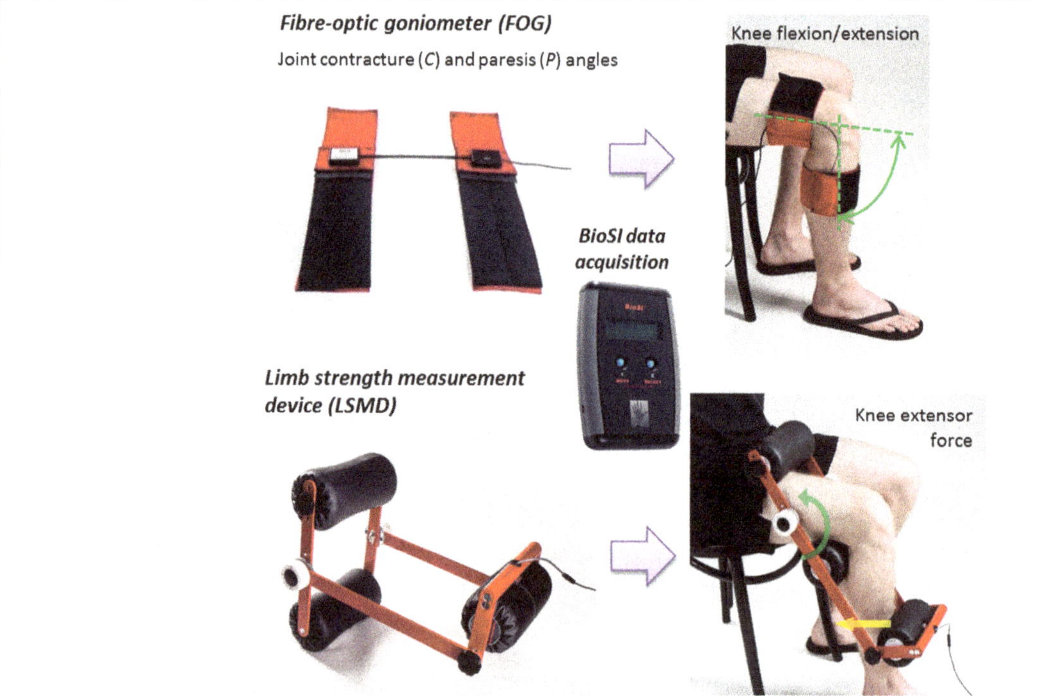


**Fig. 1** BioTone™ toolkit for lower-extremity assessment of range of motion and strength. Top: ShapeSensor™ (Measureand Inc.) fibre-optic goniometer (FOG) is inserted into Neoprene and Velcro mounting cuffs, and attached to the leg to measure knee flexion/extension angle. Bottom: Limb strength measurement device (LSMD) is shown for the knee in extensor configuration. Centre: BioSI (University of New Brunswick, Canada) data acquisition unit used to acquire sensor data and save it to the laptop computer

(bottom). The LSMD was designed to enable autonomous muscle force measurement, thus eliminating the variability inherent with hand-held dynamometry protocols or subjectivity of the manual muscle test. The LSMD was adjusted to "extensor" orientation and positioned on the leg such that the subject's knee joint formed an angle of approximately 90 deg. The subject then sat with leg supported (by therapist) and attempting to extend the knee with maximal effort. The load cell in the LSMD recorded the patient's maximal force generation. This was repeated three times with 15 s between trials. Because the "lever arm" of the LSMD (distance from joint centre to distal pad which contains the force transducer) was fixed, forces measured by the device are a proportional measure of muscle torque across participants, and were normalized by dividing by body weight.

### *Pendulum test*

For the instrumented pendulum test, the FOG was positioned as described above. To avoid interference with the thigh cuffs, EMG electrodes were placed on the lateral vastii and lateral hamstring. The reference electrode was placed on the hand. The participant was positioned in a reclining wheelchair with legs hanging freely over the edge of the seat and torso inclined to approximately 30 degrees to avoid stretching the biceps femoris. The therapist then slowly raised the lower leg to full extension (or passive extension limit) and held the leg horizontal until the participant was completely relaxed, as indicated by real-time EMG display. The participant's lower-leg was then released and allowed to oscillate until coming to rest (oscillation amplitude less than 3 deg). Pendulum tests were repeated at least three times for the right and/or left legs. In the present study, EMG was only used to confirm if a spastic contraction occurred during the test.

### Data analysis

Pendulum test metrics (RI, F1amp, Plateau angle, etc.) extracted for analysis have been described in detail elsewhere [19] and are shown in Fig. 2. Number of cycles (full oscillations) was counted between start of motion and until the oscillation amplitudes is less than 3 degrees [27]. Data from knees of participants with bilateral involvement were treated as separate data points. All statistical analyses were completed using IBM Statistical Package for Social Science (IBM SPSS, Version 23). All statistical analyses were conducted with an alpha level of .05 for statistical significance.

First we used a 2-way ANOVA to compare pendulum test metrics (RI, F1amp, E1amp, ERI, Plateau angle and Number of cycles as described in Fig. 2) across levels of extensor and flexor spasticity, coded as binary independent variables from the clinical MAS scores. Binary variables were required for testing classification models as part of the second objective.

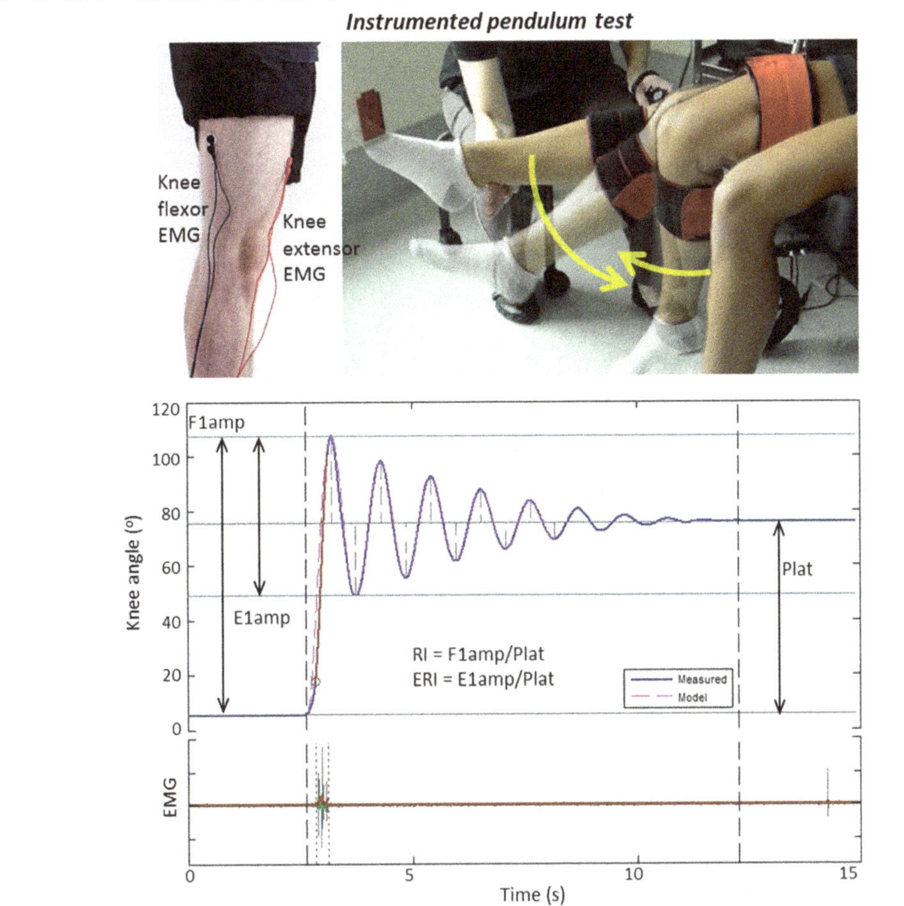


**Fig. 2** BioTone™ toolkit for lower-extremity assessment of spasticity using the pendulum test. Top: Electromyographic (EMG) electrodes are placed on extensor and flexor muscles and used to ensure muscle relation prior to the start of testing and to monitor involuntary contractions during the test, and fibre-optic goniometer (FOG) system used to measure knee angle during the pendulum test. Bottom: Pendulum test metrics include the plateau angle (Plat), first flexion amplitude (F1amp), and first extension amplitude (E1amp) which are used to compute the relaxation index (RI) and the extension relaxation index (ERI)

We created two separate models in order to determine if the above metrics can classify patients by no or any spasticity, and whether we can classify present spasticity as low or high. In the first model (Model 1), we divided knees into no spasticity (MAS = 0) and any spasticity (MAS > =1), based on their clinical assessment of knee extensor and flexor spasticity prior to the pendulum test. In the second model (Model 2), we excluded the MAS = 0 knees and re-grouped knees by low/moderate spasticity (MAS = [1,1+]) and high/severe spasticity (MAS= [2, 3]), for extensors and flexors. Although other groups could be used, the selected groupings had the best distribution and were considered the most clinically relevant.

Therefore, Model 1 had factors: *Ext1* (0 = No extensor spasticity, 1 = Any extensor spasticity) by *Flx1* (0 = No flexor spasticity, 1 = Any flexor spasticity), and Model 2 had factors: *Ext2* (0 = Low/moderate extensor spasticity, 1 = High/severe extensor spasticity) by *Flx2* (0 = Low/moderate flexor spasticity, 1 = High/severe flexor spasticity). Models were tested for main effects of extensor spasticity and flexor spasticity, and interaction effects between extensor and flexor spasticity. Interaction effects were used to answer the first research question, and the main effects analysis was used to answer the second research question, as detailed below:

#### Flexor spasticity influence on measurement of extensor spasticity

To determine if the measurement of extensor spasticity is influenced by flexor spasticity during the pendulum test we need to analyze the interaction effects of the ANOVA tests for Model 1 and Model 2. A significant interaction effect for Model 1 would suggest that *any* flexor spasticity might influence the outcome of the extensor spasticity assessed using the pendulum test. A significant interaction effect for Model 2 would suggest that the *amount* of flexor spasticity may be an important consideration for trusting the measurement of extensor spasticity using the pendulum test.

These analyses were then repeated with a variety of covariates (age, time since diagnosis/injury, passive and active extension range and isometric extensor strength) to further explore sources of variability in the pendulum test.

### *Assessment of pendulum test metrics for quantifying spasticity*

To determine which pendulum test metrics are the best indicators of clinical muscle spasticity as represented by the Modified Ashworth Scale (MAS), we first need to analyze the main effects of the ANOVA tests for Model 1 and Model 2. A significant main effect for extensor (or flexor) spasticity will indicate if the pendulum test metric is sensitive to spasticity state of the knee. For pendulum test metrics with significant main effects, we then determined which metric offered the best predictive value of clinical spasticity. This was accomplished using Receiver Operator Characteristic (ROC) analysis to determine which metrics have the best predictive potential (area under curve, AUC > .7), followed by classification analysis using logistic regressions on the binary spasticity groupings (Model 1 and Model 2) for each of the pendulum test metrics. Tests for Model 1 will determine which metric is best at discriminating any spasticity from no spasticity, and the analysis of Model 2 will determine which metric is best at grading spasticity into low/moderate spasticity versus high/severe spasticity.

## Results

Ninety-three patients (65 male and 28 female) composed of four patient cohorts (45 with ABI, 14 with MS, 12 with CP, and 22 with SCI) completed the pendulum test assessments and had MAS scores and other measures available. A total of 53 patients were measured unilaterally and 39 bilaterally, resulting in a total of 131 knees tested (56 ABI, 23 MS, 18 CP and 34 SCI). Participant demographic data (age and years since diagnosis) for the four cohorts are shown in Table 1. Also included are BioTone™ measures of contracture (passive extension range), paresis (passive – active extension range) and strength (peak isometric extensor force) for each cohort group.

MAS score for knee flexors and extensors for patient cohorts are shown in Table 2, and are grouped according to most affected and least affected side, with the total number of knees shown in the last row. The majority of flexor knees were assigned a score of 1 ($n = 31$) or 1+ ($n = 24$). Likewise, extensor knees followed a similar pattern with 33 knees being assigned a grade of 1, and 14 being assigned a grade of 1+, but with 16 being assigned a grade of 2. A total of 46 and 53 knee flexors and extensors, respectively, had a score of 0.

Pendulum test outcomes are summarized in Fig. 3, and Tables 3 and 4. Figure 3 shows pendulum test metrics against therapist rated MAS scores for knee extensors (top) and flexors (bottom). Data in Tables 3 and 4 show the marginal means of the pendulum test scores according to the factorial levels of each model. Table 3 shows results for Model 1 with factors *Ext1* and *Flx1*, and Table 4 shows results for Model 2 with factors *Ext2* and *Flx2*. Table 5 shows the resulting *p*-values generated for main effects and interaction effects from testing Model 1 (top) and Model 2 (bottom).

### Does flexor spasticity influence the pendulum test?

Statistical results in Table 5 for analysis of Model 1 revealed a significant effect of *Ext1* for all metrics ($p < 0.001$), except for Number of cycles ($p = 0.594$). No significance effect of *Flx1* was present for any metrics except for E1amp ($p = 0.013$). Additionally, there were no significant *Ext1* × *Flx1* interactions ($p > 0.05$). For Model 2, the only pendulum test metrics to show any statistically significant differences were ERI, which had a significant *Ext2* effect ($p = .012$) and *Flx2* effect ($p = .031$), and E1amp which had a significant *Ext2* effect. Similar to Model 1, there were no significant ($p > .05$) interactions observed for any of the pendulum test metrics.

Lastly, ANCOVA were conducted using both models to identify any confounding effects of age, time since injury/diagnosis, passive ROM, active-passive ROM, and extensor strength. We found that both models maintained the same significance patterns when covariates

**Table 1** Participant demographic data and BioTone™ assessment of contracture, paresis and strength

| | Gender | | Age, years | | Years Since Dx | | Contracture (Passive ROM Min, deg) | | Paresis (Passive ROM min-Active ROM min, deg) | | Flexor Strength (N) | | Knees |
|---|---|---|---|---|---|---|---|---|---|---|---|---|---|
| | M n | F n | Mean | (SD) | Mean | (SD) | Mean | (SD) | Mean | (SD) | Mean | (SD) | N |
| ABI ($n = 45$) | 35 | 10 | 52 | (16) | 4.8 | (4.8) | −6.96 | (5.25) | 7.46 | (12.9) | 258.1 | (135.7) | 56 |
| MS ($n = 14$) | 6 | 8 | 54 | (13) | 17.2 | (8.7) | −5.26 | (4.81) | 11.3 | (19.2) | 215.7 | (134.2) | 23 |
| CP ($n = 12$) | 5 | 7 | 34 | (12) | 34.3 | (11.9) | −4.28 | (7.12) | 6.50 | (5.93) | 160.8 | (45.9) | 18 |
| SCI ($n = 22$) | 19 | 3 | 45 | (14) | 6.4 | (10.5) | −6.11 | (5.17) | 13.3 | (20.2) | 153.1 | (99.4) | 34 |
| Total ($n = 93$) | 65 | 28 | 49 | (16) | 11.0 | (12.9) | −6.09 | (5.46) | 9.52 | (15.7) | 215.7 | (126.4) | 131 |

*AB* acquired brain injury including stroke; *MS* multiple sclerosis; *CP* cerebral palsy; *SCI* spinal cord injury

**Table 2** Modified Ashworth Scale scores for knee flexors and extensors from clinical examination

| | MAS score – Knee flexors | | | | | MAS score – Knee extensors | | | | |
|---|---|---|---|---|---|---|---|---|---|---|
| | 0 | 1 | 1+ | 2 | 3 | 0 | 1 | 1+ | 2 | 3 |
| | n | n | n | n | n | n | n | n | n | n |
| *Most affected side* | | | | | | | | | | |
| ABI (*n* = 34) | 20 | 8 | 2 | 2 | 2 | 14 | 11 | 4 | 4 | 1 |
| MS (*n* = 17) | 6 | 2 | 4 | 5 | 0 | 6 | 1 | 3 | 4 | 3 |
| CP (*n* = 11) | 3 | 3 | 1 | 1 | 3 | 0 | 4 | 1 | 3 | 3 |
| SCI (*n* = 20) | 6 | 5 | 3 | 2 | 4 | 7 | 8 | 3 | 2 | 0 |
| Total (*n* = 82) | 35 | 18 | 10 | 10 | 9 | 27 | 24 | 11 | 13 | 7 |
| *Least affected side* | | | | | | | | | | |
| ABI (n = 22) | 6 | 10 | 4 | 1 | 1 | 8 | 8 | 3 | 1 | 2 |
| MS (*n* = 6) | 2 | 1 | 2 | 0 | 1 | 2 | 0 | 2 | 1 | 1 |
| CP (*n* = 7) | 2 | 2 | 3 | 0 | 0 | 3 | 0 | 3 | 1 | 0 |
| SCI (n = 14) | 3 | 0 | 5 | 5 | 1 | 5 | 4 | 2 | 1 | 2 |
| Total (*n* = 49) | 13 | 13 | 14 | 6 | 3 | 26 | 9 | 3 | 3 | 4 |
| Total (*n* = 131) | 46 | 31 | 24 | 16 | 12 | 53 | 33 | 14 | 16 | 11 |

*ABI* acquired brain injury including stroke; *MS* multiple sclerosis; *CP* cerebral palsy, *SCI* spinal cord injury

were entered into the model. Therefore, covariates did not confound any effects on pendulum test metrics.

These results show that the pendulum test metrics studied were not significantly influenced by level of spasticity in the antagonistic flexor muscles, and were unaffected by variability in the covariates listed above. In addition, results showed that the pendulum test metrics studied had a good ability to discriminate between no spasticity and any spasticity of knee extensors, as grouped by the assigned MAS. However, they were not able to discriminate very well between low/moderate and high/severe spasticity.

### Which pendulum test metric(s) best predicts clinical spasticity?

We first conducted ROC analysis on each of the metrics to evaluate their ability to detect the presence of spasticity (Model 1) and level of spasticity (Model 2). AUC values from operator curves (in Fig. 4.) are presented in Table 6. AUC values >.7 indicate the metric is potentially a good predictor of the dichotomous dependent variable, and AUC of ≤.5 indicates no predictive potential (diagonal line in ROC graphs in Fig. 4).

For Model 1, most metrics were acceptable predictors of *Ext1*, having AUC > .7, with the exception of Number of cycles and Plateau angle which had AUC's of 0.66 and 0.69, respectively. Not surprisingly, Model 1 yielded poor results for predicting *Flx1* (Factors MAS = 0, and MAS = 1,1+,2,3) with AUC values < 0.7 for all pendulum test metrics. Model 2 yielded AUC values between 0.59 and 0.71 for predicting *Ext2* (Factors MAS = 1,1+ and MAS = 2,3) and all were < .7 for predicting *Flx2* (Factors; MAS = 1,1+, and MAS = 2,3). This analysis suggests that RI, ERI, F1amp, E1amp and (borderline) Plateau angle are individually acceptable predictors of extensor spasticity as graded by the MAS, but with no clear winner.

Given that E1amp and ERI are somewhat redundant in their nature of depicting contraction of the flexor muscles, and the above results suggest they have limited predictive value, both were dropped from the remaining analysis. The metrics RI and F1amp appear to be the superior metrics for predicting spasticity presence, per any MAS greater than 0. Clearly RI is a function of F1amp, and Plateau angle, as shown in Fig. 2. Therefore, we next conducted logistic regression analysis to classify extensor muscle spasticity according to Model 1 (*Ext1*) and Model 2 (*Ext2*) using RI alone, versus using F1amp and Plateau angle in the logistic regression.

These results are shown in Tables 7 and 8, respectively. For each metric the classification table is shown with resulting sensitivities and specificities as well as positive and negative predictive values (*PPV* and *NPV*) and overall classification accuracy. The logistic regression model for each metric is footnoted in the tables.

For Model 1 (Table 7), the best classification accuracy value of 77.9% was for the model using F1amp and Plateau angle simultaneously, versus using any one of them individually or using the RI term alone. This model had the highest sensitivity (.81), specificity (.7), PPV (.86) and NPV (.62). For Model 2 (Table 8.) the highest classification accuracy value was 70.9% for the model using Plateau angle alone, followed by 68.6% for the model using F1amp and Plateau angle in combination. However, it is clear for Model 2 that predictively is heavily biased toward the negative condition; that is, the classifier was better at identifying negative results (low/moderate spasticity: MAS = [1,1+]) than positive results (high/severe spasticity: MAS= [2, 3]).

These results show that F1amp and Plateau angle explained more variance in the logistic regression analysis than RI alone, for detecting the presence of spasticity per a MAS or 1 or greater (*Ext1*). However, no pendulum test metrics were suitable for discriminating between knees with high/severe spasticity and those with low/moderate spasticity (*Ext2*) as measured by MAS.

## Discussion

The Wartenberg pendulum test [16] has been around since the 1950's, but gained interest in the late 80's/early 90's as test for assessing muscle spasticity [17, 18, 23, 24]. Several studies have shown feasibility of using sensor technologies with an instrumented pendulum test [19, 21, 28] and repeatability and validity for assessing spasticity in patients has generally been reported as positive [25, 29, 30]. Although the test has some drawbacks [31], such as being sensitive to posture during the test [32, 33], clinical

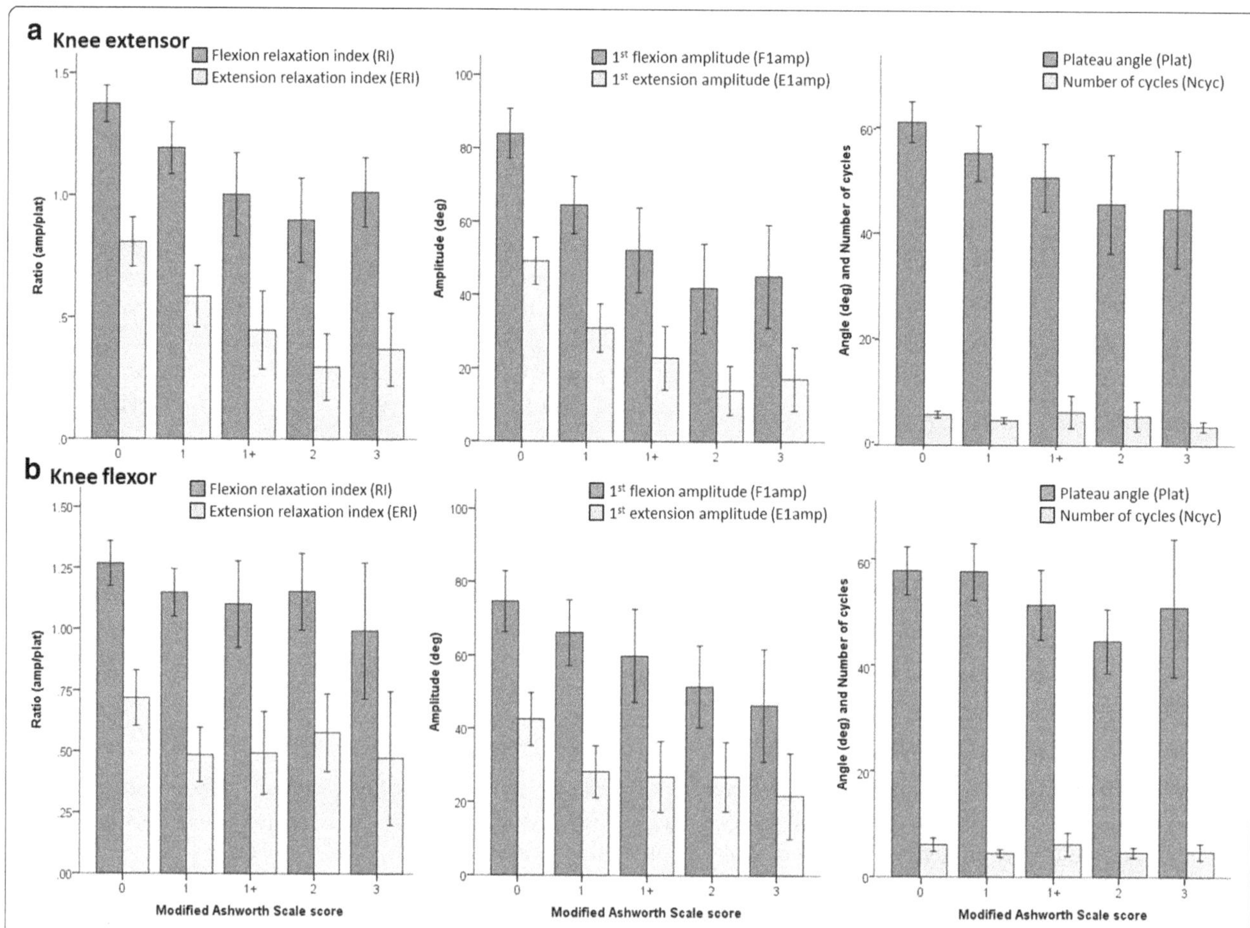


**Fig. 3** Spasticity metrics derived from the instrumented pendulum test for with standard deviation error bars for **a**. knee extensors and **b**. knee flexors. Left side panels show mean scores by MAS category flexion relaxation index (RI) and extension relaxation index (ERI) phases. Middle panels show mean scores by MAS category for 1st flexion amplitude (F1amp) and 1st extension amplitude (E1amp) phases. Right side panels show mean scores by MAS category for plateau (final resting) angle (Plat) and number of oscillation cycles (Ncyc)

evidence is increasing that the pendulum test has value for objectively quantifying spasticity of knee extensors [34–39], as well as flexors and extensor of the elbow [40, 41].

Due to the velocity-dependent nature of spasticity, the initial and second swing of the pendulum test are considered key for the detection of spasticity [17]. The maximum angular velocity of the knee occurs during the initial flexion oscillation [19, 27]. This first excursion (F1amp) results in the largest velocity of musculotendinous stretch, and has been found to be sensitive to differences in spasticity in the quadriceps [23]. Bajd and Vodovnik [17] further refined the approach by dividing F1amp by the plateau angle to quantify the relaxation index, RI.[1]

Therefore, the RI is simply F1amp adjusted for the resting position Plateau angle (hence excursion angle), thus accounting for changes in muscle length and structure that can result from ongoing spasticity, or activity induced changes in resting tonic reflex. Stillman and McMeeken [19] introduced the extension relaxation index, or ERI, along with its characteristic amplitude E1amp, describing the excursion of the first (return) extension oscillation. In theory, this metric should be sensitive to flexor spasticity, although there is very little evidence of the use of ERI/E1amp in clinical studies.

Several studies have identified F1amp and/or RI [17, 23, 27, 34, 35] as acceptable metrics for identifying spasticity of the quadriceps, although its reliability has been debated in other works [27]. While these metrics have been identified as acceptable, many patient groups with knee extensor spasticity also experience knee flexor spasticity; in our study, MAS scores for extensors and flexors of patient's knees were significantly correlated ($r = .525$, $p < .001$). To our knowledge, no other studies have examined the influence of flexor spasticity on the commonly reported pendulum test output metrics mentioned above. Our study provides convincing evidence that the degree of flexor spasticity does not have a significant impact upon commonly used pendulum test metrics F1amp and RI.

In addition, we quantified the predictive value of these metrics for detecting the presence of muscle spasticity,

**Table 3** Model 1 marginal means for factors *Ext1* and *Flx1* on pendulum test metrics (RI = relaxation index, ERI = extension relaxation index, F1amp = 1st flexion amplitude, E1amp = 1st extension amplitude, Plat = plateau angle, and Ncyc = number of oscillation cycles)

| MODEL 1 | | *Flx1*: Flexor spasticity | | | | | |
|---|---|---|---|---|---|---|---|
| *Ext1*: Extensor spasticity | | No spasticity (MAS = 0) | | | Any spasticity (MAS > 0) | | |
| | | Mean | (SD) | n | Mean | (SD) | n |
| RI | No spasticity | 1.4 | (.2) | 26 | 1.3 | (.3) | 19 |
| | Any spasticity | 1.1 | (.3) | 22 | 1.1 | (.3) | 64 |
| ERI | No spasticity | .9 | (.3) | 26 | .7 | (.3) | 19 |
| | Any spasticity | .5 | (.4) | 22 | .4 | (.3) | 64 |
| F1amp | No spasticity | 88.2 | (22.6) | 26 | 78.2 | (21.5) | 19 |
| | Any spasticity | 58.6 | (26.7) | 22 | 52.8 | (24.6) | 64 |
| E1amp | No spasticity | 54.6 | (21.6) | 26 | 41.9 | (19.6) | 19 |
| | Any spasticity | 28.3 | (21.0) | 22 | 22.1 | (17.5) | 64 |
| Plat | No spasticity | 62.4 | (13.3) | 26 | 59.1 | (11.8) | 19 |
| | Any spasticity | 52.2 | (16.7) | 22 | 49.4 | (16.0) | 64 |
| Ncyc | No spasticity | 6.3 | (2.1) | 26 | 5.2 | (2.2) | 19 |
| | Any spasticity | 5.8 | (6.3) | 22 | 4.9 | (3.5) | 64 |

and for discriminating the level of spasticity as measured by MAS. The results demonstrate that the pendulum test is a valid tool to distinguish knee extensors with spasticity (MAS > 0), from those without spasticity (MAS = 0), but that none of the metrics we analyzed were able to discriminate between knees with low/moderate (MAS = [1,1+])

**Table 4** Model 2 marginal means for factors *Ext2* and *Flx2* on pendulum test metrics (RI = relaxation index, ERI = extension relaxation index, F1amp = 1st flexion amplitude, E1amp = 1st extension amplitude, Plat = plateau angle, and Ncyc = number of oscillation cycles)

| MODEL 2 | | *Flx2*: Flexor spasticity | | | | | |
|---|---|---|---|---|---|---|---|
| *Ext2*: Extensor spasticity | | Low spasticity (MAS = [1,1+]) | | | High spasticity (MAS= [2, 3]) | | |
| | | Mean | (SD) | n | Mean | (SD) | n |
| RI | Low spasticity | 1.1 | (.3) | 30 | 1.1 | (.5) | 9 |
| | High spasticity | .8 | (.4) | 8 | 1.0 | (.3) | 17 |
| ERI | Low spasticity | .5 | (.3) | 30 | .6 | (.5) | 9 |
| | High spasticity | .2 | (.1) | 8 | .4 | (.3) | 17 |
| F1amp | Low spasticity | 60.8 | (24.5) | 30 | 48.1 | (22.2) | 9 |
| | High spasticity | 42.9 | (31.9) | 8 | 45.9 | (19.3) | 17 |
| E1amp | Low spasticity | 24.4 | (19.2) | 30 | 28.0 | (20.7) | 9 |
| | High spasticity | 12.1 | (13.1) | 8 | 19.5 | (12.8) | 17 |
| Plat | Low spasticity | 53.6 | (14.4) | 30 | 47.8 | (15.1) | 9 |
| | High spasticity | 47.1 | (21.0) | 8 | 43.8 | (16.0) | 17 |
| Ncyc | Low spasticity | 4.7 | (2.7) | 30 | 5.2 | (1.8) | 9 |
| | High spasticity | 6.9 | (7.7) | 8 | 4.1 | (1.8) | 17 |

**Table 5** Significant levels for 2-way ANOVA tests for pendulum test metrics

| *p*-values | RI | ERI | F1amp | E1amp | Plat | Ncyc |
|---|---|---|---|---|---|---|
| Model 1 | | | | | | |
| *Ext1* | < 0.001* | < 0.001* | < 0.001* | < 0.001* | 0.001* | 0.594 |
| *Flx1* | 0.264 | 0.057 | 0.098 | 0.013* | 0.304 | 0.181 |
| *Ext1* × *Flx1* | 0.679 | 0.520 | 0.654 | 0.395 | 0.948 | 0.956 |
| Model 2 | | | | | | |
| *Ext2* | 0.103 | 0.021* | 0.149 | 0.040* | 0.251 | 0.598 |
| *Flx2* | 0.319 | 0.031* | 0.477 | 0.269 | 0.319 | 0.262 |
| *Ext2* × *Flx2* | 0.238 | 0.827 | 0.259 | 0.704 | 0.776 | 0.093 |

*effect is significant at $p < .05$
Model 1: Ext1 and Flx1 are coded 0 = No spasticity vs 1 = Any spasticity
Model 2: Ext2 and Flx2 are coded 0 = Low/ moderate spasticity vs 1 = High/ severe spasticity

and high/severe (MAS= [2, 3]) spasticity. The data in Fig. 3 provide a potential explanation for this observation: note that RI and F1amp decreased in magnitude until extensor MAS = 2 then increase for MAS = 3 to levels similar as MAS = 1+. This U-shape tendency was similarly reported between RI and reflex torque from a model-based analysis [20], but was not observed in other studies [27].

Figure 3 also shows that Plateau angle had a more linear relationship with extensor MAS score. This may also explain, when classifying by Model 1 (no spasticity vs any spasticity), why the combination of F1amp and Plateau angle in the logistic regression model had better classification accuracy than did the models with RI, F1amp or Plateau angle individually. This supports the notion proposed by others [41] that F1amp corresponds to stretch-reflex hyper-excitability and Plateau angle corresponds to passive resistance to stretch of the muscle, both of which comprise (and confound [4]) the clinical presentation of spasticity when using the MAS. This finding is consistent with the study by Fowler et al. [41], who reported that variability in resting angle (our plateau angle) contributed to their finding of low reliability of the RI for children with CP; which adds further support to using F1amp and Plateau angle as separate independent variables for quantifying spasticity.

Although none of the metrics we studied passed muster for Model 2 classification (low spasticity vs high spasticity), the metric with the best overall accuracy, sensitivity and specificity (>.7) was Plateau angle, as shown in Table 8. However, the PPV was too low (.24) to be of value for reliably discriminating spasticity in the clinic, and it failed to reach AUC criteria (>.7) using the ROC analysis. It is important to note, however, that we only examined a few of the available metrics from the pendulum test. Other variables or approaches may be more sensitive to discriminating level of spasticity, such as time to first peak, first excursion velocity, and area under the pendulum curve

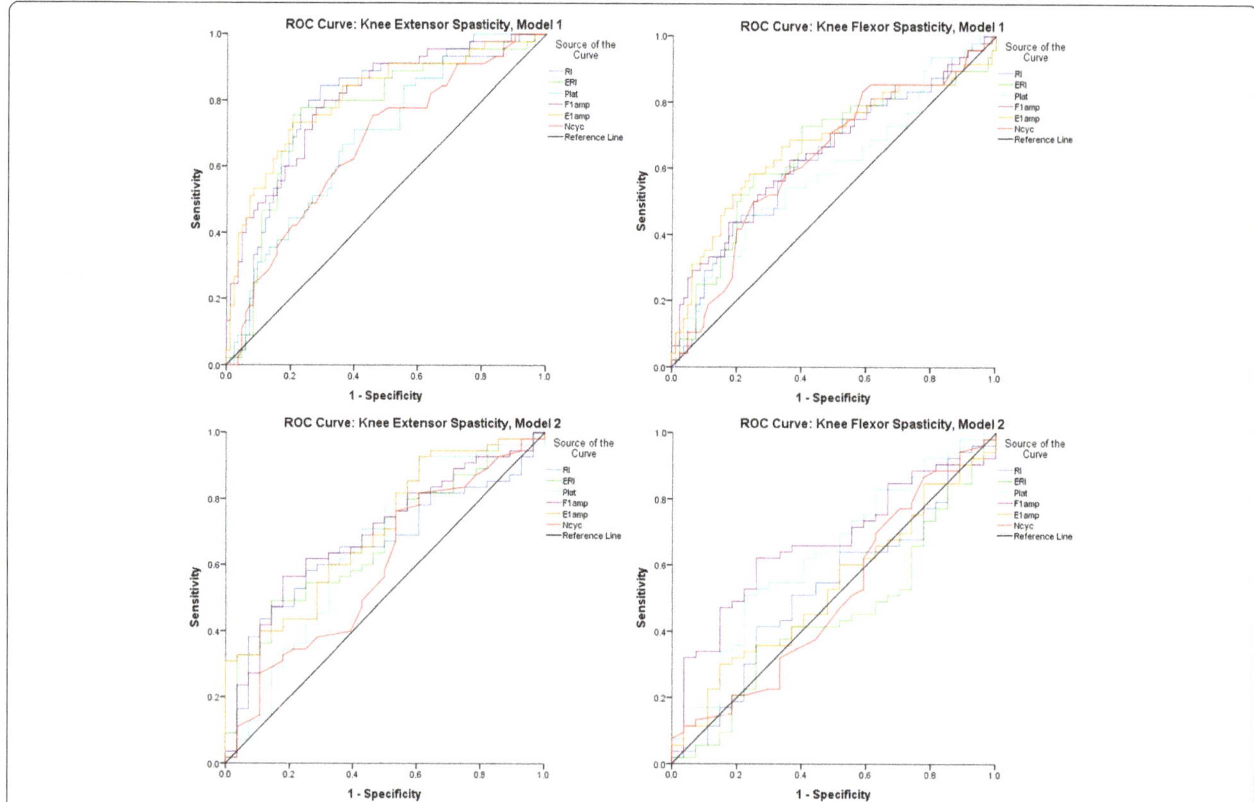


**Fig. 4** Receiver Operator Characteristic (ROC) curves for Model 1 (top panels) and Model 2 (bottom panels) for extensors (left panels) and flexors (right panels), for each of the pendulum test metrics studied

[25, 42], stiffness and damping properties [20, 36, 43], or deriving metrics from combining kinematic information with electromyographic (EMG) information [44].

Other studies have reported that number of cycles is an indicator of spasticity [17, 41]. In our study, however, number of cycles had the poorest classification performance of all metrics for both Model 1 and 2 analyses. Although our averages for number of cycle agreed with data reported by others [41], Fig. 3 shows that the number of cycles did not correspond to MAS score.

## Limitations

This study had several limitations. Firstly, the gold standard used for comparison was the Modified Ashworth Scale (MAS), which despite its ubiquity has been questioned as a meaningful measure of spasticity due to the intermingling effects of hyper-excitability of the stretch reflex and passive resistance to stretch [4, 9]. This means that the inability of the pendulum test metrics to discriminate between low/moderate and high/severe spasticity in our study may have been due to subjectivity and lack of sensitivity of the MAS scale, rather than the ability of the pendulum test to quantify the effect. Nevertheless, data from others [41] shows that the pendulum test was less reliable when muscle becomes exceedingly resistant to passive stretch, which stands to reason given the limited excursion of the limb after release in patients with severe spasticity (c.f. [20, 25]).

**Table 6** ROC analysis area under curve (AUC) values for each pendulum test metric and for Model 1 and Model 2 prediction of spasticity

| AUC (95% CI) | RI | ERI | F1amp | E1amp | Plat | Ncyc |
|---|---|---|---|---|---|---|
| Model 1 | | | | | | |
| *Ext1* | 0.784 (0.698,0.869) | 0.762 (0.672,0.852) | 0.807 (0.729,0.885) | 0.808 (0.726,0.890) | 0.691 (0.599,0.783) | 0.665 (0.567,0.762) |
| *Flx1* | 0.635 (0.534,0.736) | 0.652 (0.549,0.754) | 0.655 (0.554,0.756) | 0.675 (0.573,0.778) | 0.600 (0.496,0.70) | 0.631 (0.530,0.731) |
| Model 2 | | | | | | |
| *Ext2* | 0.659 (0.541,0.777) | 0.675 (0.557,0.792) | 0.695 (0.578,0.812) | 0.705 (0.589,0.820) | 0.655 (0.523,0.786) | 0.590 (0.459,0.721) |
| *Flx2* | 0.526 (0.392,0.661) | 0.446 (0.314,0.580) | 0.665 (0.544,0.785) | 0.531 (0.400,0.662) | 0.636 (.508,0.764) | 0.506 (0.368,0.644) |

**Table 7** Classification results using logistic regression for Model 1 prediction of extensor spasticity from pendulum test metrics: RI, F1amp, and Plat

| LR terms for Model 1 (*Ext1*) | Observed *Ext1* | Predicted *Ext1* | | Predictive value |
|---|---|---|---|---|
| | | No spasticity | Any spasticity | |
| [a]RI | No spasticity | 23 | 22 | *NPV* = .51 |
| | Any spasticity | 12 | 74 | *PPV* = .86 |
| | Overall | *Spec* = .66 | *Sens* = .77 | *CA* = 74.0% |
| [b]F1amp | No spasticity | 24 | 21 | *NPV* = .53 |
| | Any spasticity | 14 | 72 | *PPV* = .84 |
| | Overall | *Spec* = .63 | *Sens* = .77 | *CA* = 73.3% |
| [c]Plat | No spasticity | 15 | 30 | *NPV* = .33 |
| | Any spasticity | 11 | 75 | *PPV* = .87 |
| | Overall | *Spec* = .58 | *Sens* = .71 | *CA* = 68.7% |
| [d]F1amp, Plat | No spasticity | 28 | 17 | *NPV* = .62 |
| | Any spasticity | 12 | 74 | *PPV* = .86 |
| | Overall | *Spec* = .70 | *Sens* = .81 | *CA* = 77.9% |

[a]RI: Constant = 5.153; $Beta_{(RI)} = -3.646$, $p < .001$
[b]F1amp: Constant = 4.236; $Beta_{(F1amp)} = -.051$, $p < .001$
[c]Plat: Constant = 3.383; $Beta_{(Plat)} = -.049$, $p < .001$
[d]F1amp, Plat: Constant = 3.258; $Beta_{(F1amp)} = -.073$, $p < .001$; $Beta_{(Plat)} = .045$, $p = .082$

Another limitation is that the study was carried out at three different hospitals across four different cohorts of patients. Regarding the different sites, the BioTone™ software fully controlled the testing protocol (from order of testing to acquisition of all measures) and all three sites used the same version of the system and sensors, and all site personnel (physical therapists) were trained by the research team (M.J. and A.S.) when the system was deployed. Of greatest concern, would be the variability across therapists in their MAS assessments, however, this was not analyzed in the current paper. Regarding patient groups, although we did not analyze effects by patient cohort (due to the lower numbers of MS and CP) the measurement approach (clinical MAS and BioTone) was independent of neurological etiology, and it is expected that the relationships between pendulum test metrics and clinical assessment would not be influenced by this fact.

## Conclusions

The pendulum test metrics studied for quantifying knee extensor spasticity were not sensitive to level of flexor spasticity (as measured by MAS), age, years since diagnosis, or physical parameters of the limb. The pendulum test metrics RI and F1amp were good predictors of the presence of clinical spasticity in knee extensors, however, none of the metrics studied were acceptable predictors of the level of spasticity, as measured by MAS. The best logistic regression model for detecting presence of spasticity used F1amp and Plateau angle as separate input terms rather than using RI or F1amp alone. More research is needed to determine if pendulum test metrics are sensitive outcomes measures for managing problematic spasticity.

## Endnotes

[1]note that R2 was originally used by [17] but most authors now use RI as introduced by [19].

**Acknowledgements**
The authors are indebted to contributing staff and graduate students of the UNB Institute of Biomedical Engineering and Stan Cassidy Centre for Rehabilitation, Nova Scotia Rehabilitation Hospital in Halifax, and Spaulding Rehabilitation Hospital in Boston.

**Table 8** Logistic regression results for Model 2 prediction of extensor spasticity from pendulum test metrics: RI, F1amp, and Plat

| LR terms for Model 2 (*Ext2*) | Observed *Ext2* | Predicted *Ext2* | | Predictive value |
|---|---|---|---|---|
| | | Low spasticity | High spasticity | |
| [a]RI | Low spasticity | 52 | 5 | *NPV* = .91 |
| | High spasticity | 26 | 3 | *PPV* = .10 |
| | Overall | *Spec* = .66 | *Sens* = .38 | *CA* = 64.0% |
| [b]F1amp | Low spasticity | 49 | 8 | *NPV* = .86 |
| | High spasticity | 20 | 9 | *PPV* = .31 |
| | Overall | *Spec* = .71 | *Sens* = .53 | *CA* = 67.4% |
| [c]Plat | Low spasticity | 54 | 3 | *NPV* = .95 |
| | High spasticity | 22 | 7 | *PPV* = .24 |
| | Overall | *Spec* = .71 | *Sens* = .70 | *CA* = 70.9% |
| [d]F1amp, Plat | Low spasticity | 50 | 7 | *NPV* = .88 |
| | High spasticity | 20 | 9 | *PPV* = .31 |
| | Overall | *Spec* = .71 | *Sens* = .56 | *CA* = 68.6% |

[a]RI: Constant = 1.006; $Beta_{(RI)} = -1.623$, $p = .026$
[b]F1amp: Constant = .834; $Beta_{(F1amp)} = -.029$, $p = .005$
[c]Plat: Constant = 1.044; $Beta_{(Plat)} = -.035$, $p = .022$
[d]F1amp, Plat: Constant = .897; $Beta_{(F1amp)} = -.028$, $p = .072$; $Beta_{(Plat)} = -.003$, $p = .907$

**Funding**

This work was supported by the Atlantic Canada Opportunities Agency, Atlantic Innovation Fund, Project # 195180.

**Authors' contributions**

AW and CM conceived the research questions, performed the statistical analysis and wrote the manuscript draft. AS, MJ and CO collected the data and managed the project. CM and CO acquired the funding to support the project. All authors reviewed, edited and approved the final version of the manuscript.

**Competing interests**

The authors declare that they have no competing interests.

**Author details**

[1]Institute of Biomedical Engineering, University of New Brunswick, Fredericton, NB E3B 5A3, Canada. [2]Faculty of Kinesiology, University of New Brunswick, Fredericton, NB, Canada. [3]Stan Cassidy Centre for Rehabilitation, Fredericton, NB, Canada.

**References**

1. Public Health Agency of Canada. Mapping connections: Understanding of neurological conditions in Canada. Ottawa ON: Public Health Agency of Canada; 2014.
2. Logan LR. Rehabilitation techniques to maximize spasticity management. Top Stroke Rehabil. 2011;18(3):203–11.
3. Kostas JP, Ristanis SE, Soucacos PN. Botulinum a toxin for treatment of lower limb spasticity in cerebral palsy gait analysis in 49 patients. Acta Orthop Scand. 2003;74(6):749–55.
4. Malhotra S, Pandyan AD, Day CR, Jones PW, Hermens H. Spasticity, an impairment that is poorly defined and poorly measured. Clin Rehabil. 2009; 23(7):651–8.
5. Bohannon RW, Smith MB. Interrater reliability of a modified Ashworth scale of muscle spasticity. Phys Ther. 1987;67(2):206–7.
6. Mackey AH, Walt SE, Lobb G, Stott NS. Intraobserver reliability of the modified Tardieu scale in the upper limb of children with hemiplegia. Dev Med Child Neurol. 2004;46(4):267–72.
7. Ansari NN, Naghdi S, Younesian P, Shayeghan M. Inter- and intrarater reliability of the modified modified Ashworth scale in patients with knee extensor poststroke spasticity. Physiother Theory Pract. 2008;24(3):205–13.
8. Ansari NN, Naghdi S, Hasson S, Mousakhani A, Nouriyan A, Omidvar Z. Inter-rater reliability of the modified modified Ashworth scale as a clinical tool in measurements of post-stroke elbow flexor spasticity. NeuroRehabilitation. 2009;24(3):225–9.
9. Pandyan AD, Price CI, Rodgers H, Barnes MP, Johnson GR. Biomechanical examination of a commonly used measure of spasticity. Clin Biomech Bristol Avon. 2001;16(10):859–65.
10. Akpinar P, Atici A, Ozkan FU, Aktas I, Kulcu DG, Sarı A, et al. Reliability of the modified Ashworth scale and modified Tardieu scale in patients with spinal cord injuries. Spinal Cord. 2017;55:944–9.
11. Fleuren JF, Voerman GE, Erren-Wolters CV, Snoek GJ, Rietman JS, Hermens HJ, et al. Stop using the Ashworth Scale for the assessment of spasticity. J Neurol Neurosurg Psychiatry. 2009;81(1):46–52. Journal Article
12. McGibbon CA, Sexton A, Jones M, O'Connell C. Elbow spasticity during passive stretch-reflex: clinical evaluation using a wearable sensor system. J Neuroengineering Rehabil. 2013;10(1):61. -0003-10–61
13. McGibbon C, Sexton A, Jones M, O'Connell C. Quantification of elbow muscle tone from an instrumented manual stretch-reflex test. Phys Med Rehabil Res [Internet]. 2016 [cited 2017 Oct 29];1(5). Available from: http://oatext.com/Quantification-of-elbow-muscle-tone-from-an-instrumented-manual-stretch-reflex-test.php.
14. Calota A, Feldman AG, Levin MF. Spasticity measurement based on tonic stretch reflex threshold in stroke using a portable device. Clin Neurophysiol Off J Int Fed Clin Neurophysiol. 2008;119(10):2329–37.
15. Bar-On L, Aertbelien E, Wambacq H, Severijns D, Lambrecht K, Dan B, et al. A clinical measurement to quantify spasticity in children with cerebral palsy by integration of multidimensional signals. Gait Posture. 2013;38(1):141–7.
16. Wartenberg R. Pendulousness of the legs as a diagnostic test. Neurology. 1951;1(1):18–24.
17. Bajd T, Vodovnik L. Pendulum testing of spasticity. J Biomed Eng. 1984;6(1):9–16.
18. Lin DC, Rymer WZ. A quantitative analysis of pendular motion of the lower leg in spastic human subjects. IEEE Trans Biomed Eng. 1991; 38(9):906–18.
19. Stillman B, McMeeken J. A video-based version of the pendulum test: technique and normal response. Arch Phys Med Rehabil. 1995;76(2):166–76.
20. Le Cavorzin P, Hernot X, Bartier O, Carrault G, Chagneau F, Gallien P, et al. Evaluation of pendulum testing of spasticity. Ann Readaptation med Phys rev Sci Soc Francaise reeducation Fonct Readaptation Med Phys 2002;45(9):510–516.
21. Jamshidi M, Smith AW. Clinical measurement of spasticity using the pendulum test: comparison of electrogoniometric and videotape analyses. Arch Phys Med Rehabil. 1996;77(11):1129–32.
22. Yeh C-H, Hung C-Y, Wang Y-H, Hsu W-T, Chang Y-C, Yeh J-R, et al. Novel application of a Wii remote to measure spasticity with the pendulum test: proof of concept. Gait Posture. 2016;43:70–75.
23. Bajd T, Bowman B. Testing and modelling of spasticity. J Biomed Eng. 1982;4(2):90–6.
24. Bohannon RW. Variability and reliability of the pendulum test for spasticity using a Cybex II isokinetic dynamometer. Phys Ther. 1987;67(5):659–61.
25. Bohannon RW, Harrison S, Kinsella-Shaw J. Reliability and validity of pendulum test measures of spasticity obtained with the Polhemus tracking system from patients with chronic stroke. J Neuroengineering Rehabil. 2009; 6:30. (Journal Article):30–0003–6–30
26. Polman CH, Reingold SC, Banwell B, Clanet M, Cohen JA, Filippi M, et al. Diagnostic criteria for multiple sclerosis: 2010 revisions to the McDonald criteria. Ann Neurol 2011;69(2):292–302.
27. Fowler EG, Nwigwe AI, Ho TW. Sensitivity of the pendulum test for assessing spasticity in persons with cerebral palsy. Dev Med Child Neurol. 2000;42(3):182–9.
28. Valle MS, Casabona A, Sgarlata R, Garozzo R, Vinci M, Cioni M. The pendulum test as a tool to evaluate passive knee stiffness and viscosity of patients with rheumatoid arthritis. BMC Musculoskelet Disord. 2006;7:89. Journal Article
29. White H, Uhl TL, Augsburger S, Tylkowski C. Reliability of the three-dimensional pendulum test for able-bodied children and children diagnosed with cerebral palsy. Gait Posture. 2007;26(1):97–105.
30. Syczewska M, Lebiedowska MK, Pandyan AD. Quantifying repeatability of the Wartenberg pendulum test parameters in children with spasticity. J Neurosci Methods. 2009;178(2):340–4.
31. Biering-Sørensen F, Nielsen JB, Klinge K. Spasticity-assessment: a review. Spinal Cord. 2006;44(12):708–22.
32. Fleuren JF, Nederhand MJ, Hermens HJ. Influence of posture and muscle length on stretch reflex activity in poststroke patients with spasticity. Arch Phys Med Rehabil. 2006;87(7):981–8.
33. de Azevedo ERFBM, Maria RM, Alonso KC, Cliquet AJ. Posture influence on the pendulum test of spasticity in patients with spinal cord injury. Artif Organs. 2015;39(12):1033–7.
34. Szopa A, Domagalska-Szopa M, Kidoń Z, Syczewska M. Quadriceps femoris spasticity in children with cerebral palsy: measurement with the pendulum test and relationship with gait abnormalities. J Neuroengineering Rehabil 2014;11:166 (September).
35. Krause P, Szecsi J, Straube A. Changes in spastic muscle tone increase in patients with spinal cord injury using functional electrical stimulation and passive leg movements. Clin Rehabil. 2008;22(7):627–34.
36. Lotfian M, Mirbagheri MM, Kharazi MR, Dadashi F, Nourian R, Irani A, et al. Pendulum test measure correlates with gait parameters in children with cerebral palsy. Conf Proc Annu Int Conf IEEE Eng Med Biol Soc IEEE Eng Med Biol Soc Annu Conf. 2016;2016:1708–11.
37. Hofstoetter US, McKay WB, Tansey KE, Mayr W, Kern H, Minassian K. Modification of spasticity by transcutaneous spinal cord stimulation in individuals with incomplete spinal cord injury. J Spinal Cord Med. 2014;37(2):202–11.

38. Tancredo JR, Maria RM, Azevedo ER, Alonso KC, Varoto R, Cliquet Junior A. Clinical assessment of spasticity in individuals with spinal cord injury. Acta Ortop Bras. 2013;21(6):310–4.
39. Nordmark E, Anderson G. Wartenberg pendulum test: objective quantification of muscle tone in children with spastic diplegia undergoing selective dorsal rhizotomy. Dev Med Child Neurol. 2002;44(1):26–33.
40. Lin CC, Ju MS, Lin CW. The pendulum test for evaluating spasticity of the elbow joint. Arch Phys Med Rehabil. 2003;84(1):69–74.
41. Huang H-W, Ju M-S, C-CK L. Flexor and extensor muscle tone evaluated using the quantitative pendulum test in stroke and parkinsonian patients. J Clin Neurosci Off J Neurosurg Soc Australas. 2016;27:48–52.
42. Sterpi I, Caroli A, Meazza E, Maggioni G, Pistarini C, Colombo R. Lower limb spasticity assessment using an inertial sensor: a reliability study. Physiol Meas. 2013;34(11):1423–34.
43. Joghtaei M, Arab AM, Hashemi-Nasl H, Joghataei MT, Tokhi MO. Assessment of passive knee stiffness and viscosity in individuals with spinal cord injury using pendulum test. J Spinal Cord Med. 2015;38(2):170–7.
44. Yeh CH, Young HWV, Wang CY, Wang YH, Lee PL, Kang JH, et al. Quantifying spasticity with limited swinging cycles using pendulum test based on phase amplitude coupling. IEEE Trans Neural Syst Rehabil Eng `2016;24(10):1081–1088.

# 6

# A robot-based gait training therapy for pediatric population with cerebral palsy: goal setting, proposal and preliminary clinical implementation

Cristina Bayón[1], Teresa Martín-Lorenzo[2,3], Beatriz Moral-Saiz[2,4], Óscar Ramírez[1], Álvaro Pérez-Somarriba[2], Sergio Lerma-Lara[2,4], Ignacio Martínez[2] and Eduardo Rocon[1,5*]

**Abstract**

**Background:** The use of robotic trainers has increased with the aim of improving gait function in patients with limitations. Nevertheless, there is an absence of studies that deeply describe detailed guidelines of how to correctly implement robot-based treatments for gait rehabilitation. This contribution proposes an accurate robot-based training program for gait rehabilitation of pediatric population with Cerebral Palsy (CP).

**Methods:** The program is focused on the achievement of some specifications defined by the International Classification of Functioning, Disability and Health framework, Children and Youth version (ICF-CY). It is framed on 16 non-consecutive sessions where motor control, strength and power exercises of lower limbs are performed in parallel with a postural control strategy. A clinical evaluation with four pediatric patients with CP using the CPWalker robotic platform is presented.

**Results:** The preliminary evaluation with patients with CP shows improvements in several aspects as strength (74.03 ± 40.20%), mean velocity (21.46 ± 33.79%), step length (17.95 ± 20.45%) or gait performance (e.g. 66 ± 63.54% in Gross Motor Function Measure-88 items, E and D dimensions).

**Conclusions:** The improvements achieved in the short term show the importance of working strength and power functions meanwhile over-ground training with postural control. This research could serve as preliminary support for future clinical implementations in any robotic device.

**Keywords:** Robotic training, Therapy, Cerebral palsy, Exoskeleton, Rehabilitation, Gait

## Background

Gait limitation is one of the main impairments in children with Cerebral Palsy (CP) [1]. This mobility deficiency in CP is commonly the consequence of a damage of the child's Central Nervous System (CNS), and an optimal functional training is required in order to maximize the improvements [2], which will highly contribute to the enhancement of the independence and, therefore, the quality of life for both the young patient and his family [3].

The use of robotic trainers for neurorehabilitation applications has increased in the last decades, both in adulthood and childhood, and in several motor diseases [3–5]. Robot-based therapies have been developed and improved beyond reducing the clinician's effort. Currently, a new generation of robotic devices [6–8] provides means for encouraging the patients to an active participation in exercises, which are now more task specific. Both the implemented novel control strategies and

* Correspondence: e.rocon@csic.es
[1]Centre for Automation and Robotics (CAR), CSIC-UPM, Ctra Campo Real km 0.2 - La Poveda-Arganda del Rey, 28500 Madrid, Spain
[5]Universidade Federal do Espírito Santo Vitória, Brazil
Full list of author information is available at the end of the article

the modularity of new exoskeletons and gait trainers offer promising possibilities to enhance the rehabilitation outcomes by adapting the treatment to the patient's needs [9, 10]. Nevertheless, so far there is not enough evidence to ensure that classic robot-based rehabilitation provides better treatment outcomes by itself than conventional physical strategies in childhood [11]. New approaches are needed in order to improve the rehabilitation, making the robotic therapy a key feature of the change.

One of the main drawbacks for the everyday use of these technologies into the rehabilitation practice, apart from the price of these devices, is the absence of studies that describe a detailed robotic training program for gait rehabilitation. The wide variety of changes that could be applied to the parameters of robotic training therapies, makes unclear how to specify rehabilitation settings with the aim of providing a suitable solution for a large population size. Additionally, most of current studies are only focused on lower limbs strategies. However, the upper body (head and trunk movement) also influences gait function through walking balance [12], so a proper program should not ignore these features.

This manuscript presents a detailed robot-based therapy proposal for the rehabilitation of gait function in children with CP, which is based on the achievement of some specifications defined by the International Classification of Functioning, Disability and Health framework, Children and Youth version (ICF-CY) [13]. It contributes with better answers on how to implement robotic rehabilitation following defined guidance, establishing the baseline settings and subsequently tailoring the therapy to each patient.

The proposed robotic rehabilitation therapy works around a key factor: the implementation of strength and power exercises at the same time than over-ground walking guidance, performing in parallel an active head-trunk control therapy. As a result, the robot-based program recreates a situation as similar as possible to a real gait scenario, and encourages the patients to control different movements associated with gait: not only individual movements of lower limb joints but also the synergy between them while maintaining a proper posture of upper body. We hypothesize that these essential components, performed following an appropriate progression of the variables, will boost the rehabilitation of our patients.

Eventually, the robot-based therapy proposed in this manuscript is evaluated in four patients with CP in order to provide preliminary results of its application. Although the proposed robot-based treatment could be implemented in any robotic device adapted for that, in this study we selected the CPWalker training platform [7] as one scenario to test the effectiveness of the approach. This device is a novel robotic prototype with partial body weight support (PBWS) for over-ground rehabilitation of children with CP. It provides means for adapting the therapy to the user's necessities through different levels of assistance in multi-joint over-ground training. Other gait trainers, such as Lokomat [14] or LOPES [6], could have been also used to evaluate this training method, since they are also able to tailor each therapy session to the patient's needs.

## Methods

### Rehabilitation device

The CPWalker rehabilitation platform [7] is a robotic device composed by an exoskeleton linked to a walker that provides support and balance to the child during over-ground training (Fig. 1). The device is able to implement users' PBWS and allows the adaptation of exercises to the patient's capabilities by means of individual controllers for each joint, which increases the modularity of the system [7]. Each joint of the CPWalker exoskeleton can operate in a wide range of modes (see Fig. 1):

i) *Position control mode*: in this mode, the robot imposes a prescribed gait pattern to the user's lower limbs. The aim is that the patient learns the walking motion sequence correctly.
ii) *Impedance control modes*: these modes take into account the patients' collaboration. Thereby, the prescribed gait pattern should be achieved by the sum of robotic assistance and patient's cooperation. Three different modes of impedance may be executed in CPWalker (high, medium and low), which tolerate variable deviations from the programmed gait trajectories, enhancing the patients' participation and taking advantage of their residual movements through assist as needed strategies (AAN).
iii) *Zero-force control mode*: in this mode, the trajectory reference is not given, and the patient is who entirely moves the legs with a minimal resistance of the exoskeleton. It is used with patients with enough motor control (acquired with the previous modes) but poor balance, so the CPWalker provides stability and PBWS while the patient implements the gait pattern.

Within each mode, some variables of the robotic platform are updated along the treatment sessions: PBWS, gait velocity and the percentage of the total range of motion (ROM) [9]. The possibility of varying these parameters enables the customization of the therapy to the progression of each patient and gives a higher versatility for the treatment design. Additionally, the robotic platform also includes a biofeedback strategy to motivate the children to actively correct their posture during

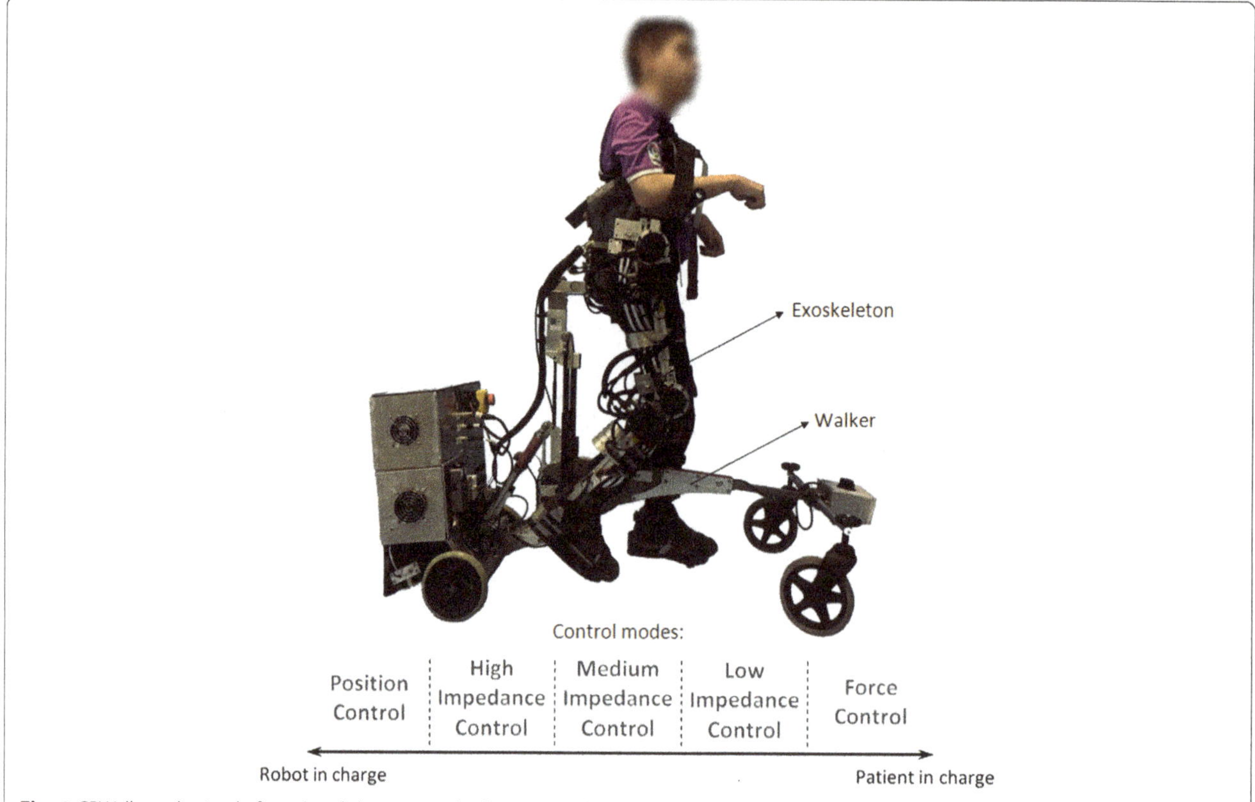


**Fig. 1** CPWalker robotic platform (*exoskeleton* to guide the patients' lower limbs and *walker* with PBWS to provide balance control during over-ground walking). The wide variety of operating modes comprehends from "robot in charge" to "patient in charge" states: i) position control mode; ii) high impedance control mode; iii) medium impedance control mode; iv) low impedance control mode; v) zero-force control mode

walking [7, 9]. The CPWalker robotic platform may be easily controlled by a clinician through an intuitive interface, which controls and monitors the exercises in real time.

The variety of control modes, the number of adjustable parameters, the ease of setting the variables individually for each joint and the possibility of implementing different strategies associated with gait simultaneously, made CPWalker an appropriate platform for testing the robotic therapy proposed in this manuscript.

### Robotic training program

In order to define the objectives of the robot-based treatment, the we adopted the conceptual framework of ICF-CY [13]. The proposal was focused on improving the principal gait-related functions derived from this international classification. Concretely, the selected goals of the ICF-CY to be achieved with the treatment and the work methodology implemented in the robotic device (in this case CPWalker), are presented in Table 1.

To achieve the goals presented in Table 1, we previously performed a systematic selection of variables based on the requirements of the National Strength and Conditioning Association (NSCA) youth training guidelines, which suggests that eccentric and explosive strength exercises should be the beginning of a proper training to ensure greater muscle power generation and the transference of gains to gait (Table 2)

According to the proposed objectives (Table 1) and complying with the NSCA youth training guidelines (Table 2), the treatment was conceptualized into two main *phases*, where the ROM, PBWS and gait velocity were the principal parameters under variation. The intention was that the patient maximized the gains acquired in the whole rehabilitation period (ideally the sum of robot-based exercises and common non-robotic therapy). A detailed description of each *phase* follows:

i) *First phase*: the main aim of this *phase* was to improve motor control, teaching the patients the correct sequence of motion and increasing strength. The patients were requested to follow the movements established by the exoskeleton with the minimal possible resistance during swing period, pushing the ground at each step and trying to keep the maximum flexion-extension values at the end of each gait cycle. Instructions were given to ensure the comprehension of normative gait patterns, and verbal encouragement in addition to direct feedback by graphics in real-time was delivered throughout the sessions.

**Table 1** Goal settings of the ICF-CY and the robot-based solutions adopted with CPWalker platform

| ICF-CY functional domain [13] | Implementation on CPWalker |
|---|---|
| Mobility and stability of joint functions (b710, b715) | The different control modes of the exoskeleton are used to guide the movement of a single or multiple joints, improving motor control. The exoskeleton also helps to maintain stability through the coordinated actions of surrounding tissues. This domain is exercised along the whole treatment with diverse robotic assistance. |
| Muscle power functions (b730) | The *second phase* of the training requires the patient to contract a muscle or muscle groups to generate the necessary force in order to start and maintain the movement with AAN strategies. The force must be maintained for a time in the extremes of the gait pattern (maximum flexion and extension) in order to reach the these maximum values in the complete range of motion. |
| Muscle endurance functions (b740) | Muscle endurance is exercised when the patient is requested to sustain a muscle contraction to finalize the required movement with AAN strategies, mainly in the extremes of the gait pattern (maximum flexion and extension). |
| Control of voluntary movement functions (b760) | The voluntary movement is implemented through the control and coordination of simple and complex movements to collaborate with AAN strategies. Lower impedance implies more patient's control. |
| Gait pattern functions (b770) | Motor control and gait pattern functions are trained through the different control modes to guide the lower limbs following prescribed gait patterns at several velocities and supports. |
| Maintaining a body position (d415) | A biofeedback strategy for postural control is used to notify the patients when they lose the correct position of the upper body. See section "Postural control". |
| Walking (d450) | Over-ground walking training is executed in all sessions with controlled PBWS, at different velocities and supports. |
| Proprioceptive functions (b260) | During random moments of the first training session, the patients perceive feeling using a mask on the eyes at the same time than the robot performs the movement for single or multiple joints with 100% of PBWS. |
| Formal relationships (d740) | Creating and maintaining patient-researcher relationship. See section "Motivation and inclusion of challenges". |

ii) *Second phase*: the main aim of this *phase* was to further train motor control and increase power in order to ensure the transference to the independent gait pattern. Aware of the sequence of movement of a normal gait pattern, the patient's contribution became an important aspect to develop neuroplasticity and preserve the gained motor control [15]. The active participation was achieved both by boosting the patient's motivation and by requiring self-activity [16]. The latter was implemented through AAN algorithms and the impedance control modes presented previously for CPWalker [7].

Figure 2 represents the schematic view of the therapy proposal. The treatment was composed of a total of 16 sessions (*first phase*: 8 sessions for strength training and motor control learning; and *second phase*: 8 sessions to transfer the gains to gait through power performance).

**Table 2** National Strength and Conditioning Association (NSCA) youth training guidelines [32]

| Adapted youth resistance training NSCA guidelines | | |
|---|---|---|
| Variables | Strength | Power |
| Muscle actions | Eccentric / Concentric | Eccentric / Concentric |
| Exercise | Single / Multi-joint | Single / Multi-joint |
| Intensity | ↑ Load / ↓ Velocity | ↓ Load / ↑ Velocity |
| Velocity | Moderate | Moderate / Fast |

Exercises were multi-joint and gait-oriented, demanding concentric-eccentric actions based on the gait phase that was being performed. During the whole treatment (*first* and *second phases*) the position of head and trunk during walking exercises were monitored, especially because these patients usually walked looking at the ground. For this purpose, the strategy for postural control of CPWalker encouraged the patients giving an acoustic feedback to when their position was inappropriate, so they could realize and rectified it by themselves. The program was reinforced with modifications on AAN levels according to the patient's evolution based on performance evaluations related to ROM, PBWS and gait velocity. Furthermore, several challenges were included to enhance the patients' motivation.

### Duration of the study

The robot-aided treatment was proposed for a whole period of 2 monthly cycles (1 month for each *phase* of the treatment) with the aim of having enough sessions to generate significant neural changes [17]. The children trained 2 non-consecutive days per week for 8 weeks (16 sessions, see Fig. 2). The sessions consisted of a 10–15 min warm-up and 60 min of over-ground exercise with CPWalker, including 3 min of independent gait as a cool-down phase.

As Fig. 2 indicates, the first 8 sessions corresponded with general motor control and strength exercises, where the

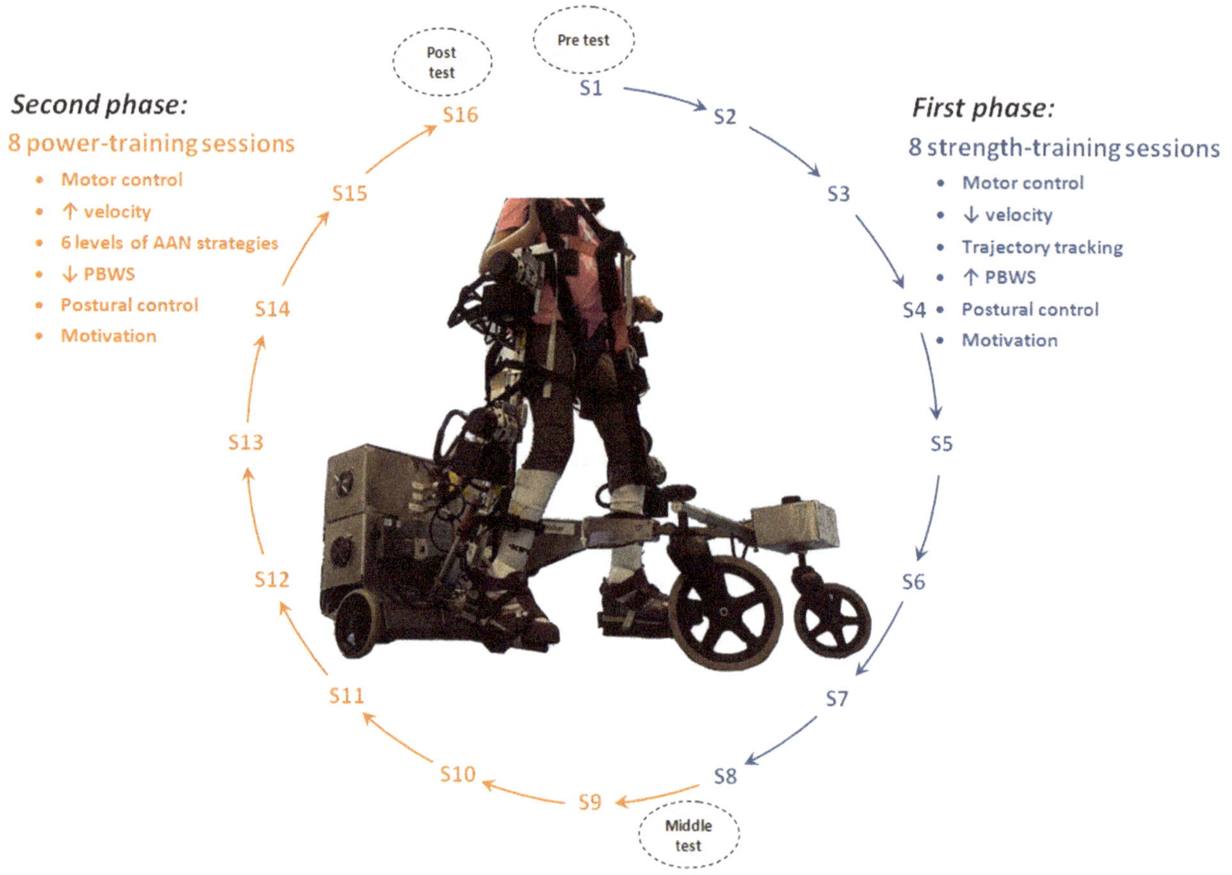

**Fig. 2** Robot-based training program overview. *First phase:* sessions S1 to S8 for strength exercises where motor control was primary trained. *Second phase:* sessions S9 to S16 for power training where the assistance was progressively decreased with the patient's progression. The improvements were assessed at three stages of analysis: before treatment begins, between both *phases* and at the end of the program (gray ellipses in the figure)

robot imposed a gait trajectory tracking. Sessions 9 to 16 were related to muscle power performance through levels of AAN strategies, where self-activity was required.

### *Training phases*

In order to individually define the training progression through the different sessions and to comply with the NSCA guidelines, the principal modifications were implemented on ROM, PBWS and gait velocity. The selected parameters for both *phases* are represented by Figs. 3 and 4 respectively. This selection was concluded in collaboration with our clinical partners, based on the evaluation of previous studies carried out with CPWalker [9, 18]. The robot-based tasks began with high assistance and PBWS, and they progressed toward greater ROM and smaller PBWS as long as the patient overcame the different levels of the sessions. Parameter variations within and between sessions were performed as long as session goals were attained and when the clinical staff agreed, based on levels of spasticity, fatigue and motor control presented in the last day. If the patient

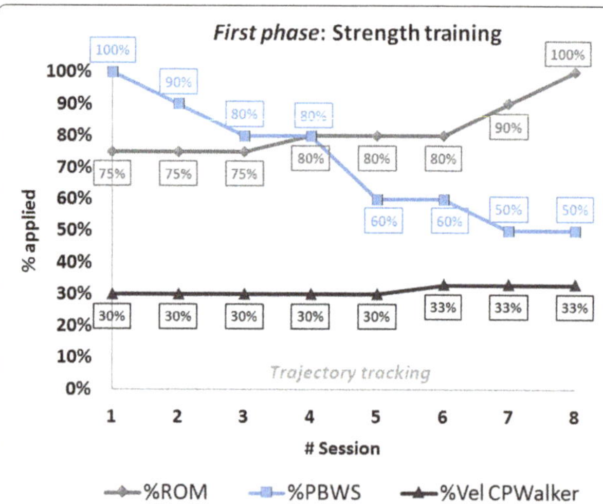


**Fig. 3** *First phase*: strength training progression values along first 8 sessions of the robot-based therapy. Trajectory tracking motion was imposed by the robot. The light-gray line represents the movement amplitude (%ROM), the blue line the changes of %PBWS and the dark-gray line is referred to gait velocity percentage for each session

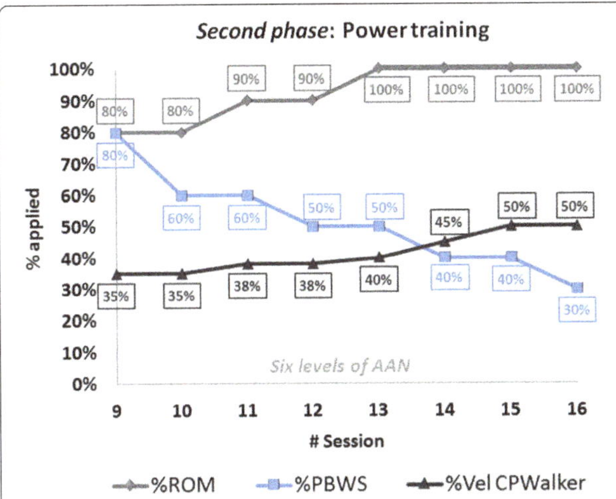


**Fig. 4** *Second phase*: power training progression values along sessions 9 to 16 of the robot-based therapy. Six different levels of assistance were selected on the exoskeleton (combining hips and knees). The light-gray line represents the movement amplitude (%ROM), the blue line the changes of %PBWS and the dark-gray line is referred to gait velocity percentage for each session

was not ready to jump to the next challenge, the session was repeated with the same last percentages.

Within the *first phase* (Fig. 3), the first session was performed with the children completely suspended (100% of PBWS) in order to adapt the users to the movements with the robot. Moreover, during random moments of this first session, the patients wore a mask on their eyes to feel the motion performance. For the rest of the *first phase* children's lower limbs were guided through a pure position control imposed by the CPWalker platform, with a gradual decrease in PBWS (Fig. 3 blue), and a gradual amplification of ROM (Fig. 3 light-grey). Gait velocity during this *first phase* was maintained around a regular and small value (Fig. 3 dark-grey). Notice that in general, only one variable at each session was varied.

The *second phase* of the training (sessions 9 to 16 in Fig. 4) presented an additional difficulty that enhanced the user's collaboration in the exercises performance through different levels of AAN strategies in the exoskeleton. The initial ROM for the *second phase* was set at 80% of the total gait pattern and reached 100% by session 13 (Fig. 4 light-grey), time at which velocity was highly increased (Fig. 4 dark-grey). Note that gait velocity for this *phase* became around double of the one achieved in the *first phase*, which is in relation to the requisites exposed in Table 2. Furthermore, PBWS declined up to 30% of weight supported by the platform (Fig. 4 blue).

It is important to highlight in Figs. 3 and 4 that the percentage of PBWS was related to the individual patient's total weight, and the percentage of ROM was applied to the total trajectory of the gait pattern programmed in the control of CPWalker [7]. The estimated changes in gait velocity are represented regarding percentages of CPWalker platform, where 100% corresponded to 0.6 m/s.

### Tailored assist as needed strategies

With the aim of enhancing the patient's participation in the *second phase* and consequently improving outcomes of the treatment, in sessions 9 to 16, position control was substituted by six adapted levels of impedance control in the joints of the exoskeleton (Table 3).

One of the main advantages of CPWalker robot was the possibility of selecting three different modes of assistance individualized per joint beyond a pure position control (high, medium and low impedances). In consonance with this, six situations (levels) were adopted following the criteria of our clinical partners in order to define the scales of difficulty in the assistance of knees and hips in the *second phase* (Table 3). Thereby, the patients were considered fit to move to the next level when they achieved a performance higher than 85% in the execution of each session, together with its corresponding parameters (ROM, PBWS and gait velocity) represented by Fig. 4. This percentage of performance was calculated comparing the real motion executed by the children and the desired gait pattern of each session.

### *Postural control*

It was important to ensure that throughout all the sessions, patients maintained a proper posture of head and trunk because it facilitates the performance of any activity of daily living, and improves the social interaction, the participation and communication [12, 19]. In this sense, the aim of our proposal was to provide biofeedback to the patients each time that they kept an incorrect position of the body during walking. With this goal, the CPWalker robot used Inertial Measurement Units (IMUs) (Technaid, Spain) to measure the rotation of head and trunk in real time, and give acoustic feedback when subjects overcame predefined maximum values selected by clinicians. In response to the acoustic feedback, patients were instructed to correct their position, time at which the acoustic feedback ceased. This strategy had been previously proved with promising results in children with spastic diplegia [9].

**Table 3** Levels of assistance in *first* and *second* training *phases*

| | *First phase* | *Second phase* | | | | | |
|---|---|---|---|---|---|---|---|
| | Level 0 | Level 1 | Level 2 | Level 3 | Level 4 | Level 5 | Level 6 |
| Hips | P | HI | MI | MI | MI | LI | LI |
| Knees | P | P | P | HI | MI | MI | LI |

The assistance on knee joint always went behind the assistance on hip, due to the knee movement during gait is performed following inertial forces, assigning to the hip movement higher importance. A higher level is implemented if the session performance (real motion versus desired pattern) is bigger than 85%

*P* position, *HI* high impedance, *MI* medium impedance, *LI* low impedance

***Motivation and inclusion of challenges***

In the field of physical rehabilitation, especially in childhood disability, the F-words (Function, Family, Fitness, Fun, Friends and Future) defined by Dr. Rosenbaum [20] become really important. It is essential to maintain a high patients' motivation because this concept could affect treatment outcomes [21]. To address this issue, we introduced challenges with goals in each session of the robot-based therapy with the aim of having a more engaged user. An example of that is a classification board where the children could follow their progression along the different sessions, and they were rewarded when the goals of each session were correctly fulfilled.

Moreover, with the same objective of enhancing motivation, the data collected with the robotic platform was explained to the patients through graphics so they could feel part of the team, and their interest in the treatment increased.

The patient's motivation was subjectively measured in each session through a scale from 0 to 10 points, with 0 being no motivation and 10 being maximum motivation.

**Metrics**

In order to objectively measure the patient's evolution and due to the lack of homogeneity among children with CP, we decided to evaluate the progression of the therapy by comparing each patient to himself, instead of maintaining a control group. We carried out some analyzes and evaluation metrics in different occasions of the study (Table 4): *during* the use of the robot, before the treatment begins (*pre*), in the *middle* and after the whole sessions (*post*).

The 10mwt was assessed for two situations: normal comfortable walking speed and maximum walking speed. Three trials were collected for each situation and subsequently, the average of the three trials was calculated.

Regarding the 6mwt, it was performed indoors along a flat corridor, where the walking course had a 30 m length and was marked every 3 m. The turnaround points were marked with cones. The patients received information about remaining time every minute, but they were not encouraged during the exercise [22]. The heart rate was also measured for each patient in two situations: resting and just after finishing the test. This parameter gives the possibility of calculating the Physiological Cost Index (PCI) after the exercise, which is used to quantify the energy expended by the patients during the exercise and their progression, [23].

3D gait analysis was recorded at 200 Hz using a motion capture system Smart-DX (BTS Bioengineering, Italy). In order to obtain gait kinetics, a set of reflective markers were placed over the skin on discrete anatomical sites according to the Helen Hayes Model [24]. Subjects walked barefoot at a self-selected speed.

Maximum isometric strength was measured in kgf with a hand-held dynamometer microFET2 (Hoggan Scientific LLC, USA). Three records were taken and averaged for each movement bilaterally (dorsiflexion, plantarflexion, knee flexion-extension, hip flexion-extension, abduction and adduction).

**Table 4** Evaluation metrics and moment of application

| Metric | Utility | When | | | |
|---|---|---|---|---|---|
| | | During | Pre | Middle | Post |
| 10-m walking test (10 mwt) [30] | Gait-speed measure | | * | * | * |
| 6-min walking test (6 mwt) [33] | Global responses involved during exercise. Walked distance and endurance | | * | * | * |
| Physiological Cost Index [23] | Energy expended | | | * | * |
| 3D gait analysis | Kinematics, spatial-temporal parameters, gait deviation index (GDI), gait profile score (GPS)... | | * | * | * |
| Maximum isometric strength with hand-held dynamometer | Maximum voluntary contraction of primary lower limbs muscles | | * | * | * |
| Selective Control Assessment of Lower Extremity (SCALE) [25] | Quantify selective voluntary motor control and its progression | | * | * | * |
| Gross Motor Function Measure (GMFM-88) dimensions D (standing) and E (walking) [26] | Evaluate changes in gross motor function according to activities as walking, running or jumping | | * | | * |
| Kinesiophobia assessment | Psychological influence of fear and pain | | * | | * |
| Gillette Functional Assessment Questionnaire (FAQ) [34] | Independent activities, locomotor abilities and user's satisfaction | | * | | * |
| ROM performance | Measure of ROM in lower limbs during robot walking | * | | | |
| Selective force | Evolution of forces executed by the user during robot walking | * | | | |
| Patient's motivation | Estimation of user's motivation in 0 to 10 scale | * | | | |

The particularity of the SCALE assessment [25] was that it was evaluated by the same physiotherapist bilaterally on three occasions (pre, middle and post), with the aim of reducing the subjective error.

The changes of GMFM-88 [26] were collected for the 88 items, but we focused on dimensions D (standing) and E (walking).

The kinesiophobia assessment consisted of a test composed of 10 questions of 1 to 4 points each. The responses were given by the patients without parents' influence.

Two FAQ questionnaires were requested: one as initial questionnaire at the beginning and the other as follow up at the end of the treatment. These surveys presented several questions for parents and others referred to children.

During the whole treatment, ROM performance and force interactions were measured for each session in order to evaluate if the patient was prepared to jump to the next stage with more difficult parameters and level of assistance.

Finally, the users' motivation was subjectively evaluated by the practitioner from 0 to 10 points for each session with the robot.

### Patients

Four children diagnosed with spastic CP affecting muscle strength and motor control of lower limbs (two male, two female, weight 44.75 ± 6.29 kg, height 1.56 ± 0.29 m and age 14.50 ± 2.38 years-old) were selected to be participants for testing the robotic training proposal (P1, P2, P3 and P4 in Table 5). The inclusion criteria for patients' recruitment followed: i) children aged 11 to 18 years suffering from spastic diplegia; ii) Gross Motor Function Classification System (GMFCS [1]) levels I to IV; iii) maximum weight 75 kg; iv) anthropometric measures of lower limbs according to the exoskeleton of CPWalker; v) capable of understanding the proposed exercises; and vi) able to signal pain or discomfort. The exclusion criteria was: i) patients who experimented concomitant treatments 3-months prior study (e.g. orthopedic surgery or botulinum toxin); ii) children with muscle-skeletal deformities or unhealed skin lesions in the lower limbs that could prevent the use of the exoskeleton; iii) patients with critical alterations of motor control as dystonia, choreoathetosis or ataxia; iv) aggressive or self-harming behaviors; and v) severe cognitive impairment. The study was carried out at "Hospital Infantil Universitario Niño Jesús", (Spain). The Local Ethical Committee of this hospital gave approval to the study (R-0032/12) and warranted its accordance with the Declaration of Helsinki. All participants and families were informed, and parental consents were obtained prior to participation (Additional file 1).

**Table 5** Patients' description

| Patient | Age | GMFCS | Weight (kg) | Height (m) | Walking support | |
|---|---|---|---|---|---|---|
| | | | | | 50 m | 500 m |
| P1 (F) | 12 | III | 40 | 1.56 | CT | WC |
| P2 (F) | 16 | II | 42 | 1.60 | Cane | Cane |
| P3 (M) | 17 | III | 54 | 1.53 | PW | WC |
| P4 (M) | 13 | II | 43 | 1.55 | – | – |

Two females (F) and two males (M) with spastic diplegia were selected. No medication 3-months prior the study was taken by the patients. The type of walking support without the aid of the robot is indicated for a distance of 50 m and 500 m: crutches-CT, wheelchair-WC, posterior walker-PW and cane

## Results

Due to personal reasons unrelated to the study, three of four patients (P1, P2 and P4) completed 15 of a possible 16 total sessions. Concretely, P1 lost the session number 8, P2 lost the number 7 and P4 the number 11. The rest of training was completed successfully.

The modifications of parameters (ROM, PBWS and gait velocity) proposed in Figs. 3 and 4, were fulfilled by all children without problems. The progressions of the levels of assistance provided by Table 3, which were tailored for each patient between sessions 9 to 16 (*power training* with AAN strategies), are represented in Fig. 5, where the maximum reached level was level 5 by P1 in the last two sessions.

### Gait speed, endurance and global responses

All patients improved the outcomes in D and E dimensions (assessed together) of the GMFM-88 scale [26] (Fig. 6a). Showing the dimensions D and E separately, the increments per patient (post-pre) were: P1 (38.48%$_D$, 34.71%$_E$), P2 (5.13%$_D$, – 1.39%$_E$), P3 (25.90%$_D$, 15.72%$_E$) and P4 (10.26%$_D$, 22.20%$_E$). Regarding these results, all patients except P2 got minimal clinically important difference (MCID) for dimensions D and E [27]. The rest of dimensions of the GMFM-88 showed in conjunction a total increment for this scale of 18.16% in P1, 3.93% in P2, 11.8% in P3 and 8.11% in P4, which represent for P1, P3 and P4 gains higher than seven percentage points, considered as positive changes by therapists.

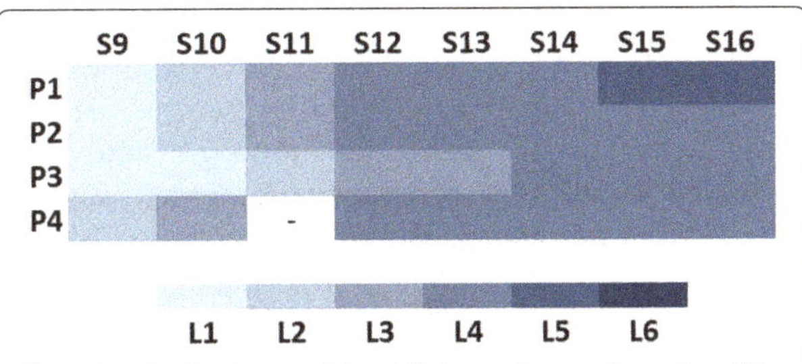


**Fig. 5** Levels of assistance (L1 to L6) depending on the patient (P1 to P4) and the AAN session (S9 to S16). The patients could jump to the next level if they achieved at least the 85% of the pattern desired for each session. The level for P4 in S11 is not represented because P4 lost this session

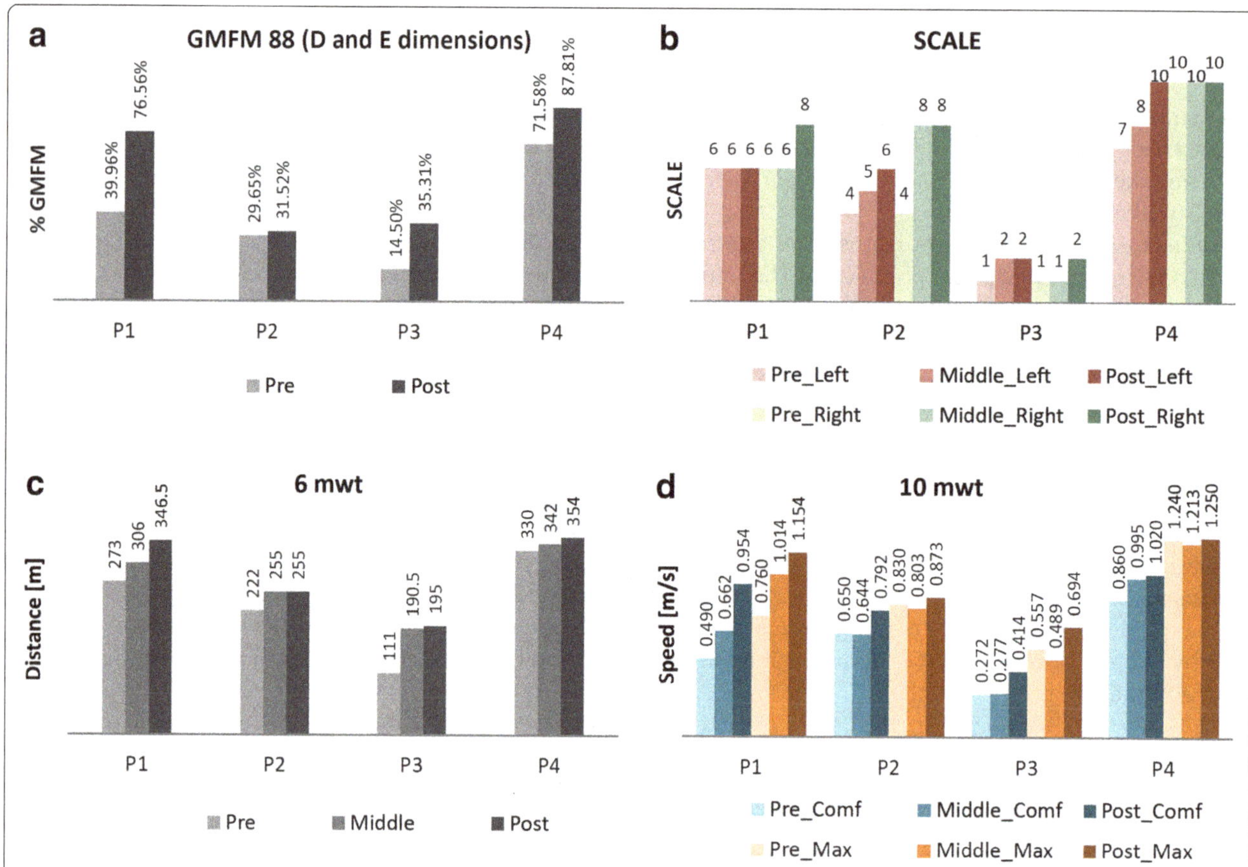


**Fig. 6** **a** Results of GMFM-88 (D and E dimensions), **b** SCALE, **c** 6mwt, and **d** 10mwt in pre, middle and post analysis for all the patients (P1 to P4). The SCALE was measured bilaterally (left and right). The 10mwt was performed in two situations: comfortable speed (Comf) and maximum speed (Max)

The SCALE assessment also showed better results at the end of the robot-based treatment (Fig. 6b), [25]. In this case, although the value for the left leg in P1 was kept same as at the beginning (Fig. 6b, red bars for P1), the rest of measures were increased or maintained as maximum (SCALE equal to 10 points). No reference of MCID has been found for SCALE assessment.

Finally, both the walked distance in the 6mwt and the walking speed in the 10mwt increased after the training period (Fig. 6c and d respectively). The two situations evaluated for the 10mwt are represented: a comfortable speed for each child (blue bars in Fig. 6d) and the same exercise at maximum speed (orange bars in Fig. 6d). More concretely, the percentage of progressions comparing post and pre-analysis in these metrics, were: P1 (6mwt: 26.92%; $10mwt_{comf}$: 94.69%; $10mwt_{max}$: 51.84%); P2 (6mwt: 14.86%; $10mwt_{comf}$: 21.85%; $10mwt_{max}$: 5.18%); P3 (6mwt: 75.68%; $10mwt_{comf}$: 52.21%; $10mwt_{max}$: 24.60%) and P4 (6mwt: 7.27%; $10mwt_{comf}$: 18.60%; $10mwt_{max}$: 0.81%).

In case of 6mwt, the MCID was established as 50 m in stroke survivors [28]. In children and adolescents with CP, the minimum detectable change (MDC), which may not represent a MCID, has been settled as 64 m for GMFCS II and 47.4 m for GMFCS III [29]. Comparing our results to these significances, P1 and P3 reached the MCID for the 6mwt with improvements of 73.5 m and 84 m respectively.

In case of 10mwt, 0.13 m/s is the MCID in spinal injuries [30]. For CP, only the MDC was found, but Thompson et al. ensured in [29] that it did not present reliability. In that sense, P1 and P3 obtained clinical significance in the $10mwt_{max}$ comparing with the MCID exposed in [30].

The Physiological Cost Index was evaluated comparing middle and post assessments during the 6mwt. All patients reduced the PCI: P1 obtained 0.75 beats/m (middle) and 0.55 beats/m (post); P2 0.89 beats/m (middle) and 0.80 beats/m (post); P3 1.57 beats/m (middle) and 1.26 beats/m (post); and P4 0.33 beats/m (middle) and 0.03 beats/m (post).

#### Strength progression

In order to quantify the patient's maximum strength performing on defined and individual movements without the robot, three measures were taken for each required motion. According to that, Fig. 7 represents the average

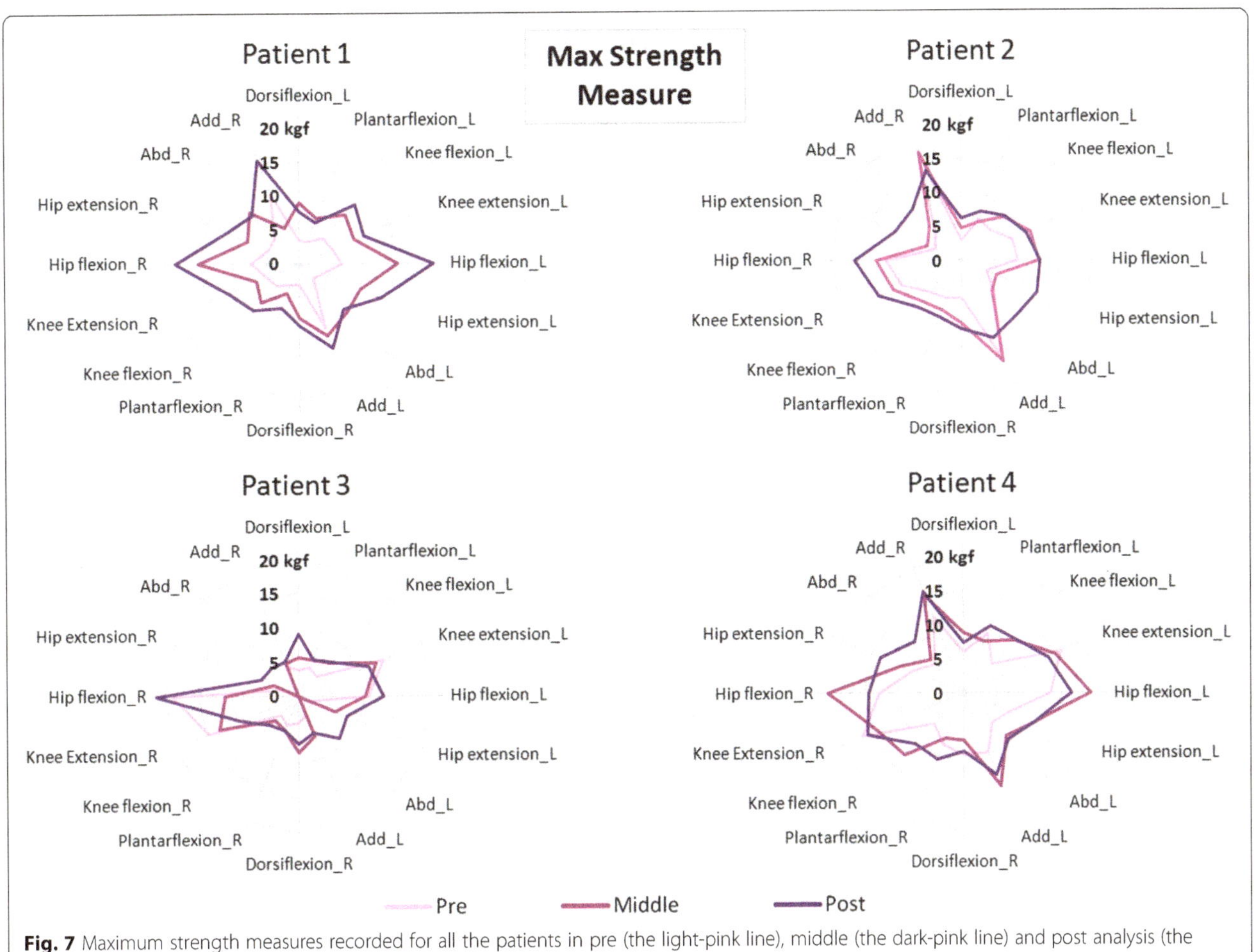


**Fig. 7** Maximum strength measures recorded for all the patients in pre (the light-pink line), middle (the dark-pink line) and post analysis (the purple line). Both legs were evaluated, right (R) and left (L)

values (in kgf) of the individualized movements recorded in pre, medium and post analyzes. In general, results from Fig. 7 show that the purple line (post assessment) covers the light-pink line (pre measures) for all the patients. In some cases, it even covers the dark-pink line (measured just after finishing the first 8 strength training sessions). This means that higher values of strength were reached by the children after the robot-based treatment. Concretely, the general improvements (including all the required movements) per patient in maximum isometric strength measure, comparing pre and post averages, were: P1: 129.77 ± 58.71%; P2: 61.39 ± 58.55%; P3: 70.54 ± 83.68% and P4: 34.41 ± 30.41%.

### Kinematics and spatiotemporal variability

The 3D kinematic analysis provided outcomes focused on gait improvements respect to normality. The GPS and GDI (Fig. 8) are accepted indexes that represent how close the patient's gait is to the desired gait. Related to these metrics and comparing pre and post analyzes, all the patients obtained better values for both sides (left and right) after the robot-based treatment. Nevertheless, they were not clinically significant, except for the right side of P1 (around 10 points in GDI). It is important to highlight that we believe that the post results in P4 could be affected by personal circumstances non-related to the study that occurred the day of the test.

Table 6 shows the values of Fig. 8 in detail and also includes some of the spatial-temporal parameters recorded during the studies. The average improvement percentages (four patients) in spatiotemporal parameters were: 21.46 ± 33.79% for mean velocity, 2.84 ± 13.96% for cadence and 17.95 ± 20.45% for step length.

### ROM performance

Although all the patients succeeded the changes in the parameters of ROM, velocity and PBWS for the different sessions, the progression in AAN levels was individualized for each subject (Fig. 5). Thereby, P1 was the most advanced, achieving level 5 as the maximum. An example of ROM performance difference between trajectory tracking and AAN strategy is represented by Fig. 9, which shows the data collected for P1 in session 3 (Fig. 9a) and session

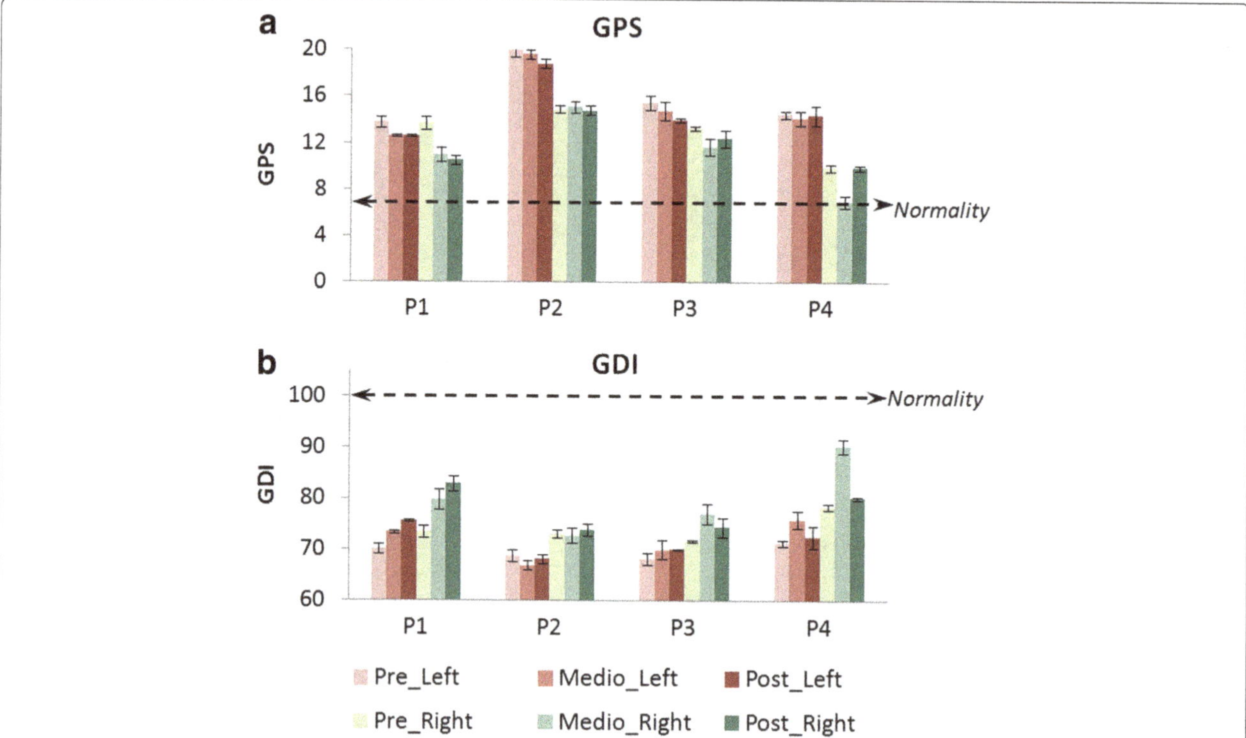


**Fig. 8** **a** GPS and **b** GDI for pre, middle and post kinematic analyzes of patients P1 to P4. The results represent means ± standard error bilaterally (left in red bars and right in green bars). Normality in GPS considers values lower than 7 points (doted-black line in (**a**)), and normality in GDI comprehends values higher than 100 points (doted-black line in (**b**))

16 (Fig. 9b). The last one was recorded within level 5 (low impedance at hips and medium impedance at knees).

### Qualitative variables

The motivation was subjectively evaluated by the practitioner who was with the children during the whole period of the study. The averaged motivation values on a scale from 0 to 10 points were: 9.4 for P1, 8.6 for P2, 9.44 for P3 and 8.87 for P4. Moreover, three of four patients decreased the kinesiophobia score after the 16 sessions.

Parents and patients filled a FAQ questionnaire at the beginning and a follow-up at the end of the treatment. Regarding parent questionnaires, results show that all of them thought that the strength and mobility were better at the end of the study thanks to the robot-based therapy. Meanwhile half of them also included the endurance as an improved variable due to the robot. One-hundred percent of parents felt satisfied towards the results of robotic therapy with CPWalker, and they ensured that they would like to do it again. The patients' opinion was very similar. They were also satisfied and, in general, they described the treatment as: "really fun", "the robot makes you feel light and independent" and "safe".

## Discussion and Conclusion

The main aim of this research was to provide a first approach to the implementation of a novel and defined robotic rehabilitation method that could cover the most important clinical aspects of the ICF-CY framework. This proposal was tested with four pediatric patients with CP, which provided us some preliminary outcomes to assess it. Although the patients' progression was evaluated without a control group, we do not consider it as a relevant limitation of the study, since the wide variety of differences among each child with CP makes interesting and even necessary to expose the improvements by comparing each patient with himself.

According to the results, the greatest benefits due to the robot-based treatment corresponded to P1 and P3, who were the most affected levels of GMFCS (III in both cases). In general, the higher values of gait speed and improved values of global responses achieved by all the children in several tests, may be in benefit of the patients' social mobility. Visual inspection of the graphics show that changes appeared after a small number of sessions (middle tests) and they were commonly increased or maintained until the post studies. The results showed statistical significance for the improvements of most of

**Table 6** Spatial-temporal parameters, GPS and GDI values for pre, middle and post analyzes in all the patients. The results were calculated taking an average of 40 steps

| Patient | Analysis | Side | Stride time (s) | Mean velocity (m/s) | Cadence (steps/min) | Step length (m) | Gait Profile Score (GPS) | Gait Deviation Index (GDI) |
|---|---|---|---|---|---|---|---|---|
| Normality | | | 0.93 ± 0.04 | 1.20 ± 0.20 | 129.60 ± 8.40 | 0.58 ± 0.06 | < 7 | > 100 |
| P1 | Pre | Left | 1.52 ± 0.12 | 0.50 ± 0.1 | 79.05 ± 7.51 | 0.39 ± 0.02 | 13.80 ± 0.50 | 70.19 ± 1 |
| | | Right | 1.54 ± 0.15 | | | 0.38 ± 0.04 | 13.70 ± 0.60 | 73.50 ± 1.13 |
| | Middle | Left | 1.81 ± 0.04 | 0.40 ± 0 | 66.20 ± 1.58 | 0.25 ± 0.18 | 12.60 ± 0.10 | 73.43 ± 0.30 |
| | | Right | 1.82 ± 0.12 | | | 0.40 ± 0.03 | 11 ± 0.60 | 79.82 ± 2 |
| | Post | Left | 1.27 ± 0 | 0.80 ± 0 | 96 ± 1.30 | 0.46 ± 0.04 | 12.60 ± 0.10 | 75.62 ± 0.16 |
| | | Right | 1.23 ± 0.03 | | | 0.56 ± 0.02 | 10.50 ± 0.40 | 82.87 ± 1.42 |
| P2 | Pre | Left | 1.42 ± 0.04 | 0.50 ± 0 | 86.70 ± 5.19 | 0.26 ± 0.01 | 20 ± 0.60 | 68.81 ± 1.23 |
| | | Right | 1.36 ± 0.11 | | | 0.39 ± 0.01 | 14.90 ± 0.30 | 73.08 ± 0.78 |
| | Middle | Left | 1.43 ± 0.08 | 0.40 ± 0 | 86.80 ± 1.86 | 0.22 ± 0.02 | 19.60 ± 0.40 | 66.93 ± 0.84 |
| | | Right | 1.34 ± 0.04 | | | 0.39 ± 0.01 | 15.10 ± 0.50 | 72.83 ± 1.41 |
| | Post | Left | 1.58 ± 0.14 | 0.40 ± 0.10 | 76 ± 7.50 | 0.25 ± 0.04 | 18.80 ± 0.40 | 68.12 ± 0.87 |
| | | Right | 1.61 ± 0.18 | | | 0.41 ± 0 | 14.80 ± 0.40 | 73.96 ± 1.21 |
| P3 | Pre | Left | 1.33 ± 0.27 | 0.30 ± 0 | 84.90 ± 9.90 | 0.27 ± 0.01 | 15.50 ± 0.60 | 68.23 ± 1.19 |
| | | Right | 1.58 ± 0.03 | | | 0.15 ± 0.03 | 13.30 ± 0.20 | 71.67 ± 0.20 |
| | Middle | Left | 1.73 ± 0.23 | 0.20 ± 0.10 | 66.20 ± 8.23 | 0.17 ± 0.12 | 14.80 ± 0.80 | 70.08 ± 1.91 |
| | | Right | 1.98 ± 0.25 | | | 0.12 ± 0.04 | 11.70 ± 0.70 | 77.08 ± 2.03 |
| | Post | Left | 1.43 ± 0.08 | 0.40 ± 0 | 86.80 ± 4.45 | 0.39 ± 0.01 | 14 ± 0.20 | 70.11 ± 0.16 |
| | | Right | 1.35 ± 0.08 | | | 0.14 ± 0.10 | 12.40 ± 0.70 | 74.46 ± 1.88 |
| P4 | Pre | Left | 1.02 ± 0.06 | 0.80 ± 0 | 121.20 ± 7.12 | 0.41 ± 0.03 | 14.50 ± 0.30 | 71.43 ± 0.57 |
| | | Right | 0.97 ± 0.06 | | | 0.37 ± 0.02 | 9.90 ± 0.30 | 78.39 ± 0.63 |
| | Middle | Left | 1 ± 0.05 | 0.90 ± 0 | 121.35 ± 4.48 | 0.44 ± 0.01 | 14.20 ± 0.60 | 75.94 ± 1.70 |
| | | Right | 0.99 ± 0.03 | | | 0.44 ± 0.01 | 7 ± 0.50 | 90.23 ± 1.42 |
| | Post | Left | 0.96 ± 0.02 | 0.90 ± 0 | 121.20 ± 1.80 | 0.46 ± 0.03 | 14.40 ± 0.80 | 72.59 ± 2.17 |
| | | Right | 1.02 ± 0.01 | | | 0.47 ± 0 | 9.90 ± 0.20 | 80.13 ± 0.33 |

the tests, reaching also clinical significance in some of them, especially for P1 and P3.

The most challenging part was the *second phase* of the training, which allowed the possibility of adapting the level of assistance depending on the patient's progression. Thereby, any subject could achieve the last level (level 6), and although the action of reaching level 4 took P3 longer than the rest of children, this patient could pass through it in the last three sessions.

It is interesting to highlight that the outcomes from isometric strength measure showed important peaks of improvement, especially for hip and knee flexion-extension, which was targeted with the CPWalker robotic platform. These higher values were observed from the middle to the post analysis. The results of the present study are difficult to compare with the scientific literature due to the lack of studies using exoskeletons for gait resistance training. However, we can establish that our results are in line with previous studies assessing conventional resistance strength training in CP [31].

In relation to 3D-kinematic analysis, as we said before, P4 suffered a non-grata personal situation the day of the post study. We believe this affected the results of this metric. Nevertheless, the whole population improved the values for spatiotemporal parameters, GDI and GPS, although some of these improvements were not clinically significant. It is important to highlight that not only the kinematics was improved, also the physiological expenditure in walking activities decreased, as illustrated by the reduction of the PCI in all patients, which means that better gait performance implies lower energy cost.

Finally, the motivation scale for the patients and the parents' satisfaction with the robot-based treatment was very high in most aspects.

The greatest achievements of this proposal come from the possibility of exercising different gait functions in an orderly way, individualized per joint and at the same time than over-ground walking. The proposed protocol could be applied to any current robotic device for gait rehabilitation making minimal changes on it: e.g. in

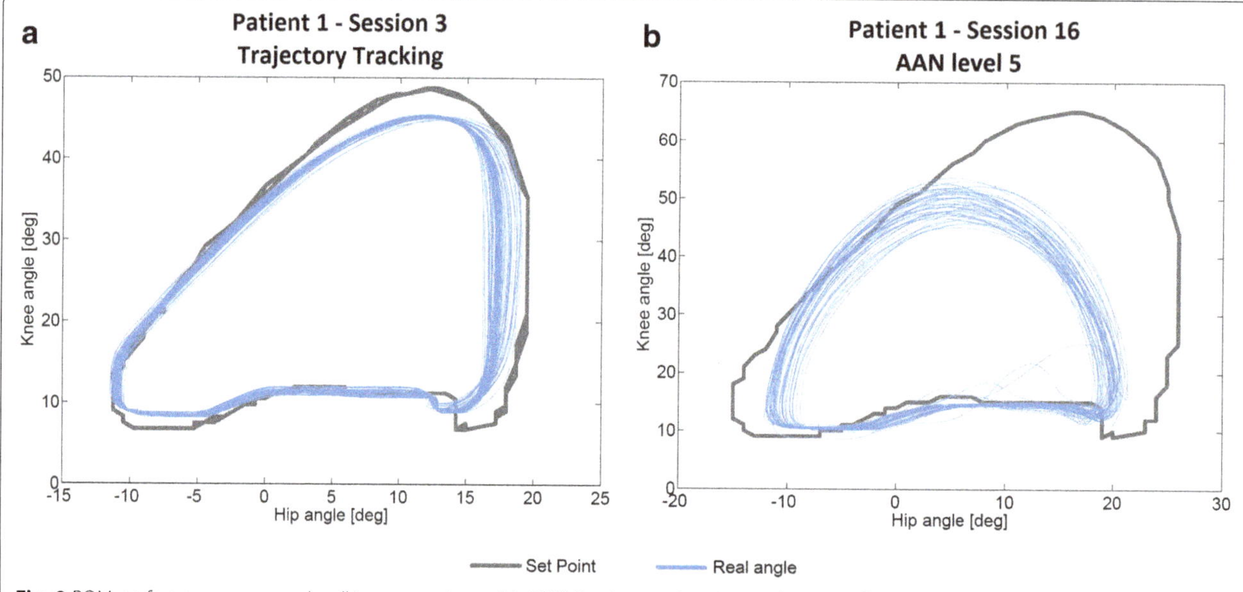


**Fig. 9** ROM performance measured walking some steps with CPWalker in two situations: **a** Session 3 for patient 1 (*first phase* of the training with trajectory tracking) and **b** Session 16 for patient 1 (*second phase* of the training through AAN strategies with level 5 of assistance (LI at hips and MI at knees))

treadmill pediatric platforms as Lokomat [14], despite the impossibility of over-ground walking, it already has different controllers, whose operation modes are close to the levels of impedance of CPWalker, ensuring the progression of the therapy into the sessions. Regarding working postural control in parallel with lower limbs training, which we considered one of the key factors of the study, it may be solved through other solutions if the selected robotic device does not have a similar strategy as described with CPWalker, but it is crucial to guarantee its compliance to get the best results of the treatment [12].

The principal limitation of this research is that we followed up only in short term, so further research with a higher population size is needed to evaluate if the improvements will be kept over time. Moreover, although other studies propose interventions on 3 non-consecutive days per week [31], the patients of the present study performed the robotic exercises during 2 non-consecutive sessions per week. This enabled them to continue their conventional therapies in parallel to the robot-based rehabilitation. The conventional therapies had been attended on a regular basis for years, so we considered to not abolish them for ethical reasons. The conventional therapies of the patients were performed 2 days a week and consisted of exercises on balance and strength focusing on quadriceps and abs. Although the patients were doing conventional and robotic therapy in parallel, we consider that the improvements achieved in this proposal are exclusively associated to the use of the CPWalker, since the patients received usual care therapies for years with no significant improvements in their functional situation. In conclusion, the method implemented with CPWalker is complementary to the common therapies, providing new possibilities to the clinical practice through robotic rehabilitation and also reaching better outcomes than conventional therapy alone.

We are currently working on new future lines in which we plan to include electromyography during the use of the robotic trainer in order to assess more objectively the patients' progressions. Electromyography data will serve to evaluate the therapy outcomes in terms of reorganization of the neural structures, which will be performed by analyzing the raw neural drive to muscles and the muscle synergies.

In a nutshell, this manuscript contributed with a defined robotic treatment that could be implemented in most of the existing rehabilitation robotic devices for lower limbs, and which we evaluated positively in four patients with CP using CPWalker.

#### Abbreviations

10mwt: 10 m walking test; 6mwt: 6 min walking test; AAN: Assist as needed; CNS: Central nervous system; CP: Cerebral Palsy; FAQ: Functional Assessment Questionnaire; GDI: Gait deviation index; GMFCS: Gross motor function classification system; GMFM: Gross motor function measure; GPS: Gait profile score; ICF-CY: International Classification of Functioning, Disability and Health framework, Children and Youth version; IMUs: Inertial measurement units; MCID: Minimal clinically important difference; MDC: Minimal detectable change; NSCA: National Strength and Conditioning Association; PBWS: Partial body weight support; PCI: Physiological cost index; ROM: Range of motion; SCALE: Selective control assessment of lower extremity

**Acknowledgements**
Authors would like to thank Biomechanical Institute of Valencia and other colleges from Spanish National Research Council and Niño Jesús Hospital for their collaboration in other aspects and achievements carried out with CPWalker robotic platform. Authors appreciate Dr. Gaebler's advice for the development of the robotic program proposal.
We would like to thank *Made for Movement* company for supporting us with *NF-Walker* device.
We also acknowledge the effort and contribution from all testing children and their families.
We acknowledge support of the publication fee by the CSIC Open Access Publication Support Initiative through its Unit of Information Resources for Research (URICI).

**Funding**
The work presented in this paper has been carried out with the financial support from the Ministerio de Economía y Competitividad of Spain, under Contract DPI2012–39133-C03–01.

**Authors' contributions**
CB collaborated in the design of the training program, developed the control algorithm of the robotic device, processed the clinical data and drafted the manuscript. TM collaborated in the design of training program proposal and reviewed the manuscript. BM collaborated in the clinical processing data derived from the study and helped with the recording of clinical metrics. OR was involved in the development of the hardware of the robotic device. APS participated in patients' recruitment and collaborated in the recording of clinical metrics. SLL participated in patients' recruitment and collaborated in the review and critique of the manuscript. IM aided in the patient's recruitment and reviewed and criticized the manuscript. ER collaborated in all facets of the contribution. All authors read and approved the final manuscript.

**Competing interests**
The authors declare that they have no competing interests.

**Author details**
[1]Centre for Automation and Robotics (CAR), CSIC-UPM, Ctra Campo Real km 0.2 - La Poveda-Arganda del Rey, 28500 Madrid, Spain. [2]Hospital Infantil Universitario Niño Jesús, Madrid, Spain. [3]Universidad Rey Juan Carlos, Madrid, Spain. [4]Universidad de La Salle, Madrid, Spain. [5]Universidade Federal do Espírito Santo Vitória, Brazil.

**References**
1. Palisano R, Rosenbaum P, Walter S, Russell D, Wood E, Galuppi B. Development and reliability of a system to classify gross motor function in children with cerebral palsy. Dev Med Child Neurol. 1997;39:214–23. PMID: 9183258.
2. Dietz V. Clinical aspects for the application of robotics in locomotor neurorehabilitation. In: Reinkensmeyer DJ, Dietz V, editors. Neurorehabilitation Technol: Springer; 2016. p. 209–21. https://doi.org/10.1007/978-3-319-28603-7_11.
3. Meyer-Heim A, a van Hedel HJ. Robot-assisted and computer-enhanced therapies for children with cerebral palsy: current state and clinical implementation. Semin Pediatr Neurol. 2013;20:139–45. https://doi.org/10.1016/j.spen.2013.06.006.
4. Bayón C, Raya R, Lara SL, Ramírez Ó, Serrano J, Rocon E. Robotic therapies for children with cerebral palsy: a systematic review. Transl Biomed. 2016;7:1–10. https://doi.org/10.21767/2172-0479.100044.
5. Diaz I, Gil J, Sanchez E. Lower-limb robotic rehabilitation: literature review and challenges, J robot. 2011.
6. van Asseldonk EHF, van der Kooij H. Robot-Aided Gait Training with LOPES. In: Dietz V, editor. Neurorehabilitation Technol. London: Springer London; 2012. p. 379–96. https://doi.org/10.1007/978-1-4471-2277-7_21.
7. Bayón C, Ramírez O, Serrano JI, Del Castillo MD, Pérez-Somarriba A, Belda-Lois JM, Martínez-Caballero I, Lerma-Lara S, Cifuentes C, Frizera A, Rocon E. Development and evaluation of a novel robotic platform for gait rehabilitation in patients with cerebral palsy: CP Walker. Robotics Autonomous Systems. 2017;91:101–14. https://doi.org/10.1016/j.robot.2016.12.015.
8. Bartenbach V, Gort M, Riener R. Concept and Design of a Modular Lower Limb Exoskeleton. In: 6th IEEE RAS/EMBS Int. Conf. Biomed. Robot. Biomechatronics; 2016. p. 649–54. https://doi.org/10.1109/BIOROB.2016.7523699.
9. Bayón C, Lerma S, Ramírez O, Serrano JI, Del Castillo MD, Raya R, Belda-Lois JM, Martínez I, Rocon E. Locomotor training through a novel robotic platform for gait rehabilitation in pediatric population: short report. J Neuroeng. Rehabil. 2016;13:98. https://doi.org/10.1186/s12984-016-0206-x.
10. Rüdt S, Moos M, Seppey S, Riener R, Marchal-Crespo L. Towards more efficient robotic gait training: a novel controller to modulate movement errors, 6th IEEE RAS/EMBS Int. Conf Biomed Robot Biomechatronics; 2016. p. 884–9.
11. Lefmann S, Russo R, Hillier S. The effectiveness of robotic-assisted gait training for paediatric gait disorders: systematic review. J Neuroeng Rehabil. 2017;14:1. https://doi.org/10.1186/s12984-016-0214-x.
12. Wallard L, Dietrich G, Kerlirzin Y, Bredin J. Robotic-assisted gait training improves walking abilities in diplegic children with cerebral palsy. Eur J Paediatr Neurol. 2017:1–8. https://doi.org/10.1016/j.ejpn.2017.01.012.
13. World Health Organization. International classification of functioning, disability and health, children and youth version: World Health Organization; 2007.
14. Aurich-Schuler T, Grob F, Van Hedel HJA, Labruyère R. Can Lokomat therapy with children and adolescents be improved? An adaptive clinical pilot trial comparing guidance force, path control, and FreeD. J Neuroeng Rehabil. 2017;14:1–14. https://doi.org/10.1186/s12984-017-0287-1.
15. Hussain S, Xie SQ, Liu G. Robot assisted treadmill training : mechanisms and training strategies. Med Eng Phys. 2011;33:527–33. https://doi.org/10.1016/j.medengphy.2010.12.010.
16. Van Asseldonk EHF, Veneman JF, Ekkelenkamp R, Buurke JH, Van Der Helm FCT, Van Der Kooij H. The effects on kinematics and muscle activity of walking in a robotic gait trainer during zero-force control. IEEE Trans Neural Syst Rehabil Eng. 2008;16:360–70. https://doi.org/10.1109/TNSRE.2008.925074.
17. Damiano DL, Abel MF. Functional outcomes for strenght training in spastic cerebral palsy. Arch Phys Med Rehabil. 1998;79:119–25.
18. Martín-Lorenzo T, Lerma-Lara S, Bayón C, Ramírez Ó, Rocon E. The CPWalker for strength training in children with spastic cerebral palsy: a training program proposal. In: Int. Conf Neurorehabilitation; 2016.
19. Dewar R, Love S, Johnston LM. Exercise interventions improve postural control in children with cerebral palsy: a systematic review. Dev Med Child Neurol. 2014; https://doi.org/10.1111/dmcn.12660.
20. Rosenbaum P, Gorter JW. The "F-words" in childhood disability: I swear this is how we should think! Child Care Health Dev. 2012;38:457–63. https://doi.org/10.1111/j.1365-2214.2011.01338.x.
21. Maclean N, Pound P. A critical review of the concept of patient motivation in the literature on physical rehabilitation. Soc Sci Med. 2000;50:495–506. PMID: 10641802. http://www.ncbi.nlm.nih.gov/pubmed/10641802. Accessed 11 Dec 2016.
22. Borg G. Psychophysical bases of perceived exertion. Med Sci Sport Exerc. 1982;14:377–81.
23. Butler P, Engelbrecht M, Major RE, Tait JH, Stallard J, Patrick JH. Physiological cost index of walking for normal children and its use as an Indicator of physical handicap. Dev Med Child Neurol. 1984;26:607–12. https://doi.org/10.1111/j.1469-8749.1984.tb04499.x..
24. Kadaba MP, Ramakrishnan HK, Wootten ME. Measurement of lower extremity kinematics during level walking. J Orthop Res. 1990;8:383–92. http://trama.deib.polimi.it/allegati/Kadaba_1990.pdf (Accessed 22 Dec 2015)
25. Fowler EG, Staudt LA, Greenberg MB, Oppenheim WL. Selective control assessment of the lower extremity (SCALE): development, validation, and interrater reliability of a clinical tool for patients with cerebral palsy. Dev Med Child Neurol. 2009;51:607–14. https://doi.org/10.1111/j.1469-8749.2008.03186.x.
26. D.J. Russell, P.L. Rosenbaum, L.M. Avery, M. Lane, Gross motor function measure (GMFM-66 and GMFM-88) user's manual, press, Cambridge University, 2002.

27. Kaphingst KA, Persky S, Lachance C. Outcome tools used for ambulatory children with cerebral palsy: responsiveness and minimum clinically important differences. Dev Med Child Neurol. 2008;50:918–25. https://doi.org/10.1080/10810730902873927.Testing.
28. Rehabilitation Measures Database, 6 Minute Walk Test, (n.d.). https://www.rehabmeasures.org (Accessed 7 May 2018).
29. Thompson P, Beath T, Bell J, Jacobson G, Phair T, Salbach NM, Wright VF. Test-retest reliability of the 10-metre fast walk test and 6-minute walk test in ambulatory school-aged children with cerebral palsy. Dev Med Child Neurol. 2008;50:370–6. https://doi.org/10.1111/j.1469-8749.2008.02048.x.
30. Rehabilitation Measures Database, Timed 10-Meter Walk Test, Timed 10-m. Walk TEST (n.d.). www.rehabmeasures.org (Accessed 11 Aug 2016).
31. Park EY, Kim WH. Meta-analysis of the effect of strengthening interventions in individuals with cerebral palsy. Res Dev Disabil. 2014;35:239–49. https://doi.org/10.1016/j.ridd.2013.10.021.
32. Faigenbaum C, Kramer AD, Blimkie W. Youth resistance training: updated position statement paper from the National Strength and conditioning association. J Strength Con Res. 2009;23:60–79. https://doi.org/10.1519/JSC.0b013e31819df407.
33. Crapo RO, Casaburi R, Coates AL, Enright PL, MacIntyre NR, McKay RT, Johnson D, Wanger JS, Zeballos RJ, Bittner V, Mottram C. ATS statement: guidelines for the six-minute walk test. Am J Respir Crit Care Med. 2002;166:111–7. https://doi.org/10.1164/rccm.166/1/111.
34. Gorton GE, Stout JL, Bagley AM, Bevans K, Novacheck TF, Tucker CA. Gillette functional assessment questionnaire 22-item skill set: factor and Rasch analyses. Dev Med Child Neurol. 2011;53:250–5. https://doi.org/10.1111/j.1469-8749.2010.03832.x.

# 7

# Using principal component analysis to reduce complex datasets produced by robotic technology in healthy participants

Michael D. Wood[1], Leif E. R. Simmatis[1], J. Gordon Boyd[1,2,3], Stephen H. Scott[1] and Jill A. Jacobson[4*]

**Abstract**

**Background:** The KINARM robot produces a granular dataset of participant performance metrics associated with proprioceptive, motor, visuospatial, and executive function. This comprehensive battery includes several behavioral tasks that each generate 9 to 20 metrics of performance. Therefore, the entire battery of tasks generates well over 100 metrics per participant, which can make clinical interpretation challenging. Therefore, we sought to reduce these multivariate data by applying principal component analysis (PCA) to increase interpretability while minimizing information loss.

**Methods:** Healthy right-hand dominant participants were assessed using a bilateral KINARM end-point robot. Subjects (*Ns* = 101–208) were assessed using 6 behavioral tasks and automated software generated 9 to 20 metrics related to the spatial and temporal aspects of subject performance. Data from these metrics were converted to Z-scores prior to PCA. The number of components was determined from scree plots and parallel analysis, with interpretability considered as a qualitative criterion. Rotation type (orthogonal vs oblique) was decided on a per task basis.

**Results:** The KINARM performance data, per task, was substantially reduced (range 67–79%), while still accounting for a large amount of variance (range 70–82%). The number of KINARM parameters reduced to 3 components for 5 out of 6 tasks and to 5 components for the sixth task. Many components were comprised of KINARM parameters with high loadings and only some cross loadings were observed, which demonstrates a strong separation of components.

**Conclusions:** Complex participant data produced by the KINARM robot can be reduced into a small number of interpretable components by using PCA. Future applications of PCA may offer potential insight into specific patterns of sensorimotor impairment among patient populations.

**Keywords:** KINARM, End-point robot, Principal component analysis, Data reduction

## Background

Robotic technology, such as KINARM (BKIN Technologies, Kingston, ON, Canada), provides objective metrics using the participants' upper limbs to assess proprioception and sensorimotor function, as well as executive function. In stroke survivors, this technology has identified subtle neurocognitive deficits not apparent on routine clinical testing [1], and various KINARM tasks have been administered to multiple patient populations (e.g., traumatic brain injury, fetal alcohol spectrum disorder) [2–5]. The KINARM behavioral battery (KINARM Standard Tests™) currently consists of 9 tasks that include automated data analysis routines. Within each task, up to 20 performance items (e.g., reaction time) are calculated, leading to the potential generation of over 100 metrics per participant. However, when assessing performance deficits that are indicative of adverse neurological outcomes among participants, this granular and complex performance output may impede deriving meaningful interpretations.

* Correspondence: jill.jacobson@queensu.ca
[4]Department of Psychology, Queen's University, 318 Craine Hall, 62 Arch Street, Kingston, ON K7L 3N6, Canada
Full list of author information is available at the end of the article

Principal component analysis (PCA) is a data reduction technique used to identify linear combinations of measured variables that account for the most overall variance in responses [6–8]. The first principal component accounts for the largest amount of variance, followed by the second, and so forth [9]. PCA is best used when the measured variables are theorized to be causal or formative indicators of the overarching construct rather than reflective or effects of it, which would be better assessed using factor analysis (FA). Because performance on the tasks determine a participant's level of functioning rather than the other way around, PCA is the more appropriate technique for the KINARM battery than is FA.

The primary objective of this study was to reduce the dimensionality of healthy participant data produced by the KINARM End-Point robot by using PCA. This analysis should increase overall interpretability by reducing redundant KINARM parameters into behaviorally meaningful components, which has the potential to demonstrate the clinical utility of PCA by capturing a minimal number of performance measures that could assist with the characterization of deficits among various patient populations.

## Methods

### Participant recruitment

Healthy participants were community-based and were randomly recruited via advertisements on lab and departmental websites, as well as in local classifieds (online and print). Trained research staff screened each adult participant (> 17 years old) to ensure that task instructions could be easily understood, no prior neurological deficits were reported, and that subjects had no prior medical conditions that could affect upper limb mobility (see Additional file 1). Once enrolled, participants were then assessed by our research staff at one of two sites: Kingston Health Sciences Centre or Queen's University, both sites in Kingston, ON, Canada. See Table 2 for detailed demographics (e.g., age, gender, education). The Queen's University and Affiliated Hospitals Health Sciences Research Ethics Board approved recruitment and assessment of these participants. Informed consent was obtained from each participant prior to the KINARM assessment.

### Robotic assessment

Participants were seated, in a height-adjustable chair that was locked in place, at a two-dimensional virtual reality system that displayed each task in the horizontal plane. As shown in Fig. 1, participants' vision of their hands and arms was occluded, their head was positioned in the center of the visual field, and visual feedback of hand position (when provided) was represented on the screen by a white circle. Participants were instructed to grasp onto the KINARM End-Point robotic handles (BKIN Technologies Ltd., Kingston, ON, Canada), permitting free movement in the horizontal plane without anti-gravity support at proximal or distal arm segments. A trained operator described each task, using a standardized script, before it was performed by the participant. Automated data collection and analysis software (Dexterit Version 3.6.2) measured and quantified subject performance. For each performance metric, the software computed a Z-score that accounted for age, sex, and handedness. Only performance metrics that could be normalized, and therefore converted to Z scores, were considered in the current analysis. We examined 6 tasks from the KINARM Standard Tests™, four tasks assessed upper-limb motor function in right-handed participants: visually guided reaching (VGR), object hit (OH), object hit and avoid (OHA), and level 1 of ball on bar (BonB). The Arm position matching (APM) task assessed upper limb proprioception, and reverse visually guided reaching (RVGR) assessed cognitive-motor function. For detailed descriptions of all tasks, see Table 1. Within each task, approximately 9–20 performance metrics were produced, participants were instructed to take breaks as needed, and the assessment took < 1 h to complete. For detailed descriptions on all task parameters, see Additional file 2 or obtain from the BKIN Technologies Ltd. KINARM manual [10]. All included participants were selected based on their ability and understanding to follow task commands. Each task was comprised of various healthy participants, as not all participants completed each task. To reduce handedness as a potential confounding variable, only right hand dominant subjects were included in the analysis.

### Principal component analysis

PCA analysis was conducted in R version 3.3.2 [11] using the psych package version 1.7.8 [12]. KINARM performance items were converted to Z-scores prior to PCA. For correlation matrices associated with each task, see Additional file 3. Scree plots and parallel analysis were examined to determine the number of components per KINARM task. Briefly, eigenvalues (variance explained by a component) were plotted in descending order to generate a scree plot [13, 14] per KINARM task. Parallel analysis was used to generate a random data set that possessed the same number of items and sample size as each KINARM task. Eigenvalues then were calculated for the random data, which also are graphed on the scree plot mentioned previously. The number of components prior to where the random data crosses the observed KINARM participant data are retained (i.e., retain the number of components that have eigenvalues larger than those from the simulated dataset) [13–16]. In PCA, if more than one component is retained, then multiple solutions for the parameters

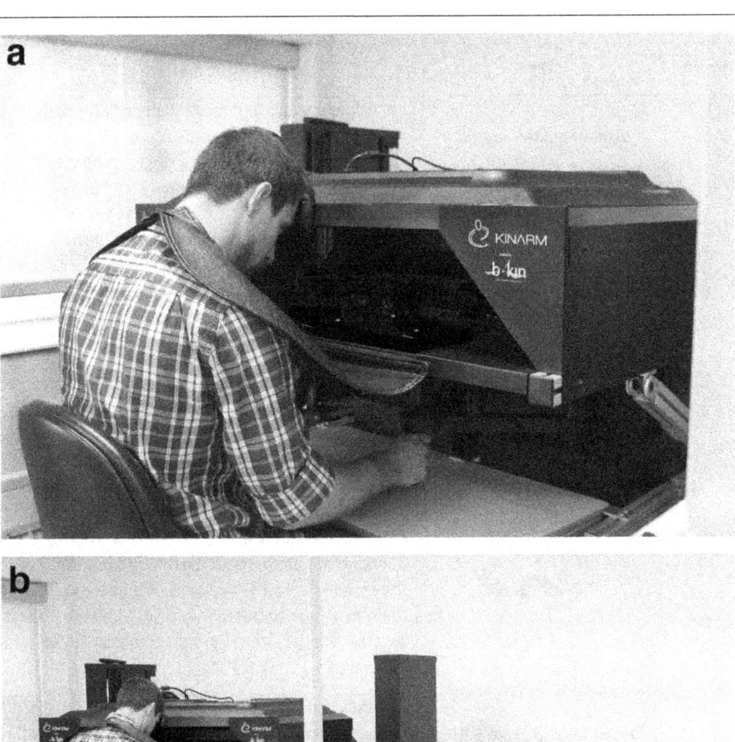


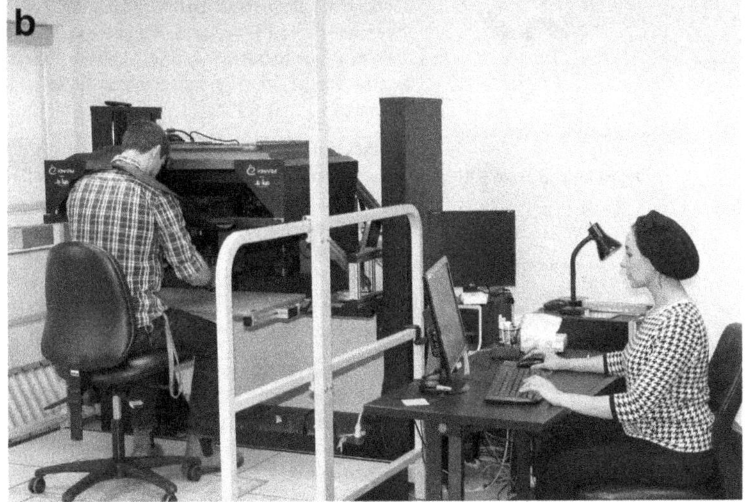


**Fig. 1** KINARM End-Point robot set-up during participant assessment

would fit the data equally well. Rotation is used to help select the best solution. An oblique rotation (oblimin) [17], which allows for components to be correlated, first was conducted for each solution. If all intercomponent correlations were low (<|0.30|), then an orthogonal rotation (varimax) [18] was used instead. A component loading was considered substantial if the loading was ≥0.40. Only participants who were right hand dominant were chosen for this analysis, and only data from their dominant hand were used for PCA. To reduce possible practice effects for those participants who completed multiple assessments, we used the participant performance metrics from the *first* KINARM assessment only (i.e., for participants who completed the same task multiple times [repeated testing], we used their performance score only from the first assessment of that task). These identified components were then analyzed for interpretability, which was defined as parameters having substantial loadings on each component and groupings of performance parameters having behaviorally meaningful conclusions (e.g., the constructs of "*total movement*" and "*posture/reaction time*" being separable). To further characterize participant performance, components and their respective parameters that loaded highly were used to broadly generate intuitive labels for each component per task rather than successively numbering components.

## Results

### Participant characteristics

The number of participants across each task varied (Ns = 101–208). Participant level of education consistently ranged from high school education to postdoctoral researcher. The percent of participants that were male ranged from 40 to 46%, and the mean age of subjects (range: 41–46 years) was fairly consistent across all

**Table 1** KINARM Task Descriptions and Domains of Assessment

| Domain | Task | Description |
|---|---|---|
| Motor | Visually guided reaching (VGR) | Participants grasp and move the right arm of the robot (visual feedback is represented as a white circle) from a central target to 4 peripheral target locations, in succession, quickly and accurately (red circles). Participants hold their position inside the red circle until a new target appears, which is always at the start position. A new target will then appear after a brief hold in the start position and participants then move to the following target. This task assesses visuomotor control of the upper limbs. |
| | Object hit (OH) | Participants control green paddles during this task and red balls fall towards them from the top of the screen. Participants are instructed to hit away as many targets (red circles) as possible, while the task continues to increase in difficulty (i.e., objects fall more frequently and with increasing velocity). When a participant makes contact with a ball, the KINARM robot produces a force that mimics the force felt by the subject as if they made contact with an actual object using a real paddle (unique to this task and OHA below). This task assesses spatial attention, sensorimotor control and planning. |
| | Object hit and avoid (OHA) | This task is similar to OH, except participants are instructed to hit only 2 shapes out of a possible 8, while avoiding distractor targets. Participants are required to inhibit movements towards distractor targets and only permit movements to the selected targets. Distractors do no provide force feedback, as described above, only the 2 target shapes. This task assesses attention, rapid sensorimotor control and inhibition. |
| | Ball on bar (BonB) | Participants hands are represented by a long white bar connecting both of their hands, with a white ball resting on top. Four targets are then presented one at a time in a diamond pattern and participants move to one of the 4 targets, at a time, in a clockwise direction. Participants are instructed to move quickly and accurately to as many target locations as possible within the 60 second time limit. This task tested the bimanual motor coordination. |
| Sensory | Arm position matching (APM) | This proprioceptive task involved the participants' dominant arm (i.e., right) being moved by the robot to one of four preprogrammed coordinates in a square shape. Participants receive no visual feedback of the arm being controlled by the robot. After the robot completes moving the participant's arm, participants have to mirror-match this position using the contralateral arm (i.e., left). |
| Cognitive | Reverse visually guided reaching (RVGR) | This task is similar to VGR; however, visual feedback is inverted. For example, an arm movement in one direction (right) would cause the visual feedback on screen to move in the opposite direction (left). This task required participants to inhibit their automatic response to reach towards the spatial goal and instead move in the opposite direction to achieve the visual goal. |

tasks. Only a small subset of participants ($n = 11$) completed all 6 KINARM tasks. For more detailed demographics per task, see Table 2.

### Data reduction

The number of KINARM performance parameters, per task, were substantially reduced (range 67–79%) by grouping related parameters into components. APM, BonB, VGR, RVGR, and OH reduced from 9 to 14 performance metrics to 3 components, which accounted for 70–79% of variance. OHA was reduced from 20 performance items to 5 components, while still accounting for 82% of the variance (Table 3).

### Principal component analysis across the KINARM battery

Descriptions of each task and their respective components are provided below as follows: component name (number of KINARM parameters contributing to that component [according to the ≥.40 load criteria mentioned previously], percent of variance explained by that component). See Table 4 for component loadings per task, all available items (i.e., metrics that could be converted to Z scores) were considered for each PCA per task. For scree plots with parallel analysis per task, see Fig. 2.

### Arm position matching

Scree plot and parallel analysis yielded three models (1-, 2-, and 3-component solutions) that warranted further examination. For the 2- and 3-component solutions, the use of an orthogonal rotation was justified, as the inter-component correlations ranged from |-0.18| to |0.16|. Both the 1- and 2-component solutions included parameters that did not load onto any of the components (several loadings <|0.40|) or had sizeable cross loadings. Although several parameters had substantial loadings (≥|0.40|) on at least one component in the 3-component solution, the 3-component solution was the most interpretable (see Table 4 for the component loadings). The 3 components were related to 1) *position accuracy* (4, 25%), 2) *movement variability (6, 24%),* and 3) *contraction/expansion* (3, 21%).

### Ball on bar

Three models (1-, 2-, and 3-component solutions) were further examined based on the findings of the scree plot and parallel analysis. For the 2- and 3- component solutions, we used an oblique rotation because inter-component correlations ranged from |0.15| to |0.46|. Most parameters had substantial loadings (≥|0.40|) on at least one component in the 3-component solution. Bar length variability and hand path bias did not load substantially onto any of the 3 components. However, the 3-component solution was the most interpretable (see Table 4 for the component loadings). The three components were associated with 1) *speed and success* (6, 37%), 2) *bar angle* (4, 19%) and 3) hand-speed maxima which is related to *movement smoothness* (3, 16%). The contribution of *speed and success* to task performance suggests that speed is highly influential for high success rates.

### Visually guided reaching

Scree plot and parallel analysis yielded three models (1-, 2-, and 3-component solutions) for examination. For the 2- and 3-component solutions, we used an orthogonal rotation (varimax) as inter-component correlations ranged from |0.06| to |0.16|. Both 1- and 2-component solutions included a number of parameters that did not load onto any of the components (loadings <|0.40|) or had sizeable cross loadings, but all parameters had substantial loadings (≥|0.40|) on at least 1 component in the 3-component solution. The 3-component solution also was the most interpretable (see Table 4 for the component loadings). The components were related to: 1) *initial*

**Table 2** Participant Demographics and Task Characteristics

| Task | N | Mean Age (range) | Male Gender (%) | Dominant Arm | Range of Education | Mean Number of Trials [SD] | Mean Task Duration in seconds [SD] |
|---|---|---|---|---|---|---|---|
| Arm Position Matching | 184 | 44.43 (18–87) | 84 (46) | Right | High school – Postdoctoral research** | 25.26 [0.98] | 88.13 [19.41] |
| Ball on Bar | 208 | 42.16 (18–87) | 84 (40) | Right | High school – Postdoctoral research*** | 1 | 62.57 [1.60] |
| Visually Guided Reaching | 200 | 42.51 (18–88) | 80 (40) | Right | High school – Postdoctoral research*** | 24 [0.07] | 102.55 [6.83] |
| Reverse Visually Guided Reaching | 101 | 41.56 (19–81) | 46 (46) | Right | High school – Post-doctoral research | 30 [0.10] | 180.27 [16.71] |
| Object Hit | 190 | 46.28 (18–87) | 86 (45) | Right | High school – Postdoctoral research** | 1 | 138.03 [0.43] |
| Object Hit and Avoid | 170 | 45.56 (18–87) | 76 (45) | Right | High school – Postdoctoral research* | 1 | 138.29 [0.51] |

For range of education, High school reflects completion of grade twelve or completion of secondary school equivalent. * indicate the number of participants that were missing data for that column. For example, *($n = 1$), **($n = 2$), *** ($n = 3$)

**Table 3** Proportion of variance explained by each component across KINARM tasks

| Task | Number of Parameters Used for PCA | Variance explained per component | | | | | Cumulative variance | % of data reduced by PCA |
|---|---|---|---|---|---|---|---|---|
| | | 1 | 2 | 3 | 4 | 5 | | |
| APM | 12 | 25% | 24% | 21% | | | 70% | 75% |
| BonB | 14 | 37% | 19% | 16% | | | 72% | 79% |
| VGR | 9 | 34% | 30% | 15% | | | 79% | 67% |
| RVGR | 11 | 38% | 24% | 12% | | | 74% | 73% |
| OH | 14 | 27% | 26% | 18% | | | 71% | 79% |
| OHA | 20 | 21% | 20% | 18% | 16% | 7% | 82% | 75% |

Grey shaded region indicates that there are no components for these tasks. *APM* Arm position matching, *BonB* Ball on bar, *VGR* Visually guided reaching, *RVGR* Reverse visually guided reaching, *OH* Object hit, *OHA* Object hit and avoid

*movement and smoothness* (5, 34%), 2) *total movement* (5, 30%), and 3) *posture and reaction time* (3, 15%).

#### Reverse visually guided reaching

Scree plot and parallel analysis yielded three models (1-, 2-, and 3-component solutions) for further examination. For the 2- and 3-component solutions, we used an orthogonal rotation, as the inter-component correlations ranged from |-0.09| to |0.13|. Both the 1- and 2-component solutions included items that did not load onto any of the components (all loadings < |0.40|), but all items had substantial loadings (≥ |0.40|) on at least one component in the 3-component solution. The 3-component solution also was the most interpretable (see Table 4 for the component loadings). The 3 components were associated with: 1) *initial movement and smoothness* (7, 38%), 2) *total movement* (5, 24%), and 3) *reaction time* (1, 12%). *Initial movement and smoothness* accounted for more variance in this task than it did in VGR. Therefore, *initial movement* is imperative for success in RVGR.

#### Object hit

According to the scree plot and parallel analysis, three models (1-, 2-, and 3-component solutions) warranted further examination. For the 2- and 3-component solutions, we used an orthogonal rotation, as the inter-component correlations ranged from |0.03| to |0.24|. Both the 1- and 2-component solutions included items that did not load onto any of the components (loadings <|0.40|) or had sizeable cross loadings, but all items had substantial loadings (≥|0.40|) on at least one component in the 3-component solution. The 3-component solution also was the most interpretable (see Table 4 for the component loadings). The 3 components were comprised of: 1) *hand bias* (6, 27%), 2) *speed and area* (5, 26%), and 3) *accuracy* (4, 18%). Hits with left and right hands cross loaded onto both *accuracy* and *hand bias*.

#### Object hit and avoid

According to the scree plot and parallel analysis, five models (1-, 2-, 3-, 4- and 5-component solutions) warranted further examination. For the 2-, 3-, 4-, and 5-component solutions, we kept the oblique rotation because the inter-component correlations ranged from |-0.12| to |0.30|. The 1-, 2-, 3-, and 4-component solutions included items that did not load onto any of the components (all loadings < |0.40|) or had substantial cross loadings, but all items had substantial loadings (> |0.40|) on at least one component in the 5-component solution, which was also the most interpretable (see Table 4 for the component loadings). The 5 components were related to: 1) *impulsivity* (6, 21%), 2) *accuracy* (6, 20%), 3) *hand bias* (6, 18%), 4) *speed and area* (5, 16%), and 5) *miss bias* (1, 7%). Three of these components (*hand bias, speed and area,* and *accuracy*) are shared with OH, whereas *miss bias* and *distractor avoidance* are components unique to OHA.

## Discussion

The KINARM robot produces granular datasets of performance metrics associated with sensory, motor, visuospatial, and executive function. The primary objective of this study was to reduce multivariate data produced by the KINARM and generate interpretable and intuitively understandable components across 6 behavioral tasks to improve characterization of participant performance. KINARM data dimensionality was substantially reduced, while still retaining a large proportion of variance. Therefore, complex participant data produced by the KINARM robot can be reduced into a small number of components that characterize participant performance.

#### Interpretability and classification of components

We initially used PCA to reduce our multivariate dataset, and we then investigated if this statistical technique identified interpretable and classifiable components of participant performance. Within a task, components explained intuitive performance metrics. For example, in APM (a task that assesses limb proprioception), the 3 components identified were distinct measures of function (*position accuracy, movement variability* and *contraction/expansion).* Furthermore, KINARM parameters were separated reproducibly across multiple tasks. For

**Table 4** Principal component loadings for KINARM behavioral battery

| Task | KINARM Items | Component Loadings | | | | |
|---|---|---|---|---|---|---|
| | | 1 | 2 | 3 | 4 | 5 |
| Arm Position Matching | Absolute Error X | **0.870** | 0.210 | – 0.040 | | |
| | Absolute Error Y | 0.320 | **0.400** | – 0.230 | | |
| | Absolute Error XY | **0.910** | 0.250 | –0.100 | | |
| | Variability X | 0.260 | **0.810** | –0.130 | | |
| | Variability Y | 0.140 | **0.750** | –0.060 | | |
| | Variability XY | 0.270 | **0.860** | –0.140 | | |
| | Contraction Expansion Ratio X | –0.120 | –0.040 | **0.890** | | |
| | Contraction Expansion Ratio Y | 0.020 | – 0.140 | **0.800** | | |
| | Contraction Expansion Ratio XY | –0.060 | – 0.050 | **0.990** | | |
| | Shift X | **–0.550** | **0.580** | –0.070 | | |
| | Shift Y | 0.250 | **–0.580** | –0.070 | | |
| | Shift XY | **0.870** | –0.030 | – 0.040 | | |
| Ball on Bar | Targets completed | **0.997** | –0.146 | 0.001 | | |
| | Mean movement time | **–0.965** | 0.066 | –0.037 | | |
| | Mean ball speed | **0.942** | 0.083 | 0.029 | | |
| | Mean right hand speed | **0.913** | 0.164 | –0.017 | | |
| | Mean left hand speed | **0.956** | 0.045 | 0.020 | | |
| | Right hand speed maxima | 0.054 | –0.012 | **0.950** | | |
| | Left hand speed maxima | 0.175 | –0.069 | **0.766** | | |
| | Mean bar angle rad | –0.067 | **0.633** | –0.045 | | |
| | Stdev bar angle rad | –0.056 | **0.801** | –0.201 | | |
| | Bar length variability* | 0.250 | 0.335 | 0.145 | | |
| | Hand speed diff | **0.473** | **0.639** | 0.108 | | |
| | Norm absolute hand speed diff | 0.027 | **0.849** | 0.166 | | |
| | Hand speed peak bias | –0.263 | 0.113 | **0.732** | | |
| | Hand path bias* | –0.098 | 0.342 | –0.191 | | |
| Visually Guided Reaching | Posture speed | –0.077 | **0.519** | **0.512** | | |
| | Reaction time | 0.078 | –0.162 | **0.863** | | |
| | Initial direction error | **0.547** | 0.291 | **0.435** | | |
| | Initial distance ratio | **–0.915** | 0.073 | – 0.096 | | |
| | Speed maxima count | **0.780** | –0.324 | –0.323 | | |
| | Min max speed difference | **0.738** | **0.568** | 0.098 | | |
| | Movement time | 0.105 | **–0.950** | – 0.035 | | |
| | Path length ratio | **0.773** | **0.485** | 0.142 | | |
| | Max speed | 0.371 | **0.847** | –0.148 | | |
| Reverse Visually Guided Reaching | Posture speed | –0.050 | **0.614** | 0.394 | | |
| | Reaction time | 0.022 | –0.016 | **0.800** | | |
| | Initial direction error | **0.818** | 0.224 | 0.055 | | |
| | Initial distance ratio | **–0.856** | 0.133 | – 0.343 | | |
| | Initial speed ratio | **–0.806** | – 0.023 | 0.316 | | |
| | Speed maxima count | **0.837** | –0.101 | 0.302 | | |
| | Min max speed difference | 0.263 | **0.721** | 0.053 | | |
| | Movement time | **0.621** | **–0.680** | 0.229 | | |

**Table 4** Principal component loadings for KINARM behavioral battery *(Continued)*

| Task | KINARM Items | Component Loadings | | | | |
|---|---|---|---|---|---|---|
| | | 1 | 2 | 3 | 4 | 5 |
| | Path length ratio | **0.547** | **0.692** | −0.013 | | |
| | Max speed | −0.248 | **0.871** | −0.303 | | |
| | Correction time | **0.788** | 0.095 | −0.330 | | |
| Object Hit | Total hits | −0.028 | 0.137 | **0.932** | | |
| | Hits with left | **−0.612** | 0.053 | **0.658** | | |
| | Hits with right | **0.545** | 0.218 | **0.758** | | |
| | Median error | −0.046 | 0.119 | **0.647** | | |
| | Miss bias[a] | −0.040 | − 0.022 | − 0.365 | | |
| | Right hand speed | 0.322 | **0.847** | 0.204 | | |
| | Movement area (right) | 0.200 | **0.881** | 0.052 | | |
| | Left hand speed | −0.148 | **0.888** | 0.132 | | |
| | Movement area (left) | −0.219 | **0.915** | 0.061 | | |
| | Hand bias of hits | **0.893** | 0.166 | 0.082 | | |
| | Hand transition | **−0.863** | − 0.008 | 0.051 | | |
| | Hand selection overlap | −0.056 | **0.612** | 0.094 | | |
| | Hand speed bias | **0.876** | 0.037 | 0.140 | | |
| | Movement area bias | **0.719** | −0.290 | −0.063 | | |
| Object Hit and Avoid | Total hits | −0.125 | **0.933** | 0.037 | 0.067 | 0.006 |
| | Hits with left | −0.092 | **0.780** | **−0.453** | 0.034 | 0.233 |
| | Hits with right | −0.061 | **0.748** | **0.483** | 0.070 | −0.210 |
| | Total Distractor hits | **0.982** | 0.009 | 0.013 | 0.036 | −0.012 |
| | Distractor hits (left) | **0.925** | 0.034 | −0.020 | −0.025 | 0.019 |
| | Distractor hits (right) | **0.896** | −0.001 | 0.033 | 0.070 | −0.027 |
| | Median error | **−0.414** | **0.419** | 0.076 | 0.073 | −0.012 |
| | Miss bias | −0.043 | 0.034 | 0.063 | 0.008 | **0.914** |
| | Right hand speed | 0.133 | 0.239 | 0.304 | **0.724** | −0.025 |
| | Movement area (right) | 0.004 | −0.054 | 0.360 | **0.854** | 0.050 |
| | Left hand speed | 0.133 | 0.215 | −0.241 | **0.764** | 0.136 |
| | Movement area (left) | −0.010 | −0.035 | − 0.327 | **0.922** | − 0.125 |
| | Hand bias of hits | −0.010 | 0.042 | **0.809** | 0.058 | −0.387 |
| | Hand transition | −0.060 | 0.030 | **−0.824** | −0.082 | − 0.390 |
| | Hand selection overlap | 0.021 | −0.291 | 0.107 | **0.568** | 0.124 |
| | Hand speed bias | 0.003 | 0.045 | **0.840** | 0.007 | −0.238 |
| | Movement area bias | 0.029 | −0.029 | **0.860** | −0.131 | 0.239 |
| | Objects hit | **0.484** | **0.856** | 0.003 | 0.037 | −0.041 |
| | Distractor proportion | **0.963** | −0.112 | −0.003 | 0.027 | −0.005 |
| | Object processing rate | −0.380 | **0.799** | 0.000 | 0.071 | 0.041 |

Bold text indicates that the component loading is substantial (≥ .40). [a]denotes that the KINARM parameter did no substantially load onto any component

example, OHA was divided into 5 components: *impulsivity, accuracy, hand bias, speed and area,* and *miss bias.* Three of these components (*accuracy, hand bias,* and *speed and area*) are comparable to the PCA results observed from the OH task, thereby providing evidence that these tasks assess similar underlying behaviors. However, despite assessing similar underlying behaviors, the variance explained by each component differs. For example, *speed and area* accounts for more variability than *accuracy,* whereas in OHA *accuracy* accounts for more variability than *speed and area.* The addition of two unique components (i.e., *miss bias* and *impulsivity*)

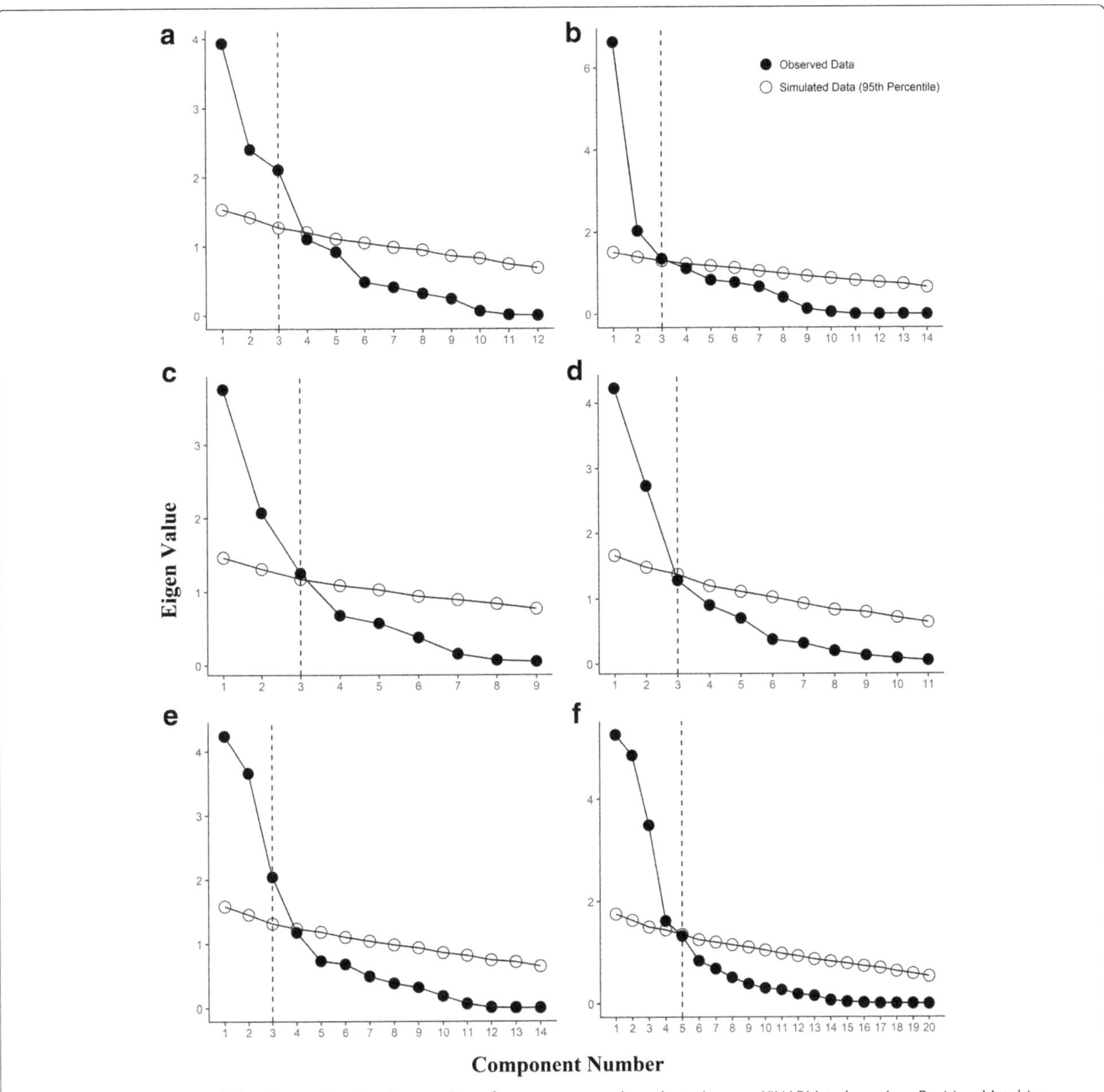


**Fig. 2** Scree plots with parallel analysis indicating the number of components to be selected across KINARM tasks. **a** Arm Position Matching; **b** Ball on Bar; **c** Visually Guided Reaching; **d** Reverse Visually Guided Reaching; **e** Object Hit; **f** Object Hit and Avoid

during the OHA may quantitatively represent the increase in cognitive load during this task. These two components that are absent from OH, and which account for 28% of overall variability, likely reflect the different impacts of impulsive movements in OH and OHA. In OH, quick and effectively impulsive movements are rewarded; quick movements with little inhibition will result in many targets being hit. In contrast, in OHA there is a penalty for impulsive movements that are not processed thoroughly prior to execution (i.e., deciding whether or not an object is a target or distractor prior to executing a movement towards it). Furthermore, behaviorally during the OH task, the subject can hit a total of 300 targets during the trial. In contrast, there are only 200 targets during OHA, plus 100 distractors. Therefore, *hand bias* and *speed and area* account for less variability in OHA, as now the subjects must inhibit their automatic responses (i.e., *impulsivity*), which also results in an increased need for improved *accuracy*. This contrast in cognitive load between tasks may anecdotally account for the difference in variance explained by the components mentioned above across OH and OHA. This result of similar inter-task grouping of parameters across related KINARM tasks was also observed during VGR and RVGR. Taken together, our results demonstrate the potential utility of PCA when applied to granular datasets, as well as the increased interpretability of KINARM parameters when categorized into behaviorally meaningful components that characterize performance.

Furthermore, parameter associations, positive or negative, strong or weak, within their respective components were logically related to participant performance. For example, the PCA of the BonB task identified the component *speed and success*. Targets completed, mean ball speed, mean right- and left-hand speed, and hand speed difference all had substantial positive loadings, whereas mean movement time had a substantial negative loading. Therefore, as participants increased their right- and left-hand speed, and thus ball speed, participants decrease the time from a target being displayed on screen to the target being reached by the participant, which results in a greater number of targets being completed throughout the task. In addition, the parameter norm absolute hand speed difference loaded substantially on *bar angle* but had a very low loading on *speed and area*. Behaviorally, this result is intuitive. If a participant reacts to a target quickly, the bar angle may become skewed if not carefully balanced, which is reflected by the substantial positive loadings of the other angle related parameters onto this component. These direction-based associations, low or high, were observed across most KINARM tasks and their respective parameters. These components will need to be further validated across KINARM platforms and patient populations. Overall, our analysis produced a consistent inter-task classification of behavioural variables (e.g., VGR and RVGR, OH and OHA), which demonstrates that these tasks assess similar underlying behaviours and further supports the validity of using this statistical technique on KINARM performance metrics. Our current analysis has substantially reduced these granular datasets into biologically plausible, interpretable, and behaviorally meaningful components.

### Cross-loading of parameters with multiple components

PCA indicated that some of the KINARM parameters cross-loaded with multiple components. For example, the APM task identified 3 components (*position accuracy, movement variability,* and *contraction/expansion*), which is fairly consistent with previous research that described these 3 variables as observed patterns of impairment after stroke [2]. Despite mostly a strong separation of these three components, shift x loaded onto both the *position accuracy* and *movement variability* components, with an inverse sign from negative to positive respectively. Shift indicates a systematic bias to move either left/right or front/back in the workspace. Therefore, in the *position accuracy* component, shift X (left/right) indicates, with a negative association, that shift X biases to the left as absolute error in the X and XY plane increases. In contrast, component 2, *movement variability*, indicates that as X shifts in the positive direction (i.e., hand moves to the right), overall variability increases.

Furthermore, certain parameters did not intuitively cross-load onto multiple components. For example, the OH task identified 3 components (*hand bias, speed and area,* and *accuracy*). Unexpectedly, the parameter total hits, which should be intuitively related to hits with left and right, did not cross-load onto the *accuracy* and *hand bias* components. However, this result demonstrates that total hits is highly related to *accuracy*, which is separable from *hand bias*. These components are not highly interrelated, as indicated by the use of orthogonal rotation, and measure separable aspects of participant performance. Therefore, *hand bias* measures the bias towards the participants limb preference, and use in space, which is separable from overall *accuracy* and the total targets being hit (e.g., right-handed participants may have biased use of their right hand, but this is separable from the number of targets that were accurately hit using either hand).

Interestingly, only three parameters did not substantially load onto any component: miss bias of OH and bar length variability and hand path bias from the BonB task. The reason for these low loadings is unclear. However, miss bias loaded highly onto a separate component for the OHA task, which may indicate that any bias of misses toward one side of the work space, or the other, is related to the increase in cognitive load during OHA. Furthermore, the low loadings for bar length variability and hand path bias may be the consequence of only analyzing level one of BonB, as this task increases with difficulty in the subsequent levels [19]. However, previous research has identified that level one of BonB identified the most performance impairments (i.e., highest number of parameters failed) in stroke participants, relative to control subjects [19]. Furthermore, the number of parameters failed tended to decrease with increasing task difficulty in successive levels, potentially due to the increased variability observed among controls, which ultimately influenced cut-off criteria used to quantify impairment among stroke participants [19]. Therefore, level one of BonB may be the most sensitive level used to detect impairment and is the most relevant for the current analysis. Future analysis may need to apply PCA to each level separately to investigate if the components are similar across all three levels. However, despite identifying some parameters that cross-load with multiple components and only a couple that did not substantially load onto any component, the majority of KINARM performance metrics did not cross-load with one another. Most components demonstrated high loadings with their respective components, which indicates a strong separation of components that quantify participant performance.

### Limitations and future directions

We excluded left-hand dominant participants, as many participants were right-handed, which means our findings may not generalize to left-handed participants. Furthermore, we could not conduct PCA across all task parameters, as participants did not complete all KINARM tasks. Therefore, we were unable to conduct PCA on broadly pooled performance scores across tasks, which confined our results to descriptions of each task only. However, the PCA has yielded task specific separations of components to further characterize sensorimotor function among healthy subjects, which may serve as a normative data set for future clinical comparisons of subject performance. In addition, two to three participants per task, except for RVGR, were missing participant data for education obtained. However, these missing data should not substantially impair our range of education. In addition, some tasks do not have Z-scores for all parameters being recorded (these metrics could not be standardized), and thus these parameters were excluded from this analysis. We conducted PCA only on the End-Point robot data, and therefore, we will need to conduct PCA on healthy participant data generated using the KINARM Exoskeleton. It is unclear how antigravity support provided by the Exoskeleton will affect participant performance and the generated components. Therefore, PCA of Exoskeleton data will be imperative to future investigations to implement PCA across KINARM platforms. Also, the current analysis was conducted in healthy participants, and future PCA will need to examine participant performance using a clinical sample, such as stroke, to validate these components and characterize performance in clinical populations to further demonstrate the utility of our current findings. Furthermore, the current analysis does not address individual or group differences in our healthy participant sample, which may complicate future comparisons between healthy subjects and clinical populations. However, using PCA across 6 KINARM tasks, which assesses a broad range of upper limb sensorimotor function, has led to a substantial reduction in the dimensionality of our data, and produced interpretable components of performance.

Therefore, PCA of KINARM data has the potential to become a valuable clinical tool. Applying PCA and identifying main sources of variability in a clinical examination can help make data from research tools, such as the KINARM, more concise and easily interpretable. Furthermore, identifying sources of strong variability may improve the detection of fluctuations in performance, which could increase the clinical relevance of robotic assessment as an evaluation tool. Therefore, analyzing data produced by a clinical population offers the potential to increase the clinical utility of the KINARM robot by maximizing interpretability of participant performance, while also minimizing information loss, to increase the characterization of performance among various populations. It is not clear if patient performance will result in similar components of performance patterns as healthy participants. Therefore, future applications of this analysis may offer potential insight into specific patterns of sensorimotor impairment among patient populations.

## Conclusions

Using PCA, granular KINARM performance data can be substantially reduced to a small number of interpretable components with minimal information loss and still accounting for a large amount of variance. To validate our current findings and to further characterize sensorimotor function and impairment, the components derived from our sample of healthy participants will serve as normative data for future comparisons to patient populations.

## Additional files

**Additional file 1:** Volunteer screening checklist. This form was administered to participants by trained researchers to screen for previous neurological and musculoskeletal complications. (PDF 53 kb)

**Additional file 2:** KINARM Standard Tests Parameter Tables. These tables provide detailed descriptions of each parameter per KINARM task. These tables were extracted from Dexterit-E 3.6 User Guide with permission from BKIN Technologies Ltd. (PDF 145 kb)

**Additional file 3:** Correlation matrices of KINARM parameters across all tasks. (DOCX 30 kb)

**Acknowledgments**

The authors would like to thank Kim Moore and Simone Appaqaq for assessing the control participants, as well as Helen Bretzke and Justin Peterson for their contribution to software implementation and database management.

**Funding**

Funding provided by an Ontario Research Foundation – Research Excellence grant and a GSK-CIHR Chair in Neuroscience to SHS. The funding agencies had no role in the design of this study and they did not have any role in data collection or analysis.

**Authors' contributions**

MDW participated in the statistical plan, data analysis and drafting of the manuscript. LERS participated in data analysis and drafting of the manuscript. SHS participated in study design and drafting of the manuscript. JGB participated in study design and drafting of the manuscript. JAJ is the primary investigator. She participated in the statistical plan, reproducing the data analysis and results, and drafting of the manuscript. All authors read and approved the final manuscript.

#### Competing interests

MDW has nothing to disclose.
LERS has nothing to disclose.
SHS is the cofounder of BKIN Technologies, the manufacturer of the KINARM device.

JGB receives a stipend from the Trillium Gift of Life Network to support his role as the Hospital Donation Support Physician.
JAJ has nothing to disclose.

#### Author details

[1]Centre for Neuroscience Studies, Queen's University, Botterell Hall, 18 Stuart St, Kingston, ON, Canada. [2]Department of Critical Care Medicine, Queen's University, Kingston, ON, Canada. [3]Department of Medicine, Queen's University, Kingston, ON, Canada. [4]Department of Psychology, Queen's University, 318 Craine Hall, 62 Arch Street, Kingston, ON K7L 3N6, Canada.

#### References

1. Coderre AM, et al. Assessment of upper-limb sensorimotor function of subacute stroke patients using visually guided reaching. Neurorehabil Neural Repair. 2010;24:528–41.
2. Dukelow SP, et al. Quantitative assessment of limb position sense following stroke. Neurorehabil Neural Repair. 2010;24:178–87.
3. Dukelow SP, Herter TM, Bagg SD, Scott SH. The independence of deficits in position sense and visually guided reaching following stroke. J Neuroeng Rehabil. 2012;9:72.
4. Debert CT, Herter TM, Scott SH, Dukelow S. Robotic assessment of sensorimotor deficits after traumatic brain injury. J Neurol Phys Ther. 2012; 36:58–67.
5. Williams L, et al. Sensory-motor deficits in children with fetal alcohol Spectrum disorder assessed using a robotic virtual reality platform. Alcohol Clin Exp Res. 2014;38:116–25.
6. Pearson K. On Lines and Planes of Closest Fit to Systems of Points in Space. Philos Mag Ser 6. 1901;2:559–72.
7. Hotelling H. Analysis of a complex of statistical variables into principal components. J Educ Psychol. 1933;24:417–41.
8. Ringner M. What is principal component analysis? Nat Biotechnol. 2008;36: 303–4.
9. Joliffe IT, Cadima J. Principal component analysis: a review and recent developments. Phil Trans R Soc A. 2016;374:20150202.
10. BKIN Technologies Ltd, Kingston, ON, Canada. Dexterit-E User Guide v 3.6 – KINARM Standard Tests Parameter Definitions. http://www.bkintechnologies.com/download/dexterit-e-user-guide-v-3-6-kinarm-standard-tests-parameter-definitions/. Accessed 2 Oct 2017.
11. Team RC. R: a language and environment for statistical computing. Vienna, Austria: R Foundation for Statistical Computing; 2016. https://www.R-project.org/
12. Revelle, W. Psych: Procedures for Personality and Psychol Res. Evanston: Northwestern University; 2017. https://cran.r-project.org/package=psych.
13. Sakaluk JK, Short SD. A methodological review of exploratory factors analysis in sexuality research: used practices, best practices, and data analysis resources. J Sex Res. 2017;54:1-9.
14. Cattell RB. The scree test for the number of factors. Multivariate Behav Res. 1966;1:245–76.
15. O'connor BP. SPSS and SAS programs for determining the number of components using parallel analysis and Velicer's MAP test. Behav Res Methods Instrum Comput. 2000;32:396–402.
16. Horn JL. A rationale and test for the number of factors in factor analysis. Psychometrika. 1965;30:179–85.
17. Jennrich RI, Sampson PF. Rotation for simple loadings. Psychometrika. 1966; 31:313–23.
18. Kaiser HF. The varimax criterion for analytic rotation in factor analysis. Psychometrika. 1958;23:187–200.
19. Lowrey CR, Jackson CPT, Bagg S, Dukelow SP, Scott SH. A novel robotic task for assessing impairments in bimanual coordination post-stroke. Int J Phys Med Rehabil. 2014;S3:002.

# 8

# Validation of wearable visual feedback for retraining foot progression angle using inertial sensors and an augmented reality headset

Angelos Karatsidis[1,2*], Rosie E. Richards[3], Jason M. Konrath[1], Josien C. van den Noort[3,4], H. Martin Schepers[1], Giovanni Bellusci[1], Jaap Harlaar[3,5] and Peter H. Veltink[2]

**Abstract**

**Background:** Gait retraining interventions using real-time biofeedback have been proposed to alter the loading across the knee joint in patients with knee osteoarthritis. Despite the demonstrated benefits of these conservative treatments, their clinical adoption is currently obstructed by the high complexity, spatial demands, and cost of optical motion capture systems. In this study we propose and evaluate a wearable visual feedback system for gait retraining of the foot progression angle (FPA).

**Methods:** The primary components of the system are inertial measurement units, which track the human movement without spatial limitations, and an augmented reality headset used to project the visual feedback in the visual field. The adapted gait protocol contained five different target angles ranging from 15 degrees toe-out to 5 degrees toe-in. Eleven healthy participants walked on an instrumented treadmill, and the protocol was performed using both an established laboratory visual feedback driven by optical motion capture, and the proposed wearable system.

**Results and conclusions:** The wearable system tracked FPA with an accuracy of 2.4 degrees RMS and ICC=0.94 across all target angles and subjects, when compared to an optical motion capture reference. In addition, the effectiveness of the biofeedback, reflected by the number of steps with FPA value ±2 degrees from the target, was found to be around 50% in both wearable and laboratory approaches. These findings demonstrate that retraining of the FPA using wearable inertial sensing and visual feedback is feasible with effectiveness matching closely an established laboratory method. The proposed wearable setup may reduce the complexity of gait retraining applications and facilitate their transfer to routine clinical practice.

**Keywords:** Foot progression angle, Inertial sensors, Real-time biofeedback, Augmented reality headset, Gait retraining, Knee osteoarthritis

## Background

Knee osteoarthritis (KOA) is a leading cause of disability in the elderly population [1]. To date, there is no cure available for the disease and treatment options are of pharmacological, surgical or biomechanical nature [2–4]. Pharmacological treatments alleviate only the symptoms (pain, discomfort and swelling), while surgical treatments usually involve total knee replacement and are only considered in severe stages of the disease [5]. Biomechanical interventions are conservative non-pharmacological treatments, which aim at decreasing or distributing the loading across the knee joint. This mechanical joint loading has been related to cartilage degeneration, pain, and disease progression [6].

A common biomechanical treatment is gait retraining. Through these treatments patients learn and gradually adopt a modified gait pattern that results in decreased loading across the knee joint. [7, 8]. The training is typically achieved by tracking the body biomechanics and using this information to drive a real-time feedback modality, such as a vibration, an audio sound or a

*Correspondence: angelos.karatsidis@xsens.com
[1]Xsens Technologies B.V, Pantheon 6, 7521 PR Enschede, The Netherlands
[2]Department of Biomedical Signals and Systems (BSS), Technical Medical Centre, University of Twente, Enschede, The Netherlands
Full list of author information is available at the end of the article

visualization pattern [9]. An advantage of gait retraining compared to other biomechanical treatments, such as use of wedge insoles, knee braces or canes, is that it does not require any additional devices to alter gait mechanics.

Knee joint loading can be quantified through the medial and lateral tibiofemoral contact forces [10]. Due to practical difficulties in measuring the internal knee contact forces in a non-invasive manner, the net knee joint moment has been considered a convenient surrogate measure [11]. However, instructing patients to decrease a complex kinetic parameter, such as the knee joint moment, in real-time, has been shown to be less effective than explicitly instructing the correct movement that achieves the decrease in the loading [12]. These findings are in line with previous studies which showed that higher reduction in the knee adduction moment can be achieved by altering related kinematic parameters, such as the foot progression angle (FPA) [13–20].

Despite the demonstrated benefits of gait retraining, it is currently not used in clinical practice [21]. One of the primary reasons impeding clinical adoption is the expensive, complex, time-consuming, and space-bound instrumentation that is required to accurately assess the biomechanical parameters. Conventionally, a gait laboratory is utilized, in which multiple cameras track the three-dimensional positions of skin-mounted passive or active markers. Next, the segment positions and orientations are assessed through computational techniques such as direct or inverse kinematics [22, 23].

An ambulatory alternative to the lab-bounded measurement systems is composed of inertial measurement units (IMUs) that can derive orientation of a sensor in space [24]. Specifically, fusion of the accelerometer, gyroscope, and magnetometer signals and incorporation of a biomechanical model and external contact updates enable consistent drift-free motion capture [25, 26]. In addition, IMU-based systems are typically low cost, low power, highly portable, minimally obstructive, easily wearable, acceptable by older adults and therefore comprise an ideal alternative to facilitate the clinical translation of movement analysis systems. Despite their potential, to date, the use of IMUs in gait retraining applications for KOA has only received limited attention [27, 28]. Exploiting the advantages of IMUs, featuring high performance and applicability, may remove the complexity of the current laboratory approaches, decrease the costs, and make treatments available to a larger number of patients.

Besides motion tracking, the second component required in gait retraining is the biofeedback. In a recent systematic review, studies using laboratory-based biofeedback to target knee joint loading either directly or indirectly were analyzed [29]. Most studies used visual feedback modes [7, 12, 20, 30–36] or multi-modal visual-tactile [37, 38], and less often solely tactile [19, 39–41] or auditory feedback [42]. Another review focusing on wearable sensing and feedback techniques reported that until recently, most studies utilizing wearable feedback incorporated primarily tactile modalities [9]. These devices are typically unobtrusive, but they act as on/off switches that can only convey binary information to the user. In addition, tactile feedback was reported to require longer training times for patients to converge to a target pattern, compared to visual feedback. Wheeler et al. [37] Wearable visual feedback was until recently challenging due to practical limitations. The conceptual and technical feasibility of wearable visual feedback for knee joint angle using two IMUs and a small screen on a smart-glass was demonstrated by Steuner et al. [43]. Recent advances in augmented reality (AR) headsets, such as the Microsoft HoloLens [44], allow the projection of virtual objects on the user's field of view, via head-worn screens. As a result, the wearable biofeedback setups can be enriched with quantitative information, which can not only convey whether the user is achieving the desired target range, but also quantify the difference from the target.

The overarching objective of this study was to develop and evaluate a wearable biofeedback system for gait retraining purposes, as an alternative to currently existing lab-bound setups. In order to achieve this, the first objective was to develop a wearable real-time visual feedback driven by FPAs calculated using input from a commercially available inertial motion capture system; utilizing accelerometers, gyroscopes, and magnetometers. We hypothesized that the proposed wearable system would provide accurate assessments in timing and magnitudes of the FPA when compared to a conventional optical motion capture laboratory setup. The second objective was to evaluate the feedback effectiveness of the wearable system reflected by the number of steps with FPA within a defined target range. We hypothesized that participants would perform equally well in achieving the desired FPAs using the wearable system, when compared to the established laboratory setup. It is envisioned that the proposed wearable setup may reduce the complexity of gait retraining and facilitate their transfer into routine clinical practice.

## Methods

### Subjects

Eleven (11) healthy volunteers (4 males, 7 females, age: 28.26 ± 4.55 years; height: 1.78 ±0.10 m; weight: 77.91 ±15.01 kg; body mass index (BMI): 24.50 ±2.52 kg/m$^2$) participated in the data collection performed at the Virtual Reality Laboratory of the VUmc Amsterdam. Subjects provided written informed consent prior to their voluntary participation in the study and after receiving detailed information about the study. Ethical approval was provided by the Scientific and Ethical Review Board (Dutch:

Vaste Commissie Wetenschap en Ethiek - VCWE) of the Faculty of Behavior & Movement Sciences, VU University Amsterdam.

### Instrumentation

Human movement analysis was performed in a Gait Real-time Analysis Interactive Lab (GRAIL, MOTEK BV, Amsterdam, NL) depicted in Fig. 1. The GRAIL system is composed of a dual-belt instrumented treadmill with two full 6D force plates beneath each belt capturing at 1000 Hz. In addition, the system features a ten-camera system tracking 22 passive reflective markers at 100 Hz (Vicon, Oxford Metrics Group, Oxford, UK). Markers were placed on the following body locations according to the lower body configuration of Human Body Model 2: anterior and posterior superior iliac spine, medial and lateral femoral epicondyle, medial and lateral malleolus, second metatarsal, fifth metatarsal, calcaneus, lateral mid-shank, and lateral mid-thigh. To enable the laboratory feedback, a semi-cylindrical screen located anterior to the treadmill was utilized that projected an immersive virtual reality environment. Integration and control of the GRAIL components is enabled by the D-Flow software package [45], and real-time biomechanical modeling was performed through the Human Body Model (HBM) software package [31, 46].

Concurrently with the GRAIL measurements, Xsens MVN Awinda inertial motion capture system (Xsens Technologies BV, Enschede, NL) [47] was used with the lower body configuration. Seven Xsens MTw IMUs with dimensions 47 x 30 x 13 mm and orientation dynamic accuracy 0.75 deg RMS for roll/pitch and 1.5 deg for heading components were used. The full scales of the measurement units are $\pm 160 m/s^2$ for the accelerometer, $\pm 2000 deg/s$ for the gyroscope, and $\pm 1.9 Gauss$ for the magnetometer [48]. Five IMUs were mounted on pelvis, thighs, and shanks using the accompanying Velcro straps and two IMUs were placed on feet by firmly tying them with the laces on each participant's own shoes. The software version of Xsens MVN Analyze 2018.0 was used to reconstruct the lower body kinematics at 60 Hz [26]. The software features consistent behavior, even at the presence of magnetic disturbances, making it suitable for use on a treadmill and any other environment regardless of its magnetic field homogeneity. Segment orientations were obtained through the software by applying the IMU-to-segment alignment, found using an a-priori-known upright pose (N-pose) performed by the subject during the calibration [49]. The second part of the calibration of Xsens MVN consisted of comfortable walking in a straight line for approximately 5 meters. The output of the Xsens MVN Analyze is three-dimensional positions and orientations of the modeled body segments, expressed in an external coordinate frame defined during the calibration [26].

To enable the wearable biofeedback, Microsoft HoloLens was used (Microsoft Corp., Redmond, WA, USA) [44]. This wearable augmented reality headset device is capable of projecting holograms (three-dimensional visualizations) in the environment of use. The biofeedback was developed as a Universal Windows Platform (UWP) application built in Unity 3D Game Engine version 5.6.2 (Unity Technologies SF, San Francisco, CA, USA), and receives kinematic input (packet size = 760 bytes) from Xsens MVN Analyze, in real-time, via User Datagram Protocol (UDP) at 30 Hz. Networking of the devices was configured via an access point (TP-Link TL-WR802N, 300MBit/s, 2.4GHz), which was connected to the computer running Xsens MVN Analyze via Ethernet and to the Microsoft HoloLens via Wi-Fi. The components comprising the wearable biofeedback system are illustrated in Fig. 2.

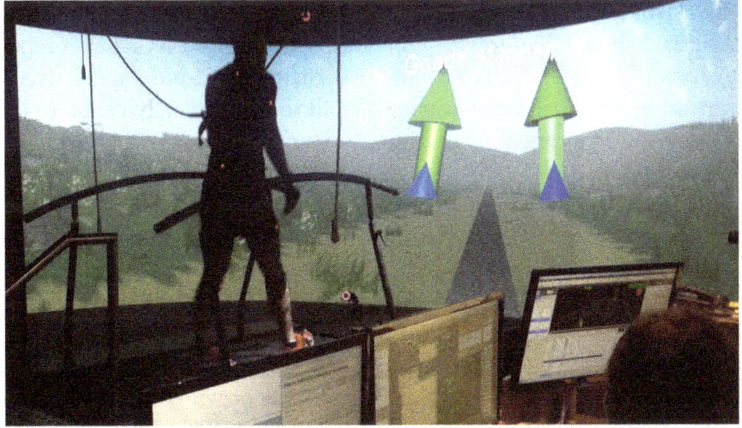

**Fig. 1** Virtual reality laboratory equipped with a GRAIL system. The subject receives a target foot progression angle (FPA) through an arrow, which changes color, from red to green, depending on the performed angle

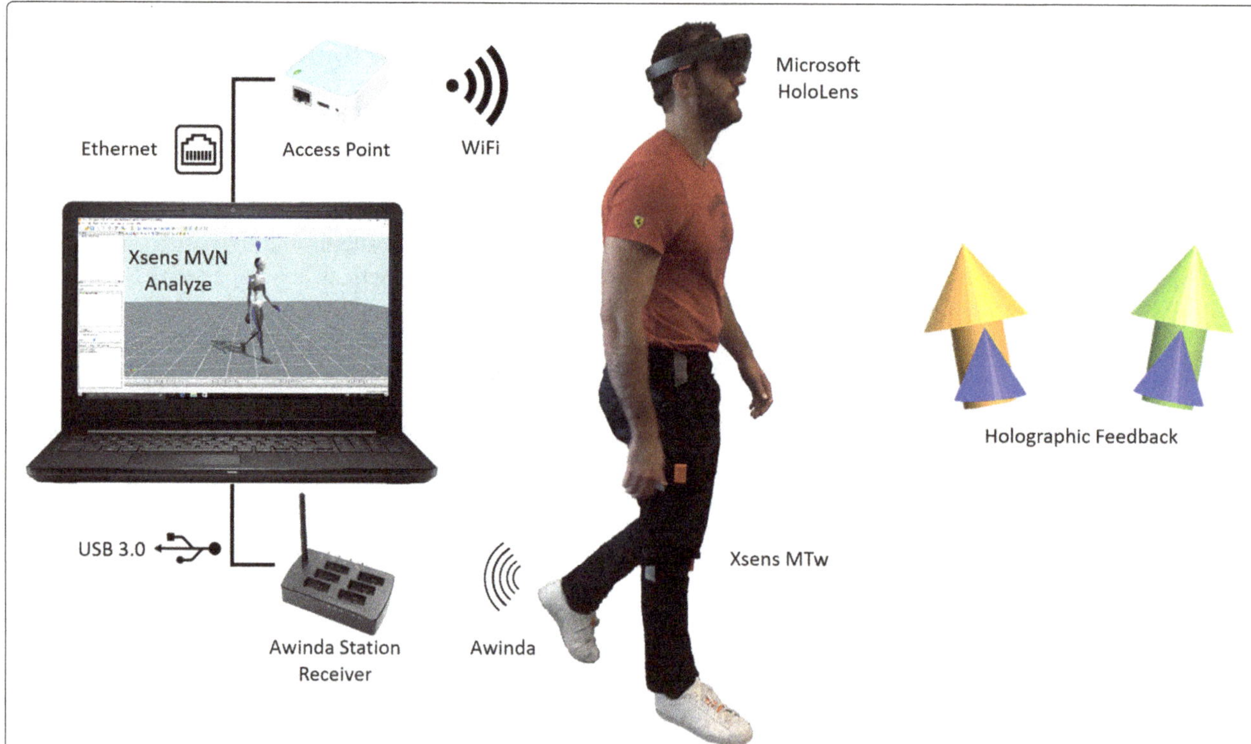


**Fig. 2** Wearable biofeedback setup. Xsens MVN Analyze receives the MTw sensor data via the Xsens Awinda Station, reconstructs the lower body kinematics, and streams via UDP. Microsoft HoloLens receives the kinematic input via Wi-Fi, calculates the foot progression angle, and updates the holographic feedback visualization

The biofeedback was visualized similarly on both laboratory and wearable screen, in accordance with a previous gait retraining study involving patients of KOA [12]. Figure 2 illustrates the wearable biofeedback setup, in which the feedback object is a blue cone with 2D orientation updated based on the FPA per step. The target object is an arrow placed behind the feedback cone, the color of which is updated depending on the agreement between the estimated and target FPA. More specifically, the color changes were based on the absolute difference between target and performed FPA: green when $|\text{FPA-target}| \leq 2°$, yellow when $2° < |\text{FPA-target}| \leq 5°$, and red when $|\text{FPA-target}| > 5°$. These are arbitrary chosen values, with the green range matching targets used in previous studies [12, 36, 41].

In the Microsoft HoloLens visualizations, billboarding and tag along features were added to the holograms to update their position, such that they would always face the user and only translate when they were entirely outside the user's field of view. These techniques ensure availability of the content at all times while minimizing the unpleasant effects of visualizations that are tighly coupled to the motion of the headset (head-locked content) [50].

For practical reasons, the Microsoft HoloLens application featured speech command capabilities to enable the initialization of the training protocol. Thus, right after the initialization of the treadmill belt, the researcher performing the experiment approached the subject who was already walking on the treadmill to provide the triggering key-phrase "go to mode zero" close to the microphone of the Microsoft HoloLens.

### Experimental procedures

Subsequently, placement of the reflective markers and IMUs was performed, followed by calibration of Xsens MVN system. As a last preparation step, participants performed a T-pose and walked a few steps on the treadmill to anatomically calibrate the optical motion capture system.

The first series of experiments comprised treadmill walking at a preselected constant speed of 1.2 m/s for 13.5 minutes. An acclimatization period with no target was provided for one minute. Next, the following five target angles were projected in a random order, for two minutes each: 15, 10, 5 degrees toe-out, 0 degrees straight toes, and 5 degrees toe-in. Visual feedback on the performed FPA was provided on each performed step on the laboratory screen. Before any new target, a 30-second active rest period was provided, during which subjects kept walking without receiving any target.

During the second series of experiments, participants wore the Microsoft HoloLens for a brief period to familiarize with the device. The complete protocol was repeated for another 13.5 minutes, by projecting the targets and IMU-driven feedback only on the wearable screen of the Microsoft HoloLens. An additional calibration step was introduced, where participants were asked to walk with their toes pointing straight for one minute to detect and reduce any heading offsets introduced by the inertial motion capture system. The average FPA estimated by the wearable system during this period was subtracted from the FPA values estimated in the rest of the trial. An example of the feedback protocol with the targeted and performed FPA values, and the various modes across the full duration of the protocol is illustrated in Fig. 3.

## Computational procedures

The inertial motion capture system outputs global positions and orientations of the tracked body segments. The toe and heel positions are used to calculate the FPA of the *i*th step, when the foot is placed approximately horizontally on the treadmill belt. Assuming heel strike at initial foot contact, entire contact of a foot is identified at the timepoint when the magnitude of heel and toe velocities is close to zero, empirically found as the magnitude of the first derivative of toe position ($\boldsymbol{p}_t$) and heel position ($\boldsymbol{p}_h$), $|\dot{\boldsymbol{p}}_t| < 0.2$ m/s and $|\dot{\boldsymbol{p}}_h| < 0.2$ m/s, respectively.

We define the foot vector for the *ith* step ($\boldsymbol{r}_{f,i}$) as the line from heel ($\boldsymbol{p}_{h,i}$) to toe ($\boldsymbol{p}_{t,i}$) during phase of entire foot contact:

$$\boldsymbol{r}_{f,i} = \boldsymbol{p}_{t,i} - \boldsymbol{p}_{h,i} \tag{1}$$

Similarly, the heading vector of the *ith* step ($\boldsymbol{r}_{w,i}$) is defined as the displacement vector between the position of the heel in two successive steps:

$$\boldsymbol{r}_{w,i} = \boldsymbol{p}_{h,i} - \boldsymbol{p}_{h,i-1} \tag{2}$$

FPA is calculated as the difference between the foot and the heading vectors projected on the transverse plane (Fig. 4), defined by anterior ($x$) and lateral ($y$) axes:

$$\theta_{FP,i} = arctan2\left(\frac{r_{w,i,x}}{r_{w,i,y}}\right) - arctan2\left(\frac{r_{f,i,x}}{r_{f,i,y}}\right) \tag{3}$$

In the laboratory system, FPA is calculated and averaged within a time interval, when the vertical ground reaction force recorded by the respective plate, is greater than a threshold of 10N. Toe and heel positions correspond to the markers placed on the second metatarsal and calcaneus, respectively and the foot vector is calculated based

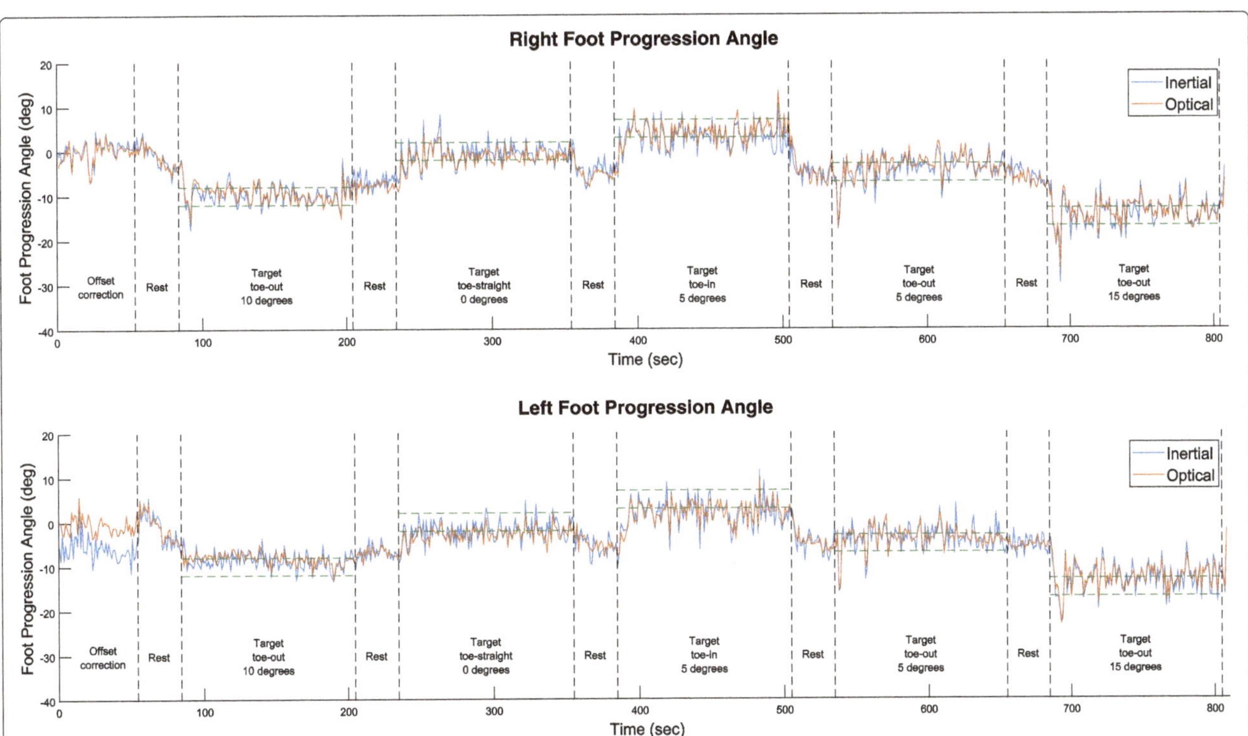


**Fig. 3** Illustration of the results for right and left foot progression angles across the whole training protocol, estimated at each entire foot contact via inertial (blue lines) or optical (red lines) motion capture input. An offset correction is calculated during the first 60 s when participants are instructed to walk with straight toes and applied after that. A unique random target is provided for 120 s (mid-point of green dashed lines indicating the ±2° good step range), after 30 s of no target (rest) period

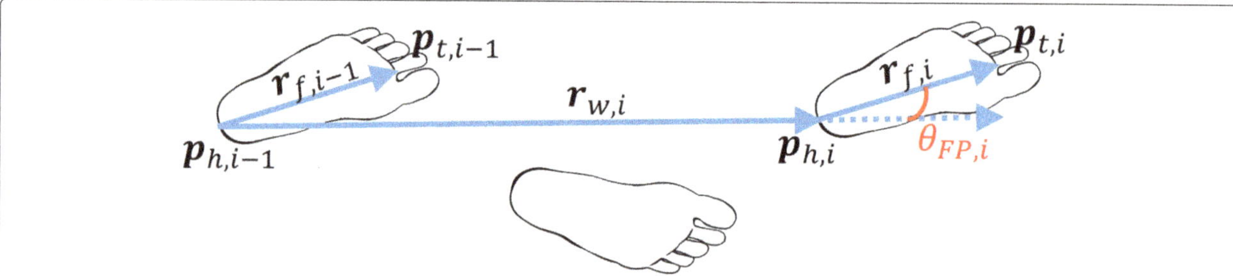


**Fig. 4** Transverse caudal view of the feet, illustrating the calculation of the foot progression angle for the right foot. The foot progression angle ($\theta_{FP}$) of the *i*th step is derived from the difference of foot vector ($r_f$) and heading vector ($r_w$). The latter two vectors are computed based on the positions of the heel ($p_h$) and toe ($p_t$) as illustrated in the figure

on Eq. 1. Differently to the wearable system, the heading vector is constant and defined as the anterior axes of the lab coordinate system aligned with the belt of the treadmill.

### Data Processing and Statistical Analysis

Data analysis focused on evaluating the performance of the wearable system in estimating FPA, in real-time versus the laboratory system, and to quantify the effectiveness of the two feedback modalities. Firstly, we compared the calculated FPA using the inertial motion capture system versus the optical motion capture system during the second series of experiments with the wearable feedback. The root-mean-squared differences were computed per target angle. Pearson's $r^2$ correlation and two-way random single measures intraclass correlation (ICC) were used to to quantify the agreement and consistency between the two estimation systems. Secondly, we examined the effectiveness of the wearable biofeedback system versus the established laboratory solution in altering the user's FPA. To quantify the effectiveness, we analyzed the number and percentage of good steps, defined as the steps with FPA within the ±2 degrees tolerance range as suggested in the literature [38, 41]. An analysis of variance (ANOVA) across all target conditions and systems, with Tukey's post-hoc analysis was performed to test whether the performed FPAs differ significantly across target conditions. Significance level was set to 0.05 and confidence interval at 95%. Data analysis was performed in MATLAB 2017a.

## Results

Correlation and Bland-Altman plots are shown in Fig. 5. Correlation coefficients were found of 0.9 for $r^2$ and of 0.94 for ICC. Accuracy analysis across all data points showed RMS difference of 2.38 degrees and level of agreement (LOA) about 4.7° between the wearable and laboratory estimates. Per target mode, RMS difference (average across subjects ± standard deviation) was found to be 2.25 ± 1.10, 2.18 ± 0.90, 2.02 ± 0.90, 2.62 ± 1.22, 1.86 ± 0.73 degrees for target FPAs of −15, −10, −5, −0, and 5 degrees, respectively.

During the first minute of the wearable feedback session, when subjects were instructed to walk with straight

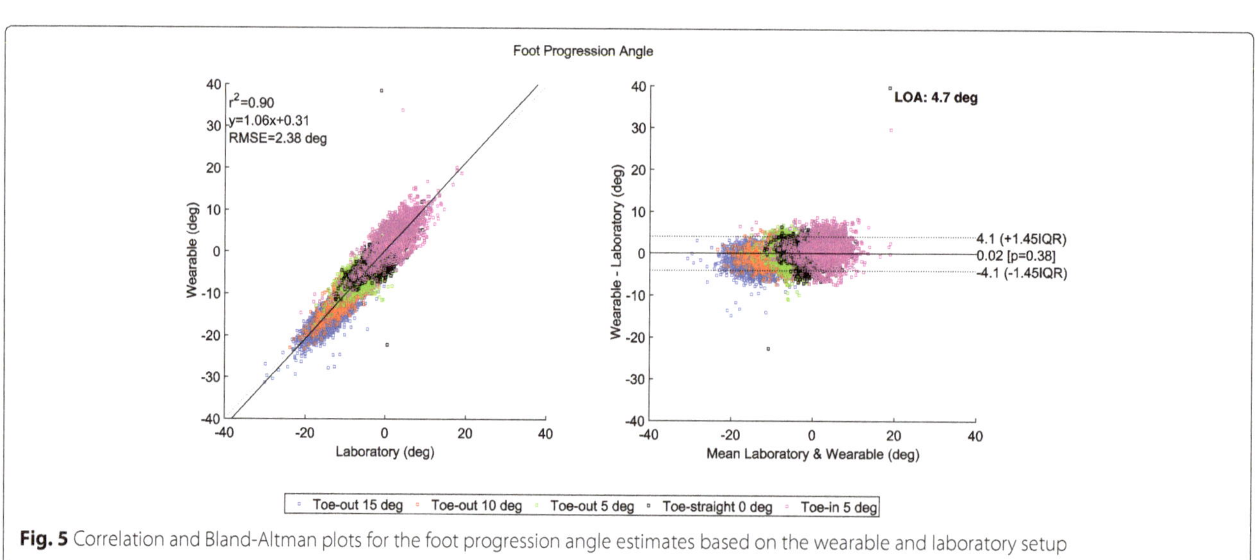


**Fig. 5** Correlation and Bland-Altman plots for the foot progression angle estimates based on the wearable and laboratory setup

toes, mean FPA values equaled $-1.61 \pm 2.47°$ as recorded by the laboratory system, and $-1.96 \pm 2.91°$ as recorded by the wearable system.

Figure 6 shows the box plots per target mode and per feedback scheme. The mean ± standard deviations of the differences from the targets for the wearable feedback were $0.48 \pm 3.75°$, $0.07 \pm 2.86°$, $-0.46 \pm 2.57°$, $-0.78 \pm 3.01°$, $-0.83 \pm 3.97°$ and for the laboratory feedback $1.04 \pm 3.44°$, $0.28 \pm 3.13°$, $-0.91 \pm 3.03°$, $-1.20 \pm 3.07°$, $-1.49 \pm 3.50°$ for $-15°$, $-10°$, $-5°$, $0°$, and $5°$ target FPA, respectively. In both systems multivariate ANOVA test showed significant differences between the FPAs of each target mode regardless of system used for the feedback, while post-hoc analysis across the five different modes showed that the FPAs during each target mode differed significantly to other target modes ($p < 0.001$).

Feedback effectiveness based on the percentage of good steps with FPA within the $\pm 2°$ range is illustrated in Fig. 7. Percentage of good steps in the laboratory feedback was $51 \pm 15\%$ across 12033 steps over all subjects and targets. Per target percentage of good steps was found to be $42.7 \pm 13.2\%$, $52.7 \pm 12.2\%$, $58.6 \pm 15.5\%$, $54.5 \pm 18.7\%$, $46.3 \pm 15.7\%$ for FPAs of $-15$, $-10$, $-5$, 0, and 5 degrees, respectively. In the case of FPA feedback provided and calculated in the wearable setup, an overall percentage of good steps $48.3 \pm 12.8\%$ across 12075 steps was found. For the aforementioned ascending order of targets, effectiveness per mode was observed to be of $39.4 \pm 9.2\%$, $54.4 \pm 14.9\%$, $54.3 \pm 11.0\%$, $51.4 \pm 15.1\%$, $42.1 \pm 13.9\%$. When the FPAs of the wearable feedback were calculated from the optical motion capture system, the percentage of good steps was found overall $51.3 \pm 13.4\%$, with individual per mode effectiveness of $45.3 \pm 7.4\%$, $56.5 \pm 17.0\%$, $56.1 \pm 10.3\%$, $53.3 \pm 19.7\%$, $45.1 \pm 12.7\%$.

## Discussion

In this study we proposed a method to perform gait retraining of the FPA using real-time biofeedback based entirely on wearable sensing and feedback modules. To our knowledge, this is the first study investigating a fully wearable visual feedback system for the purpose of retraining the FPA. Our findings demonstrated that FPA estimates derived from the inertial motion tracking input matched closely the ones from optical motion capture, with an overall RMS difference of 2.38 degrees. In addition, when incorporating a wearable augmented reality headset, the biofeedback effectiveness, based on steps within a $\pm 2°$ target range, matched closely the laboratory approach.

Our accuracy analysis depends on an optical motion capture system reference. However, previous studies have indicated that orientations of the transverse plane may also suffer from inter-trial, inter-session, and inter-observer differences. In particular, for the foot heading angle, median within-assessor reliability across four studies [51–54] was reported to provide multiple correlation coefficient of 0.55 [55]. In addition, the same systematic review discussed five other studies that

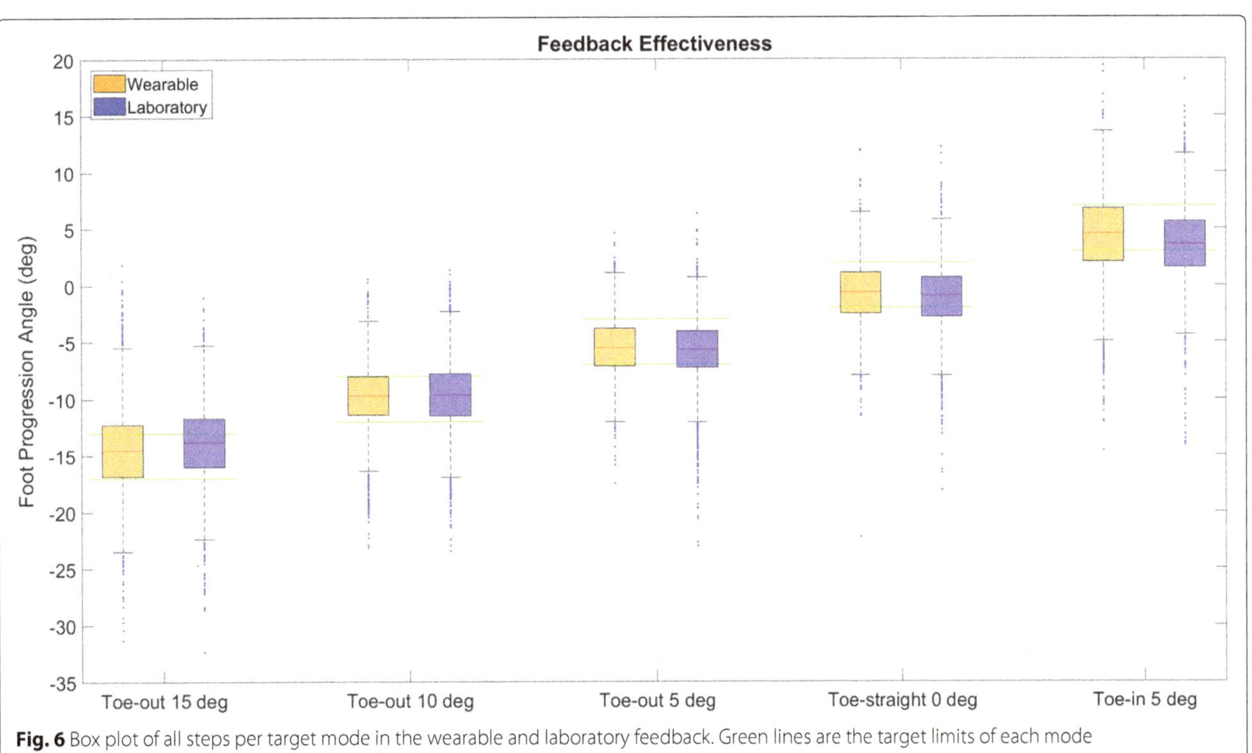


**Fig. 6** Box plot of all steps per target mode in the wearable and laboratory feedback. Green lines are the target limits of each mode

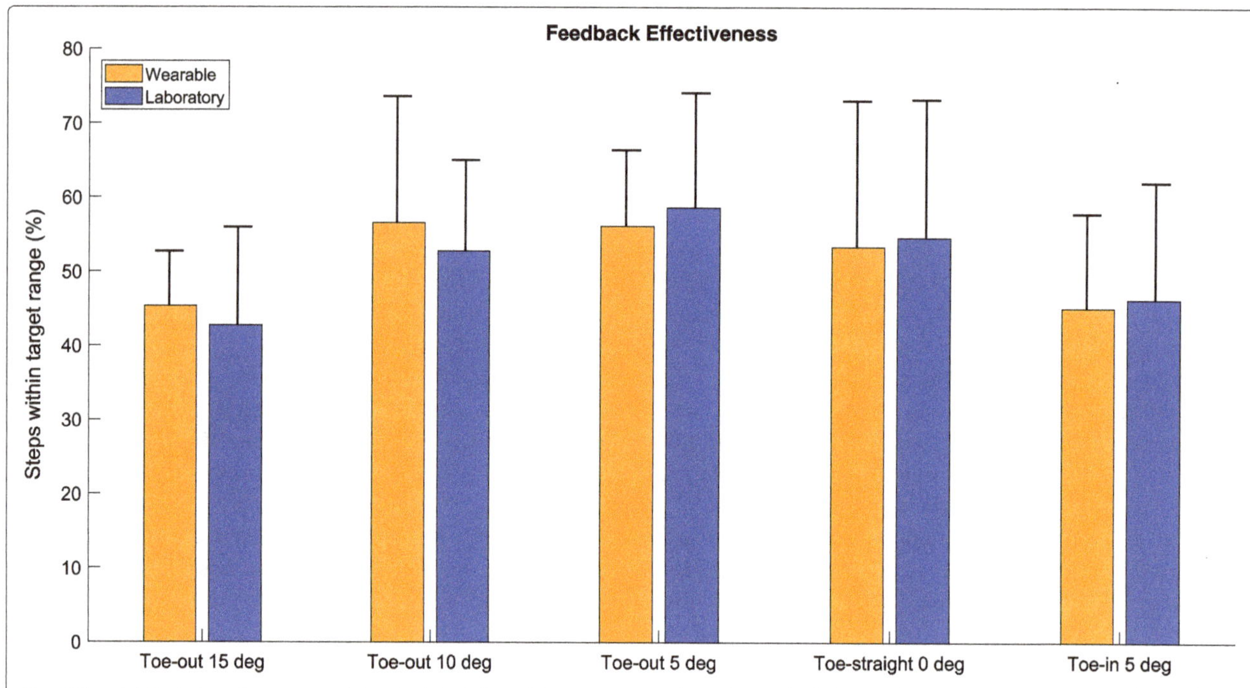


**Fig. 7** Bar plot illustrating the mean and standard deviation of the percentage of good steps (± 2 degrees from the target), across 11 subjects, for wearable (orange) and laboratory (blue) feedback setups

reported inter-assessor standard deviation of the foot progression that ranged from 2 to 5 degrees [56–60]. These literature findings suggest that error magnitudes of around 2 degrees, as found in our study, are typically found in conventional optical motion capture systems, as a result of marker placement, computational method, measurement system accuracy and resolution, or observer's experience and skills. Therefore, given that these measurement errors are considered clinically acceptable, our method is sufficiently accurate in tracking FPA for the specific application.

In our study we used a sensor set of seven IMUs required by Xsens MVN software to track both positions and orientations of the feet and other lower body segments. A question arises whether fewer IMUs would suffice for this task. Related studies have proposed a set of one sensor per foot combined with a magneto-inertial sensor fusion algorithm to derive the FPA [27, 61]. Even though those studies noted no effect of magnetic disturbances in the estimates, it has been previously shown that inertial-magnetic motion tracking is affected by the homogeneity of the magnetic field [62]. In addition, the use on a treadmill, which typically contain several electromagnetic components beneath the belt, would create a non-homogeneous magnetic field. As a result, an approach heavily relying on magnetometers would suffer from orientation drift over time. The present study used the latest version of Xsens MVN software, which provides a consistent pose of the body pose regardless of magnetic disturbances in the environment of use [26].

Calculation of the foot vector angle using both inertial and optical motion capture solutions may suffer from offsets of approximately 1–3 degrees, due to measurement and modeling error in both approaches. In inertial motion capture, offsets in the foot vector may be introduced as a result of a mismatch between the modeled and practiced N-pose used to calibrate the system. Similarly, optical motion capture may be susceptible to sub-centimeter misplacement of the markers on the second metatarsal and calcaneus. For instance, misplacements of markers on the foot may result in erratic estimation of the foot vector. In the accuracy analysis we subtracted these offsets, based on the median FPA during the first one minute of the trial. Moreover, gait event detection methods differ between both systems and may introduce differences. The wearable system relies on detection of near-zero velocity to detect contact with the ground which may be sensitive to the walking speed and style. In contrast, the laboratory system is based on force plate detection, which may be erratic in real-time in case the subject steps on the contra-lateral force plate. In the offline analysis we have corrected for these cases, by calculating the gait events based on marker velocity. Another source of difference in the laboratory system is that for convenience, the heading vector was set constant, aligned with the anterior axis of the global coordinate system matching the movement direction of the belt. In other words, contrary to the wearable system, differences in walking direction are not taken into account in the FPA calculation of the laboratory system.

A major advantage of the proposed wearable system compared to laboratory-based setups is its significantly lower cost. Fully functional virtual reality laboratories typically cost between tens to hundreds thousand dollars, depending on the type of optical motion capture systems, instrumented treadmills, and immersive environment systems. The setup proposed in this study is composed of an augmented reality headset and inertial sensors. Current price of the development version of Microsoft HoloLens is around 3000 dollars, while the cost of goods for inertial sensor components (accelerometer, gyroscope, and magnetometer) has nowadays dropped to a few tens of dollars per module. Additional costs may include the cost for networking devices, computers and software. Costs for software vary considerably and are therefore difficult to quantify, since they usually depend on development efforts, number of users and other market-driven factors [63].

Besides costs, the proposed method based on a set of IMUs and an AR headset reduces the complexity and increases the flexibility of gait retraining methods significantly compared to conventional laboratory techniques. Alternative approaches with lower cost and complexity have been previously proposed, even without the necessity for electronic equipment. For instance, mirror-based biofeedback for FPA retraining of patients with knee osteoarthritis has been investigated by Hunt et al. [33]. That study reported significantly lower performance of the mirror feedback compared to real-time visual biofeedback, with mean differences of approximately 2 degrees. However, despite the significantly lower performance of the mirror and given the high costs of laboratory-based biofeedback setups, the former was favored as an acceptable solution for clinical practice. Our work provides a method that matches closely the performance of laboratory feedback systems in terms of both tracking accuracy and feedback effectiveness, while reducing the costs and increasing the portability and potential of performing gait retraining in any environment. In addition, recent studies reported that subject-specific gait modifications decrease knee joint loading significantly more, compared to generalized targets [64]. Therefore, performance and portability of motion tracking and biofeedback are both important factors in retaining the decrease of joint loading, effectively in each individual patient.

The feedback effectiveness of both wearable and laboratory systems is affected by the arbitrary chosen value of $\pm 2°$. This has been a point of discussion also by Chen et al. [41] when applying a vibrotactile feedback setup driven by marker-based motion capture. Further research is required to identify optimal thresholds for the green, yellow and red zones of the visualization. Moreover, in this study we have used visualization of the target and feedback values as arrows and cones able to rotate. Alternative visualizations such as in [31] should be examined to find the ideal visualization method for gait retraining of the FPA with visual biofeedback.

Inertial sensing and augmented reality technologies have both advanced vastly in the last years. Their unique combination enables not only applications of rehabilitation, such as gait retraining, but could potentially be expanded to other fields, such as live entertainment and gaming using input from a user's own body motion to drive graphics in a mixed reality environment. Devices such as the Microsoft HoloLens utilize a number of sensors to derive its own position and orientation in space. All heavy computations needed for self-localization and visualization of the holograms in space are executed standalone in real-time. In addition, these kinematic estimates can be fused with inertial motion capture to correct any position drift introduced by the latter, for instance due to errors in the measured segment lengths. The downsides of the first commercially available and latest to date version of the headset are that it is rather heavy (approx. 0.5 kg), bulky, and impractical for use in daily life. Moreover, the field of view is currently narrow. Upcoming developments in augmented reality headsets are expected to improve the functionality for potentially unobtrusive daily life use in the future.

We evaluated the method on a treadmill, however, the wearable setup enables the application of gait retraining in overground walking that differs in terms of gait mechanics and metabolic energy cost [65]. As discussed previously the use of more sensors may reduce the drift over time, however, the system may not be comfortable for uses of long duration during daily living. Further research towards magnetically immune motion tracking systems that require fewer sensors is necessary, to achieve reliable orientation estimates regardless the environmental conditions or movement performed. Leveraging the increased practicality of fewer sensors and consistent performance over time could enable daily life applications, requiring continuous monitoring of important kinematic parameters.

In our study, awareness of the distance from the FPA target was considered an important advantage of visual feedback with respect to alternative modalities, such as tactile and auditory. Wheeler et al. compared both visual and tactile feedback, reporting that despite both being equally effective, visual feedback required less time from subjects to converge to the targeted gait pattern. However, whether quantitative information can actually boost the feedback effectiveness remains unknown. Another advantage of visual feedback compared to other modes are the gamification prospects, which could motivate subjects to perform the training in a game-like fashion. Nevertheless, further comparison studies between various feedback modalities should be performed to assess the most effective and most engaging type of biofeedback.

In our study we examined the wearable sensing and feedback in a group of young healthy adults, similarly to previous studies evaluating experimental technology for gait retraining [7, 28, 31, 32, 35, 37–39, 41, 42, 66]. However, the eventual application is targeted to patients of KOA who are generally older and less familiar with technology. Further studies to evaluate the applicability of the system in patients with KOA is required. Potential issues that may be met with the current setup is the inability of patients to perform the N-pose due to increased static knee varus/valgus type malalignment with an unknown effect to the FPA estimate. Manual input of the joint angles performed during the static calibration trial may be an appropriate solution for this. Nevertheless, the portability of the system could facilitate applications initially in clinical environments with the help of a medical specialist, and subsequently for home use. In particular, such setup allows for increased number of training sessions, which may result in enhanced training retention over time. Moreover, combining such home retraining system with telemedicine techniques could enable objective data for remote monitoring of the gait pattern of patients and identifying changes over time.

## Conclusion

This study investigated the feasibility, accuracy, and effectiveness of combining a commercially available inertial motion capture system and an augmented reality headset to perform gait retraining to alter the FPA. The findings proved sufficient accuracy of the FPA estimates with the ones obtained from optical motion capture. At the same time, average feedback effectiveness based on number of steps within a $\pm 2^\circ$ range from the target was found around 50% for both setups. The proposed setup is completely wearable and enables gait retraining applications in clinical settings without the need for a complex gait motion analysis laboratory. For daily monitoring of FPA, further developments towards reduced sensor setups with immunity to magnetic disturbances are recommended.

**Acknowledgments**
The authors thank the participants for volunteering for the study. Moreover, the authors thank Pakwan Suppaiboonsuk for the insightful initial exploration on ambulatory feedback during her Bachelor assignment and Kevin Selman for his help in understanding the Microsoft HoloLens capabilities during his Master thesis.

**Funding**
This study is part of KNEEMO Initial Training Network, funded by the European Union's Seventh Framework Programme for research, technological development, and demonstration under Grant Agreement No. 607510 (www.kneemo.eu).

**Authors' contributions**
Conceptualization: AK RR JK JvdN HMS GB JH PHV Data Curation: AK Formal Analysis: AK Funding Acquisition: GB JH PHV Investigation: AK RR JK Methodology: AK RR JK PHV Project Administration: GB JH PHV Resources: RR GB JH PHV Software: AK RR Supervision: JvdN HMS GB JH PHV Validation: AK Visualization: AK Writing - Original Draft Preparation: AK Writing - Review & Editing: AK RR JK JvdN HMS GB JH PHV. All authors read and approved the final manuscript.

**Competing interests**
Four of the authors are employed at Xsens Technologies B.V., the company that manufactures and sells the inertial motion capture system used in this study.

**Author details**
[1]Xsens Technologies B.V, Pantheon 6, 7521 PR Enschede, The Netherlands. [2]Department of Biomedical Signals and Systems (BSS), Technical Medical Centre, University of Twente, Enschede, The Netherlands. [3]Department of Rehabilitation Medicine, Amsterdam Movement Sciences, VU University Medical Center, Amsterdam, The Netherlands. [4]Academic Medical Center, Musculoskeletal Imaging Quantification Center (MIQC), Department of Radiology and Nuclear Medicine, Amsterdam Movement Sciences, Amsterdam, The Netherlands. [5]Department of Biomechanical Engineering, Delft University of Technology, Delft, The Netherlands.

**References**

1. Woolf AD, Pfleger B. Burden of major musculoskeletal conditions. Bull World Health Organ. 2003;81:646–56.
2. Felson DT, Lawrence RC, Hochberg MC, et al. Osteoarthritis: New insights. part 2: treatment approaches. Ann Intern Med. 2000;133(9): 726–37. https://doi.org/10.7326/0003-4819-133-9-200011070-00015. /data/journals/aim/19968/0000605-200011070-00015.pdf.
3. Conaghan PG, Dickson J, Grant RL. Care and management of osteoarthritis in adults: summary of nice guidance. BMJ. 2008;336(7642): 502–3. https://doi.org/10.1136/bmj.39490.608009.AD. http://www.bmj.com/content/336/7642/502.full.pdf.
4. Boutron I, Tubach F, Giraudeau B, Ravaud P. Methodological differences in clinical trials evaluating nonpharmacological and pharmacological treatments of hip and knee osteoarthritis. JAMA. 2003;290(8):1062–70. https://doi.org/10.1001/jama.290.8.1062. /data/journals/jama/4893/joc22095.pdf.
5. Sarzi-Puttini P, Cimmino MA, Scarpa R, Caporali R, Parazzini F, Zaninelli A, Atzeni F, Canesi B. Osteoarthritis: An overview of the disease and its treatment strategies. Semin Arthritis Rheum. 2005;35(1):1–10. https://doi.org/10.1016/j.semarthrit.2005.01.013. Osteoarthritis in General and Specialist Practice in Italy: The AMICA Study.
6. Reeves ND, Bowling FL. Conservative biomechanical strategies for knee osteoarthritis. Nat Rev Rheumatol. 2011;7(2):113–22.
7. Barrios JA, Crossley KM, Davis IS. Gait retraining to reduce the knee adduction moment through real-time visual feedback of dynamic knee alignment. J Biomech. 2010;43(11):2208–13. https://doi.org/10.1016/j.jbiomech.2010.03.040. NIHMS150003.
8. Fregly BJ, Reinbolt JA, Rooney KL, Mitchell KH, Chmielewski TL. Design of patient-specific gait modifications for knee osteoarthritis rehabilitation. IEEE Trans Biomed Eng. 2007;54(9):1687–95.

9. Shull PB, Jirattigalachote W, Hunt Ma, Cutkosky MR, Delp SL. Quantified self and human movement: A review on the clinical impact of wearable sensing and feedback for gait analysis and intervention. Gait Posture. 2014;40(1):11–9. https://doi.org/10.1016/j.gaitpost.2014.03.189.
10. Jackson B, Wluka A, Teichtahl A, Morris M, Cicuttini F. Reviewing knee osteoarthritis—a biomechanical perspective. Sci Med Sport. 2004;7(3):347–57.
11. Zhao D, Banks SA, Mitchell KH, D'Lima DD, Colwell CW, Fregly BJ. Correlation between the knee adduction torque and medical contact force for a variety of gait patterns. J Orthop Res. 2007;25(6):789–97. https://doi.org/10.1002/jor.20379.
12. Richards RE, van den Noort JC, van der Esch M, Booij MJ, Harlaar J. Effect of real-time biofeedback on peak knee adduction moment in patients with medial knee osteoarthritis: Is direct feedback effective? Clin Biomech. 2018;57:150–158. https://doi.org/10.1016/j.clinbiomech.2017.07.004.
13. Lin C-J, Lai K-A, Chou Y-L, Ho C-S. The effect of changing the foot progression angle on the knee adduction moment in normal teenagers. Gait Posture. 2001;14(2):85–91.
14. Mündermann A, Dyrby CO, Hurwitz DE, Sharma L, Andriacchi TP. Potential strategies to reduce medial compartment loading in patients with knee osteoarthritis of varying severity: reduced walking speed. Arthritis Rheumatol. 2004;50(4):1172–8.
15. Chang A, Hurwitz D, Dunlop D, Song J, Cahue S, Hayes K, Sharma L. The relationship between toe-out angle during gait and progression of medial tibiofemoral osteoarthritis. Ann Rheum Dis. 2007;66(10):1271–5. https://doi.org/10.1136/ard.2006.062927.
16. Guo M, Axe MJ, Manal K. The influence of foot progression angle on the knee adduction moment during walking and stair climbing in pain free individuals with knee osteoarthritis. Gait Posture. 2007;26(3):436–41.
17. Rutherford D, Hubley-Kozey C, Deluzio K, Stanish W, Dunbar M. Foot progression angle and the knee adduction moment: a cross-sectional investigation in knee osteoarthritis. Osteoarthr Cartil. 2008;16(8):883–9.
18. Simic M, Wrigley TV, Hinman RS, Hunt Ma, Bennell KL. Altering foot progression angle in people with medial knee osteoarthritis: The effects of varying toe-in and toe-out angles aremediated by pain and malalignment. Osteoarthr Cartil. 2013;21(9):1272–80. https://doi.org/10.1016/j.joca.2013.06.001.
19. Shull PB, Shultz R, Silder A, Dragoo JL, Besier TF, Cutkosky MR, Delp SL. Toe-in gait reduces the first peak knee adduction moment in patients with medial compartment knee osteoarthritis. J Biomech. 2013;46(1):122–8. https://doi.org/10.1016/j.jbiomech.2012.10.019.
20. Hunt MA, Takacs J. Effects of a 10-week toe-out gait modification intervention in people with medial knee osteoarthritis: A pilot, feasibility study. Osteoarthr Cartil. 2014;22(7):904–11. https://doi.org/10.1016/j.joca.2014.04.007.
21. Bergmann J, McGregor A. Body-worn sensor design: what do patients and clinicians want? Ann Biomed Eng. 2011;39(9):2299–312.
22. Kainz H, Graham D, Edwards J, Walsh HP, Maine S, Boyd RN, Lloyd DG, Modenese L, Carty CP. Reliability of four models for clinical gait analysis. Gait Posture. 2017;54:325–31.
23. Andersen MS, Damsgaard M, Rasmussen J. Kinematic analysis of over-determinate biomechanical systems. Comput Methods Biomech Biomed Eng. 2009;12(4):371–84. https://doi.org/10.1080/10255840802459412.
24. Luinge HJ, Veltink PH. Measuring orientation of human body segments using miniature gyroscopes and accelerometers. Med Biol Eng Comput. 2005;43(2):273–82. https://doi.org/10.1007/BF02345966.
25. Roetenberg D, Luinge H, Slycke P. Xsens MVN: full 6DOF human motion tracking using miniature inertial sensors. Xsens Motion Technol BV Tech Rep. 2009;1–9.
26. Schepers HM, Giuberti M, Bellusci G. Xsens MVN: Consistent Tracking of Human Motion Using Inertial Sensing. Tech Rep Xsens Technol B V. 2018;1–8.
27. Huang Y, Jirattigalachote W, Cutkosky MR, Zhu X, Shull PB. Novel Foot Progression Angle Algorithm Estimation via Foot-Worn, Magneto-Inertial Sensing. IEEE Trans Biomed Eng. 2016;63(11):2278–85. https://doi.org/10.1109/TBME.2016.2523512.
28. Xu J, Bao T, Lee UH, Kinnaird C, Carender W, Huang Y, Sienko KH, Shull PB. Configurable, wearable sensing and vibrotactile feedback system for real-time postural balance and gait training: proof-of-concept. J Neuroengineering Rehabil. 2017;14(1):102.
29. Richards R, van den Noort JC, Dekker J, Harlaar J. Gait retraining with real-time biofeedback to reduce knee adduction moment: Systematic review of effects and methods used. Arch Phys Med Rehabil. 2017;98(1):137–50. https://doi.org/10.1016/j.apmr.2016.07.006.
30. Hunt MA, Bennell KL. Predicting dynamic knee joint load with clinical measures in people with medial knee osteoarthritis. Knee. 2011;18(4):231–4. https://doi.org/10.1016/j.knee.2010.05.014.
31. van den Noort JC, Steenbrink F, Roeles S, Harlaar J. Real-time visual feedback for gait retraining: toward application in knee osteoarthritis. Med Biol Eng Comput. 2014;275–86. https://doi.org/10.1007/s11517-014-1233-z.
32. Hunt MA, Simic M, Hinman RS, Bennell KL, Wrigley TV. Feasibility of a gait retraining strategy for reducing knee joint loading: Increased trunk lean guided by real-time biofeedback. J Biomech. 2011;44(5):943–7. https://doi.org/10.1016/j.jbiomech.2010.11.027.
33. Hunt MA, Takacs J, Hart K, Massong E, Fuchko K, Biegler J. Comparison of Mirror, Raw Video, and Real-Time Visual Biofeedback for Training Toe-Out Gait in Individuals With Knee Osteoarthritis. Arch Phys Med Rehabil. 2014;95(10):1912–7. https://doi.org/10.1016/j.apmr.2014.05.016.
34. Segal NA, Glass NA, Teran-Yengle P, Singh B, Wallace RB, Yack HJ. Intensive Gait Training for Older Adults with Symptomatic Knee Osteoarthritis. Am J Phys Med Rehabil. 2015;1. https://doi.org/10.1097/PHM.0000000000000264.
35. Pizzolato C, Reggiani M, Saxby DJ, Ceseracciu E, Modenese L, Lloyd DG. Biofeedback for gait retraining based on real-time estimation of tibiofemoral joint contact forces. IEEE Trans Neural Syst Rehabil Eng. 2017;25:1612–1621.
36. Richards R, van der Esch M, van den Noort JC, Harlaar J. The learning process of gait retraining using real-time feedback in patients with medial knee osteoarthritis. Gait Posture. 2018;62:1–6.
37. Wheeler JW, Shull PB, Besier TF. Real-Time Knee Adduction Moment Feedback for Gait Retraining Through Visual and Tactile Displays. J Biomech Eng. 2011;133(4):041007. https://doi.org/10.1115/1.4003621.
38. Shull PB, Lurie KL, Cutkosky MR, Besier TF. Training multi-parameter gaits to reduce the knee adduction moment with data-driven models and haptic feedback. J Biomech. 2011;44(8):1605–9. https://doi.org/10.1016/j.jbiomech.2011.03.016.
39. Dowling AV, Fisher DS, Andriacchi TP. Gait Modification via Verbal Instruction and an Active Feedback System to Reduce Peak Knee Adduction Moment. J Biomech Eng. 2010;132(7):071007. https://doi.org/10.1115/1.4001584.
40. Shull PB, Silder A, Shultz R, Dragoo JL, Besier TF, Delp SL, Cutkosky MR. Six-week gait retraining program reduces knee adduction moment, reduces pain, and improves function for individuals with medial compartment knee osteoarthritis. J Orthop Res. 2013;31(7):1020–5. https://doi.org/10.1002/jor.22340.
41. Chen DKY, Haller M, Besier TF. Wearable lower limb haptic feedback device for retraining foot progression angle and step width. Gait Posture. 2017;55. Supplement C:177–83. https://doi.org/10.1016/j.gaitpost.2017.04.028.
42. Ferrigno C, Stoller IS, Shakoor N, Thorp LE, Wimmer MA. The feasibility of using augmented auditory feedback from a pressure detecting insole to reduce the knee adduction moment: a proof of concept study. J Biomech Eng. 2016;138(2):021014.
43. Seuter M, Opitz L, Bauer G, Hochmann D. Live-feedback from the imus: Animated 3d visualization for everyday-exercising. In: Proceedings of the 2016 ACM International Joint Conference on Pervasive and Ubiquitous Computing: Adjunct. ACM; 2016. p. 904–7.
44. The leader in Mixed Reality Technology – HoloLens. https://www.microsoft.com/en-us/hololens. Accessed 25 Aug 2017.
45. Geijtenbeek T, Steenbrink F, Otten B, Even-Zohar O. D-flow: immersive virtual reality and real-time feedback for rehabilitation. In: Proceedings of the 10th International Conference on Virtual Reality Continuum and Its Applications in Industry. ACM; 2011. p. 201–8.
46. Van den Bogert AJ, Geijtenbeek T, Even-Zohar O, Steenbrink F, Hardin EC. A real-time system for biomechanical analysis of human movement and muscle function. Med Biol Eng Comput. 2013;51(10):1069–77.
47. Xsens MVN Analyze - Products - Xsens 3D motion tracking. https://www.xsens.com/products/xsens-mvn-analyze/. Accessed 23 Dec 2017.
48. Xsens MTw Awinda - Products - Xsens 3D motion tracking. https://www.xsens.com/products/mtw-awinda/. Accessed 07 Apr 2018.
49. Xsens MVN User Manual. https://xsens.com/download/usermanual/3DBM/MVN_User_Manual.pdf. Accessed 01 May 2017.

50. Billboarding and tag-along - Windows Mixed Reality - Microsoft. https://developer.microsoft.com/en-us/windows/mixed-reality/billboarding_and_tag-along. Accessed 24 Jan 2018.
51. Growney E, Meglan D, Johnson M, Cahalan T, An K-N. Repeated measures of adult normal walking using a video tracking system. Gait Posture. 1997;6(2):147–62.
52. Kadaba M, Ramakrishnan H, Wootten M, Gainey J, Gorton G, Cochran G. Repeatability of kinematic, kinetic, and electromyographic data in normal adult gait. J Orthop Res. 1989;7(6):849–60.
53. Steinwender G, Saraph V, Scheiber S, Zwick EB, Uitz C, Hackl K. Intrasubject repeatability of gait analysis data in normal and spastic children. Clin Biomech. 2000;15(2):134–9.
54. Tsushima H, Morris ME, McGinley J. Test-retest reliability and inter-tester reliability of kinematic data from a three-dimensional gait analysis system. J Jpn Phys Ther Assoc. 2003;6(1):9–17.
55. McGinley JL, Baker R, Wolfe R, Morris ME. The reliability of three-dimensional kinematic gait measurements: a systematic review. Gait Posture. 2009;29(3):360–9.
56. Eve L, McNee A, Shortland A. Extrinsic and intrinsic variation in kinematic data from the gait of healthy adult subjects. Gait Posture. 2006;24:56–57.
57. Gorton G, Hebert D, Goode B. Assessment of the kinematic variability between twelve shriners motion analysis laboratories part 2: short-term follow up. Gait Posture. 2002;16(S1):65–6.
58. Gorton GE, Hebert DA, Gannotti ME. Assessment of the kinematic variability among 12 motion analysis laboratories. Gait Posture. 2009;29(3):398–402. https://doi.org/10.1016/j.gaitpost.2008.10.060.
59. Murphy A, McGinley J, Tirosh O. Reliability of kinematic gait measurements in adult hemiplegic stroke. Proc 12th Ann Gait Clin Mov Anal Soc. 2007.
60. Schwartz MH, Trost JP, Wervey RA. Measurement and management of errors in quantitative gait data. Gait Posture. 2004;20(2):196–203.
61. Xia H, Xu J, Wang J, Hunt MA, Shull PB. Validation of a smart shoe for estimating foot progression angle during walking gait. J Biomech. 2017;61(Supplement C):193–8. https://doi.org/10.1016/j.jbiomech.2017.07.012.
62. de Vries WHK, Veeger HEJ, Baten CTM, van der Helm FCT. Magnetic distortion in motion labs, implications for validating inertial magnetic sensors. Gait Posture. 2009;29(4):535–41. https://doi.org/10.1016/j.gaitpost.2008.12.004.
63. Hill P, Group ISBS. Practical Software Project Estimation: A Toolkit for Estimating Software Development Effort & Duration: McGraw-Hill Education; 2010. https://books.google.gr/books?id=mW6phtfWZ-EC.
64. Uhlrich SD, Silder A, Beaupre GS, Shull PB, Delp SL. Subject-specific toe-in or toe-out gait modifications reduce the larger knee adduction moment peak more than a non-personalized approach. J Biomech. 2018;66:103–10.
65. Berryman N, Gayda M, Nigam A, Juneau M, Bherer L, Bosquet L. Comparison of the metabolic energy cost of overground and treadmill walking in older adults. Eur J Appl Physiol. 2012;112(5):1613–20. https://doi.org/10.1007/s00421-011-2102-1.
66. Clark RA, Pua Y-H, Bryant AL, Hunt MA. Validity of the Microsoft Kinect for providing lateral trunk lean feedback during gait retraining. Gait Posture. 2013;38(4):1064–6. https://doi.org/10.1016/j.gaitpost.2013.03.029.

# 9

# Home-based transcranial direct current stimulation plus tracking training therapy in people with stroke

Ann Van de Winckel[1*], James R. Carey[1], Teresa A. Bisson[2], Elsa C. Hauschildt[3], Christopher D. Streib[4] and William K. Durfee[3]

**Abstract**

**Background:** Transcranial direct current stimulation (tDCS) is an effective neuromodulation adjunct to repetitive motor training in promoting motor recovery post-stroke. Finger tracking training is motor training whereby people with stroke use the impaired index finger to trace waveform-shaped lines on a monitor. Our aims were to assess the feasibility and safety of a telerehabilitation program consisting of tDCS and finger tracking training through questionnaires on ease of use, adverse symptoms, and quantitative assessments of motor function and cognition. We believe this telerehabilitation program will be safe and feasible, and may reduce patient and clinic costs.

**Methods:** Six participants with hemiplegia post-stroke [mean (SD) age was 61 (10) years; 3 women; mean (SD) time post-stroke was 5.5 (6.5) years] received five 20-min tDCS sessions and finger tracking training provided through telecommunication. Safety measurements included the Digit Span Forward Test for memory, a survey of symptoms, and the Box and Block test for motor function. We assessed feasibility by adherence to treatment and by a questionnaire on ease of equipment use. We reported descriptive statistics on all outcome measures.

**Results:** Participants completed all treatment sessions with no adverse events. Also, 83.33% of participants found the set-up easy, and all were comfortable with the devices. There was 100% adherence to the sessions and all recommended telerehabilitation.

**Conclusions:** tDCS with finger tracking training delivered through telerehabilitation was safe, feasible, and has the potential to be a cost-effective home-based therapy for post-stroke motor rehabilitation.

**Keywords:** Stroke, Neurological rehabilitation, Telerehabilitation, Transcranial direct current stimulation, Physical therapy

## Background

Post-stroke motor function deficits stem not only from neurons killed by the stroke, but also from down-regulated excitability in surviving neurons remote from the infarct [1]. This down-regulation results from deafferentation [2], exaggerated interhemispheric inhibition [3], and learned non-use [4]. Current evidence suggests that post-stroke motor rehabilitation therapies should encourage upregulating neurons and should target neuroplasticity through intensive repetitive motor practice [5, 6]. Previously, our group has examined the feasibility and efficacy of a custom finger tracking training program as a way of providing people with stroke with an engaging repetitive motor practice [7–9]. In this program, the impaired index finger is attached to an electro-goniometer, and participants repeatedly move the finger up and down to follow a target line that is drawn on the display screen. In successive runs, the

* Correspondence: avandewi@umn.edu
[1]Division of Physical Therapy; Division of Rehabilitation Science, University of Minnesota, 420 Delaware Street SE (MMC388), Minneapolis, MN 55455, USA
Full list of author information is available at the end of the article

shape, frequency and amplitude of target line is varied, which forces the participant to focus on the tracking task. In one study, we demonstrated a 23% improvement in hand function (as measured by the Box and Block test; minimal detectable change is 18% [10]) after participants with stroke completed the tracking training program [9]. While our study did not evaluate changes in activity in daily life (ADL) or quality of life (because efficacy of the treatment was not the study objective), the Box and Block test is moderately correlated ($r$ = 0.52) to activities in daily life and quality of life ($r$ = 0.59) [11]. In addition, using fMRI, we showed that training resulted in an activation transition from ipsilateral to contralateral cortical activation in the supplementary motor area, primary motor and sensory areas, and the premotor cortex [9].

Recently, others have shown that anodal transcranial direct current stimulation (tDCS) can boost the beneficial effects of motor rehabilitation, with the boost lasting for at least 3 months post-training [12]. Also, bihemispheric tDCS stimulation (anodal stimulation to excite the ipsilateral side and cathodal stimulation to downregulate the contralateral side) in combination with physical or occupational therapy has been shown to provide a significant improvement in motor function (as measured by Fugl-Meyer and Wolf Motor Function) compared to a sham group [13]. Further, a recent meta-analysis of randomized-controlled trials comparing different forms of tDCS shows that cathodal tDCS is a promising treatment option to improve ADL capacity in people with stroke [14]. Compared to transcutaneous magnetic stimulation (TMS), tDCS devices are inexpensive and easier to operate. Improvement in upper limb motor function can appear after only five tDCS sessions [15], and there are no reports of serious adverse events when tDCS has been used in human trials for periods of less than 40 min at amplitudes of less than 4 mA [16].

Moreover, tDCS stimulation task also seems beneficial for other impairments commonly seen in people post-stroke. Stimulation with tDCS applied for 20 sessions of 30 min over a 4-week period has been shown to decrease depression and improve quality of life in people after a stroke [17, 18]. Four tDCS sessions for 10 min applied over the primary and sensory cortex in eight patients with sensory impairments more than 10 months post-stroke enhanced tactile discriminative performance [19]. Breathing exercises with tDCS stimulation seems to be more effective than without stimulation in patient with chronic stroke [20], and tDCS has shown promise in treating central post-stroke pain [21]. Finally, preliminary research on the effect of tDCS combined with training on resting-state functional connectivity shows promise to better understand the mechanisms behind inter-subject variability regarding tDCS stimulation [22].

Motor functional outcomes in stroke have declined at discharge from inpatient rehabilitation facilities [23, 24], likely a result of the pressures to reduce the length of stay at inpatient rehabilitation facilities as part of a changing and increasingly complex health care climate [25, 26]. Researchers, clinicians, and administrators continue to search for solutions to facilitate and post-stroke rehabilitation after discharge. Specifically, there has been considerable interest in low-cost stroke therapies than can be administered in the home with only a modest level of supervision by clinical professionals.

Home telerehabilitation is a strategy in which rehabilitation in the patient's home is guided remotely by the therapist using telecommunication technology. If patients can safely apply tDCS to themselves at home, combining telerehabilitation with tDCS would be an easy way to boost therapy without costly therapeutic face-to-face supervision. For people with multiple sclerosis, the study of Charvet et al. (2017) provided tDCS combined with cognitive training, delivered through home telerehabilitation, and demonstrated greater improvement on cognitive measures compared to those who received just the cognitive training [27]. The authors demonstrated the feasibility of remotely supervised, at-home tDCS and established a protocol for safe and reliable delivery of tDCS for clinical studies [28]. Some evidence shows that telerehabilitation approaches are comparable to conventional rehabilitation in improving activities of daily living and motor function for stroke survivors [29, 30], and that telemedicine for stroke is cost-effective [31, 32]. A study in 99 people with stroke receiving training using telerehabilitation (either with home exercise program or robot assisted therapy with home program) demonstrated significant improvements in quality of life and depression [33].

A recent search of the literature suggests that to date, no studies combine tDCS with repetitive tracking training in a home telerehabilitation setting to determine whether the combination leads to improved motor rehabilitation in people with stroke. Therefore, the aim of this pilot project was to explore the safety, usability and feasibility of the combined system. For the tDCS treatment, we used a bihemispheric montage with cathodal tDCS stimulation to suppress the unaffected hemisphere in order to promote stroke recovery [34–37]. For the repetitive tracking training therapy, we used a finger tracking task that targets dexterity because 70% of people post-stroke are unable to use their hand with full effectiveness after stroke [38]. Safety was assessed by noting any decline of 2 points or more in the cognitive testing that persists over more than 3 days. We expect day to day variations of 1 digit. Motor decline is defined by a decline of 6 blocks on the Box and Block test due to muscle weakness. This is based on the minimal detectable change (5.5 blocks/min) [10]. The standard error of measurement is at least 2 blocks for the paretic and stronger side. We expect possible variations in muscle tone that could influence the scoring of the test. Usability was assessed through a questionnaire and by observing whether the participant, under remote supervision, could don the apparatus and complete the therapy sessions. Our intent was to set the stage for a future clinical trial to determine the efficacy of this approach.

## Methods

### Participants

Participants were recruited from a database of people with chronic stroke who had volunteered for previous post-stroke motor therapy research studies at the University of Minnesota. Inclusion criteria were: at least 6 months post-stroke; at least 10 degrees of active flexion and extension motion at the index finger; awareness of tactile sensation on the scalp; and a score of greater than or equal to 24 (normal cognition) on the Mini-Mental State Examination (MMSE) to be cognitively able to understand instructions to don and use the devices [39]. We excluded those who had a seizure within past 2 years, carried implanted medical devices incompatible with tDCS, were pregnant, had non-dental metal in the head or were not able to understand instructions on how to don and use the devices. The study was approved by the University of Minnesota IRB and all enrolled participants consented to be in the study.

### Apparatus

tDCS was applied using the StarStim Home Research Kit (NeuroElectrics, Barcelona, Spain). The StarStim system consists of a Neoprene head cap with marked positions for electrode placement, a wireless cap-mounted stimulator and a laptop control computer. Saline-soaked, 5 cm diameter sponge electrodes were used. For electrode placement, we followed a bihemispheric montage [14] involving cathodal stimulation on the unaffected hemisphere with the anode positioned at C3 and the cathode at C4 for participants with left hemisphere stroke, and vice versa for participants with right hemisphere stroke. Stimulation protocols were set by the investigator on a web-based application that communicated with the tDCS control computer. A remote access application (TeamViewer) was also installed on the control computer, as was a video conferencing application (Skype).

The repetitive finger tracking training system was a copy of what we used in our previous stroke studies [7–9]. The apparatus included an angle sensor mounted to a lightweight brace and aligned with the metacarpophalangeal (MCP) joint of the index finger, a sensor signal conditioning circuit, and a target tracking application loaded on a table computer. Figure 1 shows a participant using the apparatus during a treatment session.

### Assessment measures

We collected demographic information (age, sex, and distance to the university) for each participant. Finger and wrist flexor spasticity was measured with the Modified Ashworth Scale [40], cognitive impairment with the MMSE, physical impairment with the Upper Extremity Fugl-Meyer score [41], and handedness prior to stroke with the Edinburgh Handedness Inventory [42]. The assessments were conducted by physical therapists who were experienced in the proper application of the measurement instruments.

Motor function was assessed before and after treatment using a 60 s trial of the Box and Block Test [43], which assesses rapid grasp and release of single blocks. Cognitive function was assessed before and after treatment using the Digit Span Forward Test [44], which quantifies the largest sequence of numbers the participant can repeat without errors after being recited by the investigator.

Adverse effects of the tDCS were monitored by asking participants whether they experienced any of the following symptoms since the preceding treatment: scalp pain, headache, neck pain, dental pain, tingling, nausea, itching, burning sensation, skin redness, open lesion on skin, abnormal sleep, anxiety, difficulty concentrating, dizziness, impaired memory, altered mood, altered balance, impaired use of the strong hand, or any other problem [45].

**Fig. 1** Participant with right hemiparesis receiving transcranial direct current magnetic stimulation (tDCS) in their home simultaneous while performing the finger movement tracking task on the tracking computer (left). The tDCS computer (right) shows the supervising investigator, located off-site, who communicated with the participant through the video conferencing application, controlled the tDCS stimulator through web-based software, and controlled the tracking protocols. (Permission was obtained from the participant for the publication of this picture)

Motor function was measured before the first treatment session and after the last treatment session. Cognitive function and adverse effects were measured before each treatment session so that in the event of a cognitive decline or a report of adverse effects, further sessions would be withheld immediately.

Usability was measured by through a post-treatment questionnaire and by observing over the video link the participant interacting with the apparatus.

### Protocol

For this study we followed a protocol that aligned with the guidelines for remote tDCS application suggested by Chavret et al. [46]. These include: (1) training of staff in tDCS treatment and supervision; (2) assessment of the user's capability to participate in tDCS remotely; (3) ongoing training procedures and materials including assessments of the user and/or caregiver; (4) simple and fail-safe electrode preparation techniques and tDCS headgear; (5) strict dose control for each session; (6) ongoing monitoring to quantify compliance (device preparation, electrode saturation/placement, stimulation protocol), with corresponding corrective steps as required; (7) monitoring for treatment-emergent adverse effects; (8) procedures for discontinuation of a session or study participation including emergency failsafe procedures tailored to the treatment population's level of need. We included Guidelines 1–3 and 8 in our protocol and had questionnaires and procedures in place to identify any potential adverse events and discontinue any session in case of adverse events prior or during the session. We trained patients to address Guideline 4; the dose (Guideline 5) was controlled by the therapist; we established ongoing monitoring (Guidelines 6–7) for compliance and potential adverse events.

Our study had two scenarios for treatment sessions. Under the first scenario, treatment sessions took place at the university with the supervising investigator in one room and the participant in a separate room to simulate the condition of being at home. We did this with the first 3 participants to be close to emergency services should there be any adverse effects. Under the second scenario, the investigator was at the university and the participant was at home. For both scenarios, the investigator and participant communicated through the video conferencing application. The investigator controlled the treatment applications through the remote access application. Under both scenarios, a second investigator, the observer, was with the participant at all times. The role of the observer was to monitor for adverse events and provide immediate assistance if needed. All instruction and communication with the participant was done by investigator to accurately represent the conditions of a future home-based clinical treatment session.

Figure 2 illustrates the study design and timeline. The initial in-person session involved baseline testing followed by training in how to use the tDCS and tracking training apparatus. The training included how to don the cap so that the tDCS electrodes ended up in the correct locations. The investigator first found and marked the reference point Cz (International 10/20 system for locating scalp electrodes) on the participant's head by determining the intersection of the line between the nasion and the inion and the line connecting the left and right auricular. The cap was donned and adjusted so that the marked Cz hole in the cap was aligned with the Cz mark. The distance between the front edge of the cap and the eyebrows was noted and this served as the indicator that the cap was positioned properly in future sessions, as monitored by the participant and by the investigator.

The participant was trained to soak the sponge electrodes with saline and secure them in the C3 and C4 marked holes in the cap. Saline-soaked sponge electrodes are standard for tDCS. Sponges were wet but not dripping. The color-coded lead wires were then attached so that the anode was positioned at C3 and the cathode at C4 for participants with left hemisphere stroke, and vice versa for participants with right hemisphere stroke. During the treatment sessions, the participant was guided through these steps by the investigator over the

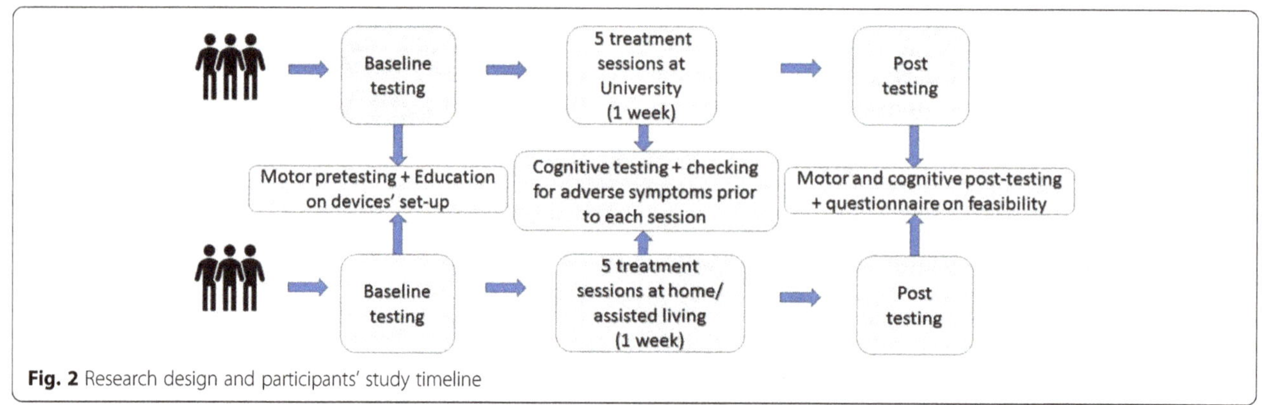


**Fig. 2** Research design and participants' study timeline

video conferencing application. Additionally, we asked the patient whether they felt some discomfort at the location of the sponges. A caretaker was called upon to look at the head and screen for redness. If the patient lived alone, the patient checked the scalp with a mirror or through taking a picture on their phone, and then inspecting the picture. The participant was also trained in how to apply the hand brace containing the angle sensor to the paretic hand so that the sensor was centered at the lateral side of MCP joint of their index finger. The complete startup procedure was repeated until the participant could re-apply these devices independently with remote guidance, if necessary, from the investigator.

Participants then had 5 treatment sessions either at the university (Fig. 2, top row) or at home (Fig. 2, bottom row). As much as possible, these 5 treatments occurred over consecutive weekdays. The investigator initiated the treatment session by calling the participant's cell phone. The investigator first conducted the Digit Span Forward Test and the survey of symptoms over the phone. The investigator then instructed the participant to turn on the tDCS control computer and the tracking training tablet computer. Once both computers were connected to the internet, the investigator could gain control using the remote access application and could communicate with the participant via video.

Next, the investigator screened the scalp for possible redness or lesion from the previous treatment by having the participant lower their head to be in view of the computer's camera and adjust their hair to give a better view of the scalp. The investigator gave further instructions until they were satisfied that all relevant parts of the scalp were assessed. As a safety check, the observer, who was in the same room as the participant, also checked the scalp. The person in the room was only an observer; not a participant. The scalp check did not influence the end result.

The participant then prepared the electrodes and donned the head cap. The investigator viewed the cap position using the video link and if needed guided the participant to make any adjustments. The investigator then remotely active the tDCS software to conduct an impedance check, and if the impedance was too high, coached the participant through taking off the cap, re-wetting the electrodes and putting the cap back on. The impedance check is automatically done by the StarStim system with the threshold set at approximately 10 kohm. Once the impedance check was passed, the investigator initiated the tDCS treatment, which was 20 min at 1.5 mA, including a 30 s ramp-up and ramp-down [13].

The investigator then shifted the participant's attention to donning the finger angle sensor and tablet running the tracking training application. The investigator initiated the application remotely and for approximately 20 min, while simultaneously receiving tDCS, the participant repeatedly extended and flexed the paretic index finger to move the computer-screen cursor as accurately as possible along various target tracks (Fig. 1). Each tracking trial was 5 to 20 s and the investigator remotely adjusted the parameters, including waveform (round, pointy or square waves), frequency (number of waves), amplitude (height of waves), polarity (how high up or down the waves went), and trial duration (how fast the cursor went), to keep the participant challenged and motivated. At the end of each trial, the screen displayed a performance score, related to an accuracy index [47], which provided further motivation to improve tracking accuracy.

At the end of 20 min of tDCS and tracking training therapy, the investigator guided the participant in removing the cap, and then checked the scalp for irritation. The observer also inspected the scalp. The investigator then thanked the participant who powered down the computers and stowed the apparatus, concluding the session for that day. Following the five treatment sessions, the participant returned to the university for the Box and Block and Digit Span Forwards post-treatment tests.

## Results

Six people with chronic stroke participated in the study (3 women; mean (standard deviation, SD) age 61 (10) years; mean (SD) time post-stroke 5.5 (6.5) years; 5 with left hemiplegia due to ischemic stroke; 1 with right hemiplegia due to hemorrhagic stroke.) Table 1 shows the demographic data and stroke characteristics for each participant. The first three participants had their treatment sessions at the university while the second three had their treatment sessions at home. Participants lived between 5 and 20 miles from the university.

All six participants completed five sessions (30 sessions total). Table 2 shows the Box and Block pre- and post-test results for the paretic and non-paretic hands, and the pre- and post-test Digit Span Forward test results; as well as the day-to day variations in the Digit Span Forward test.

There were no meaningful changes in motor or cognitive function except for Participant 2 who showed a 50% reduction in the paretic hand Box and Block Test at posttest. When questioned 2 days after the post-test, the participant stated that their spasticity typically varies throughout the day, and happened to be stronger than usual at the post-test assessment, which worsened their finger dexterity. The participant did not attribute the decrement to the tDCS and felt that their finger dexterity was at its typical level later that day.

Over the five sessions, no participants reported adverse symptoms before or after the treatment, except

**Table 1** Participant Stroke Characteristics and Treatment Information

| ID | Age (years) | Sex | Stroke (years) | Stroke Type and Location | Hemi side | Edinb. | UEFM score | MAS score | MMSE score |
|---|---|---|---|---|---|---|---|---|---|
| 1 | 46 | F | 0.75 | Ischemic, BG, insula, posteroinferior FL, TL | Left | Right | 31 | 2 | 30 |
| 2 | 60 | F | 18 | Ischemic, MCA | Left | Mixed | 38 | 2 | 30 |
| 3 | 67 | M | 6 | Ischemic, lacunar infarct of posterior lentiform nucleus, CR | Left | Right | 45 | 1 | 30 |
| 4 | 70 | M | 1.3 | Ischemic, MCA | Left | Right | 48 | 1 | 29 |
| 5 | 72 | F | 2 | Ischemic, CR | Left | Right | 51 | 1 | 29 |
| 6 | 51 | M | 5 | Hemorrhagic, Thalamus, BS | Right | Right | 55 | 0 | 29 |

*BG* Basal ganglia, *BS* brain stem, *CR* corona radiata, *Edinb* Edinburgh Handedness Inventory; *F* female, *FL* frontal love, *Hemi side* hemiplegic side, *M* male, *MCA* middle cerebral artery distribution, *UEFM* Upper Extremity Fugl-Meyer, *MAS* Modified Ashworth Scale, *MMSE* Mini-Mental State Examination, *TL* temporal lobe

for the brief, mild, tingly sensation at the electrode sites at the beginning of each treatment in all but one patient who had thick hair. A temporary feeling of tingling is expected. Neither the investigator nor the observer detected reddening of the scalp in any session.

All participants were able to don the cap and adjust to the proper location. After the initial training, they received an illustration showing the proper location of the electrodes on the cap and the cap on the head. During the initial training, landmarks (e.g. cap right above the eye brows on a particular skin line) were given to the patient. After this initial training, no prompting was needed. The observer was only in the room to ensure safety and intervention with donning and doffing the cap was not needed for any session.

From the post-treatment usability questionnaire (Table 3), 5 of the 6 participants found the set-up easy, all six were comfortable with the devices, and all 6 would recommend the telerehabilitation program to others. One participant reported difficulty in setting up the equipment and being uncomfortable in working with computer technology. This participant had good cognitive ability and did not score the lowest in motor function. The same participant, however, would recommend the treatment therapy to others and thought that the treatment was feasible.

The investigator was able to guide all of the treatment sessions without major incidents. On some occasions, the internet connection was disrupted. When this happened, the participant reconnected the computers and the trial was repeated. Temporary loss of internet

**Table 2** Cognitive and motor scores pre- and post-treatment and adverse symptoms reported before each session; and daily Digit Span testing prior to tDCS stimulation

| ID | Paretic Hand Box and Block Test (number of blocks) | | Non-paretic Hand Box and Block Test (number of blocks) | | Digit Span Forward Test (number of digits) | | | | | Adverse Symptoms Reported |
|---|---|---|---|---|---|---|---|---|---|---|
| | Baseline | Posttest | Baseline | Posttest | Baseline | | Posttest | | | |
| 1 | 4 | 3 | 70 | 73 | 15 | | 14 | | | None |
| | | | Daily Digit Span test | | 15 | 13 | 14 | 14 | 14 | |
| 2 | 10 | 5 | 50 | 49 | 6 | | 6 | | | None |
| | | | Daily Digit Span test | | 6 | 7 | 8 | 7 | 6 | |
| 3 | 17 | 21 | 52 | 56 | 9 | | 9 | | | None |
| | | | Daily Digit Span test | | 9 | 9 | 6 | 9 | 9 | |
| 4 | 15 | 19 | 52 | 55 | 8 | | 11 | | | None |
| | | | Daily Digit Span test | | 8 | 9 | 9 | 9 | 11 | |
| 5 | 27 | 28 | 72 | 71 | 10 | | 10 | | | None |
| | | | Daily Digit Span test | | 10 | 10 | 11 | 11 | 10 | |
| 6 | 36 | 38 | 67 | 78 | 12 | | 15 | | | None |
| | | | Daily Digit Span test | | 12 | 12 | 10 | 13 | 15 | |
| Mean (SD) | 18.17 (11.62) | 19.00 (13.40) | 60.05 (10.19) | 63.67 (11.79) | 10.00 (3.16) | | 10.83 (3.31) | | | |
| % improvement - Total Group | | 4.59% | | 5.23% | | | 8.33% | | | |
| - Treatment responders (#3–6) | | 11.58% | | 7.00% | | | 15.38% | | | |

**Table 3** Feasibility questionnaire: Summary of the participants' responses

| | | | | | |
|---|---|---|---|---|---|
| 1. 1. How difficult/easy was it for you to set up the equipment project? | Very Difficult | Difficult | Neutral | Easy | Very Easy |
| | 0 | 1 | 0 | 4 | 1 |
| 2. How comfortable were you at the beginning in working with computer technology? | Very uncomfortable | Uncomfortable | Neutral | Comfortable | Very Comfortable |
| | 0 | 1 | 0 | 0 | 5 |
| 3. 1. How comfortable are you now in working with computer technology for your rehabilitation? | Very uncomfortable | Uncomfortable | Neutral | Comfortable | Very Comfortable |
| | 0 | 1 | 0 | 0 | 5 |
| 4. 1. To what extent would you recommend tDCS with telerehabilitation to another person with stroke? | Definitely Not Recommend | Not Recommend | Neutral | Recommend | Definitely Recommend |
| | 0 | 0 | 0 | 2 | 4 |
| 5. 1. What is your overall opinion on the feasibility (i.e. ease and capability) of using tDCS with telerehabilitation? | Definitely Not Feasible | Doubtfully Feasible | Neutral | Somewhat Feasible | Highly Feasible |
| | 0 | 0 | 0 | 1 | 5 |

Number in each cell indicates number of participants selecting that response

connectivity occurred in eight of 30 sessions, and included Participants 1 and 3 at the university and Participant 5 who used the guest network in the community room of their assisted living facility. This happened during the setup phase only; the actual treatment phase was never affected. When internet connectivity problems occurred, the participants could always re-establish the connection after one to three attempts.

A high impedance error trigger occurred in six of the 30 treatments. The impedance check happens before the tDCS treatment is initiated, and the program would not start or continue unless safe impedance was assured. Re-wetting the electrodes solved this problem in each instance. Multiple wetting of the sponges was only needed for one participant who had thick hair.

## Discussion

This study explored the safety and feasibility of applying tDCS in combination with a finger tracking task through telerehabilitation in people with stroke. The results showed no adverse events attributable to tDCS, and all participants successfully completed the five treatment sessions. The role of the observer was minimal, as there were no safety concerns, although the presence of the observer may have influenced the participant to complete all five sessions. As shown by the results of the usability questionnaire and by the observations of the investigator, participants found the apparatus relatively easy to use, including donning the head cap and the finger sensor.

Temporary loss of internet connectivity was the main difficulty encountered; a problem that can be avoided by embedded a cellular interface into the device for internet access or by restricting the therapy to those with reliable internet connections. While Skype was used for this study, future trials will use a secure conferencing application such as VSee Messenger. A second problem was the occasional high impedance error. This is normal for electrotherapy treatments and with proper training, users can learn how to fix the error by re-soaking the electrodes with saline.

Three different electrode montages are possible for tDCS in stroke: anodal in the affected hemisphere; cathodal in the unaffected hemisphere; or combined anodal/cathodal (bihemispheric) [3, 48]. We chose the cathodal stimulation in the unaffected hemisphere to correct the interhemispheric imbalance after stroke by suppressing over-activation in the unaffected hemisphere. While safety with tDCS has been demonstrated with intensity levels up to 4 mA for less than 40 min [49, 50], the choice location of the anode and cathode has been varied over different studies [14]. Cathodal tDCS, however seems to be the most promising treatment option to improve ADL capacity in people with stroke [14].

Variations among participants, age, the time and type of stroke lesion, as well as type of training delivered in conjunction with tDCS, all can influence the effectiveness of tDCS therapy [15, 48, 51, 52]. Rabadi and Aston (2017) demonstrated large effect size improvements in motor function in eight participants with severe motor impairments after acute stroke after applying tDCS for 30 min with 3 h of inpatient rehabilitation therapy, compared to a control group who received sham tDCS and therapy [53]. Several studies have demonstrated retained improved motor abilities between 3 weeks and 3 months post-intervention in chronic stroke after a treatment that combined physical therapy with tDCS stimulation [12, 54, 55].

Unlike transcranial magnetic stimulation (TMS), tDCS technology is simple and potentially low-cost, even for cloud-connected versions. Devices for tDCS therapy have been commercialized and prices should continue to drop if tDCS therapy is adopted as a standard of care. Further, with proper training in the use of the device and occasional remote check-ins with a therapist, we believe that supervised home use of tDCS will be safe, feasible and affordable.

Telerehabilitation offers the opportunity for practice at home as well as reaching people who live remotely. Although geographic data are not available for rehabilitation services, some data are available regarding stroke centers and stroke consultations. For example, in rural areas in northeastern states only 44% of the population had access to stroke centers within 30 miles, compared to 92.3% of the population in urban areas. These states are already implementing telemedicine-enabled stroke consultation [56]. In Minnesota, physical therapists can use telemedicine in real-time or as store-and-forward system for patients under Medicaid. Doing exercises at home with regular therapist check-ins will likely increase adherence to a home exercise therapy program, and may improve the chance of improved outcomes in motor function post-stroke. Another advantage of telerehabilitation is the time and cost saved for the patient not having to drive to the rehabilitation center, pay for parking, and for one-on-one therapy services. This savings is especially relevant for patients living in rural and remote communities. With telerehabilitation, the clinic also realizes savings by reducing costs associated with room set-up, intake, supplies, and therapist time beyond that associated with the remote check-in.

Participants in this study neither declined nor improved in motor and cognitive function. Because the long-term goal is to use tDCS plus tracking training as a post-stroke therapy tool, efficacy must be demonstrated, which means showing improved motor function when compared to tracking training alone. The reason no improved in motor function occurred in this study is it is likely that each treatment session was too short and there were an insufficient number of treatment sessions. Previous finger tracking therapy studies showed improved motor skills after 18 to 20 treatment sessions, each 45 to 60 min, a far more intensive paradigm than what we used for this study [7–9]. Combining tDCS stimulation with motor learning therapy, especially one that targets precise finger control, may enhance its impact on functional recovery. Recently, more robot and virtual reality games have provided gloves or other devices for these hand and finger movements to be trained more intensively [57, 58]. Clinical trials should use multivariate models capturing different patient baseline characteristics to predict which patients would respond to treatment. This will help develop a targeted, individualized brain stimulation therapy for patients with stroke [59]. For those studies, the minimum selection criteria for participants who could benefit from the therapy would be the ability to don the tDCS electrodes and finger tracker hardware either themselves or with the aid of a caregiver, and the cognitive ability to remember the treatment procedure and safety steps. The ability to reliably place the electrodes in their proper location will depend on the particular tDCS system chosen for the study. Additionally, participants must have a minimum ability to move the finger, which for this study was set at 10 degrees, but could be set lower for future studies. In our study, the participants were cognitively high functioning, but according to Woytowicz et al. (2017), our patients are categorized with moderate to mild upper limb impairments, with visible limitations in hand use [60]. The reason that patients with severe upper limb impairment were not included is because one of our inclusion criteria stated that the patients had to be able to move at least 10 degrees with the index finger in order to perform the finger tracking movement. Therefore, we cannot generalize our findings to the total population. Further studies should investigate if patients without hand function on the affected side are able to don the tDCS cap. Our age range of patients is 46–72 years of age, with average age 61 years of age. Recent evidence suggests that 1/3 of the people with a stroke are less than 65 years old [61], but our results may not generalize to the general population of individuals with stroke.

## Conclusions

As in-clinic healthcare costs continue to rise, economical, home-based treatments for post-stroke motor rehabilitation will become increasingly significant. Based on the results of this study, a home-based post-stroke therapy that combines low-cost tDCS and tracking training can be a safe treatment option. Although the study sample size was small, the participants found the set-up easy, were comfortable with the devices and unanimously recommended the use of tDCS and finger tracking as a telerehabilitation program. We conclude that tDCS combined with finger tracking training is safe and feasible for the people with stroke. Clinical trials are needed to determine if this promising remote therapy with tDCS and finger tracking is effective.

**Acknowledgments**

We would like to thank Margaret Merline for contributing valuable technical support for this project.

**Funding**

This work was funded by an Institute for Engineering in Medicine Seed Grant at the University of Minnesota. The funding agency was not involved in the design, collection, analysis, interpretation of data, and in writing the manuscript.

**Authors' contributions**

All authors have made substantial contributions to the research design, data acquisition, data analysis or interpretation of data, to drafting the paper or reviewing it critically, and all have approved the submitted version. JC and WD designed the study; AVDW, JC, TB, EH, and CS conducted the study, including patient recruitment, data collection and monitoring. AVDW, who is the guarantor of the study, and JC decided on the analytic strategy, conducted the data analysis and prepared the manuscript draft. TB, EH, WD and CS revised it critically for important intellectual content. All authors had complete access to the study data. All authors approved the version to be published. All authors participated sufficiently in the work to take public responsibility for appropriate portions of the content.

**Competing interests**

The authors declare that they have no competing interests.

**Author details**

[1]Division of Physical Therapy; Division of Rehabilitation Science, University of Minnesota, 420 Delaware Street SE (MMC388), Minneapolis, MN 55455, USA. [2]Division of Physical Therapy, University of Minnesota, Minneapolis, MN 55455, USA. [3]Department of Mechanical Engineering, University of Minnesota, Minneapolis, MN 55455, USA. [4]Department of Neurology, University of Minnesota, Minneapolis, MN 55455, USA.

**References**

1. Khedr EM, Abdel-Fadeil MR, Farghali A, Qaid M. Role of 1 and 3 Hz repetitive transcranial magnetic stimulation on motor function recovery after acute ischaemic stroke. Eur J Neurol. 2009;16:1323–30.
2. Classen J, Schnitzler A, Binkofski F, Werhahn KJ, Kim YS, Kessler KR, et al. The motor syndrome associated with exaggerated inhibition within the primary motor cortex of patients with hemiparetic. Brain. 1997;120:605–19.
3. Murase N, Duque J, Mazzocchio R, Cohen LG. Influence of interhemispheric interactions on motor function in chronic stroke. Ann Neurol. 2004;55:400–9.
4. Taub E, Crago J, Burgio L, Groomes T, Cook EI, DeLuca S. An operant approach to rehabilitation medicine: overcoming learned nonuse by shaping. J Exp Anal Behav. 1994;61:281–93.
5. Hömberg V. Neurorehabilitation approaches to facilitate motor recovery. Handb Clin Neurol. 2013;110:161–73.
6. Bolognini N, Pascual-Leone A, Fregni F. Using non-invasive brain stimulation to augment motor training-induced plasticity. J Neuroeng Rehabil 2009;17;6:8.
7. Carey JR, Durfee WK, Bhatt E, Nagpal A, Weinstein SA, Anderson KM, et al. Comparison of finger tracking versus simple movement training via telerehabilitation to alter hand function and cortical reorganization after stroke. Neurorehabil Neural Repair. 2007;21:216–32.
8. Deng H, Durfee WK, Nuckley DJ, Rheude BS, Severson AE, Skluzacek KM, et al. Complex versus simple ankle movement training in stroke using telerehabilitation: a randomized controlled trial. Phys Ther. 2012;92:197–209.
9. Carey JR, Kimberley TJ, Lewis SM, Auerbach EJ, Dorsey L, Rundquist P, et al. Analysis of fMRI and finger tracking training in subjects with chronic stroke. Brain. 2002;125:773–88.
10. Chen HM, Chen CC, Hsueh IP, Huang SL, Hsieh CL. Test-retest reproducibility and smallest real difference of 5 hand function tests in patients with stroke. Neurorehabil Neural Repair. 2009;23(5):435–40.
11. Rehabilitation Measures Database, https://www.sralab.org/rehabilitation-measures/box-and-block-test#stroke. Accessed 12 Dec 2017.
12. Allman C, Amadi U, Winkler AM, Wilkins L, Filippini N, Kischka U, et al. Ipsilesional anodal tDCS enhances the functional benefits of rehabilitation in patients after stroke. Sci Transl Med. 2016;8:330re1.
13. Lindenberg R, Renga V, Zhu LL, Nair D, Schlaug G. Bihemispheric brain stimulation facilitates motor recovery in chronic stroke patients. Neurology. 2010;75:2176–84.
14. Elsner B, Kwakkel G, Kugler J, Mehrholz J. Transcranial direct current stimulation (tDCS) for improving capacity in activities and arm function after stroke: a network meta-analysis of randomised controlled trials. J Neuroeng Rehabil. 2017;14(1):95.
15. Kubis N. Non-invasive brain stimulation to enhance post-stroke recovery. Front Neural Circuits. 2016;10:56.
16. Bikson M, Grossman P, Thomas C, Zannou AL, Jiang J, Adnan T, Mourdoukoutas AP, Kronberg G, Truong D, Boggio P, Brunoni AR. Safety of transcranial direct current stimulation: evidence based update 2016. Brain Stimul. 2016;9(5):641–61.
17. An TG, Kim SH, Kim KU. Effect of transcranial direct current stimulation of stroke patients on depression and quality of life. J Phys Ther Sci. 2017;29:505–7.
18. Valiengo LC, Goulart AC, de Oliveira JF, Benseñor IM, Lotufo PA, Brunoni AR. Transcranial direct current stimulation for the treatment of post-stroke depression: results from a randomised, sham-controlled, double-blinded trial. J Neurol Neurosurg Psychiatry. 2017;88:170–5.
19. Fujimoto S, Kon N, Otaka Y, Yamaguchi T, Nakayama T, Kondo K, et al. Transcranial direct current stimulation over the primary and secondary somatosensory cortices transiently improves tactile spatial discrimination in stroke patients. Front Neurosci. 2016;10:128.
20. Lee DJ, Lee YS, Kim HJ, Seo TH. The effects of exercise training using transcranial direct current stimulation (tDCS) on breathing in patients with chronic stroke patients. J Phys Ther Sci. 2017;29:527–30.
21. Morishita T, Inoue T. Brain stimulation therapy for central post-stroke pain from a perspective of interhemispheric neural network remodeling. Front Hum Neurosci. 2016;10:166.
22. Lefebvre S, Dricot L, Laloux P, Desfontaines P, Evrard F, Peeters A, et al. Increased functional connectivity one week after motor learning and tDCS in stroke patients. Neuroscience. 2017;340:424–35.
23. O'Brien S. Trends in inpatient rehabilitation stroke outcomes before and after advent of the prospective payment system: a systematic review. JNPT. 2010;34:17–23.
24. O'Brien SR, Xue Y, Ingersoll G, Kelly A. Shorter length of stay is associated with worse functional outcomes for medicare beneficiaries with stroke. Phys Ther. 2013;93:1592–602.
25. Dobrez D, Heinemann AW, Deutsch A, Manheim L, Mallinson T. Impact of Medicare's prospective payment system for inpatient rehabilitation facilities on stroke patient outcomes. Am J Phys Med Rehabil. 2010;89:198–204.
26. Granger CV, Markello SJ, Graham JE, Deutsch A, Ottenbacher KJ. The uniform data system for medical rehabilitation: report of patients with stroke discharged from comprehensive medical programs in 2000-2007. Am J Phys Med Rehabil. 2009;88:961–72.
27. Charvet L, Shaw M, Dobbs B, Frontario A, Sherman K, Bikson M, et al. Remotely supervised transcranial direct current stimulation increases the benefit of at-home cognitive training in multiple sclerosis. Neuromodulation. 2017. https://doi.org/10.1111/ner.12583.
28. Kasschau M, Reisner J, Sherman K, Bikson M, Datta A, Charvet LE. Transcranial direct current stimulation is feasible for remotely supervised home delivery in multiple sclerosis. Neuromodulation. 2016;19:824–31.
29. Laver KE, Schoene D, Crotty M, George S, Lannin NA, Sherrington C. Telerehabilitation services for stroke. Cochrane Database Syst Rev. 2013;12:CD010255.
30. Chen J, Jin W, Zhang XX, Xu W, Liu XN, Ren CC. Telerehabilitation approaches for stroke patients: systematic review and meta-analysis of randomized controlled trials. J Stroke Cerebrovasc Dis. 2015;24(12):2660–8.
31. Nelson RE, Okon N, Lesko AC, Majersik JJ, Bhatt A, Baraban E. The cost-effectiveness of telestroke in the Pacific northwest region of the USA. J Telemed Telecare. 2016;22(7):413–21.
32. Dumitrascu OM, Demaerschalk BM. Telestroke. Curr Cardiol Rep. 2017;19(9):85.
33. Linder SM, Rosenfeldt AB, Bay RC, Sahu K, Wolf SL, Alberts JL. Improving quality of life and depression after stroke through Telerehabilitation. Am J Occup Ther. 2015;69(2):6902290020p1–10.
34. Boggio PS, Nunes A, Rigonatti SP, Nitsche MA, Pascual-Leone A, Fregni F. Repeated sessions of noninvasive brain DC stimulation is associated with

motor function improvement in stroke patients. Restor Neurol Neurosci. 2007;25:123–9.
35. Zimerman M, Heise KF, Hoppe J, Cohen LG, Gerloff C, Hummel FC. Modulation of training by single-session transcranial direct current stimulation to the intact motor cortex enhances motor skill acquisition of the paretic hand. Stroke. 2012;43:2185–91.
36. Stagg CJ, Bachtiar V, O'Shea J, Allman C, Bosnell RA, Kischka U, et al. Cortical activation changes underlying stimulation-induced behavioural gains in chronic stroke. Brain. 2012;135:276–84.
37. Bradnam LV, Stinear CM, Barber PA, Byblow WD. Contralesional hemisphere control of the proximal paretic upper limb following stroke. Cereb Cortex. 2012;22:2662–71.
38. Van de Winckel A, Feys H, van der Knaap S, Messerli R, Baronti F, Lehmann R, et al. Can quality of movement be measured? Rasch analysis and inter-rater reliability of the motor evaluation scale for upper extremity in stroke patients (MESUPES). Clin Rehabil. 2006;20:871–84.
39. Folstein M, Folstein S, McHugh P. "Mini-mental state:" A practical method for grading the cognitive state of patients for the clinician. J Psychiat Res. 1975;12:189–98.
40. Bohannon R, Smith M. Interrater reliability of a modified Ashworth scale of muscle spasticity. Phys Ther. 1987;67:206.
41. Fugl-Meyer A, Jaasko L, Leyman I, Olsson S, Steglind S. The post-stroke hemiplegic patient: a method for evalulation of physical performance. Scand J Rehab Med. 1975;7:13–31.
42. Oldfield R. The assessment and analysis of handedness: the Edinburgh inventory. Neuropsychologia. 1971;9:97–113.
43. Mathiowetz V, Volland G, Kashman N, Weber K. Adult norms for the box and block test of manual dexterity. Am J Occup Ther. 1985;39:386–91.
44. Wechsler D. Wechsler adult intelligence scale - third edition, Wechsler memory scale - third edition: technical manual. New York: Psychological Corporation; 2002.
45. Brunoni AR, Amadera J, Berbel B, Volz MS, Rizzerio BG, Fregni F. A systematic review on reporting and assessment of adverse effects associated with transcranial direct current stimulation. Int J Neuropsychopharmacol. 2011;14:1133–45.
46. Charvet LE, Kasschau M, Datta A, Knotkova H, Stevens MC, Alonzo A, Loo C, Krull KR, Bikson M. Remotely-supervised transcranial direct current stimulation (tDCS) for clinical trials: guidelines for technology and protocols. Front Syst Neurosci. 2015;9:26. https://doi.org/10.3389/fnsys.2015.00026.
47. Carey JR. Manual stretch: effect on finger movement control and force control in stroke subjects with spastic extrinsic finger flexor muscles. Arch Phys Med Rehabil. 1990;71:888–94.
48. Sehm B. tDCS for motor stroke: the relevance of systematic comparisons. Clin Neurophysiol. 2017;128:1367–8.
49. Chhatbar PY, Chen R, Deardorff R, Dellenbach B, Kautz SA, George MS, et al. Safety and tolerability of transcranial direct current stimulation to stroke patients - a phase I current escalation study. Brain Stimul. 2017;10:553–9.
50. Buch ER, Santarnecchi E, Antal A, Born J, Celnik PA, Classen J, et al. Effects of tDCS on motor learning and memory formation: a consensus and critical position paper. Clin Neurophysiol. 2017;128:589–603.
51. Lindenberg R, Sieg MM, Meinzer M, Nachtigall L, Flöel A. Neural correlates of unihemispheric and bihemispheric motor cortex stimulation in healthy young adults. NeuroImage. 2016;140:141–9.
52. Lefebvre S, Liew SL. Anatomical parameters of tDCS to modulate the motor system after stroke: a review. Front Neurol. 2017;8:29.
53. Rabadi MH, Aston CE. Effect of transcranial direct current stimulation on severely affected arm-hand motor function in patients after an acute ischemic stroke: a pilot randomized control trial. Am J Phys Med Rehabil. 2017;96:S178–84.
54. Goodwill AM, Teo WP, Morgan P, Daly RM, Kidgell DJ. Bihemispheric-tDCS and upper limb rehabilitation improves retention of motor function in chronic stroke: a pilot study. Front Hum Neurosci. 2016;10:258.
55. Mortensen J, Figlewski K, Andersen H. Combined transcranial direct current stimulation and home-based occupational therapy for upper limb motor impairment following intracerebral hemorrhage: a double-blind randomized controlled trial. Disabil Rehabil. 2016;38:637–43.
56. Gropen T, Magdon-Ismail Z, Day D, Melluzzo S, Schwamm LH, Group NA. Regional implementation of the stroke systems of care model: recommendations of the northeast cerebrovascular consortium. Stroke. 2009;40:1793–802.
57. Adamovich S, Fluet GG, Merians AS, Mathai A, Qiu Q. Recovery of hand function in virtual reality: training hemiparetic hand and arm together or separately. Conf Proc IEEE Eng Med Biol Soc. 2008;2008:3475–8.
58. Yeh SC, Lee SH, Chan RC, Wu Y, Zheng LR, Flynn S. The efficacy of a haptic-enhanced virtual reality system for precision grasp Acquisition in Stroke Rehabilitation. J Healthc Eng. 2017;2017:9840273.
59. Plow EB, Sankarasubramanian V, Cunningham DA, Potter-Baker K, Varnerin N, Cohen LG, Sterr A, Conforto AB, Machado AG. Models to tailor brain stimulation therapies in stroke. Neural Plast 2016;2016:4071620.
60. Woytowicz EJ, Rietschel JC, Goodman RN, Conroy SS, Sorkin JD, Whitall J, McCombe WS. Determining levels of upper extremity movement impairment by applying cluster analysis to upper extremity Fugl-Meyer assessment in chronic stroke. Arch Phys Med Rehabil. 2017;98(3):456–62.
61. Hall MJ, Levant S, DeFrances CJ. Hospitalization for stroke in U.S. hospitals, 1989–2009. NCHS data brief, No. 95. Hyattsville, MD: National Center for Health Statistics; 2012. [Online: https://www.cdc.gov/nchs/products/databriefs/db95.htm].

# 10

# Amputee perception of prosthetic ankle stiffness during locomotion

Max K. Shepherd[1,2,3*], Alejandro F. Azocar[3,4,5], Matthew J. Major[6] and Elliott J. Rouse[3,4,5]

**Abstract**

**Background:** Prosthetic feet are spring-like, and their stiffness critically affects the wearer's stability, comfort, and energetic cost of walking. Despite the importance of stiffness in ambulation, the prescription process often entails testing a limited number of prostheses, which may result in patients receiving a foot with suboptimal mechanics. To understand the resolution with which prostheses should be individually optimized, we sought to characterize below-knee prosthesis users' psychophysical sensitivity to prosthesis stiffness.

**Methods:** We used a novel variable-stiffness ankle prosthesis to measure the repeatability of user-selected preferred stiffness, and implemented a psychophysical experiment to characterize the just noticeable difference of stiffness during locomotion.

**Results:** All eight subjects with below-knee amputation exhibited high repeatability in selecting their Preferred Stiffness (mean coefficient of variation: 14.2 ± 1.7%) and were able to correctly identify a 7.7 ± 1.3% change in ankle stiffness (with 75% accuracy).

**Conclusions:** This high sensitivity suggests prosthetic foot stiffness should be tuned with a high degree of precision on an individual basis. These results also highlight the need for a pairing of new robotic prescription tools and mechanical characterizations of prosthetic feet.

**Keywords:** Prosthetics, Perception, Stiffness, Variable-stiffness

## Background

Mobility is a key predictor of quality of life for lower-limb amputees [1]. Critical aspects of mobility, such as stability and energetic cost of walking, are highly affected by the mechanical behavior of leg prostheses [2–6]. The process of matching patients with appropriate prosthetic componentry is traditionally based on experiential decisions of clinicians. Consequently, selection of a specific component might be generally acceptable to a prosthesis user, but limit achievement of their full rehabilitation potential [7].

A common device prescribed to persons with lower limb loss is the modern energy storage and return (ESR) prosthetic foot, which intends to mimic the spring-like behavior of the biological ankle-foot complex during gait. ESR feet deflect during mid-stance, storing energy which is subsequently returned during terminal stance phase to help propel the body forward [8]. The effectiveness of springs in mimicking the function of the biological ankle is well supported by research on the quasi-stiffness and stiffness of the ankle during level-ground walking, as well as the time-tested success of spring-like commercial ankle-foot prostheses [9–11]. For ESR feet, stiffness—the ratio between applied force and associated deflection—is perhaps the most defining mechanical characteristic, and its effects on gait mechanics, metabolic cost of walking, socket comfort, and limb loading are well studied [2–6, 8].

Despite the importance of stiffness in locomotion, prosthetic manufacturers do not offer quantitative stiffness information—only qualitative scales associated with body weight and activity level—making cross-comparison between feet exceedingly difficult. The prescription process is further hindered by the excessive time required to test each prosthesis, as each tested device must be individually

* Correspondence: maxshep@u.northwestern.edu
[1]Shirley Ryan AbilityLab, Room 11-1414, 355 E Erie St, Chicago, IL 60611, USA
[2]Department of Biomedical Engineering, Northwestern University, 663 Clark St, Evanston, IL 60208, USA
Full list of author information is available at the end of the article

donned, aligned, adapted to, assessed, and doffed. As a result, it is not uncommon for patients to pilot only one or two prostheses; furthermore, the down-time required to switch prostheses hinders direct comparison of their impact on patients.

With the advent of computer-controlled active and quasi-passive / semi-active prostheses, new tools for improving the prescription process are on the horizon. Caputo et al. have proposed using a powered prosthesis emulator to adjust mechanical behavior of the prosthesis in real-time, thereby quickly simulating the mechanics of different prosthetic feet [12]. With such a device, the optimal foot behavior could be determined through various biomechanical objectives, such as minimizing metabolic cost or gait asymmetry. However, it is more likely that prosthetists and patients will work together, making fine adjustments to improve less-quantifiable metrics such as perceived levels of comfort, effort, and stability. With a patient-in-the-loop approach, prosthetists may adjust device parameters and ask patients for their preference, or patients may make the adjustments themselves [13]. However, the ability of patients to perceive and communicate changes in prosthesis properties—stiffness in particular—has not been quantified, despite being incorporated into the shared decision process of clinical fitting and alignment.

The resolution with which below-knee prosthesis users can discriminate changes in prosthesis stiffness has important implications for the balance of patient and clinician voices in the shared-decision prescription process. If, for instance, patients are highly sensitive to small changes in stiffness, and exhibit high repeatability in their preferences, then clinicians can rely more heavily on patient feedback. This could be particularly true if patient sensitivity to stiffness proves to be of substantially higher resolution than clinicians' perception of the corresponding gait changes. If, conversely, patients exhibit poor sensitivity, or have difficulty finding or communicating their preference, then prosthetists may need to emphasize alternative metrics for assessment.

In this experiment, we sought to determine the sensitivity of below-knee prosthesis users to prosthetic ankle stiffness. Specifically, we assessed 1) the repeatability (variability) of user-selected preferred stiffness, and 2) the difference threshold, or Just Noticeable Difference (JND), for stiffness during locomotion. Additionally, we sought to determine if simple measures, such as self-reported mobility assessment and peripheral vibration sensing, accurately predicted perception of stiffness. Our results provide a rigorous yet patient-centric perspective on the importance of optimizing stiffness on an individual basis, and provide valuable insight into the potential for new prescription methods using robotic devices.

## Methods

### Overview

To perform this experiment, we used a custom Variable Stiffness Prosthetic Ankle-Foot (VSPA Foot), which has the distinct advantages of providing known torque-angle mechanics, while enabling adjustment of stiffness in real-time, between steps during locomotion [14]. First, subjects walked on a treadmill while wearing the VSPA Foot, and used a dial to repeatedly select their preferred stiffness. We then quantified their ability to perceive changes in stiffness by making small changes to stiffness and asking whether the patient could identify if the stiffness increased or decreased. A psychometric curve was then fit to these data to determine subjects' difference threshold of stiffness perception. Preliminary results from a portion of this study were presented at the 2018 International Conference on Biomedical Robotics and Biomechatronics.

### Variable-stiffness ankle-foot prosthesis

The VSPA Foot (modified from Shepherd and Rouse, 2017 [14]) was used to perform this study (Fig. 1a). The VSPA Foot can change its stiffness by over an order of magnitude, mimicking the mechanical behavior of a range of ESR feet.[1] It employs an actively repositionable simple support beneath a leaf spring to change stiffness, and a cam-based transmission to create a custom torque-angle curve (Fig. 1b). Modifications to the VSPA Foot design included a stiffer aluminum chassis and inclusion of a 6AL-4 V titanium spring, rather than the fiberglass spring implemented in the original design; both of these modifications were implemented to increase the maximum attainable stiffness of the foot. In addition, the cam-follower was integrated into the titanium spring design, which enabled a more symmetric change to the dorsiflexion and plantarflexion mechanics as foot stiffness was increased or decreased.

The cam profile was designed to create a linear torque-angle relationship at the ankle joint in both plantarflexion and dorsiflexion, providing the mechanics of a simple torsional spring at the ankle joint. Plantarflexion stiffness (the stiffness when the ankle is plantarflexed from neutral, occurring immediately following heel strike) was designed to be 33% of dorsiflexion stiffness. This ratio follows work completed by Major et al. [2] based on characterization of common ESR feet, and from subject feedback during pilot testing.

The achievable stiffness levels were measured using a custom rotary dynamometer and an onboard encoder on the ankle axis (Fig. 1b). Hysteresis is minimal; less than 5% of energy stored is lost. For all tests, when describing stiffness quantitatively, we refer only to *dorsiflexion* stiffness, as the ankle is primarily spent dorsiflexed from neutral during stance phase.

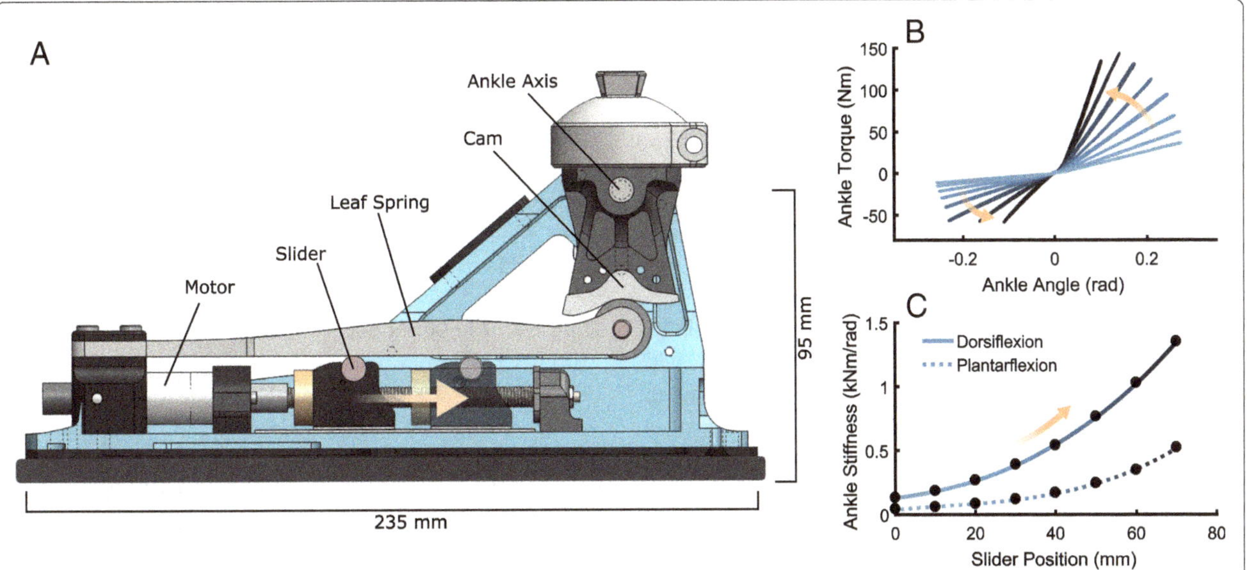


**Fig. 1 a** Modified version of the VSPA Foot [14]. As the slider is actively repositioned towards the cam, the stiffness increases (orange arrows). The cam was designed to create constant stiffness (a linear torque-angle relationship). **b** Experimental characterization of select torque vs. angle curves across the range of slider positions. **c** Ankle stiffness, defined as a function of slider position with a cubic fit. Dorsiflexion stiffness (filled line) is considered the nominal stiffness for this study

The relationship between dorsiflexion stiffness and slider position is described by a cubic fit to the characterization data (Fig. 1c).

The ankle could be commanded to change stiffness by either the experimenter, who could communicate with the onboard computer (*Raspberry Pi Zero*, Raspberry Pi Foundation, Cambridgeshire UK) over WiFi, or directly by the subject, using a small dial which could rotate indefinitely and had no absolute reference. The ankle was worn without a shoe to avoid additional compliance and confounding factors [15]. Instead, a 6 mm piece of shoe material (*SolFlex crepe*, Shore A durometer 50–55, SoleTech, Nahant, MA, US) was epoxied to the bottom of the foot to aid in shock absorption, and a thin tread (*Diamond*, durometer 50–55, SoleTech) was epoxied below that shoe material for traction.

**Participants**

Eight unilateral below-knee amputees were recruited to participate in the study (Table 1). Participants were at least nine months post-amputation, able to walk one km on a treadmill over the course of 45 min, 18–65 years old, 60–95 kg, and primarily used an ESR or hydraulic foot at home. None of the subjects had skin irritation on the affected leg or comorbidities that would affect their gait biomechanics. Informed consent was obtained by each subject prior to testing, and the study was approved by the Northwestern University Institutional Review Board.

Prior to the walking experiments, anthropometric data were taken, and several tests were administered for possible correlations with stiffness perception. Each subject's height, weight, side of amputation, take-home prosthesis

**Table 1** Subject Information

| Subject | Age (yrs) | Height (m) | Weight (kg) | Time since amputation (yrs) | Residual limb length (cm) | Customary prosthesis | Amputation etiology | PLUS-M raw score (T-Score) |
|---|---|---|---|---|---|---|---|---|
| 1 | 23 | 1.88 | 86.2 | 3 | 14 | Össur *Vari-Flex XC* | Traumatic | 60 (71.4) |
| 2 | 41 | 1.52 | 76.2 | 4 | 15 | Freedom Inn. *Senator* | Dysvascular | 58 (64.5) |
| 3 | 41 | 1.70 | 54.4 | 14 | 12 | Össur *Elation* | Traumatic | 43 (49.1) |
| 4 | 46 | 1.85 | 86.0 | 26 | 24 | College Park *Velocity* | Traumatic | 60 (71.4) |
| 5 | 33 | 1.75 | 72.5 | 13 | 15 | College Park *Soleus* | Traumatic | 36 (44.5) |
| 6 | 24 | 1.65 | 61.2 | 1 | 14 | Össur *Pro-Flex LP* | Traumatic | 54 (58.4) |
| 7 | 35 | 1.83 | 90.0 | 15 | 14 | Ability Dyn. *Rush Foot* | Dysvascular | 53 (57.3) |
| 8 | 54 | 1.78 | 84.0 | 1 | 15 | Endolite *Echelon* | Dysvascular | 54 (58.4) |

type, and length of affected limb (mid-patellar tendon to distal end) were recorded (Table 1). Peripheral vibration threshold was assessed with a Rydel-Seiffer Tuning Fork (US Neurologicals, LLC, Poulsbo, Washington US) on three bony prominences with three trials per location: amputation-side tibial tuberosity, sound-side tibial tuberosity, and sound-side medial malleolus. All trials occurred with the subject seated with eyes closed, and leg extended, resting on a stool. The scores for the three locations were averaged to create an overall vibration-sensing score. Finally, the Prosthetic Limb Users Survey of Mobility (PLUS-M) self-report instrument was administered, with the maximum score of 60 and higher scores suggestive of higher mobility [16].

#### Fitting and familiarization

Subjects first walked on the treadmill with their customary prosthesis for several minutes to familiarize themselves with the treadmill and laboratory space. The customary prosthesis was then removed, and the VSPA Foot was fit and aligned to their customary socket by a certified prosthetist. The subjects then walked on the treadmill with the VSPA Foot set to an initial stiffness given by $k = 7.5 \times$ Mass (kg), a standardized setting established from preliminary experiments. The prosthetist stopped the treadmill as necessary to adjust the alignment. During this process, the speed of the treadmill was slowly increased by the subject to 1.0 m/s, which was the nominal experiment speed. All experiments were performed at 1.0 m/s, except by Subjects 2, 4, and 6, who felt uncomfortable at this speed and elected to walk at 0.87 m/s. This speed is slightly less than is typically reported for self-selected walking speed, but was selected to ensure stability during trials in which the stiffness was changed between steps. When the prosthetist and subject were content with the alignment, and the subject expressed comfort walking with the prosthesis at the experiment speed, the treadmill was stopped for a short break before Experiment 1.

#### Experiment 1: Preferred stiffness

After familiarization and alignment, the subjects walked on the treadmill and were asked to "select the stiffness that you find to be the most comfortable at this walking speed." To change foot stiffness, subjects were free to rotate the dial at any time during the gait cycle; however, active stiffness adjustments were made only during the swing phase of gait. Subjects were encouraged to explore a range of stiffness levels until the prosthesis felt uncomfortably stiff or uncomfortably soft. After selecting their preferred stiffness (PS), the treadmill was stopped, and the researcher "reseeded" the stiffness by pseudo-randomly increasing or decreasing it by 25%. The subjects repeated the process of selecting a PS for a total of five trials. The process typically took longer (approximately 10 min) for the first trial, and less time in subsequent trials, as the subjects learned the effects of changing stiffness and became more confident in their choices. Subject 2 did not fully understand the instructions and had to be retested on a different day.[2] We characterized each subject's repeatability of selecting their PS with Coefficient of Variation ($CV_{PS}$), which is the ratio of the standard deviation (unbiased estimation) to the mean.

#### Experiment 2: Stiffness perception

Stiffness perception was assessed on the same day as Experiment 1, following a short break. Stiffness perception was determined by changing the stiffness of the prosthesis while subjects walked on the treadmill; specifically, we implemented a two-interval forced-choice (2IFC) task and the method of constant stimuli [17, 18]. Subjects were presented with sequential pairs of stiffness values and asked whether the second stimulus was "stiffer" or "less stiff" than the first. Each pair of stimuli constituted a single trial, and consisted of a *reference* stiffness and a *comparison* stiffness. The reference stiffness corresponded to each subject's mean preferred stiffness (PS) and was constant across all trials. The comparison stiffness varied between trials, and was equal to the PS scaled by a value in the set: {0.80, 0.85, 0.90, 0.95, 1.0, 1.05, 1.10, 1.15, 1.20}. This range was informed by preliminary experiments, with the intent of generating both incorrect answers for the smallest changes and very high accuracy for the largest changes. Each comparison stiffness was presented eight times, for a total of 72 trials. The reference stiffness was presented first on half of the trials, and second on the other half. The order of stiffness values was set before the experiment. The trials were pseudo-randomized; to avoid destabilizing the subjects, the trials were randomly re-shuffled until all consecutive stiffness values were within 25% of each other.

Prosthesis stiffness was changed by the experimenter approximately every 6–8 strides, and the change was indicated to subjects by an audible tone. The number of strides was chosen to allow subjects ample time to sense and report the stiffness changes, while minimizing the amount of time spent walking. Subjects only responded after the transition from the first stimulus to the second stimulus (i.e., they did not respond when stiffness was changed between trials). Subjects were allowed to take breaks whenever necessary to avoid fatigue bias. Before the official trials began, subjects were trained with stiffness changes of 15–25%, until subjects reported that they clearly understood the paradigm.

For each stiffness value, the proportion ($P$) of trials judged stiffer than the reference was used to calculate the psychometric function, a sigmoidal function which describes the relationship between stiffness change and probability of perceiving that change as an increase in stiffness. Specifically, a logistic function was fit using a maximum likelihood criterion, with lapse rate fixed at 0.02. The lapse rate accounts for trials in which subjects respond independently of stimulus level (e.g., due to a lapse of attention), and reduces biases on the parameters that determine the psychometric function [17]. The comparison stiffness levels corresponding to $P = 0.25$ and $P = 0.75$, denoted $X_{0.25}$ and $X_{0.75}$, are used to calculate the threshold of perception. This threshold is commonly termed the Just Noticeable Difference (JND), and is specifically calculated as:

$$JND = \frac{X_{0.75} - X_{0.25}}{2}$$

In this study, the JND represents the smallest *percent* change in stiffness that can reliably (with 75% accuracy) be identified as having been an increase or decrease in stiffness.[3] Lower JND values correspond to better perception, since a smaller change in stiffness can be perceived.

At the end of the experiment, subjects were asked to describe the strategies they used to sense changes in stiffness. As mentioned previously, Subject 2 did not originally understand the instructions for selecting preferred stiffness, so their JND trials were performed with a reference stiffness of 571.5 Nm/rad (weight normalized: 7.5 Nm/rad/kg) and preferred stiffness was reassessed on a second day.

### Kinematics

To assess the sensitivity of ankle kinematics to ankle stiffness, and specifically the magnitude of biomechanical changes associated with the JND, ankle kinematic data were recorded in separate walking trials from the onboard ankle encoder. Subjects walked on the treadmill for 90 s at each of five stiffness levels (order randomized): the mean preferred stiffness modified by 0%, ± 10%, and ± 20%.

The kinematic data were recorded at 30 Hz by an onboard 14-bit absolute encoder, zero-lag filtered with a fourth-order, 7.5-Hz Butterworth filter in post-processing, and then up-sampled to 100 Hz with a spline interpolation. An angle of zero corresponded with the average angle during swing. The last 20 of each subject's strides were segmented and averaged. The kinematics were then averaged across subjects at each of the tested stiffness levels. An estimate of the kinematics at the preferred stiffness ± JND was created with a linear interpolation from the ankle kinematics at the closest measured stiffness levels.

### Analysis

Each fitted psychometric function was ensured to have an acceptable goodness of fit [17, 19]. Means and standard errors were calculated across participants for both JND and $CV_{PS}$. Neither JND nor $CV_{PS}$ were normally distributed ($p < 0.05$, one sample K-S test), therefore standard errors were calculated with bootstrapping ($n = 10{,}000$ resamples, s.e.m. calculated as standard deviation of the bootstrapped means).

To test whether there was a relationship between stiffness perception and consistency of preferred stiffness selection, individual JNDs were linearly regressed on $CV_{PS}$. JNDs were also regressed on PLUS-M T-scores and Vibration sensing score (averaged scores from the three tested anatomical locations) to see if either simple test may be used to predict perception.

## Results

### Experiment 1: Preferred stiffness

Subjects' self-selected preferred stiffness (Fig. 2) ranged from 3.8 to 10.7 Nm/rad/kg (or 341.1 to 811.4 Nm/rad,-not weight-normalized). The mean weight-normalized preferred stiffness was 6.7 ± 2.0 Nm/rad/kg (mean ± s.d.). The mean within-subjects coefficient of variability ($CV_{PS}$), which serves as a measure of the subjects' consistency of selection, was 14.2 ± 1.7%.

### Experiment 2: Threshold of stiffness perception

Psychometric functions were fit to the data for all eight subjects (Fig. 3), and individual JNDs ranged from 3.7 to 13.6%. The mean JND was 7.7 ± 1.3%.

The individual JNDs were regressed against three predictors: preferred stiffness variability ($CV_{PS}$), PLUS-M score, and vibration sensing score (Fig. 4). There was a significant ($p < 0.001$) positive correlation between $CV_{PS}$ and JND. There was a non-significant ($p = 0.25$) negative trend between self-reported mobility (PLUS-M score) and JND; removing Subject 2 due to acquiescence bias (see endnote 2), this correlation became significant ($p = 0.04$).

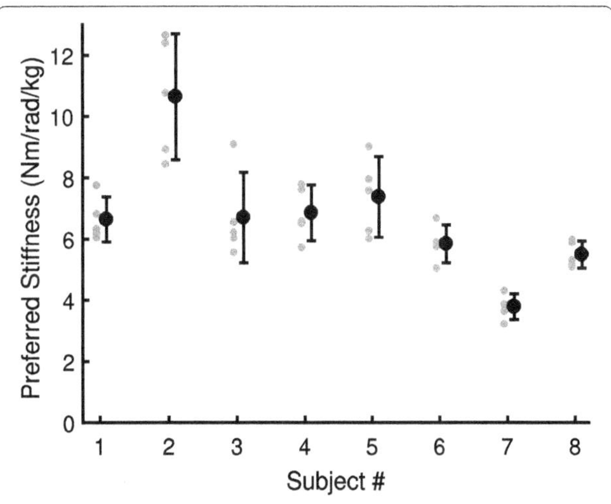


**Fig. 2** Preferred stiffness for each subject. Individual trials are shown next to each subject's mean (error bars: SD)

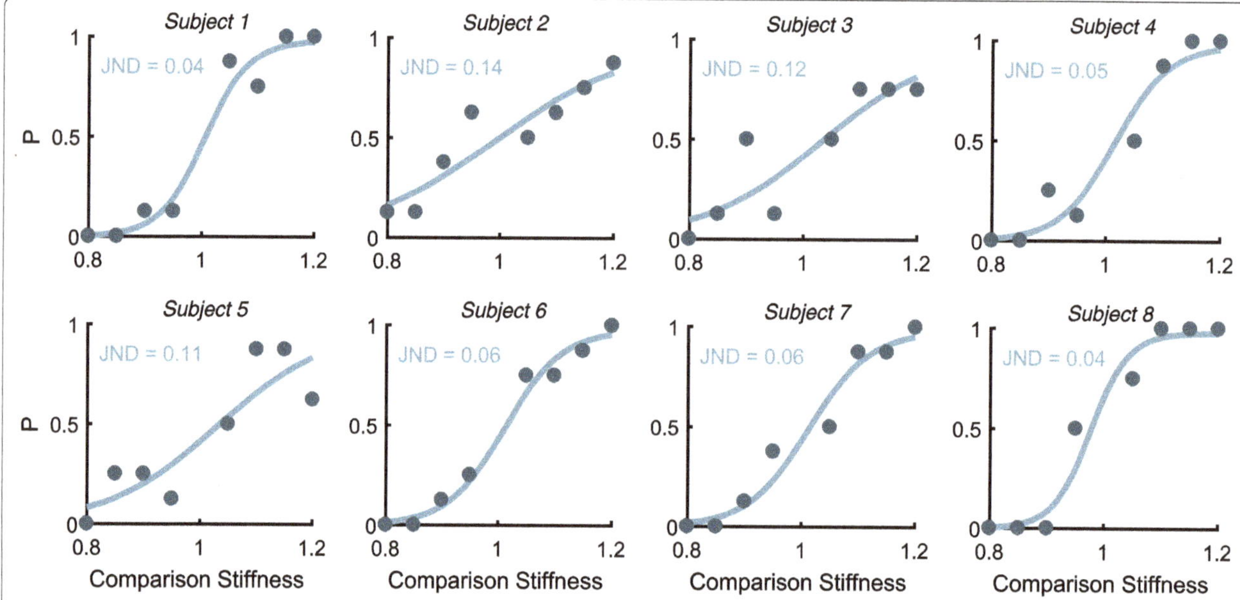


**Fig. 3** Individual psychometric curves and JNDs. The Comparison Stiffness (x-axis) is a fraction of each subject's preferred stiffness (the reference stiffness), and the y-axis label *P* represents the proportion of trials judged stiffer than the reference

As expected, the dysvascular subjects tended to have worse vibration sensing, but there was not a significant correlation between vibration sensing score and JND ($p = 0.85$).

### Kinematics

For all subjects, lower stiffness levels resulted in increased ankle range of motion (Fig. 5). An estimate of the kinematics at the JND above and below the PS is shown in blue; the change in peak dorsiflexion associated with the JND was 0.77°. The kinematics are not representative of those seen during Experiment 2, as the short time spent at each level did not allow subjects adequate time to adapt their gait. They are instead presented to demonstrate the magnitude of kinematic changes during steady-state gait associated with the threshold of stiffness perception.

## Discussion

The purpose of this study was to determine the sensitivity of below-knee prosthesis users to changes in prosthetic ankle stiffness during walking by measuring the repeatability of user-selected preferred stiffness and difference threshold for stiffness. The mean JND of 7.7% is smaller than for some other physical tasks, including able-bodied subjects' ability to sense the stiffness of a spring placed in parallel with their ankle (11.6%, [20]) or elbow (16%, [21]) while in a seated position. Stiffness sensing in these studies relied on integrating the sensations of torque and angle, which are likely dominated by feedback from local mechanoreceptors. Stiffness changes during ambulation, in contrast, may cause global kinematic and spatial orientation changes that can be sensed by the vestibular system. Specifically, at low stiffnesses, most subjects described the forward propulsion as

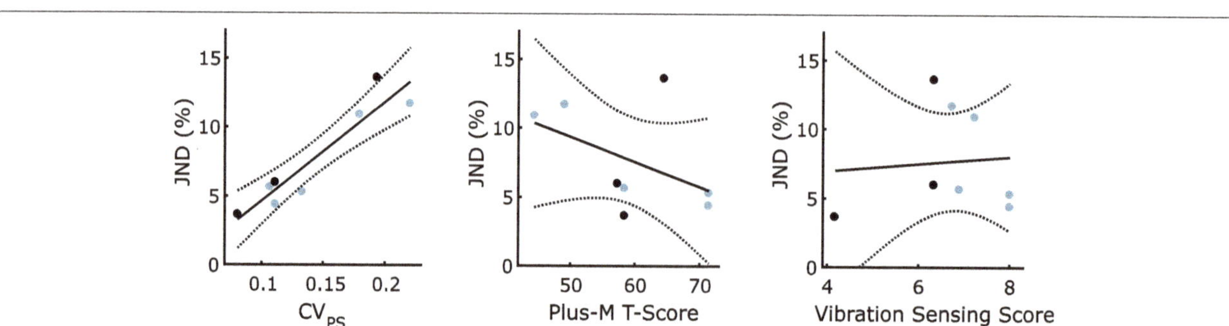


**Fig. 4** (Top) Just Noticeable Difference (JND) vs. Preferred Stiffness Variability. (Middle) JND vs. Prosthesis Limb Users Survey of Mobility (PLUS-M) T-Score. (Bottom) JND vs. Vibration Sensing Score; higher score indicates higher sensitivity to vibration. The three darker points denote the three dysvascular subjects

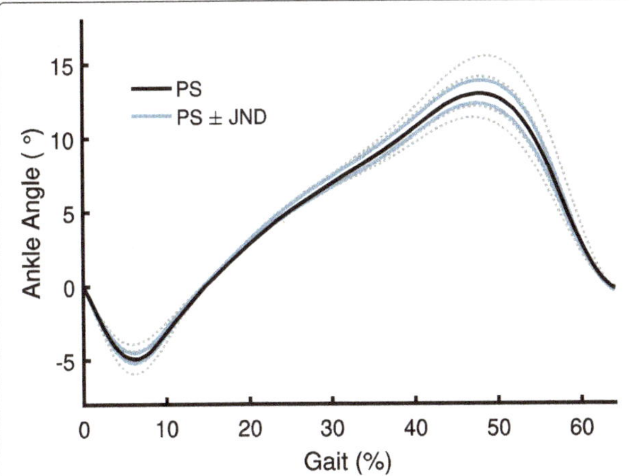


**Fig. 5** Pooled ankle kinematics of stance phase at five tested conditions: PS (black line), PS ± 10% and ± 20% (dotted gray lines). Also shown are estimates of the ankle kinematics at the PS ± JND (blue lines), which are linearly interpolated from the nearest measured stiffness values. Increasing range of motion corresponds with decreasing stiffness

inadequate, or the feeling of drop-off effect after entering midstance. Some subjects reported feeling like the stiffer setting made them stand taller during stance phase. Two subjects indicated feeling increased pressure between their knee and socket, but only at the highest stiffness levels.

There was large between-subject variability in JND, which was highly correlated with the repeatability of preferred stiffness selection ($CV_{PS}$). Specifically, subjects with high repeatability in selecting preferred stiffness could perceive smaller changes in stiffness, suggesting that the ability to consistently select a preferred stiffness may depend largely on the ability to sense it (Fig. 4). As expected, vibration sensing threshold was worse in the three dysvascular subjects, but our results suggest that sensitivity to prosthetic ankle stiffness may not be dependent on residuum vibration sensation. Further research should consider other sensory factors that may contribute to prosthesis stiffness detection.

### Implications for prescription

Our results suggest that patients' ability to effectively communicate stiffness preference is a critical piece missing in the current prescription process. The relatively low $CV_{PS}$ of 14.2% provides a frame of reference for the degree of specificity expected in patient preference. Similarly, the between-subjects variability in weight-normalized PS highlights the need to pilot more than the manufacturer-recommended feet, which are based on weight and activity level, and which would have been almost the same for all eight subjects.

Whether or not a patient's preferred stiffness is close to the 'ideal' stiffness is unknown. There are a multitude of quantitative and qualitative factors that are likely to be optimized in combination by either the prosthetist or patient, such as step length symmetry, limb loading, roll-over, or socket comfort [7]. Moreover, what a patient prefers in a clinical setting may not be the most comfortable, stable, or energetically optimal outside the clinic or long-term; adaptation of gait can continue for weeks past prescription [22]. Nevertheless, our results point to a need to more formally incorporate patient feedback of stiffness into the prescription process.

There are several limitations in the current prescription process that hinder the ability for subjects to sense and report preference for stiffness. First, it is challenging to efficiently test a range of commercial prosthetic feet in the clinic. The excessive time required to change feet, variability in alignment, and subtle differences in mechanical behavior make direct comparison of feet difficult. As a result, patients are unlikely to develop an intuitive sense of stiffness as a modifiable variable. We found that when introduced to stiffness as a controllable variable which could be adjusted instantaneously and continuously, most subjects internalized its effects, and settled on a small range of stiffnesses that they found to be most comfortable.

An approach to prescription based on systematically optimizing foot mechanics and alignment could only be feasible using robotic tools. Prosthetists and patients could work together to systematically and efficiently explore a set of mechanical variables, such as stiffness and alignment. This approach has been advocated by Caputo et al., who developed a tethered robotic platform which can emulate the mechanics of various prosthetic feet, both passive and active [12, 13].

For the described clinical optimization process to be effective, it would also have to be paired with standardized characterization of prosthetic foot behavior, which is not routinely provided by manufacturers [23]. Currently, prosthetists will either use their experience to guide them as to which feet might have a more appropriate stiffness, or make small adjustments to alignment, which has some effects similar to varying stiffness [23]. This relationship has not been studied directly, but Hansen has suggested that for feet with different mechanics, prosthetists change alignment to conform to an ideal roll-over shape [24]. Adjusting alignment is not a perfect substitute to changing ankle stiffness; while both affect the radius of curvature of the effective rocker, other roll-over characteristics vary substantially with alignment, such as arc length and horizontal displacement of the center of curvature [25].

### Implications for research

The high sensitivity of prosthesis users characterized here also has implications for how stiffness should be used as a manipulated variable in research. Several researchers have studied the effects of prosthetic ankle-foot stiffness on metrics such as stability, metabolic cost, and muscle activity. Researchers have often attempted to match the tested stiffness levels to that of common prostheses. However, the chosen variation in stiffness—which typically consists of only two or three discrete levels—has spanned a factor of 2.0 [26], 2.25 [27], 3.5 [6], and 6.2 [3]. The results of our study—both the repeatability of preferred stiffness and the sensitivity to stiffness changes—suggest that researchers may be testing unrealistic ranges of stiffness, and future studies should test a narrower window of stiffness, centered around subjects' preferred stiffness.

This experiment was enabled by the highly predictable, repeatable, and accurate behavior of the VSPA Foot as a research tool. It is unlikely that this experiment would have been possible with a fully-powered prosthesis controlled to emulate springs, as these devices typically have noise levels in their torque control approaching or exceeding the lowest JNDs measured in our experiment [12, 28]. With 0.1-mm positioning accuracy for the stiffness-varying mechanics, the VSPA Foot can make very subtle changes, and because the behavior is intrinsically passive, the motion is guaranteed to be repeatable and noise-free. Researchers seeking to answer similar questions should consider using similar quasi-passive devices that can actively modify their intrinsically passive mechanics, to mitigate the confounding effects of torque-tracking error in fully powered devices.

### Limitations

A critical element of this study was training the subjects to conceptualize "stiffer" vs. "softer/less stiff." Most subjects easily internalized the concept when they turned the dial and immediately felt the stiffness change. Both Subject #3 and Subject #5 had challenges with short-term memory loss due to the nature of their pathology, and Subject #3 requested re-training twice during the JND portion of the experiment. It seems evident that, in a clinical setting, some patients would be able to quickly grasp the concept and provide feedback of their preferences, whereas others may need more training. Similarly, the results of Experiment 2 (specifically, the ability to directionally identify changes in stiffness) were likely dependent on learning the sensations and effects of stiffness during the exploration in Experiment 1.

The pool of subjects is not a representative sample of all lower-limb amputees, due to the inclusion criteria (specifically the 90 kg upper weight limit and ability to walk for 45 min). It is possible that lower-level ambulators (below the Medicare Classification Level K3) will be less sensitive to stiffness, due to a variety of factors such as less experience walking with a prosthesis, or comorbidities such as peripheral neuropathy or excessive adipose tissue in the residuum.

Preferred stiffness is likely dependent on several factors which were controlled in our experiment, but may vary in a clinical setting. It is unknown how changes to alignment affect preferred stiffness; this is a potential avenue for future study. Similarly, we elected not to use a shoe on the prosthetic foot to facilitate complete control of stiffness adjustments, but shoes can add substantial series compliance [15], which would likely affect both preferred stiffness and JND. Stiffness preference was assessed at a steady speed on a treadmill, and may differ during over-ground walking with frequent stops. Finally, foot length may affect preferred stiffness, and may have a substantial effect on the kinematics [29].

Due to the mechanics of the VSPA-Foot, prosthesis plantarflexion and dorsiflexion stiffness could not be changed independently. We made the assumption that subjects would be most sensitive to dorsiflexion stiffness, and thus deemed dorsiflexion stiffness the nominal stiffness in the experiments. Feedback from subjects regarding their strategy confirmed this, but it is possible that we would have received different results with independent variation in stiffness or a different ratio.

## Conclusions

Below-knee amputees were able to sense a 7.7% difference in prosthetic ankle stiffness, and demonstrated high repeatability in selecting their preferred stiffness (coefficient of variability of 14.2%). This information substantiates the argument that that new tools are needed to allow patients and prosthetists to quickly locate an optimal stiffness based on shared decisions, and that standardized systems for measuring and reporting prosthesis stiffness should be adopted by manufacturers, researchers, and clinicians. Additionally, these results should inform future studies investigating the effects of prosthesis stiffness, providing a weight-normalized reference stiffness and a feasible range of stiffness levels to test.

## Endnotes

[1]While the exact mechanics of ESR prosthetic feet are complex, a common, simplified model consists of a solid foot attached to the shank via a rotational spring, articulating around an estimated ankle axis. Damping is present to varying degrees, but elasticity (stiffness) dominates the mechanics; for example, in ESR feet without a shock absorber (i.e., no damping), only 5–15% of the energy stored is lost [30].

[2]We believe Subject 2 exhibited acquiescence bias, in which they were biased to report what they thought the experimenters wanted to hear. Specifically, during the preferred stiffness test, they turned the knob a small amount, without exploring the range, and claimed that

the stiffness was their favorite. Accordingly, their results depended almost entirely on the random reseeding of stiffness between trials. Similarly, we believe their self-assessment of mobility in the Plus-M score was highly biased to show high activity level, which is in contradiction to their conversational anecdotes regarding their mobility, and our prosthetists' assessments. We asked them to return on a second day to redo the preferred stiffness test, and made it clear that we wanted them to explore the range of stiffness levels (until it was uncomfortable) before selecting their preferred stiffness for each trial.

[3]The JND is traditionally in the same units as the stimulus (e.g., Nm/rad), and is then normalized by the average stimulus intensity to calculate the Weber Fraction. However, since every subject had a different reference stiffness (their preferred stiffness), we chose to test their sensitivity as a *percentage* of their preferred stiffness (as opposed to our comparison stimuli being, for example, ± 20 Nm/rad for all subjects). Thus, in this experiment, the JND is equivalent to the Weber Fraction.

**Acknowledgments**
We would like to thank all participants and Graci Finco, CPO, for fitting and aligning the prosthesis.

**Funding**
We acknowledge financial support from the U.S. Army CDMRP PRORP under Award No. W81XWH-17-1-0704, the American Heart Association under Award No. 16PRE31160018, NSF Graduate Research Fellowship under Grant No. DGE-1324585.

**Authors' contributions**
MKS and AFA designed the study protocol with input from MJM and EJR. MKS and EJR developed the experimental prosthesis. MKS performed the experiments. All authors analyzed the data. MKS wrote the manuscript, with input from AFA, MJM, and EJR. All authors approved of the final version.

**Competing interests**
MS and ER hold pending patents on the technology (prosthesis) used in this study.

**Author details**
[1]Shirley Ryan AbilityLab, Room 11-1414, 355 E Erie St, Chicago, IL 60611, USA. [2]Department of Biomedical Engineering, Northwestern University, 663 Clark St, Evanston, IL 60208, USA. [3]Neurobionics Lab, University of Michigan, Ann Arbor, MI 48109, USA. [4]Department of Mechanical Engineering, University of Michigan, Ann Arbor, MI 48109, USA. [5]Robotics Institute, University of Michigan, Ann Arbor, MI 48109, USA. [6]Northwestern University Feinberg School of Medicine, Department of Physical Medicine and Rehabilitation, 710 North Lake Shore Drive, #1022, Chicago, IL 60611, USA.

**References**

1. Wurdeman SR, Stevens PM, Campbell JH. Mobility analysis of AmpuTees (MAAT I): quality of life and satisfaction are strongly related to mobility for patients with a lower limb prosthesis. Prosthetics Orthot Int. 2017;42(5):498-503.
2. Major MJ, Twiste M, Kenney LPJ, Howard D. The effects of prosthetic ankle stiffness on stability of gait in people with transtibial amputation. J Rehabil Res Dev. 2016;53(6):839–52.
3. Zelik KE, et al. Systematic variation of prosthetic foot spring affects center-of-mass mechanics and metabolic cost during walking. IEEE Trans Neural Syst Rehabil. Eng. 2011;19(4):411–9.
4. Fey NP, Klute GK, Neptune RR. Optimization of prosthetic foot stiffness to reduce metabolic cost and intact knee loading during below-knee amputee walking: a theoretical study. J Biomech Eng. 2012;134(11):111005.
5. Adamczyk PG, Roland M, Hahn ME. Sensitivity of biomechanical outcomes to independent variations of hindfoot and forefoot stiffness in foot prostheses. Hum Mov Sci. 2017;54:154–71.
6. Major MJ, Twiste M, Kenney LPJ, Howard D. The effects of prosthetic ankle stiffness on ankle and knee kinematics, prosthetic limb loading, and net metabolic cost of trans-tibial amputee gait. Clin Biomech. 2014;29(1):98–104.
7. Major MJ, Fey NP. Considering passive mechanical properties and patient user motor performance in lower limb prosthesis design optimization to enhance rehabilitation outcomes. Phys Ther Rev. 2017;3196:1–15.
8. Klute GK, Kallfelz CF, Czerniecki JM. Mechanical properties of prosthetic limbs: adapting to the patient. J Rehabil Res Dev. 2001;38(3):299–307.
9. Hansen AH, Childress DS, Miff SC, Gard SA, Mesplay KP. The human ankle during walking: implications for design of biomimetic ankle prostheses. J Biomech. 2004;37(10):1467–74.
10. Rouse EJ, Hargrove LJ, Perreault EJ, Kuiken TA. Estimation of human ankle impedance during the stance phase of walking. IEEE Trans Neural Syst Rehabil Eng. 2014;22(4):870–8.
11. Rouse EJ, Gregg RD, Hargrove LJ, Sensinger JW. The difference between stiffness and quasi-stiffness in the context of biomechanical modeling. IEEE Trans Biomed Eng. 2013;60(2):562–8.
12. Caputo JM, Collins SH. A universal ankle-foot prosthesis emulator for human locomotion experiments. J Biomech Eng. 2014;136(3):035002.
13. Caputo JM. Informing ankle-foot prosthesis design and prescription through systematic experimentation with a tethered robotic prosthesis. ProQuest Diss Theses. 2015:103-122. https://doi.org/10.13140/RG.2.1.3510.9603.
14. Shepherd M, Rouse E. The VSPA foot: a quasi-passive ankle-foot prosthesis with continuously variable stiffness. IEEE Trans Neural Syst Rehabil Eng. 2017;25(12):2375–86.
15. Major MJ, Scham J, Orendurff M. The effects of common footwear on stance-phase mechanical properties of the prosthetic foot-shoe system. Prosthetics Orthot Int. 2017;42(2):198-207.
16. "Prosthetic Limb Users Survey of Mobility (PLUS-M); 12-Item Short Form." [Online]. Available: http://plus-m.org/index.html. Accessed 01 Jan 2018.
17. F. Kingdom and N. Prins, Psychophysics: a practical introduction, 1. 2009.
18. Gescheider GA. Psychophysics: method, theory, and application. 2nd ed: L. Erlbaum Associates; Hillsdale. 1985.
19. Wichmann FA, Hill NJ. The psychometric function: I. fitting, sampling, and goodness of fit. Percept Psychophys. 2001;63(8):1293–313.
20. Azocar AF, Rouse EJ. Stiffness perception during active ankle and knee movement. IEEE Trans Biomed Eng. 2017;64(12):2949–56.
21. Jones LA, Hunter IW. A perceptual analysis of stiffness: Exp Brain Res; 1990; 79:150-156.
22. Ray SF, Wurdeman SR, Takahashi KZ. Prosthetic energy return during walking increases after 3 weeks of adaptation to a new device. J Neuroeng Rehabil. 2018;15(1):6.
23. Major MJ, Kenney LPJ, Twiste M, Howard D. Stance phase mechanical characterization of transtibial prostheses distal to the socket: a review. J Rehabil Res Dev. 2012;49(6):815.

24. Hansen AH, Childress DS, Knox EH. Prosthetic foot roll-over shapes with implications for alignment of trans-tibial prostheses. Prosthetics Orthot Int. 2000;24(3):205–15.
25. Hansen A. Effects of alignment on the roll-over shapes of prosthetic feet. Prosthetics Orthot Int. 2008;32(4):390–402.
26. Ventura JD, Klute GK, Neptune RR. The effects of prosthetic ankle dorsiflexion and energy return on below-knee amputee leg loading. Clin Biomech. 2011;26(3):298–303.
27. Fey NP, Klute GK, Neptune RR. The influence of energy storage and return foot stiffness on walking mechanics and muscle activity in below-knee amputees. Clin Biomech. 2011;26(10):1025–32.
28. Eilenberg MF, Geyer H, Herr H. Control of a powered ankle-foot prosthesis based on a neuromuscular model. IEEE Trans Neural Syst Rehabil Eng. 2010; 18(2):164–73.
29. Hansen AH, Meier MR, Sessoms PH, Childress DS. The effects of prosthetic foot roll-over shape arc length on the gait of trans-tibial prosthesis users. Prosthetics Orthot Int. 2006;30(3):286–99.
30. Webber CM, Kaufman K. Instantaneous stiffness and hysteresis of dynamic elastic response prosthetic feet. Prosthetics Orthot Int. 2017;41(5):463–8.

# 11

# Impact of antagonistic muscle co-contraction on in vivo knee contact forces

Adam Trepczynski[1*], Ines Kutzner[1], Verena Schwachmeyer[1], Markus O. Heller[2], Tilman Pfitzner[3,4] and Georg N. Duda[1]

**Abstract**

**Background:** The onset and progression of osteoarthritis, but also the wear and loosening of the components of an artificial joint, are commonly associated with mechanical overloading of the structures. Knowledge of the mechanical forces acting at the joints, together with an understanding of the key factors that can alter them, are critical to develop effective treatments for restoring joint function. While static anatomy is usually the clinical focus, less is known about the impact of dynamic factors, such as individual muscle recruitment, on joint contact forces.

**Methods:** In this study, instrumented knee implants provided accurate in vivo tibio-femoral contact forces in a unique cohort of 9 patients, which were used as input for subject specific musculoskeletal models, to quantify the individual muscle forces during walking and stair negotiation.

**Results:** Even between patients with a very similar self-selected gait speed, the total tibio-femoral peak forces varied 1.7-fold, but had only weak correlation with static alignment (varus/valgus). In some patients, muscle co-contraction of quadriceps and gastrocnemii during walking added up to 1 bodyweight (~ 50%) to the peak tibio-femoral contact force during late stance. The greatest impact of co-contraction was observed in the late stance phase of stair ascent, with an increase of the peak tibio-femoral contact force by up to 1.7 bodyweight (66%).

**Conclusions:** Treatment of diseased and failed joints should therefore not only be restricted to anatomical reconstruction of static limb axes alignment. The dynamic activation of muscles, as a key modifier of lower limb biomechanics, should also be taken into account and thus also represents a promising target for restoring function, patient mobility, and preventing future joint failure.

**Keywords:** Muscle co-contraction, Musculoskeletal loading conditions; in vivo joint forces, Knee osteoarthritis

## Background

Joint-related disability represents a substantial global health burden and is a frequent consequence of osteoarthritis (OA), in which bone shape is altered and cartilage is lost, causing pain and preventing smooth motion of the joints. OA and failure of total joint replacement (TJR), later in life, are often thought to be caused by excessive mechanical loading [1]. Key determinants of joint loading in clinical assessment are measures of the skeletal anatomy, typically obtained from static radiographs, such as the angle between the long axes of the tibia and femur. Here, the varus/valgus angle quantifies whether a patient has bow-legged (varus) or knock-legged (valgus) knees, and the extent of varus/valgus mal-alignment [2]. It has been suggested that deviations from ideal knee alignment result in local mechanical overload and eventual joint failure, which explains the historical focus on restoring physiologically normal alignment [3]. However, to understand the mechanical failure of joints, it is essential to unravel the interplay of bones and other passive structures with the active muscles, which together create the musculoskeletal (MS) loading conditions that eventually lead to extreme joint contact forces (JCFs) [4, 5].

The most direct knowledge regarding JCFs is provided by in vivo measurements with instrumented joint implants [6–9], which show substantial intra-individual

* Correspondence: adam.trepczynski@charite.de
[1]Julius Wolff Institute, Charité – Universitätsmedizin Berlin, corporate member of Freie Universität Berlin, Humboldt-Universität zu Berlin, and Berlin Institute of Health, Augustenburger Platz 1, 13353 Berlin, Germany
Full list of author information is available at the end of the article

differences in the peak joint loads [7, 9]. Muscle forces dominate JCFs [10], and the variability of JCF magnitudes during the same task may thus result from variations in the structure of the muscle-tendon units, but may also arise from variations in muscle strength, muscle recruitment patterns, or the presence and extent of co-contraction of antagonist muscle groups. OA patients show increased levels of co-contraction compared to healthy controls [11–13], which may result from the disease, but could also be its primary cause, or at least a factor driving its progression [14]. Even after OA was treated with knee arthroplasty, increased levels of co-contraction have been reported [15], and related to muscular insufficiencies [16]. Co-contraction can also occur during rehabilitation such as during functional electrical stimulation, and lead to underestimation of the forces produced by the stimulated muscles in analyses based on the total muscle torque [17].

Musculoskeletal models can be used to predict all muscle forces and joint loading conditions, but require extensive validation against in vivo data [18]. While musculoskeletal models have been validated against in vivo measured data in a smaller cohort before [19–23], we are not aware of evidence suggesting that such models can consistently capture the variation in potentially sub-optimal muscle activation. Therefore, to obtain a reliable quantitative description of antagonist muscle co-contraction, we constrained personalized musculoskeletal models with synchronized tibio-femoral contact forces (TFCF), measured in vivo in a unique cohort of 9 patients with instrumented knee implants (Fig. 1) [24]. We then used the constrained musculoskeletal models to identify and quantify how muscle recruitment impacts joint contact forces, and thereby modulates them from minimizing the sum of squared muscle stresses – a criterion which best matched the in vivo forces among eight commonly utilized optimization criteria [20, 23, 25–29].

## Materials and methods

### Subjects and gait analysis

This study uses data from in vivo measurements of knee contact forces previously reported [10, 30]. Gait analyses were performed on 9 TKR patients (6 male, aged 70 ± 5 years, body mass 90.5 ± 12.6 kg, body height 1.72 ± 0.04 m), assessed 26 ± 13 months post-operatively (Fig. 1).

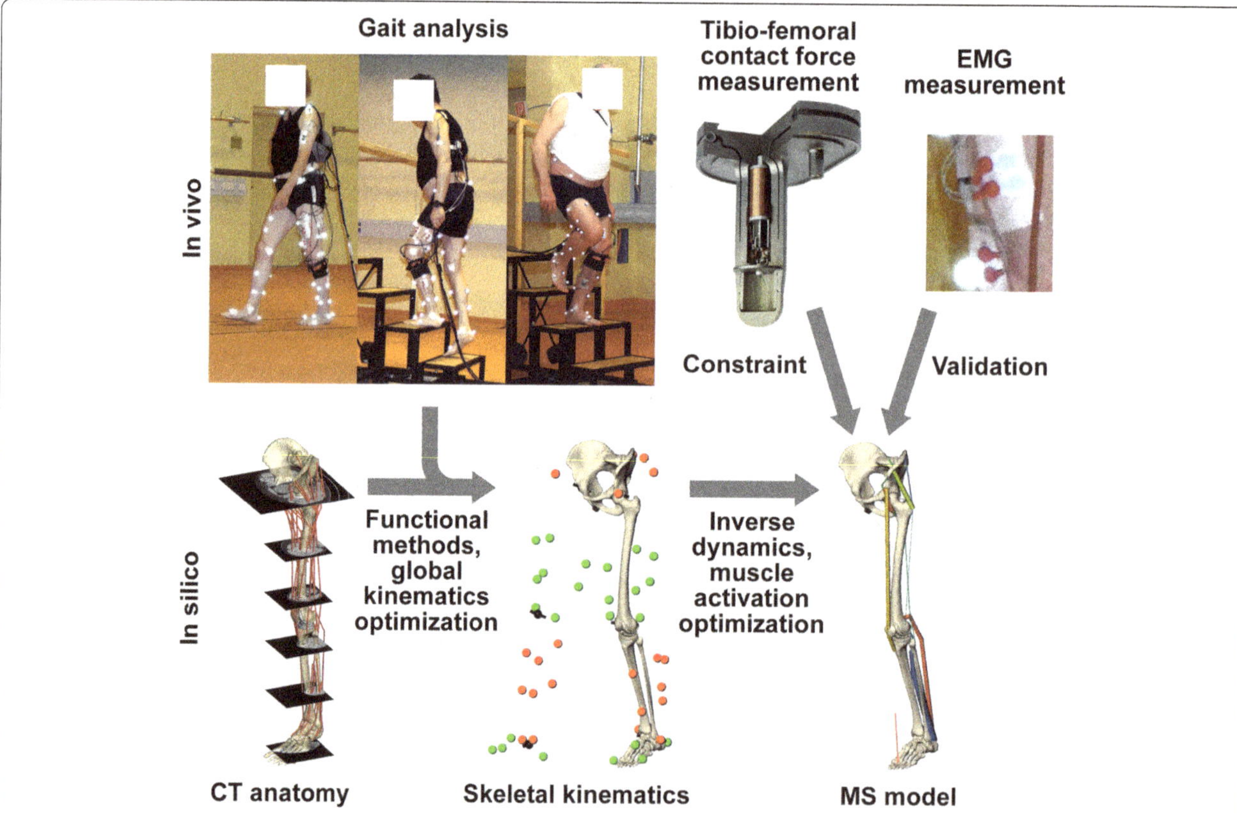


**Fig. 1** Overview of the methodology employed to assess internal loading, by combining measurements and musculoskeletal modelling. Gait analyses were performed during walking and stair negotiation, with simultaneous measurement of the in vivo tibio-femoral forces and surface EMG of the main muscles. Patient specific skeletal anatomy from CT was used to adapt reference muscles geometries, and then combined with functionally determined joint centres/axes to obtain the skeletal kinematics. The resulting musculoskeletal models were constrained to match the in vivo forces and verified using the EMG measurements

Each subject obtained an instrumented knee implant based on a clinically proven design (INNEX, Zimmer GmbH, Winterthur, Switzerland) (Fig. 1) [24]. The instrumented implant allowed for in vivo measurement of the 3 force and 3 moment components of the contact load acting on the tibial tray (at approximately 100 Hz). The magnitude of the resultant JCF was calculated and is referred to as "in vivo TFCF" in this article. The ground reaction forces were captured by 2 force plates (AMTI, Watertown, USA) at 960 Hz, while 3D kinematics of the lower limbs were measured at 120 Hz using reflective markers (Vicon, Oxford, UK). Within a single measurement three activities were investigated: level walking, stair ascent (20 cm step height, 26 cm step run) and stair descent. Across all nine subjects, a total of 169 trials were captured. A standardized protocol with the same order of activities allowed for sufficient rest periods to prevent fatigue.

From post-operative CT data (with in plane resolution 0.4–0.5 mm and slice thickness 1–2 mm), three parameters characterizing the positioning of the tibial component were determined: the tilt of the tibial component in the coronal plane relative to the long axis of the tibia (range 3.4° valgus to 4.5° varus), the component's posterior slope (3.1° to 11.4°), and its rotation (11.9° internal to 7.6° external). The mechanical axis angle (MAA), defined as the angle between the tibial and femoral mechanical axes, was derived from static coronal plane X-rays (4.5 valgus to 7.0° varus). In order to investigate the influence of alignment on the internal loads at the knee joint, the peak in vivo TFCF of each trial compared to the signed and absolute values of the aforementioned four static parameters using linear regression and one-way ANOVA (SPSS 22.0, IBM Corp., Armonk, NY). Here, the coefficient of determination ($R^2$) was used as a measure to quantify the variability in the internal forces explained by parameters of static limb alignment. The ANOVA *p*-value was used to examine the statistical significance of potential correlations.

#### Musculoskeletal modelling

A detailed description of the employed musculoskeletal modelling approach can be found in an earlier article [23], hence only a short description follows. Patient-specific bone anatomies were derived from post-operative CT scans, and combined with 51 reference muscle lines of action and wrapping points based on the Visible Human dataset (Fig. 1). The subsequent musculoskeletal analysis was performed with custom code written in C++, using optimization routines from the KNITRO (active set algorithm) and NAG (e04ucc) libraries. Functional methods were used to localize the dynamic joint centres and axes as previously reported [31–34]. The patient-specific skeletons were then aligned with the functional joint centres and axes using an inverse kinematics approach, to derive the skeletal kinematics and the muscle lines of action for the leg with the instrumented knee implant (Fig. 1). The hip and the ankle joint had 3 rotational degrees of freedom (DoF), while the tibio-femoral (TF) joint used the distal femur geometry to constrain the 6 DoF transformation between tibia and femur. Patello-femoral (PF) kinematics were determined based on a geometrical model driven by the TF kinematics under the assumption of constant patellar tendon length [23].

The joints and their local coordinate systems of the patient specific skeleton were defined as follows: hip centre – the centre of a sphere fitted to the femoral head geometry; hip proximal/distal (P/D) axis – perpendicular to the plane formed by the anterior superior iliac spines and the midpoint between the posterior superior iliac spines; hip anterior/posterior (A/P) axis – perpendicular to P/D axis and the line though both hip centres, hip medio/lateral (M/L) axis – perpendicular to P/D and A/P hip axes; knee centre – midpoint between the femoral epicondyles; knee P/D axis – line through knee and hip centres; knee A/P axis – perpendicular to knee P/D axis and the line though both femoral epicondyles; knee M/L axis – perpendicular to P/D and A/P knee axes; ankle centre – centre of circle though three points placed along the trochlea tali; ankle P/D axis – line through ankle centre and intercondylar eminence; ankle A/P axis – perpendicular to ankle P/D axis and the line though both malleoli; ankle M/L axis – perpendicular to P/D and A/P ankle axes; leg axis - the line connecting the of hip and ankle centres.

Previous studies have shown that muscle moments balance only ~ 1/3 of the external knee adduction moment (EAM) during walking and stair climbing, while the remaining ~ 2/3 are balanced passively by the contact moment resulting from asymmetrical distribution of the axial load among the condyles [10]. To determine the required muscles moment the in the frontal plane, the joint reaction force from inverse dynamics was distributed among the condyles to balance as much of the EAM as possible. During the subsequent muscle optimization, the muscles moments were only required to account for the portion of the EAM, which could not be balanced by the medio-lateral distribution of the joint reaction force from inverse dynamics. To verify this approach, the total axial muscle force from optimization was split evenly among both condyles, and added to the joint reaction forces from inverse dynamics. When compared to in vivo measured the medio-lateral distribution out approach predicted the medial ratio ($F_{med}$ / ($F_{med} + F_{lat}$)) with an error of 0.08 ± 0.05 (mean ± SD) in the constrained model over all time points of stance phase.

In order to determine the optimization criterion that best reproduces peak in vivo TFCFs, the contact forces throughout the entire stance phase were computed using

8 commonly utilized optimization criteria [20, 23, 25–29]: sum of muscle forces, sum of muscle stresses (linear, squared, cubed for both), and sum of JCFs (linear and squared) at the hip, knee and ankle (Additional file 1: Figure S1). In order to make the model predictions consistent with the individual total muscle forces levels, the in vivo measured TFCF was introduced as boundary condition into the optimization by combining the minimisation of the **S**um of **M**uscle **S**tresses **S**quared (*SMSS*) with the minimization of the **E**rror in the **TF** contact force prediction (*ETF*):

$$ETF = |F_{model} - F_{\text{in vivo}}|$$

yielding the new constrained optimization criterion (*COC*) as the weighted sum of the two components:

$$COC = nSMSS + w \times nETF^2$$

where *nSMSS* and *nETF* are normalized values of *SMSS* and *ETF*, while *w* is a weighting factor. The normalized values *nSMSS* and *nETF* were computed by dividing *SMSS* and *ETF* by their respective values that would result from a maximal contraction of all muscles. Note that for the ETF term the normalization factor is determined individually for each given time point. The choice of the value of *w* was informed by the maximal *ETF* within each trial, expressed as function of *w*: $ETF_{max}(w)$. For a value of $w = 10$ the mean slope of $ETF_{max}(w)$ for each activity and patient dropped to 5% or less of the initial slope at $w = 0$, meaning that a further increase of *w* would not reduce $ETF_{max}$ in a relevant manner (Additional file 1: Figure S2). The difference between COC and SMSS in terms of TFCFs and individual muscle forces was computed for all three activities, at the two peaks of the in vivo TFCF.

Based on the muscle forces from the model, the amount of co-contraction was quantified based on the work of Rudolph et al. [35] by the co-contraction index (CCI) defined as:

$$\text{CCI} = (F_1 + F_2) \times \text{MIN}(F_1, F_2) / \text{MAX}(F_1, F_2)$$

where $F_1$ and $F_2$ are the total forces of two antagonistic muscle groups.

### EMG processing and validation of muscle activities

To validate the predicted muscle activation patterns against the muscle activation of the major knee muscles, EMG signals were recorded for six of the nine patients at 9600 Hz using a measurement system by Biovision (Wertheim, Germany) and processed using MATLAB (version 7.7.0, MathWorks, Inc., MA). The frequencies relevant to muscle activation (10-500 Hz) were extracted using a zero-phase Butterworth filter, after which the signals were rectified and smoothed using a two-pass sliding average with a 0.1 s window. The processed EMG signals were aggregated into 3 groups (vasti, gastrocnemii and hamstrings) by adding the medial and lateral signals within each group. The forces predicted by the *COC*-model were added similarly: vasti = vastus med. + vastus lat.; gastrocnemii = gastrocnemius med. + gastrocnemius lat.; hamstrings = semitendinosus + semimembranosus + biceps femoris. For a comparison between the resulting EMG envelopes and the muscle forces predicted by *COC*-model, the signal of each muscle group in both data sets was normalized to its maximum during all stance phases of the same patient and activity [36]. The activation state of the muscle groups was identified from the normalized data using a 50% threshold of the maximal value [37]. The temporal agreement was quantified as the fraction of the time during which the on/off state was consistent between EMG and the model.

## Results

### Weak correlations between static anatomy and in vivo peak tibio-femoral contact forces

Peak TFCF magnitudes varied substantially across the nine subjects. The ranges for the mean (±SD) peak TFCFs were 2.05 ± 0.10–3.48 ± 0.11 BW (bodyweight) for walking, 2.30 ± 0.06–4.39 ± 0.26 BW for stair ascent, and 2.95 ± 0.06–4.37 ± 0.21 BW for stair descent. Linear regression did not imply strong relationships between the mean in vivo peak TFCFs and any of the implantation angles of the tibial component or the mechanical axis angle (MAA) of the leg. The coefficient of determination ($R^2$) was never higher than 0.3 and the ANOVA *p*-value was never below 0.14.

### Constraining the model with in vivo data

Among the eight initially investigated optimization criteria, the best overall match of peak TFCF between model and measurement was achieved when the sum of squared muscle stresses was minimized (*SSMS*), with mean relative errors of the peak TFCF predictions ($|F_{model} - F_{\text{in vivo}}| \;/\; F_{\text{in vivo}}$) of 16 ± 14% for walking, 14 ± 11% for stair ascent and 13 ± 9% stair descent (Additional file 1: Figure S1). When minimization of squared muscle stresses was combined with matching the in vivo TFCF in the constrained optimization criterion (*COC*), these relative errors were reduced to 4 ± 6%, 1 ± 1% and 3 ± 4% of the in vivo TFCF respectively for walking, stair ascent and stair descent (Fig. 2a). The largest remaining errors occurred during walking for K9L (21 ± 5%) and stair descent for K8L (11 ± 3%). The predicted peak hip contact force changed from 3.63 ± 0.31BW, 3.46 ± 0.49BW and 3.46 ± 0.45BW for SSMS to 3.70 ± 0.28BW, 3.63 ± 0.54BW and 3.60 ± 0.49BW for *COC*, respectively for walking, stair ascent and stair descent.

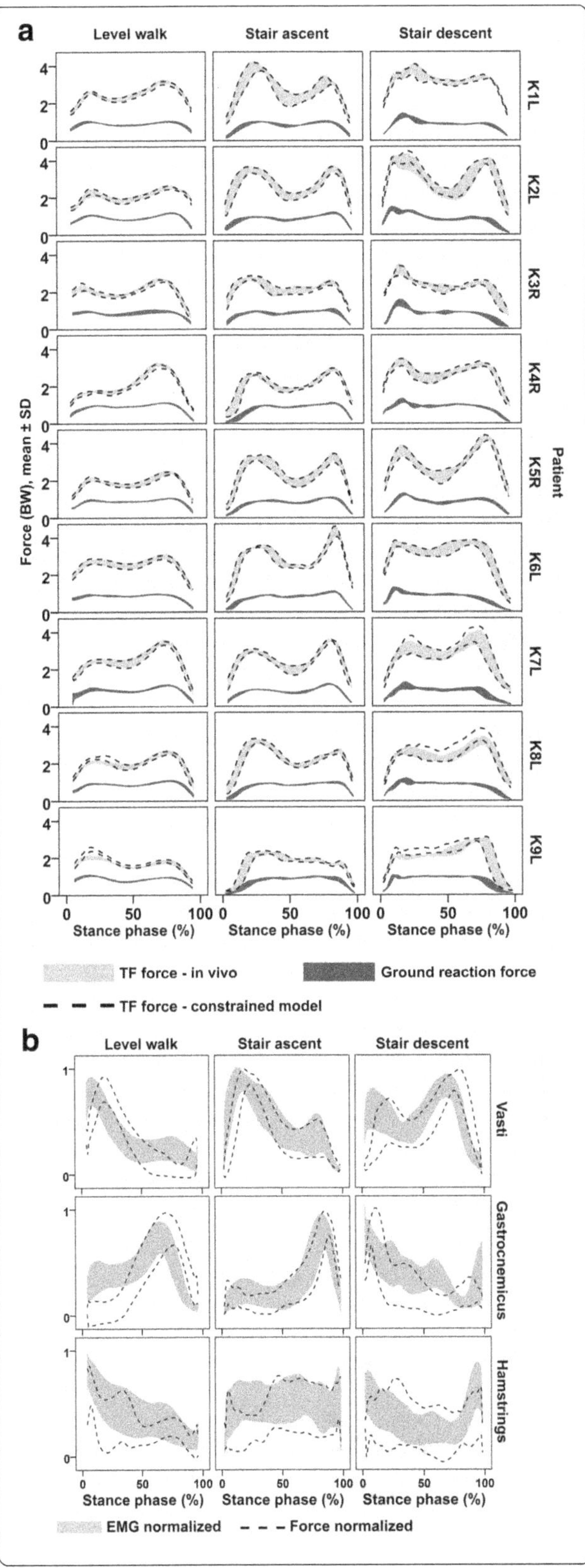


**Fig. 2** Comparison of constrained model results and measurements **a** Tibio-femoral forces measured in vivo (light grey), tibio-femoral forces from the constrained model (between dashed lines), and the measured ground reaction forces (dark grey).The indicated range covers the mean ± 1SD as determined from repeated trials. **b** Comparison of the EMG data (recorded in 6 of the 9 patients) to the computed muscle forces from the constrained model, normalized by the maximum of each patient and activity

For the six patients with available EMG data, the muscle activation profiles determined from *COC* were generally in good qualitative agreement with the EMG patterns (Fig. 2b). On average the activation states of the EMG and the model were in agreement 77% of the total stance phase. The weakest activation state agreement was observed for the hamstrings during stair ascent with 62 ± 8% (mean ± SD) of the time, ranging from 52 to 71% across patients. The best agreement was found for the gastrocnemii during stair ascent with 89 ± 6% of the time, ranging from 78 to 95% across patients (Table 1).

### Constrained muscle forces at the knee

Compared to SMSS model the COC model increased TFCF by up to 1.70 ± 0.14BW (2nd peak of stair ascent, patient K6L), and reduced it by up to 0.63 ± 0.10BW (1st peak during walking, patient K9L). While TFCF decreased in most patients during the 1st peaks of walking and stair ascent, it usually increased at the other investigated four TF force peaks (Fig. 3a). The reductions of TFCF were mainly achieved by reducing the co-contraction of the quadriceps and the hamstrings by up to 0.59 ± 0.05BW (1st peak during stair ascent, patient K1L), which coincided with increased forces of the muscles crossing only the hip. The increases of TFCF were mainly due to an increase in co-contraction of the quadriceps and the gastrocnemii by up to 1.45 ± 0.31BW (2nd peak of stair ascent,

**Table 1** The fraction of time during which the predicted muscle activation state matches the EMG derived activation state

| | | Fraction of time with activation match | |
|---|---|---|---|
| Muscle group | Activity | mean ± SD | range across patients |
| Vasti | Level walk | 83 ± 06% | 76–91% |
| | Stair ascent | 82 ± 03% | 79–86% |
| | Stair descent | 66 ± 09% | 57–83% |
| Gastrocnemii | Level walk | 69 ± 08% | 60–79% |
| | Stair ascent | 89 ± 06% | 78–95% |
| | Stair descent | 80 ± 06% | 72–90% |
| Hamstrings | Level walk | 79 ± 09% | 67–89% |
| | Stair ascent | 62 ± 08% | 52–71% |
| | Stair descent | 69 ± 17% | 54–92% |

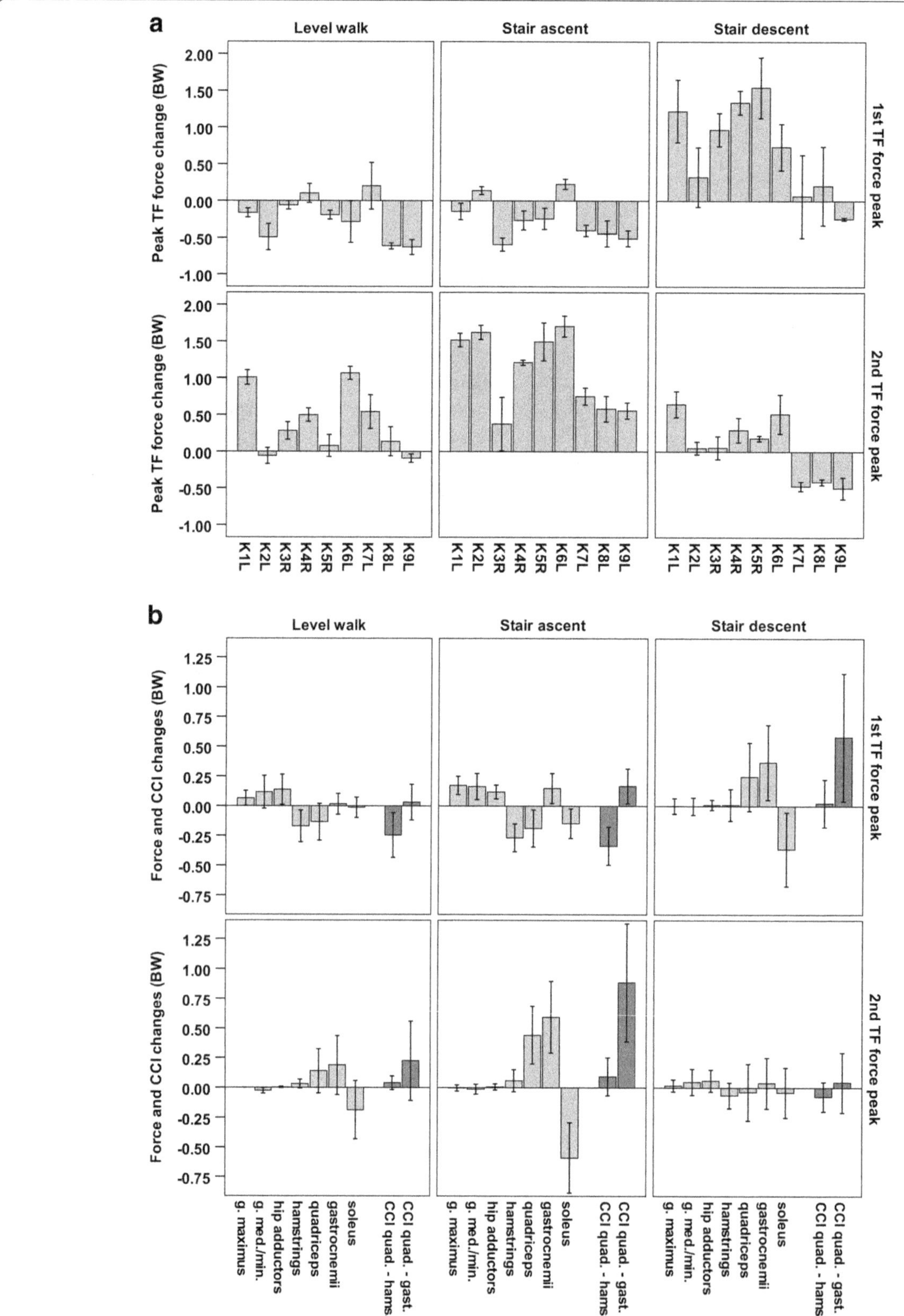


**Fig. 3** Changes in the internal forces introduced by constraining the model to match in vivo TF forces (COC), compared to minimizing the sum of muscle stresses squared (SMSS). **a** The changes in TFCF per patient. **b** The most relevant changes of forces by individual muscle groups (light grey), and the changes in co-contraction (CCI) of the knee flexors and extensors (dark grey), averaged for all patients (mean ± SD)

patient K2L), which coincided with decreased soleus activity (Fig. 3b).

## Discussion

Degeneration of natural and failure of replaced joints is, among other factors, frequently associated with mechanical overloading [14, 38–41]. Frequently, the joint axis as seen in X-rays is assumed to be the key determinant of mechanical forces acting at the joint and surgical strategies are judged by the quality of a so called "mechanical" or "anatomical" axis alignment. However, the forces acting in vivo and their determinants are widely unknown. From instrumented implant measurements, the in vivo JCFs are known to be multiples of bodyweight, even during normal walking. Such high forces are the direct consequence of muscle activity contributing up to 3/4 of the total TFCF [10]. Substantial variation in the in vivo JCFs for a given physical task suggests a high variability, not only in muscle recruitment patterns but also in the total forces exerted by the muscles [7, 9]. However, previous attempts to understand the exact role of the muscles in mechanical overload have been challenged by the difficulty to quantify *in vivo* muscle forces. Whilst EMG measurements can provide information about the temporal activation patterns of a subset of superficial muscles, there is no trivial, direct link between muscle activation levels and muscle forces [42].

Musculoskeletal modelling can provide a prediction of joint and muscle forces, but such modelling approaches have not been shown to predict TF forces accurately across more than 2 subjects [23, 43]. The presented analyses confirm that conventional approaches to determine muscle forces, aimed at efficient muscle employment [26, 44], cannot predict the individual level of antagonist muscle co-contraction. In these conventional approaches, co-contraction is avoided as a suboptimal strategy, requiring alternative methods to predict muscle co-contraction [42, 45]. We therefore used the largest available group of 9 patients with instrumented knee implants [24] to constrain personalized musculoskeletal models, and to quantify the individual amount of muscle forces resulting from co-contraction during 3 activities of daily living. To our knowledge, this is the first time that *in vivo* contact forces and in silico modelling were matched for 9 patients and multiple activities with varying knee flexion ranges. The large range of in vivo measured TF contact forces [9] within our cohort, with 1.7-fold variation even at similar gait speeds, provides a significantly wider representation of inter-subject variability than in any previous analysis. Importantly, our analyses show that this TF force variability could not be explained with differences in MAA or tibial implantation alignment, parameters that have previously been a key focus of research and clinical management of knee OA [46, 47].

The comparison of the analyses constrained by the in vivo joint contact forces (COC) to analysis strategy using commonly employed optimality criteria to estimate JCFs (SMSS) revealed that additional muscle co-contraction can increase the peak of TFCF by up to 66% in late stance during stair ascent, and by up to 56% in late stance during walking (Fig. 3a). These increases of the TFCF were the result of up to ~1BW additional force produced by the co-contraction of the gastrocnemii and the quadriceps. The variations in joint contact forces of peak forces during walking provide an indication of how far off results computed under the assumption of "optimal" muscle activation can be from reality, and underlines the decisive role of co-contraction for JCFs. Moreover, the muscle activation patterns found when constraining the solution by the in vivo measured TFCF resulted in changes to the contact forces also across the hip joint. Failure to accurately capture the manner in which muscles are activated is therefore likely to results in errors in the internal musculoskeletal loading conditions throughout the entire lower limb. While the SMSS model prefers to activate the soleus during the late stance push off, the in vivo strategy in most subjects appears to be the synergetic use of the gastrocnemii and the quadriceps. Such a strategy could allow the quadriceps to indirectly contribute to the plantar flexion moment, at the cost of higher TFCF. On the other hand, the in vivo based model resulted in lower TFCFs during the early stance phase of walking and stair ascent, by employing less co-contraction of the hamstrings and the quadriceps than the SMSS model.

Current approaches to the treatment of joint injury and degeneration primarily focus on passive structures and their effect on load distribution and joint function [46, 47]. Our findings indicate that even an "optimally" reconstructed joint axis does not necessarily prevent mechanical overloading of the joint. To avoid such overloading, the muscle activity level and their control have to be taken into account. Antagonist muscle co-contraction can also reduce tensile stresses in bones under bending loads and thus reduce fracture risk [48]. Furthermore, while co-contraction does not contribute to balancing the external moment, it can substantially stiffen the joint and thus it more stable against unexpected changes in the external load [49, 50]. A need or desire for increased joint stability at the expense of movement efficiency may thus explain increased levels of co-contraction. How the individual muscle co-contraction can be assessed, and if an intra-operative ligament tensioning or balancing could help to adjust a patients' individual soft tissue mechanics needs to be further researched. Also, it remains unclear why some subjects require higher levels of joint stabilization by muscle co-contraction, and how such individual systematic "overloading" of joints could be modulated or prevented.

The approach to quantify the muscle forces based on in vivo TFCFs as used in this study has certain limitations. Whilst the introduction of the in vivo TFCFs as boundary conditions to the optimization problem reduces the available solution space, and thereby eliminates a considerable proportion of solutions not consistent with the overall extent of muscle forces acting at the joint, there is no guarantee that the muscle activation pattern identified in this manner is indeed the very pattern utilized by a subject. However, we believe that using the optimization criterion with the least TFCF prediction error, and constraining it further to match the actual in vivo TFCF, provides the best currently possible estimation of muscle forces at the knee. Further studies which, additionally to the in vivo JCFs, also use comprehensive EMG measurements as boundary conditions to further constrain the solution space could provide even more realistic muscle force estimations [51]. While our small cohort of TKR patients cannot represent all habitual and cultural variations of human locomotion, it is still the largest group worldwide in which in vivo measurements of the TFCF are possible. Similarly to TKR patients, increased levels of co-contraction have also been previously reported in patients with advanced stages of OA [11–13, 15, 52, 53]. Despite its limited size, our cohort represents a wide range of TF loading levels and co-contraction, enabling an unprecedented quantitative analysis of the influence of muscle-co-contraction on internal loading conditions at the knee.

## Conclusions

In conclusion, we found that muscle forces dominate joint loading during dynamic activities, and that individual variations in antagonist muscle co-contraction can lead to substantially different JCFs. Treatment and rehabilitation of diseased and failed joints should thus not only consider the alignment of static limb axes, but also consider the dynamic activation of muscles as a promising new target for restoring function, patient mobility and preventing future joint failure. Computational models can be instrumental for realizing that vision, but need to consider the large variability in muscle activation strategies employed by TJR patients as demonstrated in this study.

## Additional file

**Additional file 1: Figure S1.** Tibio-femoral joint contact force (TFCF) prediction errors as function of the 9 optimization criteria used in the study. (Top) RMS errors (mean ± SD) of the prediction throughout the stance phase (top). (Bottom) Absolute difference of predicted and measured peak TFCFs. The constrained model (*COC*) uses the combination of squared muscle stresses and in vivo TFCFs, which was used to quantify the co-contraction. **Figure S2.** Maximal errors in TFCF prediction per trial ($ETF_{max}$, mean ± SD) as function of the weight w for the constraint enforcing the measured TFCF in the *COC* optimization criterion. The indicated $w = 10$ was the value at which the mean slope of $ETF_{max}(w)$ dropped to below 5% of its initial value at $w = 0$, and which was thus used in the subsequent analyses. At a value of $w = 10$ the mean ± SD of the maximum error of the TFCF was 0.16 ± 0.13BW for level walking, 0.10 ± 0.05BW for stair ascent and 0.29 ± 0.22BW for stair descent. (DOC 163 kb)

**Acknowledgements**
The authors would like to thank Dr. Alison Agres for help in formulating the manuscript.

**Funding**
This work was supported by the German Research Foundation (DFG) (Be 804/18–1), the Federal Ministry of Education and Research (OVERLOAD-PrevOP, 01EC1408A), the Open Access Publication Fund of Charité – Universitätsmedizin Berlin and the OrthoLoad Club. The funders had no role in study design, data collection and analysis, decision to publish or preparation of the manuscript.

**Authors' contributions**
AT, GND generated the study concept, performed the analyses and drafted the manuscript with input from all authors. IK, VS, AT, performed in vivo experiments, processed and analysed the data; contributed to writing the manuscript. TP, MOH contributed fundamental ideas, especially from the clinical perspective, helped to design the study and to formulate the paper. All authors were involved in critically revising the manuscript and approved the submitted version.

**Competing interests**
T. Pfitzner reports speaker fees from Link, Smith & Nephew and consultancy fees from DepuySynthes.

**Author details**
[1]Julius Wolff Institute, Charité – Universitätsmedizin Berlin, corporate member of Freie Universität Berlin, Humboldt-Universität zu Berlin, and Berlin Institute of Health, Augustenburger Platz 1, 13353 Berlin, Germany. [2]Bioengineering Sciences Research Group, Faculty of Engineering and Physical Sciences, University of Southampton, Southampton, UK. [3]Center for Musculoskeletal Surgery, Charité – Universitätsmedizin Berlin, corporate member of Freie Universität Berlin, Humboldt-Universität zu Berlin, and Berlin Institute of Health, Berlin, Germany. [4]Clinic for Adult Hip and Knee Reconstruction, Vivantes Spandau Hospital, Berlin, Germany.

**References**

1. Andriacchi TP. Osteoarthritis: Probing knee OA as a system responding to a stimulus. Nat Rev Rheumatol. 2012;8(7):371–2.
2. Paley D. Principles of deformity correction: Springer Berlin Heidelberg; 2002.
3. Kettelkamp DB, Chao EY. A method for quantitative analysis of medial and lateral compression forces at the knee during standing. Clin Orthop Relat Res. 1972;83:202–13.

4. Bennell KL, Wrigley TV, Hunt MA, Lim BW, Hinman RS. Update on the role of muscle in the genesis and management of knee osteoarthritis. Rheum Dis Clin N Am. 2013;39(1):145–76.
5. Wilson DR, McWalter EJ, Johnston JD. The measurement of joint mechanics and their role in osteoarthritis genesis and progression. Rheum Dis Clin N Am. 2008;34(3):605–22.
6. Bergmann G, Deuretzbacher G, Heller M, Graichen F, Rohlmann A, Strauss J, et al. Hip contact forces and gait patterns from routine activities. J Biomech. 2001;34(7):859–71.
7. Damm P, Schwachmeyer V, Dymke J, Bender A, Bergmann G. In vivo hip joint loads during three methods of walking with forearm crutches. Clin Biomech (Bristol, Avon). 2013;28(5):530–5.
8. D'Lima DD, Patil S, Steklov N, Slamin JE, Colwell CW Jr. Tibial forces measured in vivo after total knee arthroplasty. J Arthroplast. 2006;21(2):255–62.
9. Kutzner I, Heinlein B, Graichen F, Bender A, Rohlmann A, Halder A, et al. Loading of the knee joint during activities of daily living measured in vivo in five subjects. J Biomech. 2010;43(11):2164–73.
10. Trepczynski A, Kutzner I, Bergmann G, Taylor WR, Heller MO. Modulation of the relationship between external knee adduction moments and medial joint contact forces across subjects and activities. Arthritis Rheumatol. 2014; 66(5):1218–27.
11. Hubley-Kozey CL, Hill NA, Rutherford DJ, Dunbar MJ, Stanish WD. Co-activation differences in lower limb muscles between asymptomatic controls and those with varying degrees of knee osteoarthritis during walking (vol 24, pg 407, 2009). Clin Biomech. 2009;24(6):529.
12. Heiden TL, Lloyd DG, Ackland TR. Knee joint kinematics, kinetics and muscle co-contraction in knee osteoarthritis patient gait. Clin Biomech (Bristol, Avon). 2009;24(10):833–41.
13. Schmitt LC, Rudolph KS. Influences on knee movement strategies during walking in persons with medial knee osteoarthritis. Arthritis Rheum. 2007;57(6):1018–26.
14. Felson DT. Osteoarthritis as a disease of mechanics. Osteoarthr Cartil. 2013; 21(1):10–5.
15. Benedetti MG, Catani F, Bilotta TW, Marcacci M, Mariani E, Giannini S. Muscle activation pattern and gait biomechanics after total knee replacement. Clin Biomech (Bristol, Avon). 2003;18(9):871–6.
16. Yoshida Y, Mizner RL, Snyder-Mackler L. Association between long-term quadriceps weakness and early walking muscle co-contraction after total knee arthroplasty. Knee. 2013;20(6):426–31.
17. Levin O, Mizrahi J, Isakov E. Transcutaneous FES of the paralyzed quadriceps: is knee torque affected by unintended activation of the hamstrings? J Electromyogr Kinesiol. 2000;10(1):47–58.
18. Hicks JL, Uchida TK, Seth A, Rajagopal A, Delp SL. Is my model good enough? Best practices for verification and validation of musculoskeletal models and simulations of movement. J Biomech Eng. 2015;137(2):020905.
19. D'Lima DD, Fregly BJ, Patil S, Steklov N, Colwell CW. Knee joint forces: prediction, measurement, and significance. P I Mech Eng H. 2012;226(H2):95–102.
20. Heller MO, Bergmann G, Deuretzbacher G, Dürselen L, Pohl M, Claes L, et al. Musculo-skeletal loading conditions at the hip during walking and stair climbing. J Biomech. 2001;34(7):883–93.
21. Lin YC, Walter JP, Banks SA, Pandy MG, Fregly BJ. Simultaneous prediction of muscle and contact forces in the knee during gait. J Biomech. 2010;43(5):945–52.
22. Stansfield BW, Nicol AC, Paul JP, Kelly IG, Graichen F, Bergmann G. Direct comparison of calculated hip joint contact forces with those measured using instrumented implants. An evaluation of a three-dimensional mathematical model of the lower limb. J Biomech. 2003;36(7):929–36.
23. Trepczynski A, Kutzner I, Kornaropoulos E, Taylor WR, Duda GN, Bergmann G, et al. Patellofemoral joint contact forces during activities with high knee flexion. J Orthop Res. 2012;30(3):408–15.
24. Heinlein B, Graichen F, Bender A, Rohlmann A, Bergmann G. Design, calibration and pre-clinical testing of an instrumented tibial tray. J Biomech. 2007;40(Supplement 1):S4–S10.
25. Li G, Kaufman KR, Chao EY, Rubash HE. Prediction of antagonistic muscle forces using inverse dynamic optimization during flexion/extension of the knee. J Biomech Eng. 1999;121(3):316–22.
26. Crowninshield RD, Brand RA. A physiologically based criterion of muscle force prediction in locomotion. J Biomech. 1981;14(11):793–801.
27. Jackson M, Benkhemis I, Begon M, Sardain P, Vallee C, Lacouture P. Identifying the criterion spontaneously minimized during the take-off phase of a sub-maximal long jump through optimal synthesis. Multibody Syst Dyn. 2012;28(3):225–37.
28. Herzog W, Binding P. Cocontraction of pairs of antagonistic muscles: analytical solution for planar static nonlinear optimization approaches. Math Biosci. 1993;118(1):83–95.
29. Pedotti A, Krishnan VV, Stark L. Optimization of muscle-force sequencing in human locomotion. Math Biosci. 1978;38(1–2):57–76.
30. Kutzner I, Trepczynski A, Heller MO, Bergmann G. Knee adduction moment and medial contact force--facts about their correlation during gait. PLoS One. 2013;8(12):e81036.
31. Ehrig RM, Heller MO, Kratzenstein S, Duda GN, Trepczynski A, Taylor WR. The SCoRE residual: a quality index to assess the accuracy of joint estimations. J Biomech. 2011;44(7):1400–4.
32. Ehrig RM, Taylor WR, Duda GN, Heller MO. A survey of formal methods for determining the Centre of rotation of ball joints. J Biomech. 2006;39(15): 2798–809.
33. Ehrig RM, Taylor WR, Duda GN, Heller MO. A survey of formal methods for determining functional joint axes. J Biomech. 2007;40(10):2150–7.
34. Taylor WR, Kornaropoulos EI, Duda GN, Kratzenstein S, Ehrig RM, Arampatzis A, et al. Repeatability and reproducibility of OSSCA, a functional approach for assessing the kinematics of the lower limb. Gait Posture. 2010;32(2):231–6.
35. Rudolph KS, Axe MJ, Snyder-Mackler L. Dynamic stability after ACL injury: who can hop? Knee Surg Sports Traumatol Arthrosc. 2000;8(5):262–9.
36. Bolgla LA, Uhl TL. Reliability of electromyographic normalization methods for evaluating the hip musculature. J Electromyogr Kinesiol. 2007;17(1):102–11.
37. Howard RM, Conway R, Harrison AJ. Muscle activation sequencing of leg muscles during linear glide shot putting. Sports Biomech. 2017;16(4):463–84.
38. Bennell KL, Creaby MW, Wrigley TV, Bowles KA, Hinman RS, Cicuttini F, et al. Bone marrow lesions are related to dynamic knee loading in medial knee osteoarthritis. Ann Rheum Dis. 2010;69(6):1151–4.
39. Miyazaki T, Wada M, Kawahara H, Sato M, Baba H, Shimada S. Dynamic load at baseline can predict radiographic disease progression in medial compartment knee osteoarthritis. Ann Rheum Dis. 2002;61(7):617–22.
40. Neogi T, Nevitt M, Niu J, Sharma L, Roemer F, Guermazi A, et al. Subchondral bone attrition may be a reflection of compartment-specific mechanical load: the MOST study. Ann Rheum Dis. 2010;69(5):841–4.
41. Wimmer MA, Andriacchi TP, Natarajan RN, Loos J, Karlhuber M, Petermann J, et al. A striated pattern of wear in ultrahigh-molecular-weight polyethylene components of miller-Galante total knee arthroplasty. J Arthroplast. 1998;13(1):8–16.
42. Buchanan TS, Lloyd DG, Manal K, Besier TF. Neuromusculoskeletal modeling: estimation of muscle forces and joint moments and movements from measurements of neural command. J Appl Biomech. 2004;20(4):367–95.
43. Fregly BJ, Besier TF, Lloyd DG, Delp SL, Banks SA, Pandy MG, et al. Grand challenge competition to predict in vivo knee loads. J Orthop Res. 2012; 30(4):503–13.
44. Kaufman KR, An KW, Litchy WJ, Chao EY. Physiological prediction of muscle forces--I. theoretical formulation. Neuroscience. 1991;40(3):781–92.
45. Forster E, Simon U, Augat P, Claes L. Extension of a state-of-the-art optimization criterion to predict co-contraction. J Biomech. 2004;37(4):577–81.
46. Benjamin J. Component alignment in total knee arthroplasty. Instr Course Lect. 2006;55:405–12.
47. Sikorski JM. Alignment in total knee replacement. J Bone Joint Surg British Vol. 2008;90(9):1121–7.
48. Mizrahi J, Verbitsky O, Isakov E. Fatigue-related loading imbalance on the shank in running: a possible factor in stress fractures. Ann Biomed Eng. 2000;28(4):463–9.
49. Hogan N. Adaptive-control of mechanical impedance by Coactivation of antagonist muscles. Ieee T Automat Contr. 1984;29(8):681–90.
50. Heitmann S, Ferns N, Breakspear M. Muscle co-contraction modulates damping and joint stability in a three-link biomechanical limb. Front Neurorobot. 2011;5:5.
51. Mizrahi J. The role of electromyograms in resolving musculoskeletal interactions in able-bodied and disabled human individuals. In: Mizrahi J, editor. Advances in applied electromyography: IntechOpen; 2011. p. 3–24.
52. Fallah-Yakhdani HR, Abbasi-Bafghi H, Meijer OG, Bruijn SM, van den Dikkenberg N, Benedetti MG, et al. Determinants of co-contraction during walking before and after arthroplasty for knee osteoarthritis. Clin Biomech (Bristol, Avon). 2012;27(5):485–94.
53. Zeni JA, Rudolph K, Higginson JS. Alterations in quadriceps and hamstrings coordination in persons with medial compartment knee osteoarthritis. J Electromyogr Kinesiol. 2010;20(1):148–54.

# 12

# Overground walking patterns after chronic incomplete spinal cord injury show distinct response patterns to unloading

Christopher Schmidt Easthope[1*], Luca Renato Traini[1], Lea Awai[1,2], Martina Franz[1], Georg Rauter[3], Armin Curt[1] and Marc Bolliger[1]

**Abstract**

**Background:** Body weight support (BWS) is often provided to incomplete spinal cord injury (iSCI) patients during rehabilitation to enable gait training before full weight-bearing is recovered. Emerging robotic devices enable BWS during overground walking, increasing task-specificity of the locomotor training. However, in contrast to a treadmill setting, there is little information on how unloading is integrated into overground locomotion. We investigated the effect of a transparent multi-directional BWS system on overground walking patterns at different levels of unloading in individuals with chronic iSCI (CiSCI) compared to controls.

**Methods:** Kinematics of 12 CiSCI were analyzed at six different BWS levels from 0 to 50% body weight unloading during overground walking at $2kmh^{-1}$ and compared to speed-matched controls.

**Results:** In controls, temporal parameters, single joint trajectories, and intralimb coordination responded proportionally to the level of unloading, while spatial parameters remained unaffected. In CiSCI, unloading induced similar changes in temporal parameters. CiSCI, however, did not adapt their intralimb coordination or single joint trajectories to the level of unloading.

**Conclusions:** The findings revealed that continuous, dynamic unloading during overground walking results in subtle and proportional gait adjustments corresponding to changes in body load. CiSCI demonstrated diminished responses in specific domains of gait, indicating that their altered neural processing impeded the adjustment to environmental constraints. CiSCI retain their movement patterns under overground unloading, indicating that this is a viable locomotor therapy tool that may also offer a potential window on the diminished neural control of intralimb coordination.

**Keywords:** Body weight support, Gait pattern, Spinal cord injury, Walking, Unloading

## Background

In lower limb rehabilitation of patients with central nervous system (CNS) disorders such as incomplete spinal cord injury (iSCI), gait training with or without body weight support (BWS) is an established intervention [1, 2]. Evidence for recovery induced by locomotor training has been extensively demonstrated in animal models [3, 4]. Underlying this recovery is the neuroplasticity of spinal networks: Repetitive activation of task-specific input-output relationships that result in successful function favor remodeling, leading to neural adaptation in response to the injury [5, 6]. In humans with iSCI, there are numerous published investigations that indicate that gait training leads to accelerated recovery of walking function [7–9]. Compared to the extensive gains observed in animals however, the functional improvements in humans remain limited [10]. This is attributed to various factors, including species-specific anatomy and neurophysiology [11], low sensitivity of outcome measures [12] and suboptimal dosing/timing of the intervention [13]. In coherence with the concept of exploiting the neuroplasticity of residual circuits, it is established that training must be task-specific, active and tailored to the individual patient's capacity [5].

Recent experiments in rats with iSCI have demonstrated that larger functional gains in basic and skilled

* Correspondence: chris.schmidt@balgrist.ch
[1]Spinal Cord Injury Center, Balgrist University Hospital, Forchstrasse 340, CH-8008 Zürich, Switzerland
Full list of author information is available at the end of the article

locomotion are achieved when training is conducted overground as opposed to training on a treadmill [14]. This implies that the mechanisms driving neural recovery are more actively challenged during overground locomotion training, potentially due to subtle differences on the treadmill including the lack of goal direction and the more passive nature of hip extension (especially during fixed-point unloading) [15, 16]. A case study [17] and more recent review [7] conclude that, in order to be maximally task-specific, neurological locomotor rehabilitation should be optimally conducted overground [5]. While this is readily possible with patients who can support their own body weight, severely affected patients are dependent on BWS. These insights have inspired the development of different ceiling-mounted overground BWS systems [18, 19]. The design of overground BWS systems is challenging however, and the reduction of collateral interaction forces, which potentially impede a patient's walking pattern and body dynamics leading to distorted afferent and efferent signaling, remains a central point [20]. Low transparency of the BWS system in medio-lateral direction can for instance result in a pendulum-like motion around the point of support during walking [21] influencing the regulation of body sway [22, 23].

For effective rehabilitation, patients should hence train overground walking with a device that offers support and safety while allowing natural motion without perturbing forces in addition to the necessary BWS. In this light, we developed the FLOAT, a cable robot that provides overground BWS with minimal interaction forces to allow unimpeded locomotor training [24, 25]. Briefly, the patient is attached to a central node that is suspended via 4 cables, each deflected over a rail-mounted passive trolley and actuated by its own motor. By tensing all 4 cables, vertical BWS is generated. By tensing the anterior cables more than the posterior, additional forward forces can be applied. Constant and smooth BWS is supplied by sensing the forces acting on each cable at the node and relaying this information to a real-time processor that adapts the individual motor torques. In this way, a constant vertical (and/or forward, and/or sideward) force can be generated. When the node is moved by the patient, the distribution of forces at the sensors changes and the node position is updated to restore equilibrium. This leads to low interaction forces both in the first prototype [24] used in this study and a second, commercialized version [26] (The FLOAT, LME, Switzerland).

Unloading in a treadmill setting has been previously described to induce kinematic modifications including reduced relative stance phase, step length, and hip and knee ranges of motion [27–32]. Experiments investigating kinematics in healthy cohorts walking overground using a cable-propelled gantry system [33–35] and an overhead support system (FLOATv1) [25] have led to similar conclusions. However, it is unknown how unloading during overground walking affects gait patterns in individuals with iSCI. While there is inherent robot-human interaction that can potentially distort the gait pattern in a similar manner to controls, unloading also intrinsically engenders reduced cutaneous and loading afferents through the reduced contact forces [36]. Loading afferents are one of the main drivers for spinal networks that have long been considered especially important for chronic iSCI gait [37–39]. A reduction of the loading signal may therefore result in a degradation of the stereotyped gait pattern that is less dependent on neural circuits rostral to the lesion site [40–42]. If overground unloading is to be a useful tool in gait rehabilitation for those patients that retain volitional movement capacity but lack the strength to support their own body weight, it must be ensured that the gait pattern employed during unloading, i.e. with reduced loading information, remains physiological.

This study aimed at describing the kinematic changes induced by different levels of unloading during overground locomotion in an iSCI population. Specifically, we were interested if controlled, continuous, dynamic unloading would lead to changes in individuals with chronic iSCI (CiSCI) walking patterns and how these changes would compare to those previously reported from an able-bodied control group [25]. It was hypothesized that CiSCI show greater changes in their walking patterns in response to body unloading compared to those reported in controls.

## Methods

Fifteen CiSCI who were community ambulators were recruited and gave written informed consent prior to participating in this study. Inclusion criteria were an iSCI at least 3 months prior to inclusion and the ability to walk 10 m without or with only minor (e.g. stick) walking aids in under 20 s. CiSCI were excluded if they had any secondary neurological diagnosis, chronic pain, skin inflammation, or orthopedic or cardiovascular diagnosis or if they were still in active therapy. Control data was extracted from a previously established database of control subjects [25]. The study was approved by the local ethics committee of the Canton of Zurich (KEK Nr. PB_2016–00228) and was conducted in accordance with the Declaration of Helsinki.

To provide BWS, an in-house developed cable robot was employed [24]. The continuous, dynamic, vertical BWS was scaled to each subject's body weight. Gait kinematics were recorded using a passive infrared optical motion capture system (T-series, Vicon, UK), sampling at 200 Hz.

The CiSCI underwent a physical examination by a physician to determine their functional status prior to enrolling into the study. This encompassed walking measures in form of a ten-meter walking test (10mWT) to assess maximal walking speed, functional measures in terms of motor (lower extremity motor scores [LEMS]) and sensory scores (American Spinal Injury Association [ASIA] Impairment Scale [AIS]), and proprioceptive testing. Proprioception tests included assessment of vibration sense using a graded Rydel-Seiffer tuning fork at the Hallux dorsi, the Malleolus and the Patella on both sides and scoring the patient on a scale of 0 to 8 [43]. Furthermore, position sense of the Hallux was recorded on a scale of 0 to 2 [44] and a Romberg test [45] was performed. Information about the latencies of tibial nerve sensory evoked potentials (SEPs) and latencies of motor evoked potentials of the tibialis anterior (MEPs) were retrieved from current clinical records to improve patient characterization.

Prior to the experimental conditions, all participants were weighed and characteristic anatomical measures retrieved to enable accurate calculation of joint center positions [46, 47]. They then were equipped with a BWS harness that was modified to not obscure the pelvis landmarks and a passive reflective marker set was attached to these and other bony landmarks (Plug-in-Gait v3.0, Vicon, UK). For the remainder of the experiment, participants were attached to the BWS-system for unloading.

Participants were required to walk 20 steps in multiple passes of the 2 × 8 m workspace at six different clinically relevant BWS levels in random order (baseline (0%); 10%; 20%; 30%; 40%; 50% BWS). At the baseline level (0%), a small unloading force (~ 4 kg) was applied to enable fluid tracking of the node. Walking speed was acoustically indicated at 0.56 m/s (2 km/h) with a tolerance of 0.14 m/s (0.5 km/h) as this represented the minimal inclusion speed for the CiSCI group: A low tone was given when participants were walking too fast and a high tone when they were too slow. The speed range was employed especially so that CiSCI could walk through the workspace without being constantly corrected due to within-stride fluctuations of node velocity. For each new BWS level, familiarization passes were conducted until speed variability declined to the point that no acoustic warnings were triggered during a pass (no more than 2 passes per BWS level were necessary for any patient).

### Analysis

Marker position data were filtered using Woltring's quintic spline [48] with a laboratory-specific squared mean standard error of 15mm$^2$. Data were then synthesized with anatomical measurements to define segments and joint center locations enabling the calculation of joint angles using a clinical biomechanical model (Plug-in-Gait 3.0, Vicon, UK). Heel strike and toe off events were set manually by a single analyst using heel and toe marker trajectories and velocities. Subsequently the angle and position data were segmented into steps (heel strike to ipsilateral heel strike), time-normalized via linear interpolation to 500 points, and characteristic step parameters were extracted using custom Matlab routines (2016b, the Mathworks, USA). Step parameters included: speed, step length, step width, relative double support phase, step time, and joint (hip, knee, and ankle) ranges of motion. Ranges of motion of the center of mass (CoM) and of trunk inclination were extracted in antero-posterior (AP) and medio-lateral (ML) direction to investigate effects of the BWS system and harness on walking posture. To enable the comparison of waveforms independently of shifts in relative stance phase, joint motion patterns and intralimb coordination were separately analyzed using a stance-swing linear normalization. Joint motion patterns were averaged over all steps in one condition into a characteristic waveform and then compared on the group level. Intralimb coordination was depicted as correspondence between two joint angles in form of cyclograms (hip-knee, knee-ankle). Differences in cyclogram shape between groups and conditions were described by calculating the square root of the sum of squared distances (SSD) from the mean cyclogram at baseline BWS in the control group. Intra-subject consistency in intralimb coordination was quantified using the angular component of coefficient of correspondence (ACC). These approaches have previously been successfully used to sensitively and reliably discern changes in locomotor behavior of iSCI populations [49–51].

CiSCI data was assigned to more and less affected sides based upon a summation of sensory and motor scores for each of the lower limbs. Laterality, however, was mainly determined by the sensory component, as the motor scores were in all cases almost symmetrical. Reference data from healthy controls was averaged for both sides. All side-dependent results are reported for the more affected limb unless the limbs showed strikingly different patterns.

### Statistical analysis

All statistics for time-discrete parameters were calculated in SPSS (v24, IBM Corp., USA). Age, height and weight were normality-verified and compared between groups with unpaired t-tests. All other parameters were verified for normality (Shapiro-Wilk), homogeneity of variances (Levene's), and sphericity (Mauchly), and subsequently compared between BWS levels using a two-way ANCOVA (group x BWS) with one repeated factor (BWS) and age as a covariate. If a significant interaction or BWS main effect was detected, the factor group was fixed and the model repeated. In all cases, simple a-priori contrasts to baseline with a Sidak correction were performed and are reported when eligible. Continuous data was compared in Matlab analogously using a two-way ANOVA (group x BWS) with one repeated factor (BWS) from the spm1d package (v0.4.0; www.spm1d.org) for one-dimensional (time-continuous) data [52, 53]. Main group BWS effects were investigated via paired t-tests with appropriate Bonferroni correction

($p = 0.05/5$) within each group. All results are reported as means ± standard deviation (SD).

## Results

### Population

From the recruited population, 12 CiSCI (age: 51 ± 14 yrs., height: 176 ± 7 cm, weight: 78 ± 17 kg, 3 females) completed the full protocol and were included in the analysis (Table 1). The three excluded CiSCI could not manage the baseline condition of walking 20 strides at 0.56 $ms^{-1}$ without unloading within the tolerance range. Previously published control data [25] of 18 subjects (age: 29 ± 5 yrs., height: 174 ± 9 cm, weight: 71 ± 12 kg, 9 females) was used as a comparator for CiSCI responses to unloading. While age was significantly different between groups ($t(1,28) = 6.35$; $p = 0.000$), height ($t(1,28) = 0.69$; $p = 0.458$) and weight ($t(1,28) = 1.32$; $p = 0.458$) were similar. In comparison to clinical reference values [54], tibial SEP (46 ± 5 ms (CiSCI); 41 ± 3 ms (clinical reference data for mean group height)) and MEP (32 ± 3 ms (CiSCI); 28 ± 2 ms (clinical reference data for mean group height)) latencies were slightly increased in CiSCI. Specifically, in mean, CiSCI SEP latencies were 1.75 ± 1.5 SDs and patient MEP latencies were 2.1 ± 1.9 SDs above their height matched reference values. Walking speed ranged from 0.42–0.67 m/s in CiSCI (55 ± 22% of individual maximal velocity) and 0.5–0.62 m/s in the control group (23 ± 7% of reported maximal velocity [55]). Unloading induced a simple main effect on walking speed ($F(5,130) = 11.110$, $p = 0.000$) in the presence of a group x BWS interaction effect ($F(5,130) = 6.529$, $p = 0.000$). Both groups reduced their speed with increasing unloading ($F(5,50) = 5.867$, $p = 0.000$ (CiSCI); $F(5,85) = 3.431$, $p = 0.007$ (controls)), however this affected CiSCI more strongly than controls (Table 2).

### Spatio-temporal parameters

CiSCI adapted their temporal parameters, namely step time and double support phase to increasing BWS similarly to controls (Table 2). A simple main effect of BWS was present in step time ($F(5,130) = 6.514$, $p = 0.000$) and relative double support phase ($F(5,130) = 36.528$, $p = 0.000$), however there was no interaction effect between group and BWS ($F(5,130) = 0.479$, $p = 0.778$ and $F(5,130) = 1.354$, $p = 0.245$ respectively) (Table 2). Step length showed a simple main effect of BWS ($F(5,130) = 3.066$, $p = 0.012$) and an interaction effect ($F(5,130) = 3.116$, $p = 0.011$), with controls increasing their step length ($F(5,85) = 7.477$, $p = 0.000$) while CiSCI did not ($F(5,50) = 0.916$, $p = 0.478$). Step width on the other hand did not show a simple main effect of BWS ($F(5,130) = 1.756$, $p = 0.126$), however a group x BWS interaction was present ($F(5,130) = 2.435$, $p = 0.038$). Controls developed a local minimal at 10–20%BWS ($F(5,85) = 3.053$, $p = 0.014$), while CiSCI step width remained unchanged ($F(5,50) = 1.491$, $p = 0.209$).

### Walking posture

Walking posture was assessed using CoM motion and trunk sway in AP and ML directions (Table 2). In AP, trunk sway showed a simple main BWS effect ($F(5,130) = 5.353$, $p = 0.000$) and an interaction effect ($F(5,130) = 2.877$, $p = 0.017$). Controls reduced AP trunk sway ($F(5,85) = 16.366$, $p = 0.000$), while this was not detected in CiSCI ($F(5,50) = 0.637$, $p = 0.673$). AP CoM motion did not show a simple

**Table 1** CiSCI characteristics

| | Characteristics | | | | | | | ASIA | | | Sensory | | | SEP | | MEP | |
|---|---|---|---|---|---|---|---|---|---|---|---|---|---|---|---|---|---|
| # | Sex | Age | Lesion Level | TSI [yrs] | Height [cm] | Weight [kg] | 10mWT [s] | LEMS (50) | LT (56) | PP (56) | POS (4) | VIB (48) | RB (2) | L [ms] | R [ms] | L [ms] | R [ms] |
| 01 | m | 35 | C2 | 1.16 | 177 | 74.7 | 6.97 | 49 | 34 | 42 | 4 | 40 | 0 | 45 | 45 | 32 | 31 |
| 02 | m | 68 | C4 | 0.5 | 184 | 99.4 | 8.91 | 47 | 56 | 52 | 3 | 12 | 1 | 42 | 41 | 29 | 29 |
| 03 | m | 64 | C2 | 11 | 172 | 88.8 | 9.15 | 48 | 40 | 44 | 4 | 12 | 2 | 52 | 52 | 43 | 41 |
| 04 | m | 29 | C6 | 1.25 | 176 | 63.7 | 17.21 | 42 | 40 | 33 | 4 | 28 | 2 | 41 | 43 | 30 | 31 |
| 05 | f | 64 | C6 | 4.3 | 166 | 48.5 | 9.27 | 46 | 55 | 52 | 3 | 46 | 1 | 46 | 46 | 34 | 35 |
| 06 | m | 44 | C2 | 13 | 188 | 96.8 | 10.17 | 47 | 11 | 10 | 3 | 12 | 2 | 55 | 52 | 32 | 33 |
| 07 | f | 46 | T8 | 0.3 | 171 | 94.8 | 5.78 | 47 | 32 | 20 | 4 | 48 | 0 | 41 | 41 | 29 | 29 |
| 08 | f | 52 | T9 | 7.25 | 177 | 81.1 | 13.09 | 40 | 52 | 52 | 3 | 28 | 2 | 44 | 47 | 32 | 31 |
| 09 | m | 68 | C7 | 0.4 | 167 | 58.2 | 6.47 | 49 | 56 | 56 | 4 | 42 | 0 | 45 | 42 | NA | NA |
| 10 | m | 38 | C6 | 8.25 | 182 | 71.7 | 7.31 | 45 | 30 | 13 | 4 | 34 | 0 | 56 | 54 | 32 | 33 |
| 11 | m | 48 | C2 | 13.2 | 177 | 77 | 6.82 | 48 | 56 | 56 | 4 | 38 | 1 | 43 | 42 | 29 | 29 |
| 12 | m | 63 | T9 | 6.1 | 17 | 80 | 17.41 | 43 | 35 | 15 | 3 | 3 | 2 | 53 | 47 | 35 | 32 |

*Characteristics of the individuals with iSCI included in the analysis, especially the time since injury (TSI) and the ten-meter walking test (10mWT) as a measure of walking function. This also includes the summation of the left and right lower extremity motor scores (LEMS), light touch scores (LT), and pin prick scores (PP) along with position sense (POS), vibration sense (VIB) and Romberg (RB) scores. Latencies from sensory evoked potentials (SEP) and cortical motor evoked potentials (MEP) are reported for the tibial nerve (NA: Not assessed). Where appropriate, units and maximal possible scores are indicated in parenthesis*

**Table 2** *Gait Parameters*

| | Parameter | Group | 0% BW | 10% BW | 20% BW | 30% BW | 40% BW | 50% BW | Group BWS effect | Simple BWS effect | Group x BWS |
|---|---|---|---|---|---|---|---|---|---|---|---|
| Spatio-temporal | Step Length [mm] | Control | 435 (36) | 438 (43) | 443 (46) | **456 (48)** | 444 (46) | **462 (51)** | F(5,85) = 7.477, $p = 0.000$ | F(5,130) = 3.066, $p = 0.012$ | F(5,130) = 3.116, $p = 0.011$ |
| | | CiSCI | 418 (85) | 431 (76) | 423 (69) | 421 (73) | 416 (79) | 420 (81) | F(5,50) = 0.916, $p = 0.478$ | | |
| | Step Width [mm] | Control | 63 (27) | **59 (25)** | **57 (24)** | 60 (27) | 66 (30) | 70 (31) | F(5,85) = 3.053, $p = 0.014$ | F(5,130) = 1.756, $p = 0.126$ | F(5,130) = 2.435, $p = 0.038$ |
| | | CiSCI | 100 (28) | 91 (25) | 93 (31) | 84 (25) | 88 (28) | 84 (31) | F(5,50) = 1.491, $p = 0.209$ | | |
| | Step Time [s] | Control | 0.75 (0.06) | 0.75 (0.07) | **0.76 (0.08)** | **0.78 (0.09)** | **0.79 (0.09)** | **0.82 (0.10)** | F(5,85) = 11.342, $p = 0.000$ | F(5,130) = 6.514, $p = 0.000$ | F(5,130) = 0.479, $p = 0.778$ |
| | | CiSCI | 0.68 (0.09) | 0.71 (0.13) | 0.69 (0.11) | 0.74 (0.10) | 0.78 (0.13) | 0.79 (0.12) | F(5,50) = 1.453, $p = 0.222$ | | |
| | Double Support Phase [%] | Control | 0.29 (0.02) | 0.29 (0.02) | **0.27 (0.03)** | **0.25 (0.03)** | **0.23 (0.03)** | **0.2 (0.05)** | F(5,85) = 49.294, $p = 0.000$ | F(5,130) = 36.528, $p = 0.000$ | F(5,130) = 1.354, $p = 0.245$ |
| | | CiSCI | 0.33 (0.05) | 0.33 (0.05) | 0.3 (0.04) | 0.3 (0.04) | **0.28 (0.06)** | **0.26 (0.05)** | F(5,50) = 6.805, p = 0.000 | | |
| | Speed [km/h] | Control | 2 (0.07) | 2 (0.1) | 2 (0.1) | 2 (0.1) | 2 (0.1) | 2 (0.1) | F(5,85) = 3.431, $p = 0.007$ | F(5,130) = 11.110, $p = 0.000$ | F(5,130) = 6.529, $p = 0.000$ |
| | | CiSCI | 2.1 (0.2) | 2.1 (0.2) | 2.1 (0.2) | 2 (0.2) | 1.9 (0.2) | **1.9 (0.2)** | F(5,50) = 5.867, $p = 0.000$ | | |
| Posture | A-P Sway [°] | Control | 23 (5) | 22 (6) | **19 (4)** | **18 (4)** | **16 (4)** | **17 (4)** | F(5,85) = 16.366, $p = 0.000$ | F(5,130) = 5.353, $p = 0.000$ | F(5,130) = 2.877, $p = 0.017$ |
| | | CiSCI | 30 (14) | 29 (14) | 28 (12) | 25 (8) | 26 (8) | 24 (9) | F(5,50) = 0.637, $p = 0.673$ | | |
| | CoM AP motion [mm] | Control | 860 (70) | 870 (83) | 880 (88) | **910 (95)** | 890 (90) | **920 (98)** | F(5,85) = 6.561, $p = 0.000$ | F(5,130) = 1.786, $p = 0.119$ | F(5,130) = 4.911, $p = 0.000$ |
| | | CiSCI | 820 (130) | 840 (120) | 830 (120) | 840 (120) | 820 (150) | 820 (150) | F(5,50) = 1.471, $p = 0.216$ | | |
| | M-L Sway [°] | Control | 20 (6) | 20 (6) | 22 (6) | **23 (7)** | **24 (5)** | **24 (7)** | F(5,85) = 4.768, $p = 0.001$ | F(5,130) = 5.333, $p = 0.000$ | F(5,130) = 1.066, $p = 0.382$ |
| | | CiSCI | 27 (11) | 26 (11) | 30 (12) | 27 (9) | 30 (11) | 29 (12) | F(5,50) = 2.173, $p = 0.072$ | | |
| | CoM ML motion [mm] | Control | 43 (9) | **40 (10)** | **39 (10)** | 43 (15) | **37 (9)** | **38 (13)** | F(5,85) = 2.760, $p = 0.023$ | F(5,130) = 7.016, $p = 0.000$ | F(5,130) = 1.568, $p = 0.173$ |
| | | CiSCI | 57 (12) | 52 (15) | **49 (12)** | 49 (16) | **44 (13)** | **41 (13)** | F(5,50) = 4.601, $p = 0.002$ | | |
| Single Joint | Hip ROM [°] | Control | 32 (3) | 33 (3) | 32 (3) | 32 (4) | **30 (3)** | **30 (4)** | F(5,85) = 6.397, $p = 0.000$ | F(5,130) = 3.794, $p = 0.003$ | F(5,130) = 0.684, $p = 0.636$ |
| | | CiSCI | 34 (6.6) | 34 (6.8) | 32 (6.4) | 34 (7.6) | 33 (7.4) | 31 (7.2) | F(5,50) = 0.901, $p = 0.488$ | | |
| | Knee ROM [°] | Control | 47 (5) | 47 (5) | **46 (4)** | 47 (5) | **46 (5)** | 46 (5) | F(5,85) = 3.113, $p = 0.013$ | F(5,130) = 8.920, $p = 0.000$ | F(5,130) = 1.168, $p = 0.328$ |
| | | CiSCI | 50 (9) | 50 (9) | 49 (8) | 48 (10) | 47 (9) | **45 (10)** | F(5,50) = 5.355, $p = 0.001$ | | |
| | Ankle ROM [°] | Control | 22 (4) | 22 (4) | 22 (4) | 23 (4) | 23 (7) | **26 (7)** | F(5,85) = 6.020, $p = 0.000$ | F(5,130) = 3.774, $p = 0.000$ | F(5,130) = 1.328, $p = 0.256$ |
| | | CiSCI | 20 (5) | 21 (5) | 20 (4) | 24 (10) | 24 (12) | 24 (11) | F(5,50) = 0.853, $p = 0.519$ | | |
| Intralimb | SSD Hip-Knee [A.U.] | Control | 4.9 (1.9) | 4.9 (1.4) | 5.4 (1.9) | **6.2 (1.7)** | **8.4 (3.1)** | **12 (3.5)** | F(5,85) = 43.405, $p = 0.000$ | F(5,130) = 20.440, $p = 0.000$ | F(5,130) = 8.479, $p = 0.000$ |
| | | CiSCI | 8.3 (3.6) | 8.7 (3.9) | 7.8 (3.5) | 8.7 (3.2) | 9.4 (5.1) | 9.8 (4.9) | F(5,50) = 1.255, $p = 0.298$ | | |
| | SSD Knee-Ankle [A.U.] | Control | 4.5 (1.5) | 4.6 (1.3) | 5.4 (2) | **6.5 (2.1)** | **8.9 (3.8)** | **12 (4.2)** | F(5,85) = 31.107, $p = 0.000$ | F(5,130) = 16.615, $p = 0.000$ | F(5,130) = 7.203, $p = 0.000$ |
| | | CiSCI | 10 (4.9) | 11 (4.9) | 7.7 (2.7) | 10 (4.5) | 11 (5.2) | 11 (4.9) | F(5,50) = 2.374, $p = 0.052$ | | |
| | ACC Hip-Knee [A.U.] | Control | 0.97 (0.01) | 0.97 (0.01) | 0.97 (0.01) | 0.97 (0.01) | **0.96 (0.02)** | **0.95 (0.02)** | F(5,85) = 8.685, $p = 0.000$ | F(5,130) = 6.326, $p = 0.000$ | F(5,130) = 0.292, $p = 0.917$ |
| | | CiSCI | 0.96 (0.03) | 0.96 (0.03) | 0.95 (0.03) | 0.95 (0.04) | 0.95 (0.03) | 0.94 (0.03) | F(5,50) = 1.263, $p = 0.295$ | | |
| | ACC Knee-Ankle [A.U.] | Control | 0.95 (0.01) | 0.95 (0.02) | **0.94 (0.02)** | **0.94 (0.02)** | **0.92 (0.03)** | **0.9 (0.04)** | F(5,85) = 14.584, $p = 0.000$ | F(5,130) = 18.104, $p = 0.000$ | F(5,130) = 0.345, $p = 0.885$ |
| | | CiSCI | 0.93 (0.04) | 0.93 (0.04) | 0.91 (0.04) | 0.91 (0.06) | **0.88 (0.05)** | **0.87 (0.06)** | F(5,50) = 5.742, $p = 0.000$ | | |

*Mean gait parameter values (mean(SD)) in control and CiSCI cohorts at different BWS levels. F-statistics for simple BWS effect and group x BWS interaction are reported from the 2-way ANCOVA with one repeated factor and age as a covariate. The group-split F-statistics are reported as group effects and in case a significant effect was identified, a-priori simple contrasts to group baseline (0% BW) with a Sidak correction are indicated through* ***bold*** *formatting ($p < 0.05$). For CiSCIs, when parameters are side-dependent, only the result of the more affected side is reported unless results diverged significantly between more and less affected side* main BWS effect (F(5,130) = 1.786, $p = 0.119$), however an interaction was present (F(5,130) = 4.911, $p = 0.000$). Controls increased AP Com (F(5,85) = 6.561, $p = 0.000$), while this remained unchanged in CiSCI(F(5,50) = 1.471, $p = 0.216$). In medio-lateral direction, a simple main effect on trunk sway (F(5,130) = 5.333, $p = 0.000$) and CoM motion (F(5,130) = 7.016, $p = 0.000$) was detected, however there was no significant interaction effect (F(5,130) = 1.066, $p = 0.382$ and F(5,130) = 1.568, $p = 0.173$ respectively).

## Ranges of motion

Ranges of motion (ROMs) of lower extremity joints in the sagittal plane responded similarly to unloading (Table 2, Fig. 1). A simple main BWS effect was detected for hip (F(5,130) = 3.794, $p = 0.003$), knee (F(5,130) = 8.920, $p = 0.000$) and ankle (F(5,130) = 3.774, $p = 0.000$) joints, however there were no interaction effects (F(5,130) = 0.684, $p = 0.636$; F(5,130) = 1.168, $p = 0.328$; F(5,130) = 1.328, $p = 0.256$ respectively).

## Joint motion patterns

Unloading induced simple main effects in joint motion patterns of the hip (F* = 3.152, $p < 0.044$), knee (F* = 3.152, $p < 0.043$) and ankle (F* = 3.567, $p < 0.025$). At the knee these effects were centered around the toe-off event, while in the ankle they were mainly during the stance phase and in early swing (Fig. 1). Significant interaction effects in the joint waveforms were present at the hip level (F* = 3.152, $p < 0.039$), and the ankle level (F* = 3.567, $p < 0.012$). but not at the knee (F* = 3.152, $p > 0.05$). The interaction effect for the hip joint was strongest during the swing phase and in early

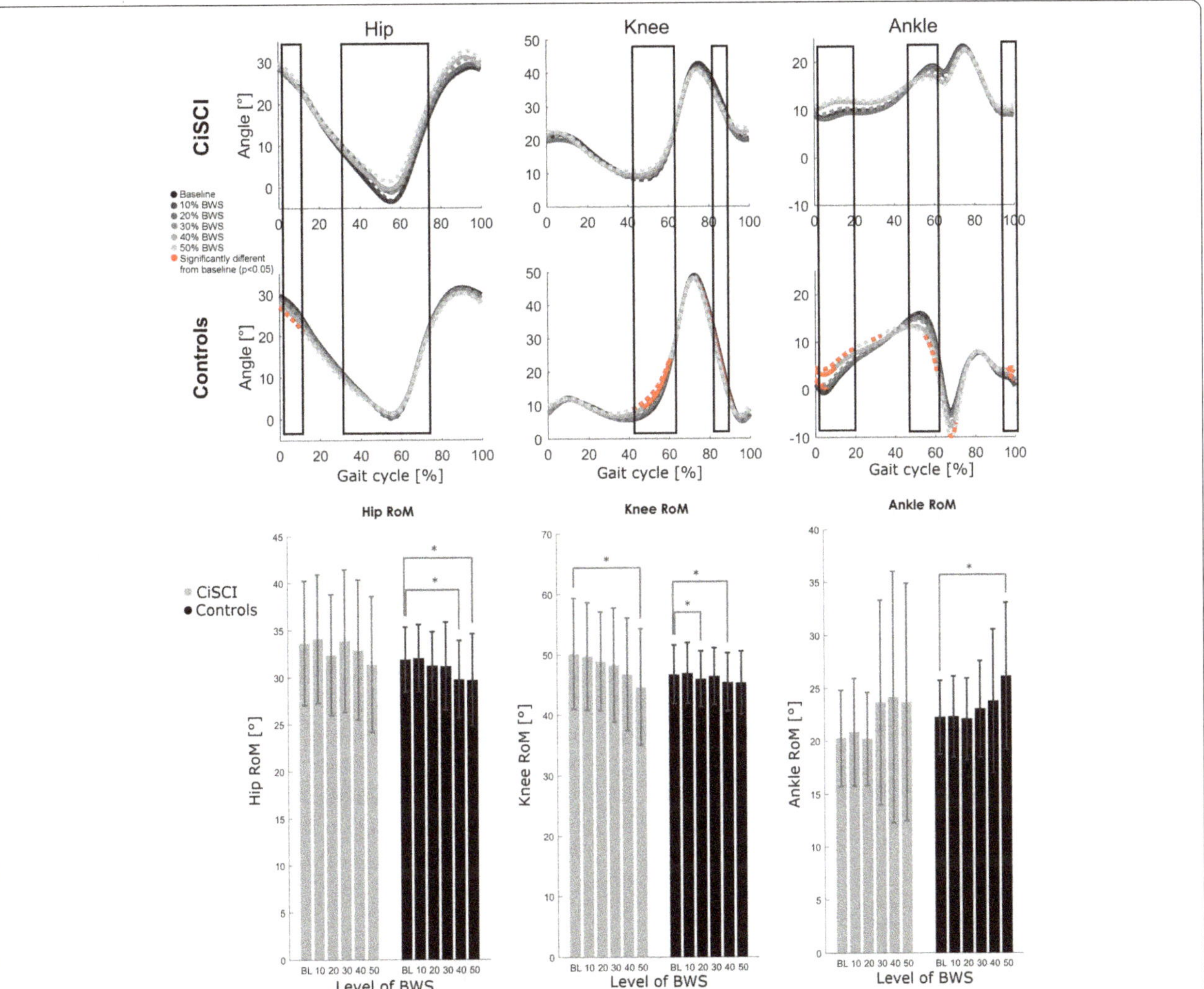


**Fig. 1** Lower limb angular waveforms & ranges of motion. Angular waveforms of hip, knee and ankle joints in the sagittal plane in controls (middle) and iSCI patients (top) at 10%, 30% and 50% BWS with increasing shades of gray. Simple main effects of BWS are indicated by the boxed areas for each joint. Significant deviation from the baseline condition within each group is indicated with red coloring for the affected section of the gait cycle. Range of motion (bottom) is depicted for hip, knee and ankle joints (sagittal plane) for patient (gray) and control (black) populations at all levels of BWS. Bars indicate group means with 1SD. Significance is detailed as *: $p < 0.05$

stance, while the ankle interaction was most prominent in late stance.

### Cyclograms

Intralimb coordination quantified via SSD showed a simple main BWS effect for both hip-knee ($F(5,130) = 20.440$, $p = 0.000$) and knee-ankle ($F(5,130) = 7.203$, $p = 0.000$) couplets. An interaction effect was present in both parameters ($F(5,130) = 8.479$, $p = 0.000$ and $F(5,130) = 7.203$, $p = 0.000$ respectively). While SSD increased with increasing unloading in controls ($F(5,85) = 43.405$, $p = 0.000$; $F(5,85) = 31.107$, $p = 0.000$), no patient response was detected ($F(5,50) = 1.255$, $p = 0.298$; $F(5,50) = 2.374$, $p = 0.052$). Shape consistency determined via ACC demonstrated a simple main BWS effect for hip-knee ($F(5,130) = 6.326$, $p = 0.000$) and knee-ankle ($F(5,130) = 18.104$, $p = 0.000$) couplets, however no interaction effects were detected ($F(5,130) = 0.292$, $p = 0.917$ and $F(5,130) = 0.345$, $p = 0.885$ respectively).

## Discussion

Application of BWS constitutes a promising approach in the rehabilitation of walking in patients suffering from CNS disorders such as iSCI or stroke. Within the focal group of patients that have voluntary control over their lower limbs, however lack the ability to ambulate freely, the FLOAT can provide transparent BWS with minimal interaction forces. Applying up to 50% BWS to CiSCI during overground walking resulted in subtle changes of walking kinematics that are largely comparable to those observed in healthy controls. While these BWS-induced changes were evident in spatio-temporal parameters such as step length and relative stance phase duration, joint motion patterns and intralimb coordination were modulated to a lower degree in iSCI than in controls. These results are contrary to our initial expectations that BWS would have a larger effect on the walking kinematics of CiSCI compared to controls. This indicates that using overground BWS for up to 20% BW unloading has no discernable effect on walking patterns and even high unloadings of 50% BW do not fundamentally distort walking kinematics. Furthermore, the differences in modulation of intra-limb coordination may provide an opportunity to specifically target this domain in iSCI rehabilitation.

### Spatio-temporal adaptation

As one of the salient characteristics of gait, the adjustment of step length and step time are prerequisites in handling changes in walking conditions. With increased unloading, step time is progressively increased in both controls and iSCI subjects. Similar observations from the literature have not resulted in a conclusive explanation overground [33, 56] or on the treadmill [27, 29–31, 57]. One possible explanation hinges on the dynamic similarity framework, assuming that the Froude number, an anatomy-independent measure of velocity dependent on leg length and gravitation, has an optimal value of around 0.25 for walking [57, 58].

$$Fr = \frac{\frac{m^*v^2}{h}}{m^*g} = \frac{v^2}{g^*h} = \frac{(sl^*sf)^2}{g^*h}$$

*Fr: Froude number, m: mass[kg], v: velocity[ms⁻¹], h: leg length[m], g: gravitational constant[ms⁻²], sl: step length[m], sf: step frequency[Hz].*

Reduced "gravity" through unloading while maintaining walking velocity should not disturb the relationship between step length and stride frequency. Applying unloading at the trunk, however, results in an interesting interaction: gravity acting on the swing leg remains at a normal level (9.81 $ms^{-2}$), while stance phase dynamics are subjected to reduced gravity plus potential robot interaction forces [59]. This typically results in a small but robust [57] reduction of the duty factor, driven mainly through an increase of step time [27, 29–31, 33, 56] while maintaining stance time. To maintain a given speed, relative stride length must be increased, disturbing the developmentally stable relationships between stride length, frequency and walking speed [60]. This behavior is adopted by control subjects. CiSCI on the other hand, display the reduced duty cycle, however they do not modulate step length as appropriately as controls, also leading to the slight reduction in walking speed registered at 50% BWS (Table 2).

Changes in step width were inconclusive in both controls and CiSCI; although an interaction between BWS and group was detected, there was no simple main effect of unloading. This could be due to interaction with the robot in medio-lateral direction, alluding to the complex interaction between BWS, lateral stability, and robot interaction forces [21, 23]. Transparency analysis leads us to somewhat discount the pendulum effect of overhead suspension [26], however the interplay between vertical unloading and step width and other parameters associated with walking stability merits further investigation in healthy subjects. CiSCI retained a large variability in step width throughout all unloading conditions, potentially masking any meaningful responses to unloading. Mean step width trended toward reduction from a pathologically high baseline, hinting that unloading enables CiSCI to walk with a narrower base of support.

### Walking posture effects

We quantified CoM displacement and trunk sway to understand how transparent BWS affects posture during walking and if this is different between CiSCI and controls. The simple main effect of BWS on ML CoM motion and AP and ML trunk sway indicates that the BWS system has some effect on walking posture and trunk control. As the unloading increases, trunk sway increases in both planes while ML motion of the CoM is reduced, and AP in controls remains unchanged. CiSCI adopt a different response pattern in both AP parameters, perhaps due to the shorter initial step length which would require less momentum transfer from the trunk to the pelvis to induce the weight shift. CiSCI also demonstrated a greater AP trunk sway in the initial condition, perhaps compensating for changes in the peak force generation of the trunk muscles. Furthermore, walking speed was reduced to a slightly greater degree in CiSCI under unloading. In summary, CiSCI may be more susceptible to the dynamics of the robot and the constraints of the harness – especially in directions where higher accelerations are necessary [25]. The overhead design and harness location potentially forced a more upright posture and allowed less forward transfer of weight to the leading foot.

The changes in medio-lateral parameters with increased unloading were reflected in both groups. This contrasted with step width, which demonstrated a local minimum at 20%BWS for controls and progressive, but non-significant reduction in CiSCI. This indicates that both groups increasingly opt to retain their CoM closer to the medial edge of their base of support with increasing BWS, alluding to a complex interaction between unloading and frontal plane dynamics [61]. Also, the superior-inferior motion speed of the BWS robot further complexifies the interaction, as the tension is not perfectly equal at all time points. This leads to subtle changes in support levels and momentum transfers in different phases of gait that likely influence the spatio-temporal structure of walking stability, however these interactions are challenging to accurately quantify.

#### Joint motion patterns

With increasing unloading, we observed increasing temporal shifts of the angle-time traces in both CiSCI and controls. To adequately compare joint motion patterns, stance and swing phase were normalized and interpolated separately to remove temporal effects while preserving time-rank and amplitude information. This enabled the detection of differences in trajectory shape induced through unloading. While the knee joints showed a simple main unloading effect, especially around the toe-off event and during stance, there was no interaction effect. This indicates that in the investigated sample we could not detect different knee strategies between CiSCI and controls in response to the unloading. At the hip joint, unloading induced a simple main effect centered around the toe-off event and there was a strong interaction effect especially during swing and following heel strike. This interaction effect may be driven by the shorter step length of the CiSCI coupled with the changes in walking posture [62]. At the ankle joint, an interaction effect was present especially in late stance where controls emphasized their push off motion while CiSCI showed no adaptation. Unloading CiSCI in the BWS system induced no detectable change in joint motion patterns, while controls exhibited subtle adaptations especially in the ankle joint at high unloading levels. In controls, the changes in ankle joint motion could be interpreted as a task-specific adaptation, necessary to maintain a given speed while optimizing other, salient gait determinants [58, 63, 64]. That we did not detect these adaptations in our cohort of primarily sensory-affected CiSCI could be linked to changes in the integration of load information into the efferent command structure [65, 66]. Models of spinal network activity during gait indicate that load- and shear-sensitive mechanoreceptors on the foot sole along with Golgi tendon organs and muscle spindle afferents contribute significantly to the successful regulation of gait [39, 66]. However, following iSCI, afferent information processing is altered on many levels [67, 68] and is partially replaced with surrogate, redundant information stemming from non-impacted sources [69] such as visual control [70, 71].

#### Intralimb coordination

Intralimb coordination patterns have been repeatedly reported to be sensitive readouts of locomotor control in iSCI subjects [49–51]. Knowledge from upper limb [72] and lower limb [73] experiments in humans leads us to interpret multi-joint coordination as a product of proprioceptive integration in the spinal cord at different levels combined with supraspinal efferent drive [74]. Intralimb coordination patterns can be quantified as the form difference from a reference shape (SSD; a form of procrustean shape analysis) [49]. Variability of these patterns can be captured by the angular component of the coefficient of correspondence (ACC; a specialized form of vector coding [51]), which describes the mean dispersion of all sequential point pairs in the cycle. Increased variability in the movement coupling of segments is detected as a decrease in the ACC (range: 0–1). Increased movement variability, especially concerning the linkage of adjacent intralimb segments, can be interpreted as increased neural noise in the generation of synergistic muscle activation patterns. Under unloading, control subjects showed progressive changes in intralimb coupling through increased shape difference and variability (Table 2 and Fig. 2) in both couplets (hip-knee and knee-ankle). In CiSCI, however, SSD remained unchanged in both proximal and distal couplets. No interaction effect was detected for ACC of both couplets, however unloading resulted in a higher coupling variability in both groups. Patient's lack of modulation of intralimb coupling under unloading may provide access to investigating and challenging this aspect after iSCI.

In synthesis, CiSCI and controls demonstrated similar responses to unloading in terms of spatio-temporal and

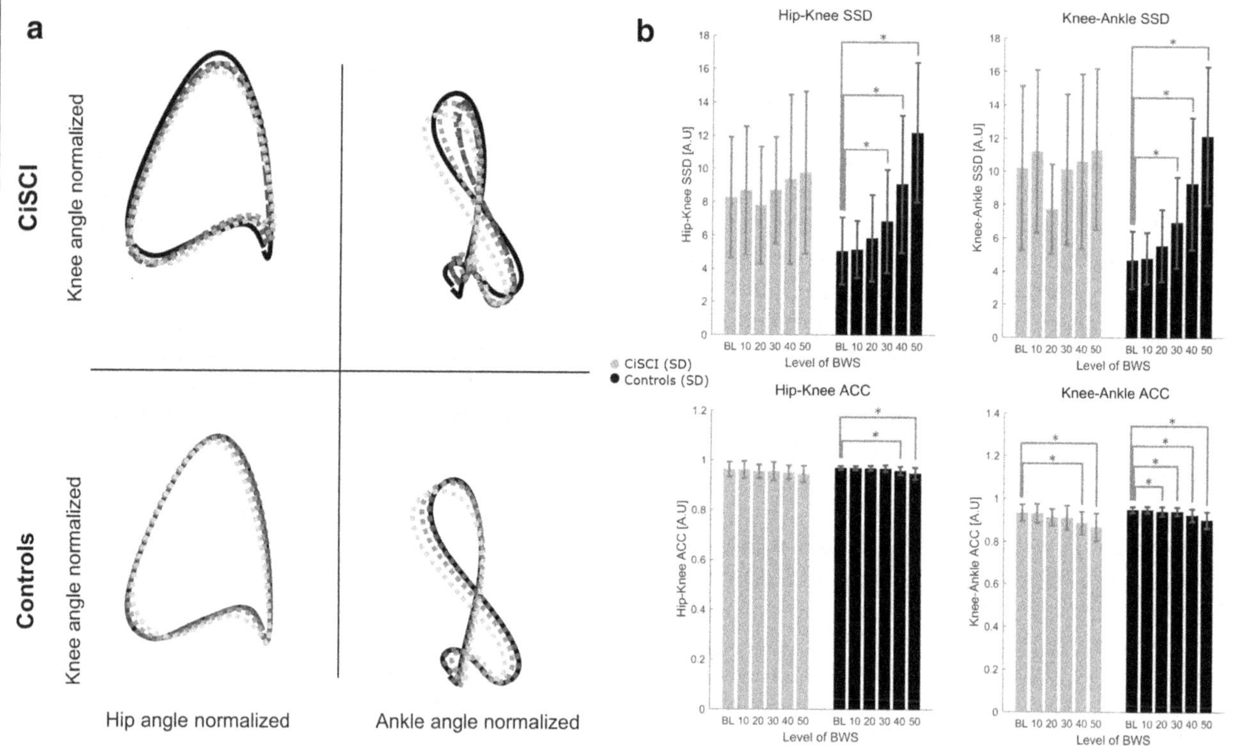


**Fig. 2** Intralimb coordination in controls and patients under BWS. **a** Response of intralimb coordination patterns to unloading in patients and controls. As a reference, mean baseline control data is shown as a continuous black line. Patient and control mean responses are depicted as dotted lines at 10%, 30% and 50% BWS in increasing shades of gray. Mean patient baseline data is shown as a dashed line. **b** Quantification metrics of the intralimb coordination. Unloading results in a change in SSD in controls (black) for both hip-knee and knee-ankle joint couplets. This coordination pattern remains unaffected in patients (gray) in both couplets. Shape consistency (ACC) is increasingly degraded in both groups with rising unloading. Bars indicate group means with 1SD. Significance is detailed as *: $p < 0.05$

walking posture parameters. CiSCI, however, retained their baseline intralimb coordination while controls modified this in response to unloading, optimizing their joint coupling patterns to adapt to the unloading while maintaining a given speed. It seems that in a healthy CNS, intralimb patterns are mediated not just at a spinal level, but also integrate a strong influence of supraspinal centers [75]. Once altered following CNS injury however, the coupling of segments has proven remarkably resilient to change. In iSCI patients, for instance, Awai and Curt [12] reported unchanged intralimb walking pattern shapes in iSCI patients throughout their rehabilitation, despite increases in walking speed and reductions in intralimb variability. Similarly, Tepavac and Field-Fote [51] reported improved consistency, yet no systematic change of shape in intralimb coupling following peroneal stimulation coupled with training in 14 iSCI patients. Analogous observations have been made in stroke survivors with lower limb impairment [76–78].

These observations could be driven by alteration of efferent drive induced by changes in supraspinal processing, including the partial loss and replacement of afferent signals. Alternatively, the coupling of segments may be mainly encoded by rhythmic spinal networks and therewith be difficult to modulate via efferent drive, especially when this drive is impaired through a lesion. Both of these two models indicate that when optimizing gait phenotype, especially intralimb coordination, to environmental constraints or when encountering novel tasks, the altered processing present in CiSCI cannot fully compensate for the degraded afferent information. Within this perspective, the response profile of CiSCI to unloading; adapted spatio-temporal parameters, unchanged joint motion patterns and interlimb coordination, may allude to the hierarchy of control in locomotion [79, 80]. Here, neural resources are reserved for task-critical parameters such as spatio-temporal organization of gait and balance control while less critical parameters such as intralimb coordination are preempted from being modulated. Any improvement of function is thought to stem from the adaptation of higher level processes while the newly established motor equivalent primitives remain invariant [81, 82]. However, to our knowledge, there has been no training paradigm that specifically targeted intralimb coordination following iSCI. Unloading might be a unique training pathway to target this rather specific deficit.

### Clinical relevance

Applying transparent overhead BWS to iSCI subjects during overground walking has negligible impact on CiSCI walking kinematics. This indicates that for CiSCI that retain volitional control of their limbs, but are not capable of supporting their own bodyweight, transparent BWS represents an opportunity to commence with safe, supported, unconstrained overground walking at an early time point. As patients improve, the amount of support can be reduced, and additional, perturbing forces can be applied by the therapist to maintain a challenging training environment. Furthermore, the device can be used for support while training other activities of daily life, such as stair climbing, obstacle crossing, balancing, curve walking, retrieving objects from the floor, sit to stand transitioning, etc. Overground BWS training remains only one of many rehabilitation tools, however it enables a transition from a treadmill rehabilitation environment to a more real-world setting.

### Outlook

Our conclusions are based upon a relatively small and heterogeneous sample of iSCI subjects with mainly sensory impairments due to the constraint of being able to walk 2 km/h without walking aids for approximately one hour of measurement. This walking speed was chosen as it is a viable speed for CiSCI and is close to the threshold for unsupervised indoor ambulation for CiSCI (0.6 m/s) [83]. Furthermore, at this velocity, controls still produce rhythmic, symmetrical gait, although it is far slower (~ 50%) than their typical age- and gender-specific walking velocity. A familiarization period was allowed at each BWS level and CiSCI and controls walked the target speed consistently during the experiment without transgressing the tolerance range, indicating sufficient acclimatization. While it would have been interesting to also analyze preferred walking speed or even the effects of different walking speeds, this would have led to a very long assessment protocol for the CiSCI. Baseline walking measures differed between the groups, and there was a significant age difference. Inclusion of age as a covariate in the statistical model however indicates that the adaption of gait patterns to BWS was not powered by this factor. The different responses between CiSCI and controls warrant further investigation in a wider range of individuals with CNS disorders walking at multiple speeds. Three CiSCI had TSIs of less than one year and would hence be considered closer to sub-acute than chronic. These individuals were included as they had achieved a functional plateau due to the relatively mild nature of their injuries. Functionally they ranked among the better individuals in the cohort. The investigated cohort of patients with a wide range of different impairments and TSIs provides some indication of a generalizable effect, however how this manifests specifically to level, severity and chronicity of impairment can only be determined by using more specific inclusion criteria. This limits the description of potential mechanisms in the current experiment. Going forward, we are interested in evaluating more precisely defined cohorts and better disentangling the mechanical effects from the biological responses to unloading.

## Conclusions

In synopsis, applying transparent BWS during overground walking in CiSCI from 30 to 50% bodyweight induces changes in temporal gait parameters and medio-lateral balance control, without disrupting the gait pattern or the intralimb coordination. These results are comparable to observations in controls, however controls subtly adapted their intralimb coordination in dependence on the BWS level. The predominantly sensory affected CiSCI cohort did not modulate their gait in a similar manner. Building upon these results, overground walking with transparent BWS may be a suitable extension of the locomotor therapist's toolbox in that it can provide walking support without disrupting gait patterns in a CiSCI population. Forward-looking, investigating the adaptation to unloading may provide a window on the residual adaptive capacity of individuals with gait impairments during locomotion. While critical endpoint control parameters may behave almost normally, the underlying joint motion patterns and intralimb coordination probe deeper into the neural framework of locomotion and may provide more sensitive readouts of locomotor capacity.

**Acknowledgements**
Not applicable.

**Funding**
This project was funded by the Swiss Federal Commission of Technology and Innovation (CTI), Grant number 17567.2 PFLS-LS, the Clinical Research Priority Program CRPP NeuroRehab of the University of Zurich UZH and Balgrist Campus AG (Swiss Center for Clinical Movement Analysis, SCMA).

**Authors' contributions**
MB, GR, and AC designed the study and assisted in data interpretation and manuscript preparation. MF and LRT recruited participants, conducted the measurements and prepared the data. LA conducted data analysis and assisted in manuscript preparation. CES assisted in data collection and preparation, conducted data analysis, and wrote the manuscript. All authors read and approved the final manuscript.

**Competing interests**
The authors declare that they have no competing interests.

**Author details**
[1]Spinal Cord Injury Center, Balgrist University Hospital, Forchstrasse 340, CH-8008 Zürich, Switzerland. [2]Sobell Department of Motor Neuroscience and Movement Disorders, University College London, London, UK. [3]BIROMED-Lab, Department of Biomedical Engineering, University Basel, Basel, Switzerland.

**References**

1. Behrman AL, Bowden MG, Nair PM. Neuroplasticity after spinal cord injury and training: an emerging paradigm shift in rehabilitation and walking recovery. Phys Ther. 2006;86:1406–25.
2. Smith AC, Knikou M. A review on locomotor training after spinal cord injury: reorganization of spinal neuronal circuits and recovery of motor function. Neural Plast. 2016;2016:1216258.
3. Grillner S, Wallén P. Central pattern generators for locomotion, with special reference to vertebrates. Annu Rev Neurosci. 1985;8:233–61.
4. Rossignol S, Frigon A. Recovery of locomotion after spinal cord injury: some facts and mechanisms. Annu Rev Neurosci. 2011;34:413–40.
5. Edgerton VR, Roy RR. Robotic training and spinal cord plasticity. Brain Res. 2009;78:4–12.
6. Knikou M. Plasticity of corticospinal neural control after locomotor training in human spinal cord injury. Neural Plast. 2012;2012:254948.
7. Wessels M, Lucas C, Eriks-Hoogland I, De Groot S. Body weight-supported gait training for restoration of walking in people with an incomplete spinal cord injury: a systematic review. J Rehabil Med. 2010;42:513–9.
8. Yang JF, Musselman KE, Livingstone D, Brunton K, Hendricks G, Hill D, et al. Repetitive mass practice or focused precise practice for retraining walking after incomplete spinal cord injury? A pilot randomized clinical trial. Neurorehabil Neural Repair. 2014;28:314–24.
9. Harkema SJ, Hillyer J, Schmidt-Read M, Ardolino E, Sisto SA, Behrman AL. Locomotor training: as a treatment of spinal cord injury and in the progression of neurologic rehabilitation. Arch Phys Med Rehabil Elsevier Inc. 2012;93:1588–97.
10. Dobkin BH, Duncan PW. Should body weight-supported treadmill training and robotic-assistive steppers for locomotor training trot back to the starting gate? Neurorehabil Neural Repair. 2012;26:308–17.
11. Friedli L, Rosenzweig ES, Barraud Q, Schubert M, Dominici N, Awai L, et al. Pronounced species divergence in corticospinal tract reorganization and functional recovery after lateralized spinal cord injury favors primates. Sci Transl Med. 2015;7:302ra134.
12. Awai L, Curt A. Locomotor recovery in spinal cord injury - insights beyond walking speed and distance. J Neurotrauma. 2016;33:1428–35.
13. Field-Fote EC, Roach KE. Influence of a locomotor training approach on walking speed and distance in people with chronic spinal cord injury: a randomized clinical trial. Phys Ther. 2011;91:48–60.
14. van den Brand R, Heutschi J, Barraud Q, DiGiovanna J, Bartholdi K, Huerlimann M, et al. Restoring voluntary control of locomotion after paralyzing spinal cord injury. Science (80- ). 2012;336:1182–5.
15. Alton F, Baldey L, Caplan S, Morrissey MC. A kinematic comparison of overground and treadmill walking. Clin Biomech. 1998;13:434–40.
16. Chiu SL, Chang CC, Chou LS. Inter-joint coordination of overground versus treadmill walking in young adults. Gait posture. Elsevier BV. 2015;41:316–8.
17. Musselman KE, Fouad K, Misiaszek JE, Yang JF. Training of walking skills overground and on the treadmill: case series on individuals with incomplete spinal cord injury. Phys Ther. 2009;89:601–11.
18. Hidler J, Brennan D, Black I, Nichols D, Brady K, Nef T. ZeroG: overground gait and balance training system. J Rehabil Res Dev. 2011;48:287–98.
19. Plooij M, Keller U, Sterke B, Komi S, Vallery H, von Zitzewitz J. Design of RYSEN: an intrinsically safe and low-power three-dimensional Overground body weight support. IEEE Robot Autom Lett. 2018;3:2253–60.
20. Mignardot JB, Le Goff CG, Van Den Brand R, Capogrosso M, Fumeaux N, Vallery H, et al. A multidirectional gravity-assist algorithm that enhances locomotor control in patients with stroke or spinal cord injury. Sci Transl Med. 2017;9.
21. Dragunas AC, Gordon KE. Body weight support impacts lateral stability during treadmill walking. J Biomech. 2016;49:2662–8.
22. Bauby C, Kuo A. Active control of lateral balance in human walking. J Biomech. 2000;33:1433–40.
23. Dean JC, Alexander NB, Kuo AD. The effect of lateral stabilization on walking in young and old adults. IEEE Trans Biomed Eng. 2007;54:1919–26.
24. Vallery H, Lutz P, von Zitzewitz J, Rauter G, Fritschi M, Everarts C, et al. Multidirectional transparent support for overground gait training. IEEE Int Conf Rehabil Robot. 2013;2013:6650512.
25. Awai L, Franz M, Easthope CS, Vallery H, Curt A, Bolliger M. Preserved gait kinematics during controlled body unloading. J Neuroeng Rehabil. 2017;14:25–35.
26. Bannwart M, Bolliger M, Lutz P, Gantner M, Rauter G. Systematic analysis of transparency in the gait rehabilitation device the FLOAT. In: 14th Int Conf control autom robot Vis; 2016.
27. Ruckstuhl H, Kho J, Weed M, Wilkinson MW, Hargens AR. Comparing two devices of suspended treadmill walking by varying body unloading and Froude number. Gait Posture. 2009;30:446–51.
28. Lewek MD. The influence of body weight support on ankle mechanics during treadmill walking. J Biomech. 2011;44:128–33.
29. van Hedel HJ a, Tomatis L, Müller R. Modulation of leg muscle activity and gait kinematics by walking speed and bodyweight unloading. Gait Posture. 2006;24:35–45.
30. Threlkeld A, Cooper L, Monger B. Temporospatial and kinematic gait alterations during treadmill walking with body weight suspension. Gait Posture. 2003;17:235–45.
31. Aaslund MK, Moe-Nilssen R. Treadmill walking with body weight support effect of treadmill, harness and body weight support systems. Gait Posture. 2008;28:303–8.
32. Apte S, Plooij M, Vallery H. Influence of body weight unloading on human gait characteristics: a systematic review. J Neuroeng Rehabil. 2018;15:53.
33. Fischer AG, Wolf A. Assessment of the effects of body weight unloading on overground gait biomechanical parameters. Clin Biomech. 2015;30:454–61.
34. Fischer AG, Wolf A. Body weight unloading modifications on frontal plane joint moments, impulses and Center of Pressure during overground gait. Clin Biomech. 2016;39:77–83.
35. Fischer AG, Debbi EM, Wolf A. Effects of body weight unloading on electromyographic activity during overground walking. J Electromyogr Kinesiol. 2015;25:709–14.
36. Sylos-Labini F, Lacquaniti F, Ivanenko YP. Human locomotion under reduced gravity conditions: biomechanical and neurophysiological considerations. Biomed Res. 2014;2014:547242.
37. Dietz V. Body weight supported gait training: from laboratory to clinical setting. Brain Res Bull. 2009;78:I–VI.
38. MacKay-Lyons M. Central pattern generation of locomotion: a review of the evidence. Phys Ther. 2002;82:69–83.
39. Dietz V. Proprioception and locomotor disorders. Nat Rev Neurosci. 2002;3: 781–90.
40. Ivanenko YP, Grasso R, Macellari V, Lacquaniti F, Ogihara N, Oku T, et al. Control of foot trajectory in human Locomotion : role of ground contact forces in simulated reduced gravity. J Neurophysiol. 2002;87:3070–89.
41. Harkema SJ, Hurley SL, Patel UK, Requejo PS, Bruce H, Edgerton VR, et al. Human lumbosacral spinal cord interprets loading during stepping. J Neurophysiol. 1997;77:797–811.
42. Beres-Jones JA, Harkema SJ. The human spinal cord interprets velocity-dependent afferent input during stepping. Brain. 2004;127:2232–46.
43. Martina ISJ, van Koningsveld R, Schmitz PIM, van der Meche FGA, van Doorn PA. Measuring vibration threshold with a graduated tuning fork in normal aging and in patients with polyneuropathy. European Inflammatory Neuropathy Cause and Treatment (INCAT) group. J Neurol Neurosurg Psychiatry. 1998;65:743–7.

44. Gilman S. Joint position sense and vibration sense: anatomical organisation and assessment. J Neurol Neurosurg Psychiatry. 2002;73:473–7.
45. Lanska DJ, Goetz CG. Romberg's sign: development, adoption, and adaptation in the 19th century. Neurology. 2000;55:1201–6.
46. Davis R, Ounpuu S, Tyburski D, Gage J. A gait analysis data collection and reduction technique. Hum Mov Sci. 1991;10:575–87.
47. Ramakrishnan HK, Wootten M, Kadaba M. On the estimation of three dimensional joint angular motion in gait analysis. 35th Annu meet Orthop res Soc. Las Vegas; 1989.
48. Woltring HJ. A Fortran package for generalized, cross-validatory spline smoothing and differentiation. Adv Eng Softw. 1986;8:104–13.
49. Awai L, Curt A. Intralimb coordination as a sensitive indicator of motor-control impairment after spinal cord injury. Front Hum Neurosci. 2014;8:148.
50. Awai L, Bolliger M, Ferguson AR, Courtine G, Curt A. Influence of spinal cord integrity on gait control in human spinal cord injury. Neurorehabil Neural Repair. 2016;30:562–72.
51. Field-Fote EC, Tepavac D. Improved Intralimb coordination in people with incomplete spinal cord injury following training with body weight support and electrical stimulation. Phys Ther. 2002;82:707–15.
52. Pataky TC. One-dimensional statistical parametric mapping in python. Comput Methods Biomech Biomed Engin. 2012;15:295–301.
53. Pataky TC, Robinson MA, Vanrenterghem J. Region-of-interest analyses of one-dimensional biomechanical trajectories: bridging 0D and 1D theory, augmenting statistical power. PeerJ. 2016;4:e2652.
54. Vogel P. Kursbuch Klinische Neurophysiologie: EMG-ENG-evozierte Potentiale. 2nd Edition. Georg Thieme Verlag; 2006.
55. Bohannon RWR. Comfortable and maximum walking speed of adults aged 20—79 years: reference values and determinants. Age Ageing. 1997;26:15–9.
56. Patino M, Goncalves A, Monteiro B, Santos I, Barela A, Barela J. Kinematic, kinetic and electromyographic characteristics of young adults walking with and without partial body weight support. Rev Bras Fisioter. 2007;11:19–25.
57. Donelan JM, Kram R. The effect of reduced gravity on the kinematics of human walking: a test of the dynamic similarity hypothesis for locomotion. J Exp Biol. 1997;200:3193–201.
58. Alexander R. Estimates of speeds of dinosaurs. Nature. 1976;261:129–30.
59. Raichlen D a, Pontzer H, Shapiro LJ. A new look at the dynamic similarity hypothesis: the importance of swing phase. Biol Open. 2013;2:1032–6.
60. Grieve DW, Gear RJ. The relationships between length of stride , step frequency , time of swing and speed of walking for children and adults. Ergonomics. 1966;9:379–99.
61. Everarts C, Vallery H, Bolliger M, Ronsse R. Adaptive position anticipation in a support robot for overground gait training enhances transparency. IEEE Int Conf Rehabil Robot. 2013;2013:6650483.
62. Kepple T, Siegel K, Stanhope SJ. Relative contributions of the lower extremity joint moments to forward progression and support during gait. Gait Posture. 1997;6(1):1–8.
63. Zarrugh MY, Todd FN, Ralston HJ. Optimization of energy expenditure during level walking. Eur J Appl Physiol Occup Physiol. 1974;33:293–306.
64. Anderson FC, Pandy MG. Dynamic optimization of human walking. J Biomech Eng. 2001;123:381–90.
65. Todorov E. Optimality principles in sensorimotor control. Nat Neurosci. 2004; 7:907–15.
66. Windhorst U. Muscle proprioceptive feedback and spinal networks. Brain Res Bull. 2007;73:155–202.
67. Raineteau O, Schwab ME. Plasticity of motor systems after incomplete spinal cord injury. Nat Rev Neurosci. 2001;2:263–73.
68. Hubli M, Dietz V, Bolliger M. Influence of spinal reflexes on the locomotor pattern after spinal cord injury. Gait Posture. 2011;34:409–14.
69. Stein BE, Stanford TR, Rowland BA. Development of multisensory integration from the perspective of the individual neuron. Nat Rev Neurosci. 2014;15: 520–35.
70. Schubert M, Prokop T, Brocke F, Berger W. Visual kinesthesia and locomotion in Parkinson's disease. Mov Disord. 2005;20:141–50.
71. Malik RN, Cote R, Lam T. Sensorimotor integration of vision and proprioception for obstacle crossing in ambulatory individuals with spinal cord injury. J Neurophysiol. 2017;117:36–46.
72. Ghez C, Sainburg R. Proprioceptive control of interjoint coordination. Can J Physiol Pharmacol. 1995;73:273–84.
73. Chiu SL, Chou LS. Effect of walking speed on inter-joint coordination differs between young and elderly adults. J Biomech. 2012;45:275–80.
74. Ivanenko YP, d'Avella A, Poppele RE, Lacquaniti F. On the origin of planar covariation of elevation angles during human locomotion. J Neurophysiol. 2008;99:1890–8.
75. Faisal AA, Selen LPJ, Wolpert DM. Noise in the nervous system. Nat Rev Neurosci. 2008;9:292–303.
76. Den Otter AR, Geurts ACH, Mulder T, Duysens J. Gait recovery is not associated with changes in the temporal patterning of muscle activity during treadmill walking in patients with post-stroke hemiparesis. Clin Neurophysiol. 2006;117:4–15.
77. Lewek MD, Cruz TH, Moore JL, Roth HR, Dhaher YY, Hornby TG. Allowing Intralimb kinematic variability during locomotor training Poststroke improves kinematic consistency: a subgroup analysis from a randomized clinical trial. Phys Ther. 2009;89:829–39.
78. Rinaldi LA, Monaco V. Spatio-temporal parameters and intralimb coordination patterns describing hemiparetic locomotion at controlled speed. J Neuroeng Rehabil. 2013;10:53.
79. Poppele R, Bosco G. Sophisticated spinal contributions to motor control. Trends Neurosci. 2003;26:269–76.
80. Grasso R, Ivanenko YP, Zago M, Molinari M, Scivoletto G, Castellano V, et al. Distributed plasticity of locomotor pattern generators in spinal cord injured patients. Brain. 2004;127:1019–34.
81. Ivanenko YP, Poppele RE, Lacquaniti F. Motor control programs and walking. Neuroscientist. 2006;12:339–48.
82. Sulzer JS, Gordon KE, Dhaher YY, Peshkin MA, Patton JL. Preswing knee flexion assistance is coupled with hip abduction in people with stiff-knee gait after stroke. Stroke. 2010;41:1709–14.
83. van Hedel HJA. EM-SCI study group. Gait speed in relation to categories of functional ambulation after spinal cord injury. Neurorehabil Neural Repair. 2008;23:343–50.

# 13

# Identification of forearm skin zones with similar muscle activation patterns during activities of daily living

Néstor J. Jarque-Bou*, Margarita Vergara, Joaquín L. Sancho-Bru, Alba Roda-Sales and Verónica Gracia-Ibáñez

## Abstract

**Background:** A deeper knowledge of the activity of the forearm muscles during activities of daily living (ADL) could help to better understand the role of those muscles and allow clinicians to treat motor dysfunctions more effectively and thus improve patients' ability to perform activities of daily living.

**Methods:** In this work, we recorded sEMG activity from 30 spots distributed over the skin of the whole forearm of six subjects during the performance of 21 representative simulated ADL from the Sollerman Hand Function Test. Functional principal component analysis and hierarchical cluster analysis (HCA) were used to identify forearm spots with similar muscle activation patterns.

**Results:** The best classification of spots with similar activity in simulated ADL consisted in seven muscular-anatomically coherent groups: (1) wrist flexion and ulnar deviation; (2) wrist flexion and radial deviation; (3) digit flexion; (4) thumb extension and abduction/adduction; (5) finger extension; (6) wrist extension and ulnar deviation; and (7) wrist extension and radial deviation.

**Conclusion:** The number of sEMG sensors could be reduced from 30 to 7 without losing any relevant information, using them as representative spots of the muscular activity of the forearm in simulated ADL. This may help to assess muscle function in rehabilitation while also simplifying the complexity of prosthesis control.

**Keywords:** Activities of daily living, Clustering analysis, Electromyography, Electrode placement, Forearm muscles, Functional principal component analysis , Myoelectric prostheses, Rehabilitation, Sollerman hand function test

## Background

The forearm and the hand are connected by the wrist, together forming a functional unit in which there are about 30 muscles that work in concert in a highly complex way [1], allowing a wide range of activities to be performed, many of them with a high level of precision. When a disease or injury to any part of the hand or forearm take place, our manipulation and grip capacities are reduced and, therefore, carrying out the most common activities in our daily life can become a serious problem.

Surface electromyography (sEMG) is a non-invasive approach that allows both structural and functional characterization of the neuromuscular system, as well as making it possible to quantify variations of this system in different situations. sEMG is applied in many fields such as motor control of human movement, myoelectric control of prosthetic and orthotic devices, and rehabilitation [2–5]. In rehabilitation, sEMG has been used to monitor the effects of rehabilitation techniques [5, 6], rehabilitation devices [7], and therapeutic exercises [8]. In this line, some studies have proposed the addition of sEMG analysis to clinical assessments to provide quantitative measures of therapy outcomes for people with motor impairment and physical disability in the upper limb, such as stroke [9] or hemiplegia [10]. In those studies, the amplitude of the sEMG signal is used to provide quantitative measures, as patients have a reduced neuromuscular amplitude [5, 9, 11]. However, there is no consensus on the muscles or areas of the forearm that must be measured in each case. This fact

* Correspondence: jarque@uji.es
Department of Mechanical Engineering and Construction, Universitat Jaume I, Avinguda Vicent Sos Baynat, s/n., 12071 Castellón, Spain

may lead to the recording of repeated muscle activity information and/or missing information about some muscles.

The central nervous system must control a structure that is in general vastly more complex than necessary to execute any particular task. In this sense, muscular activation patterns or muscular synergies [12, 13] appeared as a possible way to allow the analysis of muscular control during task execution. Some studies have described muscle patterns or synergies during certain postures [14], grasps [15] or hand movements [16, 17], during particular actions [18–22], and for individually tailored rehabilitation protocols [23]. Examining whether muscle synergies change following rehabilitation may provide an assessment of interventions that improve patients' motor function and therefore their ability to perform activities of daily living (ADL). A patient's muscle synergy profile may allow clinicians to treat motor dysfunctions more effectively by organizing patients into subclasses and tailoring the treatment to each patient's specific deficit [24].

Myoelectric prostheses are usually controlled with sEMG [25–27]. However, one quarter to one third of all amputees reject self-powered prostheses [28] because they feel a non-close-to-natural operation [29] that leads to long and discouraging training periods. A deeper knowledge of the activity of the forearm muscles during different functional tasks could help in the development of more intuitively controlled prostheses, capable of replicating the intact hand movements by using sEMG signals from the residual muscles, similarly to when they had their own intact hand. In this sense, identification of repeatable patterns of muscle activity (muscular synergies) across multiple muscle sites rather than relying on independent EMG signals has been shown to enable more natural, reliable control of myoelectric prostheses [30]. However, the electrodes are usually placed on the remaining muscles of the forearm regardless of the muscle's pre-amputation role [31], favoring a non-close-to-natural operation. To overcome this hindrance, some authors have studied the accuracy of pattern recognition of myoelectric control, by evaluating the number and locations of sEMG channels [30, 32, 33]. Daley et al. [30] distributed the electrodes around the whole forearm in a grid formation during twelve wrist and hand motions and five different grasps in healthy and amputee subjects. Their results suggest that most of the neural information available could be extracted with a greatly reduced number of electrodes. However, the optimal location of electrodes and their consistency across subjects were not reported, and the tasks used for the analysis were not representative of ADL. Upper limb tasks should be goal-oriented and of a standardized nature to obtain consistent performance [34]. Some previous works have evaluated the performance of similar SEMG classification systems by using the NINAPRO database [35–37]. However, the sEMG data contained in this database was recorded while performing non-goal-oriented actions. Clinical tests, such as the Sollerman Hand Function test (SHFT) [38], are often used to evaluate and track functional recovery of the upper extremity, providing insight into functional performance. These tests typically evaluate this functional performance using tasks simulating ADL.

In this work, we recorded sEMG activity from different areas of the whole forearm skin during the performance of the ADL included in the SHFT, a standardized test for hand function evaluation. The aim was to identify skin zones with similar muscle activation patterns in order to determine the minimum number of electrodes required to characterize the muscle activity during simulated ADL without losing any relevant information.

## Methods

### Subjects and tasks

Six able-bodied subjects (3 males and 3 females, with average (SD) age of 34.5 (8.2) years) gave their informed written consent before participating in this study, approved by the ethics committee of our University.

Twenty-one simulated ADL were selected, all of them tasks included in the SHFT. Some of the simulated ADL were adapted in order to ensure their repeatability. The activity 14 from the original SHFT was separated in two different activities. In the first one, the piece of paper was folded (A14) and in the second one the folded paper was inserted into an envelope (A15). Others were slightly adapted, for example, reducing the number of objects to be manipulated (just one zipper, one coin, one button, etc.) or indicating the direction and the degrees to be turned in each case (Fig. 1, Table 1). Each

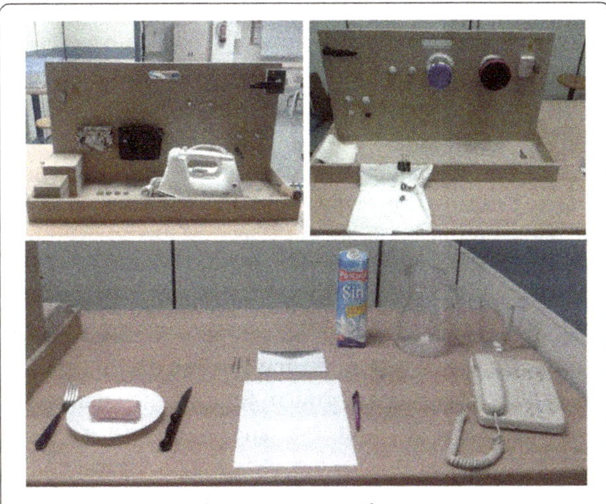

**Fig. 1** Sollerman Hand Function Test configuration

**Table 1** Description of the standardized ADL

| ADLs | DESCRIPTION |
|---|---|
| A1 | Take a key, insert it in a lock, turn it counter-clockwise 180° and leave the key in |
| A2 | Collect a coin and insert it into a change purse |
| A3 | Open and close a zipper |
| A4 | Remove the coin from the change purse and leave it on the table |
| A5 | Catch and move two wooden cubes of different sizes |
| A6 | Lift and move an iron from one marked point to another |
| A7 | Take a screwdriver and turn a screw clockwise 180° with it |
| A8 | Take a nut and bolt and turn the bolt until it is completely inserted inside the nut |
| A9 | Unscrew a lid and leave it on the table |
| A10 | Pass a button through a buttonhole with the help of both hands |
| A11 | Take a knife with the right hand and a fork with the left hand and split a piece of clay (sitting) |
| A12 | Take a bandage and put it on your left arm until the elbow |
| A13 | Pick up a pen from the table, write the Spanish word "SOL" and put the pen back on the table (sitting) |
| A14 | Fold a piece of paper with both hands and insert it into a box |
| A15 | Insert the folded paper (from A14) into an envelope and leave it on the table |
| A16 | Take a clip and put it on the flap of the envelope |
| A17 | Pick up the phone, put it to your ear and hang up the phone |
| A18 | Turn a door handle 30° |
| A19 | Pour 1 L of water from a carton into a jug |
| A20 | Pour water from the jug into the glass up to a marked point |
| A21 | Pour the water from the glass back into the jug |

simulated ADL started and ended with the body and arms in the same posture (arms relaxed at the side of the body, in the case that the person was standing, or arms resting in a relaxed position on the table if they were sitting). Precise instructions were provided for each task by the researcher administering the SHFT, including details such as the angle of rotation of the key (A1), the position of the coin (A2 & A4), the angle of rotation of the door handle (A18) or the amount of water to be poured (A20). The subjects could practice each task as many times as necessary in advance so as to become familiar with its performance before the sEMG recording.

**Electrode placement**

A grid was drawn on the forearm, covering its entire surface, by using five easily identifiable and highly reproducible anatomical landmarks (Fig. 2, Table 2). The grid defined 30 different spots (Fig. 2) and was drawn with the subject sitting comfortably, the elbow resting on the table, the arm flexed 90° with respect to the forearm, and the palm of the hand facing the subject. First, three longitudinal lines were drawn joining the anatomical landmarks 1–4, 2–3 and 2–5, and two transverse lines joining landmarks 1–2 and 3–4-5. Each of the longitudinal lines was divided into four equal parts that were used to draw the internal transverse lines of the grid.

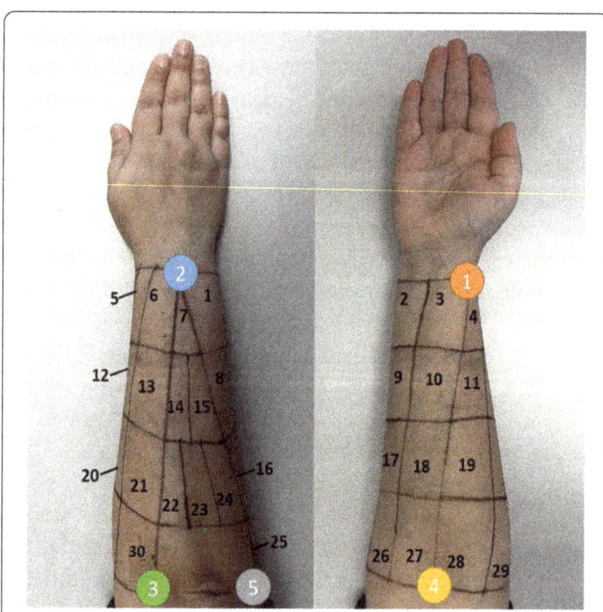

**Fig. 2** Grid with the 30 spots for sEMG recording, and anatomical landmarks used

**Table 2** Anatomical landmarks

| Number | Landmark |
|---|---|
| 1 | Styloid processes of the radius |
| 2 | Ulna head |
| 3 | Medial epicondyle of the humerus |
| 4 | Center point of the elbow |
| 5 | Humeral lateral epicondyle |

Transverse lines 1–2 and 4–5 in the cubital region were divided into three equidistant parts, and the same was applied to lines 1–2 and 4–3 in the radial region. These subdivisions were used to draw the corresponding longitudinal lines. Finally, the area formed by landmarks 2-3-5 was divided so that there were three equal spots in the most proximal region (spots 22, 23 and 24), a single spot in the most distal region (spot 7), and two spots in the intermediate region (spots 14 and 15), as shown in Fig. 2. The sensors were placed in the centre of each area and were set out in a longitudinal direction, since most of the muscles of the forearm are quite aligned to it [1]. Before placing the electrodes, the hair was removed by shaving and the skin was cleaning by using alcohol.

## Data acquisition

Muscle activity was recorded with an 8-channel sEMG Biometrics Ltd. device, with a sampling frequency of 1000 Hz. Technical information of the electrodes is shown in Fig. 3. Recording sEMG from the 30 spots of the grid required distributing the measurements over five different sessions. The spots chosen for each session (the same for each subject) were selected to be as widely distributed as possible, so that electrodes were placed in spots in not close to each other. Each simulated ADL was recorded twice within each session. It was also checked that the duration of the same simulated ADL was similar in each session. In order to check the repeatability of the simulated ADLs among sessions, all sessions recorded sEMG from spot 30.

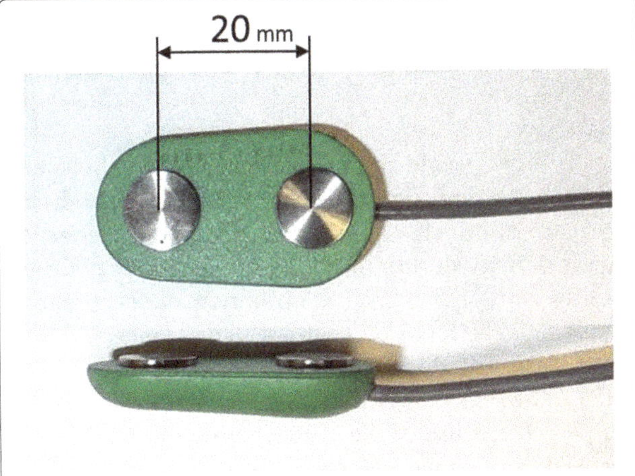


**Fig. 3** sEMG Electrodes (SX230) used: integral dry reusable electrodes with a gain of 1000, bandwidth between 20 Hz – 460 Hz and noise less than 5 μV

The sEMG records were filtered with a 4th-order bandpass filter between 25 and 500 Hz, rectified, filtered by a 4th-order low pass filter at 8 Hz, and smoothed by Gaussian smoothing. To determine muscle activity, sEMG records were normalized with the maximal values from seven records of maximum voluntary contraction (MVC), measured in each session: flexion and extension of the wrist, flexion and extension of the fingers, pronation of the forearm, ulnar deviation of the wrist, and elbow flexion. MVCs for the different movements were recorded without the use of any external device. In a comfortable posture, the subjects were asked to exert maximum effort (recorded with a dynamometer) without the help of other muscles than those of the forearm and hand. In addition, the beginning and end of the sEMG records were cut to eliminate the first and last moments in which muscle activity did not exceed 5% in any of the spots. Finally, to make records comparable in duration, all the records were interpolated to 1000 data.

## Data analysis

i) ADL repeatability: The confidence intervals (CI) of the 10 muscular activity records (2 repetitions × 5 sessions) of spot 30 were computed for each simulated ADL (1000 values of CI per ADL × 6 subjects). Statistics per ADL were considered for the analysis of repeatability.

ii) Reduction of temporary muscle activity data: The data from all repetitions at each spot for each simulated ADL were averaged, and used in the subsequent analysis, so that a total of 3780 signals (21 ADL × 6 subjects × 30 spots) of 1000 temporary data were considered. The Principal Component Analysis (PCA) was one of the first methods adapted to functional data. The fundamental idea of this extension is to conserve all the benefits of the PCA as a tool for the reduction of the dimension, conditioning them to functional data. Thus, the Functional Principal Component Analysis (FPCA) [39] could provide a basis for creating a new set of variables that incorporate the character and nature of the original functions into a smaller number of new variables. In this case, FPCA was applied to condense the temporary muscle activation data into a smaller number of parameters, in order to make the comparison of temporal patterns feasible. First, for each subject, the signals for the 21 simulated ADL from each spot were concatenated. Subsequently, each of these signals was normalized to 1 using their maximum value, in order to keep only the temporal patterns to be analyzed. Then, the data of all the subjects were concatenated. In this way, the 3780 signals were transformed into 30 functions of 126,000 data, one function per spot. FPCA was applied then to these 30 functions, extracting the functional principal components (FPCs) that explained at least 90% of the cumulative variance. Each of the 30 original functions can be expressed therefore as a common mean function plus a linear combination of the FPCs, with specific coefficients for each original function, hereinafter called reduced variables (RVs), as they condense the 126,000 data into a small number of coefficients.

iii) Similarity of patterns of muscle activation: Conglomerate or Hierarchical clustering analysis [40] is a multivariate technique that allows the classification of elements in groups or clusters, so that each element is very similar to those in its own conglomerate according to some specific selection criteria. In this case, Hierarchical clustering analysis, with the Euclidean distance as the distance criterion and Ward's method as the linkage criterion, was applied to the RVs from the 30 spots to group spots with similar muscle activation patterns. The resulting dendrogram with the spots organized in branches was displayed, and the desired number of clusters was identified by observing the distances in each step. When the distance between the clustered groups in a step becomes high in comparison to the previous steps, the elements or clusters grouped are not so close and so the grouping of the previous step may be more appropriate. The resulting groups of spots with a similar activation pattern were described. Additionally, root mean square (RMS) values of all signals measured at each spot were computed for analysis of representativeness, in order to choose a specific spot as being representative of each group.

## Results

i) ADL repeatability: Statistics of the CI of the muscular activity records at spot 30 for each simulated ADL are shown in Table 3 as a percentage of MVC. In general, the results showed a good level of repeatability of the different simulated ADL, with mean CI values of 3.33% of muscular activity and SD values of 2.22%. The most unfavorable results were for ADL 9, with mean CI of 6% (SD 4.2%) of MVC.

ii) Reduction of temporary muscle activity data: 17 FPCs explained 91% of the total variance. RVs are displayed in Fig. 4 as a rectangular array of colored cells defined by the values of the RVs.

iii) Similarity of patterns of muscle activation: Fig. 5 shows the dendrogram obtained from the hierarchical clustering with the distance between grouped elements. The dendrogram shows different possible groupings, depending on the linkage distances. The resulting groups for the case of seven clusters are displayed with different colors in Fig. 6.

**Table 3** CI Statistics: mean and standard deviation (SD)

| ADL | mean | SD |
|---|---|---|
| 1 | 2.6% | 2.3% |
| 2 | 1.9% | 1.4% |
| 3 | 1.7% | 1.4% |
| 4 | 3.6% | 3.2% |
| 5 | 4.3% | 2.5% |
| 6 | 5.0% | 3.1% |
| 7 | 3.9% | 2.6% |
| 8 | 2.9% | 2.8% |
| 9 | 6.0% | 4.2% |
| 10 | 2.7% | 1.6% |
| 11 | 4.1% | 2.2% |
| 12 | 3.9% | 2.7% |
| 13 | 2.9% | 1.1% |
| 14 | 3.1% | 1.6% |
| 15 | 3.7% | 2.5% |
| 16 | 2.1% | 1.3% |
| 17 | 3.8% | 2.3% |
| 18 | 2.7% | 1.5% |
| 19 | 2.2% | 2.1% |
| 20 | 4.0% | 2.6% |
| 21 | 2.9% | 1.6% |

Figure 7 shows the RMS values of the muscular activity measured at each spot, to be considered when choosing a specific spot as representative of a group. Spots 18 and 28 show a low level of RMS values compared to the other spots, reaching less than 5% of RMS of the muscular activity.

## Discussion

In this work we identified forearm areas with similar muscle activation patterns by means of FPCA and cluster analysis, which could be used to characterize the muscle activity during relevant simulated ADLs. Consequently, one of the main contribution of the proposed approach is the focus on goal-directed actions.

The tasks used in this study were taken from the SHFT. The selection of the tasks was originally based on the percentage use of the most common handgrips during ADL and hence reflects an accurate representativeness of hand function in day-to-day life. Muscle activity on 30 different forearm spots was obtained for each of the 21 representative simulated ADLs by merging sEMG

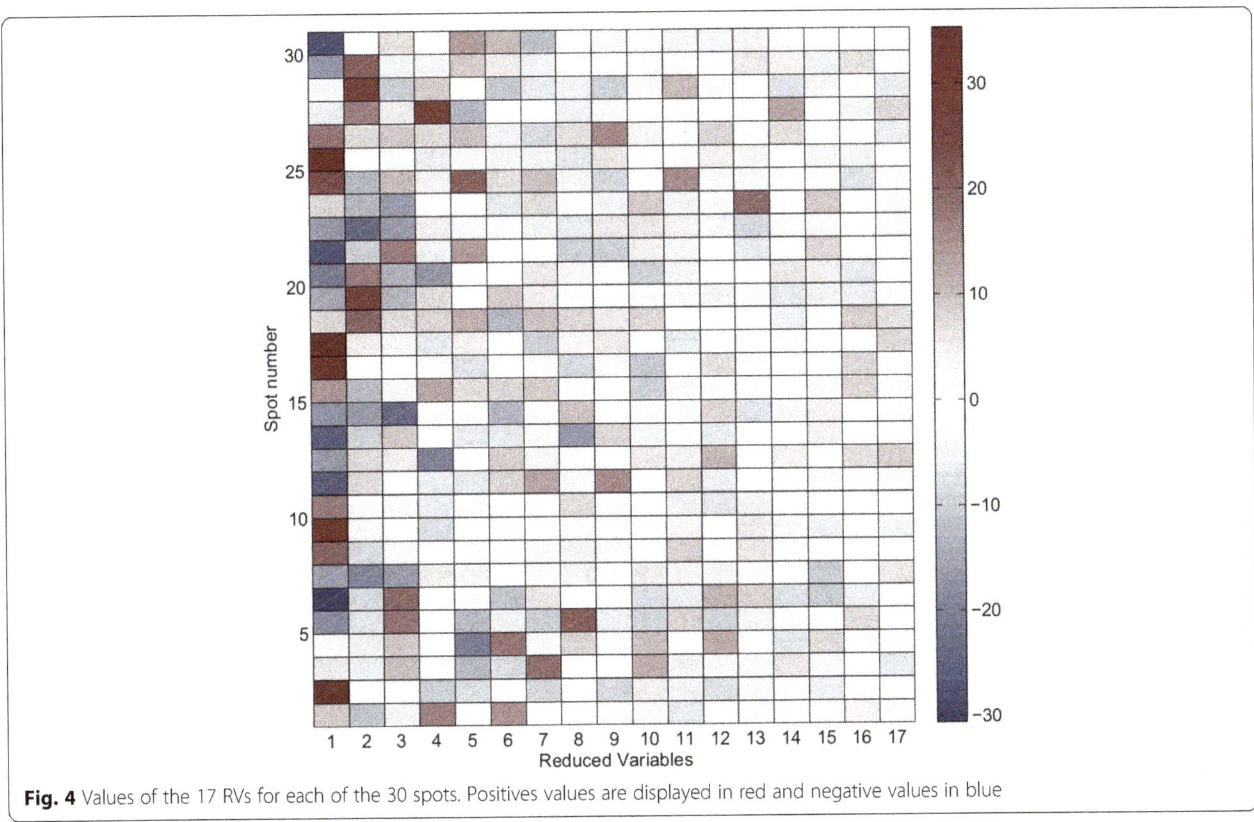


**Fig. 4** Values of the 17 RVs for each of the 30 spots. Positives values are displayed in red and negative values in blue

signals recorded in five different sessions This was possible because of the high repeatability observed for the simulated ADLs considered (mean CI values at spot 30 of 3.33% of MVC, SD 2.22%), thanks to the proposed standardization. Unscrewing a lid and leaving it on the Table (ADL 9) was the least repeatable activity, with a mean CI value of 6% of MVC.

Seventeen RVs explained 91% of the total variance. This new set of variables provided the same information but in a more easily interpretable way: each of the 30 original functions of 126,000 temporary data was expressed as a common mean function plus a linear combination of 17 FPCs, with specific coefficients (RVs) for each original function. These 17 RVs obtained from the FPCA have been used in a hierarchical cluster analysis to group spots with similar activation patterns.

From the observation of the distances in each step in the hierarchical cluster analysis, seven groups may be established based on the moment in which the distance between the clustered groups in a step becomes high in comparison to the previous steps. Therefore, we could reduce the number of EMG sensors from 30 to 7 without losing any relevant muscle activity information during the performance of simulated ADL. This is very interesting, as there are 20 muscles superimposed on each other in the forearm, thus making it practically impossible to isolate the sEMG signal of each muscle. Protocols of muscular function assessment of the forearm in rehabilitation can benefit from this. The measurement of the amplitude of the sEMG signals from the seven resulting groups may provide a reasonably complete quantitative picture of the patient's rehabilitation outcome. Furthermore, given that the number of available electrodes could be limited and that a smaller number of EMG signals may simplify the assessment, knowing the similarity of the sEMG signals with different levels of hierarchy, as provided with the method used, is also quite interesting. This also applies for prosthesis control, where the number of EMG signals to be used is limited by the complexity of the controller.

When observing the dendrogram obtained from the cluster analysis, different numbers of clusters could be chosen. Two main hierarchy levels can be observed: one involves the flexor muscles of the wrist and digits (spot groups 1 to 3), and the other includes their extensor muscles (spot groups 4 to 7). In the next hierarchy level, flexors bifurcate into wrist flexors (groups 1 and 2) and digit flexors (group 3); it is known that digit flexors also contribute to wrist flexion [41]. And something similar occurs for extensors, which bifurcate into digit extensors (groups 4 and 5) and wrist extensors (groups 6 and 7). Therefore, in the case of using two signals for prosthesis control, it would be logical to consider the two main hierarchy levels for an intuitive control in carrying out

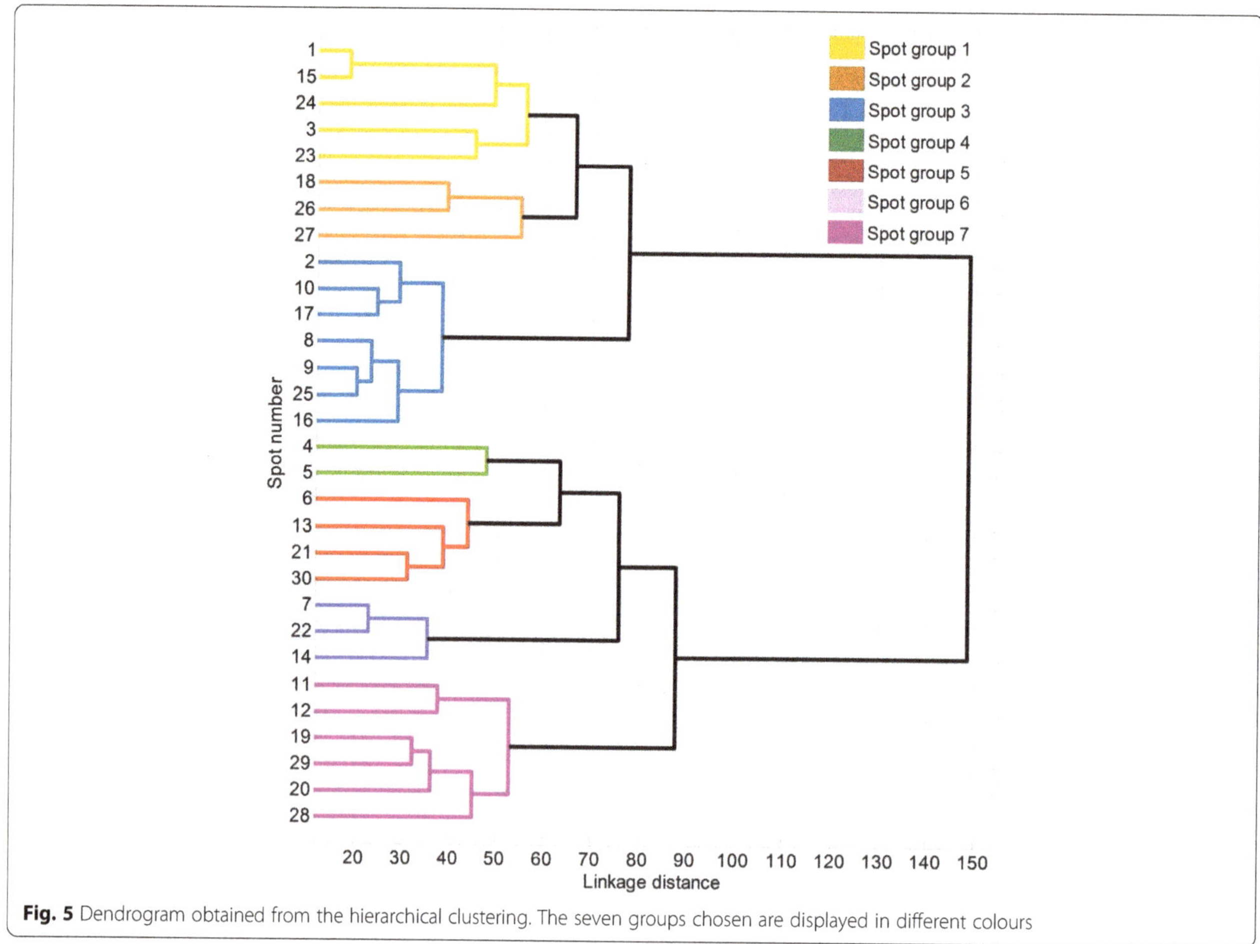


**Fig. 5** Dendrogram obtained from the hierarchical clustering. The seven groups chosen are displayed in different colours

ADL, associating the sEMG signals to the performance of flexion and extension movements of the prosthesis. In the case of using four control signals, we could use the next hierarchy level described, differentiating between hand and wrist flexors and extensors. However, the higher the number of signals is, the more difficult it will become to separate them, and using groupings that are different from those resulting from the dendrogram will not be so intuitive.

If we were to divide the classification a step further we would get the seven groups described in the results section: Group 1 is defined by spots 1, 3, 15, 23 and 24. Anatomically, it could be recording mainly the muscle activity of the flexor carpi ulnaris (FCU). The FCU is a very powerful muscle that acts as a flexor and ulnar deviator of the wrist [41], being responsible for stabilizing it during activities such as slicing meat and using a hammer. Group 2 is defined by spots 18, 26, and 27, and could be recording the muscle activity of the flexor carpi radialis (FCR) and palmaris longus (PL). Both FCR and PL are reported as wrist flexors, and FCR also as a radial deviator [41]. Group 3 is defined by spots 2, 8, 9, 10, 16, 17 and 25, and could be recording the muscle activity of the digit flexor muscles: flexor digitorum superficialis (FDS) and profundus (FDP), and flexor pollicis longus (FPL). The FDS and FDP muscles are finger flexors [42], and are located on the same forearm area, but at different depths. FPL is a flexor of all three joints of the thumb, and is the only thumb interphalangeal (IP) joint flexor [43]. Group 4 is defined by spots 3 and 4, placed next to the thumb. This group could be recording the muscle activity of the abductor pollicis longus (APL), and extensor pollicis longus (EPL) and brevis (EPB). EPB supports the extension of the metacarpophalangeal (MCP) joint of the thumb, APL participates in the abduction and extension of the thumb carpometacarpal (CMC) joint, and EPL extends the three thumb joints, and adducts the thumb CMC joint [41]. Moreover, all these muscles contribute to wrist radial deviation [41]. Group 5 is defined by spots 6, 13, 21 and 30, and could be recording the muscle activity of the extensor digitorum communis (EDC). The EDC is the primary extensor of the MCP joints of the fingers, although it also contributes to the extension of the proximal (PIP) and distal (DIP) interphalangeal joints of the fingers [41]. Group 6 is defined by spots 7, 14 and 22, and could be

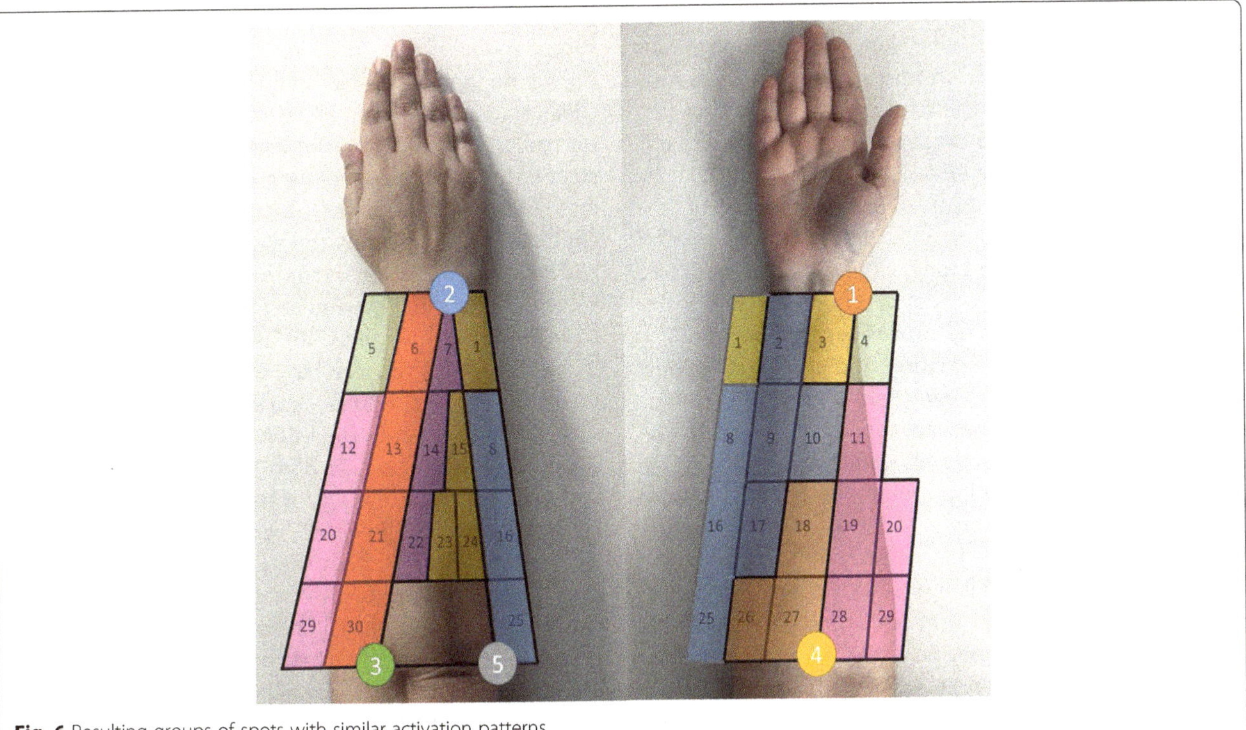


**Fig. 6** Resulting groups of spots with similar activation patterns

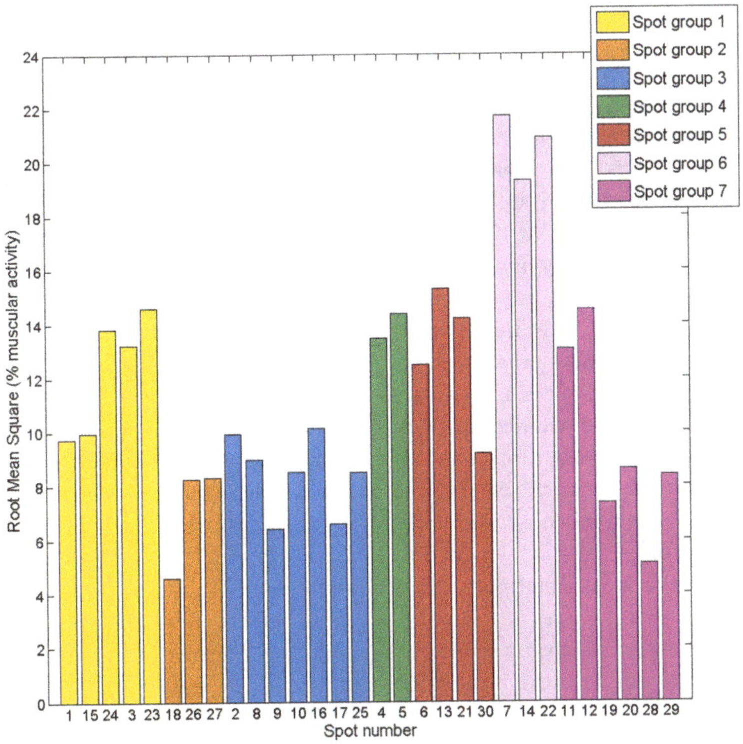

**Fig. 7** RMS of the muscular activity values of each spot. The bars are colored according to the resulting groups

recording the muscle activity of the extensor carpi ulnaris (ECU). ECU has the largest moment arm for ulnar deviation and plays an important role in supporting the distal radioulnar joint [43], the joint that enhances the manipulating skills of the hand [44]. Finally, Group 7 is defined by spots 11, 12, 19, 20, 28 and 29, and could be recording the muscle activity of the brachioradialis (BR), pronator teres (PT), extensor carpi radialis brevis (ECRB) and longus (ECRL). The main role of the ECRB and ECRL is wrist extension and radial deviation [43]. However, some studies suggested that ECRL and ECRB could have pronation and supination elbow moments [45]. BR and PT contribute to elbow flexion as well as forearm pronation and supination. Moving forward at the hierarchical level would not make sense, since we would be selecting groups of points where the distance between the spots grouped in the next step is low compared to the previous steps, and the number of sEMG signals to be used is limited.

Therefore, we propose using seven groups of spots for characterizing the muscular activity of the forearm during simulated ADL, in order to substantially reduce the number of spots to be registered, and to maintain muscular-anatomical coherence. The signals from these seven spots would be related to seven different movements: (group 1) wrist flexion and ulnar deviation; (group 2) wrist flexion and radial deviation; (group 3) digit flexion; (group 4) thumb extension and abduction/adduction; (group 5) finger extension; (group 6) wrist extension and ulnar deviation; and (group 7) wrist extension and radial deviation.

When assessing muscle function in rehabilitation, some spots may be unavailable because of the simultaneous usage of other equipment, such as sensors for kinematics measurement. When choosing the representative spots of each group to be recorded from among those available, the percentage of muscle activity measured (Fig. 7) should be taken into account. Selecting the spots with the highest percentage of muscle activity may be more reliable, since, at these spots, the muscles recorded could be more superficial or the muscle area recorded could be more centered. In particular, spots with a very low level of muscle activity should be avoided, such as spots 18 and 28. The greatest muscle activity was observed in the most distal part of the forearm (Fig. 7). However, more proximal spots could be chosen, depending on the availability, except for group 4, which is composed of two spots placed in the most distal part of the forearm.

FPCA is one of the most popular multivariate analysis techniques for extracting information from functional data, reducing the dimensions of a data set in which there are a large number of interrelated variables, while still holding as much of the total variation as possible [39]. While FPCA results in dimension reduction, FPCA vector scores can be used for clustering different functions/components using standard clustering methods. Clustering is one of the most frequently used techniques for partitioning a dataset into subgroups that contain instances that are similar to each other while being clearly dissimilar to those of the other groups. In a functional context, clustering helps to identify representative curve patterns and individuals who are very likely to be involved in the same or similar processes. Other methods reported in the literature [46] used for task identification in prosthesis control are based on the segmentation of the EMG signals into a series of windows, in which some commonly used time-domain feature sets (such as Mean Absolute Value or Zero Crossings) are extracted and used for motion classification. Our method is similar, but applied to the entire signal for all tasks and subjects, and extracting from the FPCA the features of the signal that holds as much of the total possible variation and using them for muscular classification.

The EMG recordings have been carried out with an 8-channel sEMG device, which required the repetition of the same activities for each subject in five different sessions to allow the measurement of the high number of spots chosen. Although reproducibility error has been checked to be small for the spot 30 between sessions, the use of high-density surface EMG might be considered in upcoming works.

The current study has been limited to six healthy, able-bodied subjects, and the results could be verified in further studies with a higher number of subjects and including impaired subjects. However, muscular groupings obtained in this study may be used as a first approximation, and may be used as guide for future validation for subjects with hand impairments or amputees.

As future work, studies could be conducted to relate EMG to kinematics during the performance of ADL, by using these seven spots as representative of the muscular activity of the forearm in ADL. A further step would be to evaluate kinematics and muscular synergies during specific functional tasks.

## Conclusions

This study aimed to identify skin zones with similar muscle activation patterns in order to determine the minimum number of electrodes required to characterize the muscle activity during simulated ADL without losing any relevant information. The results indicate that we could reduce the number of sEMG sensors from 30 to 7 and use them as representative spots of the muscular activity of the forearm in simulated ADL. The simulated ADL performed are included in the SHFT, which uses the most common handgrips in day-to-day life. Hence, the simulated ADL chosen in this study may reflect an accurate representativeness of the hand function. This result may help to assess muscle function in rehabilitation as well as simplify the complexity of prosthesis control.

### Abbreviations

ADL: Activities of daily living; APL: Abductor policis longus; BR: Brachioradialis; CI: Confidence intervals; CMC: Carpometacarpal; DIP: Distal interphalangeal; ECRB: Extensor carpi radialis brevis; ECRL: Extensor carpi radialis longus; ECU: Extensor carpi ulnaris; EDC: Extensor digitorum communis; EPB: Extensor policis brevis; EPL: Extensor policis longus; FCR: Flexor carpi radialis; FCU: Flexor carpi ulnaris; FDP: Flexor digitorum profundus; FDS: Flexor digitorum superficialis; FPCA: Functional principal component analysis; FPCs: Functional principal components; HCA: Hierarchical clustering analysis; IP: Interphalangeal; MCP: Metacarpophalangeal; MVC: Maximum voluntary contractions; PIP: Proximal interphalangeal; PL: Palmaris longus; PT: Pronator teres; RMS: Root mean square; RVs: Reduced variables; SD: Standard deviation; sEMG: Surface electromyography; SENIAM: Surface electromyography for the non-invasive assessment of muscles; SHFT: Sollerman hand function test

### Acknowledgements

The authors are grateful for funding from the Spanish MINECO and the European Union (FEDER funds) through project DPI2014-52095-P and FPI grant BES-2015-072480.

### Author contributions

NJB formulated the research question and the experimental design, collected and analyzed data, and wrote the manuscript. JSB and MV formulated the research question and experimental design, supervised the data collection and analysis, and reviewed and revised manuscript. VGI and ARS collaborated in the data collection and wrote the manuscript. All authors read and approved the final manuscript.

### Funding

The contents of this paper were developed under a grant from the Spanish MINECO and the European Union (FEDER funds) through project DPI2014–52095-P and FPI grant BES-2015-072480.

### Competing interest

The authors declare that they have no competing interests.

### References

1. Yu H-L, Chase RA (Robert A, Strauch B. Atlas of hand anatomy and clinical implications. Mosby; 2004.
2. Kumar D, Poosapadi Arjunan S, Singh V. Towards identification of finger flexions using single channel surface electromyography – able bodied and amputee subjects. J Neuroeng Rehabil BioMed Central. 2013;10:50.
3. Rojas-Martínez M, Mañanas MA, Alonso JF. High-density surface EMG maps from upper-arm and forearm muscles. J Neuroeng Rehabil. 2012;9:85.
4. Castellini C, Fiorilla AE, Sandini G. Multi-subject / daily-life activity EMG-based control of mechanical hands. J Neuroeng Rehabil. 2009;6:41.
5. Hogrel J-Y. Clinical applications of surface electromyography in neuromuscular disorders. Neurophysiol Clin Neurophysiol Elsevier Masson. 2005;35:59–71.
6. Vinstrup J, Calatayud J, Jakobsen MD, Sundstrup E, Jørgensen JR, Casaña J, et al. Hand strengthening exercises in chronic stroke patients: dose-response evaluation using electromyography. J Hand Ther. 2018;31:111–21.
7. Nam C, Rong W, Li W, Xie Y, Hu X, Zheng Y. The effects of upper-limb training assisted with an electromyography-driven neuromuscular electrical stimulation robotic hand on chronic stroke. Front Neurol. Frontiers Media SA. 2017;8:679.
8. Landis J, Keselman I, Murphy CN. Comparison of electromyographic (EMG) activity of selected forearm muscles during low grade resistance therapeutic exercises in individuals diagnosed with lateral epicondylitis. Work. 2005;24: 85–91.
9. Hesam-Shariati N, Trinh T, Thompson-Butel AG, Shiner CT, McNulty PA. A longitudinal electromyography study of complex movements in Poststroke therapy. 1: heterogeneous changes despite consistent improvements in clinical assessments. Front Neurol. 2017;8:340.
10. Chae J, Yang G, Park BK, Labatia I. Muscle weakness and Cocontraction in upper limb hemiparesis: relationship to motor impairment and physical disability. Neurorehabil Neural Repair. 2002;16:241–8.
11. Brorsson S, Nilsdotter A, Thorstensson C, Bremander A. Differences in muscle activity during hand-dexterity tasks between women with arthritis and a healthy reference group. BMC Musculoskelet Disord. 2014;15:154.
12. Santello M, Baud-Bovy G, Jörntell H. Neural bases of hand synergies. Front Comput Neurosci. 2013;7:23.
13. Johnston JA, Winges SA, Santello M. Neural control of hand muscles during prehension. Adv Exp Med Biol. 2009;629:577–96.
14. Coscia M, Cheung VC, Tropea P, Koenig A, Monaco V, Bennis C, et al. The effect of arm weight support on upper limb muscle synergies during reaching movements. J Neuroeng Rehabil. BioMed Central. 2014;11:22.
15. Zariffa J, Steeves J, Pai DK. Changes in hand muscle synergies in subjects with spinal cord injury: characterization and functional implications. J Spinal Cord Med. 2012;35:310–8.
16. Weiss EJ, Flanders M. Muscular and postural synergies of the human hand. J Neurophysiol American Physiological Society. 2004;92:523–35.
17. Winges SA, Furuya S, Faber NJ, Flanders M. Patterns of muscle activity for digital coarticulation. J Neurophysiol. 2013;110:230–42.
18. D'Andola M, Cesqui B, Portone A, Fernandez L, Lacquaniti F, d'Avella A. Spatiotemporal characteristics of muscle patterns for ball catching. Front Comput Neurosci. 2013;7:107.
19. Ertan H, Kentel B, Tümer ST, Korkusuz F. Activation patterns in forearm muscles during archery shooting. Hum Mov Sci. 2003;22:37–45.
20. VencesBrito AM, Rodrigues Ferreira MA, Cortes N, Fernandes O, Pezarat-Correia P. Kinematic and electromyographic analyses of a karate punch. J Electromyogr Kinesiol. 2011;21:1023–9.
21. Ertan H. Muscular activation patterns of the bow arm in recurve archery. J Sci Med Sport. 2009;12:357–60.
22. Ricci FPFM, Santiago PRP, Zampar AC, Pinola LN, Fonseca M de CR. Upper extremity coordination strategies depending on task demand during a basic daily activity. Gait Posture. 2015;42:472–8.
23. Cheung VCK, Turolla A, Agostini M, Silvoni S, Bennis C, Kasi P, et al. Muscle synergy patterns as physiological markers of motor cortical damage. Proc Natl Acad Sci. 2012;109:14652–6.
24. Safavynia SA, Torres-Oviedo G, Ting LH. Muscle synergies: implications for clinical evaluation and rehabilitation of movement. Top Spinal Cord Inj Rehabil NIH Public Access. 2011;17:16–24.
25. Atzori M, Muller H, Baechler M. Recognition of hand movements in a trans-radial amputated subject by sEMG. 2013 IEEE 13th Int Conf Rehabil robot. IEEE. 2013:1–5.
26. Castellini C, Van Der Smagt P. Surface EMG in advanced hand prosthetics. Biol Cybern. 2009;100:35–47.
27. Farina D, Ning Jiang N, Rehbaum H, Holobar A, Graimann B, Dietl H, et al. The extraction of neural information from the surface EMG for the control of upper-limb prostheses: emerging avenues and challenges. IEEE Trans Neural Syst Rehabil Eng. 2014;22:797–809.
28. Engdahl SM, Christie BP, Kelly B, Davis A, Chestek CA, Gates DH. Surveying the interest of individuals with upper limb loss in novel prosthetic control techniques. J Neuroeng Rehabil. 2015;12.
29. Zuo KJ, Olson JL. The evolution of functional hand replacement: from iron prostheses to hand transplantation. Plast Surg (Oakv) SAGE Publications; 2014;22:44–51.
30. Daley H, Englehart K, Hargrove L, Kuruganti U. High density electromyography data of normally limbed and transradial amputee subjects for multifunction prosthetic control. J Electromyogr Kinesiol. 2012;22:478–84.

31. Schultz AE, Clinical Review KTA. Current concepts neural interfaces for control of upper limb prostheses: the state of the art and future possibilities. PM R. 2011;3:55–67.
32. He Huang H, Ping Zhou P, Guanglin Li G, Kuiken TA. An analysis of EMG electrode configuration for targeted muscle Reinnervation based neural machine Interface. IEEE Trans Neural Syst Rehabil Eng. 2008;16:37–45.
33. Tkach DC, Young AJ, Smith LH, Hargrove LJ. Performance of pattern recognition myoelectric control using a generic electrode grid with targeted muscle Reinnervation patients. 2012 Annu Int Conf IEEE Eng med biol Soc. IEEE. 2012:4319–23.
34. Hebert JS, Lewicke J, Williams TR, Vette AH. Normative data for modified box and blocks test measuring upper-limb function via motion capture. J Rehabil Res Dev. 2014;51:918–32.
35. Barmpakos D, Strimpakos N, Karkanis SA, Pattichis C. Towards a versatile surface electromyography classification system. Cham: Springer; 2016. p. 33–6.
36. Baldacchino T, Jacobs WR, Anderson SR, Worden K, Rowson J. Simultaneous Force Regression and Movement Classification of Fingers via Surface EMG within a Unified Bayesian Framework. Front Bioeng Biotechnol. Frontiers Media SA. 2018;6:13.
37. Koch P, Phan H, Maass M, Katzberg F, Mertins A. Early prediction of future hand movements using sEMG data. 2017 39th Annu Int Conf IEEE Eng med biol Soc. IEEE; 2017. p. 54–57.
38. Sollerman C, Ejeskar A. Sollerman hand function test: a standardised method and its use in tetraplegic patients. Scand J Plast Reconstr Surg Hand Surg. 1995;29:167–76.
39. Ullah S, Finch CF. Applications of functional data analysis: A systematic review. BMC Med Res Methodol. 2013;13:43.
40. Hair JF, Black WC, Babin BJ, Anderson R.E. Multivariate Data Analysis. 7th ed. Pearson Prentice Hall; 2010.
41. Oatis CA. Kinesiology : the mechanics and pathomechanics of human movement: Lippincott Williams & Wilkins; 2009.
42. Bender LF. Muscles Alive: Their Functions Revealed by Electromyography. JAMA. 1967;201(4):277.
43. Brand PW, Hollister AM. Clinical mechanics of the hand. 3rd Editio. St. Louis: Mosby Publishing; 1999.
44. Almquist EE. Evolution of the distal radioulnar joint. Clin Orthop Relat Res. 1992:5–13.
45. Bremer AK, Sennwald GR, Favre P, Jacob HAC. Moment arms of forearm rotators. Clin Biomech. 2006;21:683–91.
46. Geng Y, Zhang X, Zhang Y-T, Li G. A novel channel selection method for multiple motion classification using high-density electromyography. Biomed Eng Online. 2014;13:102.

# 14

# Assessment of a markerless motion analysis system for manual wheelchair application

Jacob Rammer[1,2,3*], Brooke Slavens[6], Joseph Krzak[4,5], Jack Winters[7], Susan Riedel[1,2,3] and Gerald Harris[1,2,3,4]

**Abstract**

**Background:** Wheelchair biomechanics research advances accessibility and clinical care for manual wheelchair users. Standardized outcome assessments are vital tools for tracking progress, but there is a strong need for more quantitative methods. A system offering kinematic, quantitative detection, with the ease of use of a standardized outcome assessment, would be optimal for repeated, longitudinal assessment of manual wheelchair users' therapeutic progress, but has yet to be offered.

**Results:** This work evaluates a markerless motion analysis system for manual wheelchair mobility in clinical, community, and home settings. This system includes Microsoft® Kinect® 2.0 sensors, OpenSim musculoskeletal modeling, and an automated detection, processing, and training interface. The system is designed to be cost-effective, easily used by caregivers, and capable of detecting key kinematic metrics involved in manual wheelchair propulsion. The primary technical advancements in this research are the software components necessary to detect and process the upper extremity kinematics during manual wheelchair propulsion, along with integration of the components into a complete system. The study defines and evaluates an adaptable systems methodology for processing kinematic data using motion capture technology and open-source musculoskeletal models to assess wheelchair propulsion pattern and biomechanics, and characterizes its accuracy, sensitivity and repeatability. Inter-trial repeatability of spatiotemporal parameters, joint range of motion, and musculotendon excursion were all found to be significantly correlated ($p < 0.05$).

**Conclusions:** The system is recommended for use in clinical settings for frequent wheelchair propulsion assessment, provided the limitations in precision are considered. The motion capture-model software bridge methodology could be applied in the future to any motion-capture system or specific application, broadening access to detailed kinematics while reducing assessment time and cost.

**Keywords:** Manual wheelchair, Markerless motion capture, Musculoskeletal models, Pediatric rehabilitation

## Introduction

There are several current methods that have been successfully applied to study certain aspects of wheelchair propulsion outcomes and biomechanics. Laboratory motion analysis [1–3] is precise and detailed, yet costly and time-consuming, especially for repeated, frequent longitudinal assessments. Inertial Measurement Units (IMUs) are easier to use, but, when wrist-applied as typical practice [4] lack detailed joint kinematics at the shoulder, a key joint in assessing injury risk for manual wheelchair users. Leving et al. [5] has shown promise for IMUs to detect spatiotemporal parameters of motion, including activity level, but does not describe movement kinematics. Similarly, van der Slikke et al. [6–9] have shown accurate measurements by IMUs in speed, displacement, and acceleration metrics applied to wheelchair sports. The referenced IMU studies do not address the kinematics of individual joints, rather focusing on gross movement of the wheelchair and spatiotemporal aspects of arm movement for propulsion. Instrumented wheels, whether commercially available like the SmartWheel [10] or modified from bicycle wheel power meters [11] provide power and torque output at the pushrim, but require a motion capture system to obtain

* Correspondence: jacob.rammer@marquette.edu; Jacob.rammer@marquette.edu
[1]Orthopaedic and Rehabilitation Engineering Center (OREC), Marquette University, Olin Engineering Suite 323, Milwaukee, WI 53201-1881, USA
[2]Department of Biomedical Engineering, Marquette University, Olin Engineering Suite 323, Milwaukee, WI 53201-1881, USA
Full list of author information is available at the end of the article

kinematics. Standardized outcome measures like the Wheelchair Propulsion Test (WPT) and Wheelchair Skills Test (WST) use trained observers and standard protocols to assess function [12], but lack quantitative, kinematic data.

Based on the available solutions on the market, there is a significant need for development in this area, targeted toward physical and occupational therapists. The proposed technology quantitatively evaluates manual wheelchair mobility in a timely manner and outside of the motion analysis laboratory. The system output includes spatiotemporal parameters, joint and muscle kinematics, and propulsion pattern. The spatiotemporal parameters, which can tell the clinician how propulsion speed, cadence, and time change during a patient's rehabilitation process, indicate efficiency to the clinician. It also includes joint range of motion, which is useful in determining which joints are being utilized, and whether the patient is progressing toward a more effective strategy, Propulsion pattern is additionally provided as a qualitative (pattern) and quantitative (size of the pattern) metric – useful for assessing changes in response to therapy, and progress toward smoother and more effective propulsion. This monitoring could provide clinicians with quantitative data to indicate whether a patient is stable or deviating from an appropriate pattern during the course of care. This has a potential to be an early indicator of injury risk, as cadence and propulsion pattern were identified as predictors of injury risk in manual wheelchair users with spinal cord injury [13]. Several technological options have recently become available to make this development possible, including the Kinect® for motion capture, OpenSim for musculoskeletal biomechanics, and the Personal Wheelchair Platform to support the wheelchair and simulate overground resistance. Each technological element of the system will be introduced and discussed separately.

### Microsoft® Kinect®

The Microsoft® Kinect® is a markerless motion capture sensor designed and marketed for the consumer gaming market. It uses infrared depth sensing to capture 3-dimensional imaging and real-time algorithms to process skeletal position. The validity and research applicability of the Kinect has been widely debated in current literature, as summarized in Table 1. Several studies ([14–17], and [18]) have compared the Kinect against motion capture systems and have reported that the Kinect-detected

**Table 1** Microsoft Kinect in Upper Extremity Clinical Applications [49–68]

| Reference | Description | Key Results |
|---|---|---|
| [14] | Assessment of validity of Kinect v1.0 against marker-based motion capture; 48 normal subjects; upper and lower extremity | Similar reproducibility; different ROM detection for the lower extremity but similar results for shoulder abduction (±3°) and elbow flexion (±11°) |
| ([45, 46]; [47, 48]) | Assessment of validity of Kinect v2 for postural control and balance against marker-based motion capture; 30 normal subjects; | High reliability and concurrent validity for balance assessment (trunk, upper and lower extremity kinematics) |
| [15] | Direct comparison of Kinect against Vicon ® clinical motion capture | Kinect detection is accurate, one order of magnitude less precise than Vicon |
| [16] | Kinect vs. Vicon for gross and fine movements (controlled study of Parkinson's disease); movements included whole-body coordinated movements and shoulder flexion/abduction targeted movements | Kinect is highly accurate for gross movement detection, less for smaller hand movements; repeatable measurements ($r > 0.9$); high interclass correlation for gross extremity/body movements; low correlation for fine hand movements |
| [19] | Shoulder-specific validity and reliability of Kinect; 10 normal subjects; shoulder joint (flexion, abduction, rotation) assessed in static poses with Kinect, marker based motion analysis, and goniometer; the Kinect was tested both in anterior and sagittal view with insignificant difference in ICC | High reliability, but LOA greater than ±5°, up to 7° for shoulder abduction; Kinect shoulder measurement is most accurate in flexion (high ICC with valid measurements), and least accurate at abduction approaching 90°; note that the analysis focused on extents of motion, not the entire range of motion |
| [20] | Shoulder ROM by Kinect vs. goniometry; 15 normal and 12 with adhesive capsulitis of the shoulder; Active ROM compared between standard goniometry and Kinect | High ICC; Kinect is repeatable for shoulder ROM measurements (ICCs: 0.91 flexion, 0.94 abduction; 0.91 external rotation); Kinect accurately measures 3D shoulder ROM |
| [21] | Test-retest repeatability of Kinect for UE, both 12 healthy and 18 stroke subjects; focus on shoulder and elbow kinematics, and spatiotemporal metrics | Study showed acceptable repeatability and sensitivity in both populations; Shoulder and elbow angle measurements all showed greater than 0.9 ICC, indicating repeatability |
| [17] | Accuracy and reliability of Kinect v2 for clinical measurements – compared with Vicon; 19 normal subjects; spatial range of motion of arm movements evaluated | Most parameters ICC > 0.7; no systematic bias; all joints of the UE and torso detected by Kinect had Pearson correlation > 0.9 against Vicon; concurrent Kinect and Vicon used |
| [18] | Kinect (anterior) vs. Vicon; 20 normal subjects; balance and arm sway; Kinect and Vicon data collected separately, analyzed for variance in movement patterns and marker positions | Study found that broad movements of the upper extremities had > 90% accuracy, finer hand movements lower accuracy; activities are standardized (game-directed) for comparison between the systems |

*ICC* Interclass Correlation Coefficient, *ROM* Range of Motion, *LOA* Limits of Agreement; most studies use Kinect in anterior position, noted if different data is reproducible, accurate for gross movement detection but not finer movements, and approximately one order of magnitude lower precision than the laboratory standard marker-based systems. Studies focusing on specific aspects of detection found that shoulder kinematics and range of motion are reliable [19, 20] and test-retest reliability is acceptable in both healthy and stroke patients [21]. Based on these findings, we focus on shoulder biomechanics in our analysis.

Specifically focusing on the elbow and shoulder movements most relevant to manual wheelchair propulsion, several studies have addressed accuracy and reliability of the Kinect for this use. Comparing the shoulder kinematics from Kinect to laboratory motion capture, Bonnechere et al. found that ROM detection is within 3 degrees for shoulder abduction and 11 degrees for the elbow, with the Kinect sensor positioned anterior to the subjects. Huber et al., addressed all ranges of shoulder movement in three axes, and found that the Kinect is most valid in flexion (throughout the range of motion), with an ICC of 0.95 when compared to laboratory calibrated measures, and least accurate in extreme abduction approaching 90 degrees, with ICC of 0.76. Overall analysis of these results in terms of minimum detectable difference demonstrates differences by joint motion, with 7° at the shoulder [14], and 11° at the elbow [19]. However, these results also show that these measures are repeatable with high correlation coefficients [20], which suggests that the data that ultimately is processed from the Kinect is able to detect kinematic changes, even if the measurement accuracy is less than laboratory-grade systems.

In terms of manual wheelchair propulsion, the most important movement of the shoulder joint is in flexion, and there is no extreme abduction, so these results suggest that the Kinect is adequate in the ranges of motion applicable to manual wheelchair use. Huber et al. also compared shoulder flexion with the Kinect positioned anteriorly and laterally, and found similar ICC (0.85 and 0.84, respectively) between the positions. This provides a basis for the experimental assessment contained in this work, assessing detection accuracy within the specific workspace of manual wheelchair use and camera positioning applied.

In past work [22], technical evaluation of the system using goniometry revealed key findings regarding the capabilities of the system. The broad movements of the elbow demonstrate more precision in detection than the finer movements of the hand, a result expected due to the limited resolution of the Kinect. Detection accuracy when comparing Kinect-detected and goniometric measurements is significant enough to allow differentiation between angles of the joints, and provides sufficient kinematic data for clinical decision-making. Overall, this work indicates that the Kinect is accurate in detecting ROM and joint position of the upper extremities, with a reduced precision of approximately one order of magnitude relative to laboratory systems, and higher accuracy and precision in the proximal joints relative to the distal joints. For the purposes of this development, the Kinect adequately provides the desired level of quantitative data, but the Kinect's limitations must be accounted for when interpreting that data.

#### OpenSim musculoskeletal model

OpenSim is a free, open-source software package that allows users to develop musculoskeletal models and perform biomechanical analysis [23]. The OpenSim software (National Center for Simulation in Rehabilitation Research) and specific upper extremity model used were chosen over other alternatives (including SIMM, Any-Body, and other OpenSim models) in line with the primary project goals of cost-effectiveness, research validity and acceptance in the literature, and integration into assessment software. OpenSim has gained a significant following in scientific literature, with many studies published using the software. OpenSim is also computationally efficient, while providing sufficient data to be appropriate for this application. Given that the system is open-source, it is also easily integrated into the automated assessment software. Several upper extremity models are available that are applicable to the study of wheelchair propulsion biomechanics. Holzbaur et al. [24] developed a complete model designed to accurately represent musculoskeletal structure. The validated model was later refined [25] and enhanced for improved functionality. The newer model also incorporates scapular kinematics, and a simplified coordinate system for enhanced computational efficiency, and thus was applied for this work.

#### Stationary wheelchair propulsion platform

Roller platforms and similar ergometer devices are often used in wheelchair propulsion research, placing the wheelchair in a fixed position during analysis. This is important because it allows the wheelchair user to reach a steady-state propulsion. Some laboratories [13, 26] develop research-specific systems tailored to their needs. For instance, van der Woude et al. [2] describe a custom-developed motor-driven treadmill combined with a weight-and-pulley system to provide resistance, which they use in parallel with motion capture, energetics, and instrumented wheels. Recent development has led to the Personal Wheelchair Platform [27], which provides a safe, stable, laterally independent, and calibrated platform for manual wheelchair propulsion research. The platform is designed based on a dynamic model to provide resistance consistent with overground propulsion, and does not control or limit testing conditions. The design is entirely mechanical and, if used with the same wheelchair, would produce the same results.

## Methods

The purpose of this study was to develop and evaluate a markerless wheelchair propulsion biomechanical assessment system based on the actual needs of clinicians and wheelchair users, focused on shoulder and upper extremity kinematics. The resulting design integrates consumer technology with open-source musculoskeletal modeling technology, considering the important value and technical limitations of each component, to produce a markerless wheelchair propulsion analysis platform. The system was designed around three components: the Microsoft Kinect sensor, a stationary roller platform, and OpenSim musculoskeletal model. The system was configured with the subject and wheelchair in a stationary position on a roller platform, with Microsoft Kinect sensors placed anteriorly (for recording the static trial) and laterally on each side (for recording dynamic trials), as illustrated in Fig. 1.

The system needs two Kinect sensors for minimum operation. The Kinect produces the clearest tracking results when the primary motion is perpendicular to the sensor's line of sight. Thus, for the static trial an anterior positioned sensor is used to detect the subject in standard anatomical position, while for dynamic trials the lateral cameras are used, since sagittal plane motion is the primary action of wheelchair propulsion. The laterally-positioned Kinects also minimize occlusion of wheelchair components and body parts, allowing the sensors to maintain their view of all upper extremity segments throughout the propulsion cycle.

The Personal Wheelchair Platform (Fig. 2) supports the wheelchair, constrains its lateral motion, allows asymmetric propulsion, and provides adjustable resistance to simulate overground propulsion based on user anthropometry. Maintaining the wheelchair in a static position is key to using the markerless technology effectively. Maintaining the static position allows consistent accuracy of the kinematics and, most importantly, allows the subject to continually propel forward rather than making repeated turns within a laboratory setting. This supports community and home applications.

The final major component of the system is an OpenSim-based musculoskeletal model, developed and validated for upper extremity kinematics [25]. The model employed a virtual marker set compatible with the automated algorithms that interpret data from the Kinect sensors. The model is iteratively fitted to the motion data, and to increase the simplicity and speed of the computations, the model is used in its unilateral configuration, with each upper extremity computed separately. Key kinematic data outputs from the model include triaxial joint kinematics of the arms and trunk, and musculotendon lengths.

The Microsoft Kinect produces skeletal position, which is recorded in real-time from the sensors during the evaluation, and is subsequently input to the OpenSim musculoskeletal model. The software package was developed using MATLAB, which can interface with both the Kinect software and OpenSim modeling package when appropriately configured. Figure 3 illustrates the process, and is divided into three distinct processing phases.

Phase 1 of the process imports and filters the skeletal position data, acquired from both static and dynamic trials. A low-pass Butterworth filter, with cutoff frequency of 10 Hz, removes unwanted noise from the position data. The OpenSim model has been modified to include a custom virtual marker set that interfaces with data acquired by the sensors. The final step in Phase 1 converts position data, consisting of

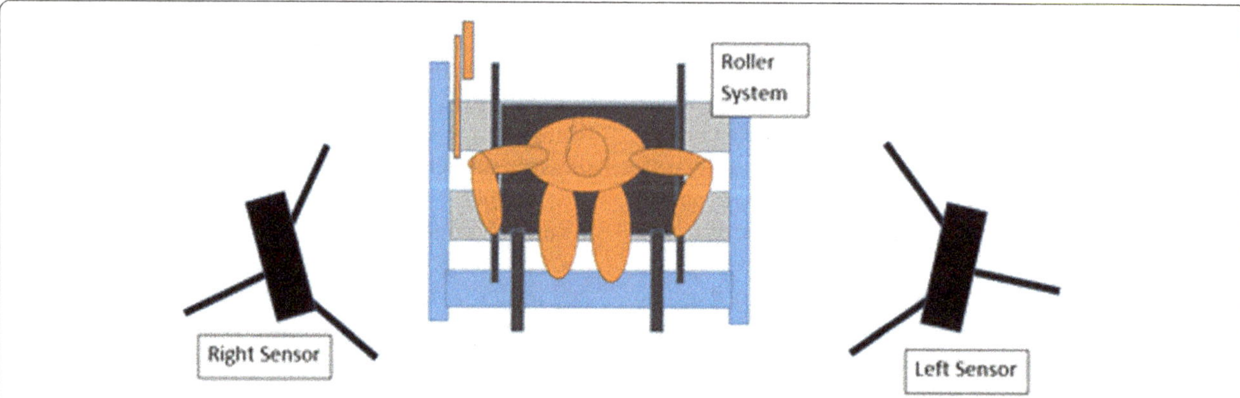


**Fig. 1** Conceptual Design and Configuration of the Markerless Wheelchair Analysis System. Subject is stationary on roller system, with a single Kinect sensor positioned in the center, anterior to the subject (for static trial), and two Kinect sensors positioned laterally, to the left and right of the subject (for dynamic trials) – the sensors are moved between trials and a total of two are needed

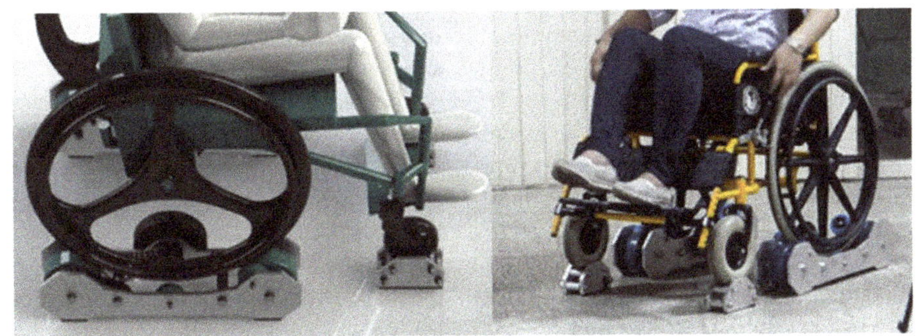

**Fig. 2** Personal Wheelchair Platform. Used to support the wheelchair and provide anthropometrically correct resistance

joint center locations and segment quaternion orientations, to virtual marker positions.

Phase 2 then includes a second set of algorithms to process the position data, automatically identifying individual propulsions from the data series, and selecting the ten most similar and consistent propulsions from those identified. The data is then divided into twenty individual data sets comprised of ten propulsions each on the left and right sides and is ready for OpenSim processing. Each individual trial is processed separately,

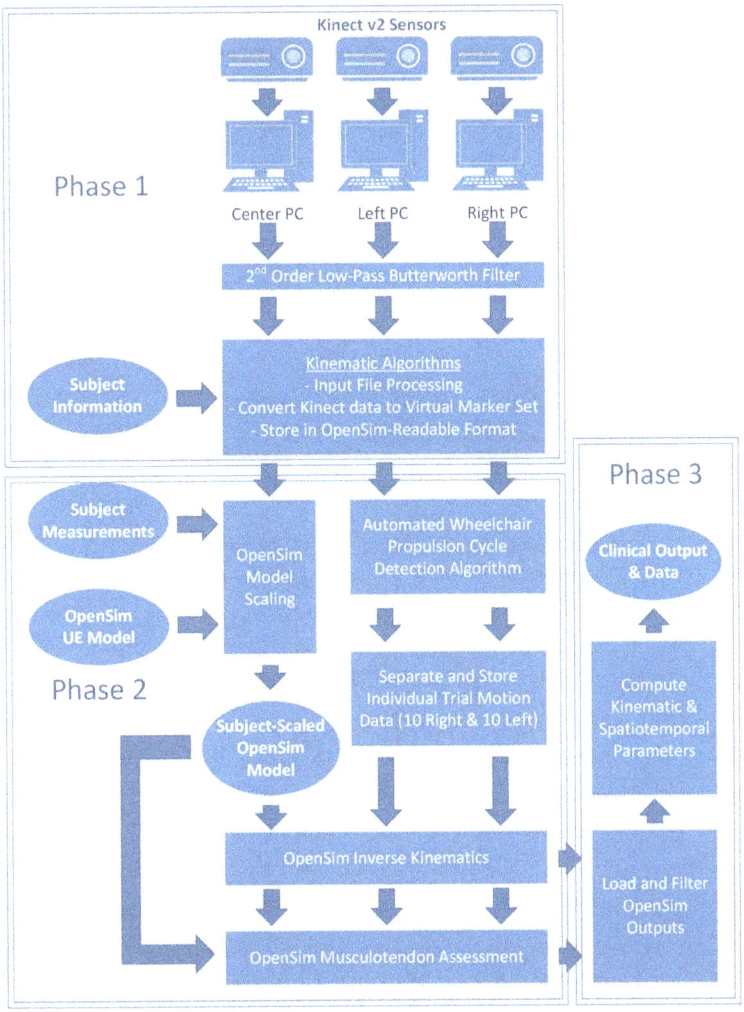

**Fig. 3** Block Diagram of Markerless Kinematic Processing Algorithm. Phases 1, 2, and 3 of processing referenced in text are denoted by boxed regions

producing an individual set of kinematic data for each trial. OpenSim processing is conducted by customized MATLAB algorithms. First, the model is scaled using data from the static trial and measurements provided by the evaluator. The scaling is proportional and uses the anatomical scaling capability of OpenSim, adequate for gross kinematic analysis in clinical research. The joint kinematics and musculotendon lengths from the static trial are recorded as the baseline, normal values. Next, each dynamic trial is processed using the subject-specific model. The iterative inverse kinematics method fits the model to the motion data at each time point. Then, muscle analysis is conducted, using geometric mapping to compute the musculotendon length changes.

Phase 3 in Fig. 3 computes spatiotemporal parameters, joint ranges of motion, and musculotendon excursions, and average and standard deviation values for each parameter. A formatted output is created in MATLAB to display all relevant parameters and outputs of the evaluation. This output is displayed automatically on-screen and saved as an image file for printing. All raw and processed data and parameters are saved in a MATLAB archival data file for future research and processing.

The clinical wheelchair propulsion analysis output from the markerless system is two printable pages, created by the automated software. This format mimics reports produced for clinical gait analysis with marker-based systems and includes both kinematic plots and spatiotemporal parameter data in a standardized, easily interpreted format for clinical use. This work expands on the contributions to the field made by de Groot et al. [28]'s WHEEL-I system and the OptiPush instrumented wheel, described in Kwarciak et al. [29]. Figure 4 gives an example of the kinematic outputs for a 15-year-old subject with spina bifida, provides kinematic plots of the joint motion of each key upper extremity joint and thoracic motion. Each plot of upper extremity joint motion presents the left (blue) and right (red) kinematics, with thin lines representing individual trials and thick lines representing the mean of all trials. The vertical blue and red lines on each plot indicate the point when the hand leaves the pushrim, which identifies the transition from propulsion phase to recovery phase. The first segment, from 0% to the vertical line, is the propulsion phase, where the hand is in contact with, and actively propelling, the pushrim. The second segment, from the vertical line to 100%, is the recovery phase, where the hand returns to its starting position. The transition between push and recovery phase is computed with an automated MATLAB script, which compares the hand's position with the known semicircular arc of the

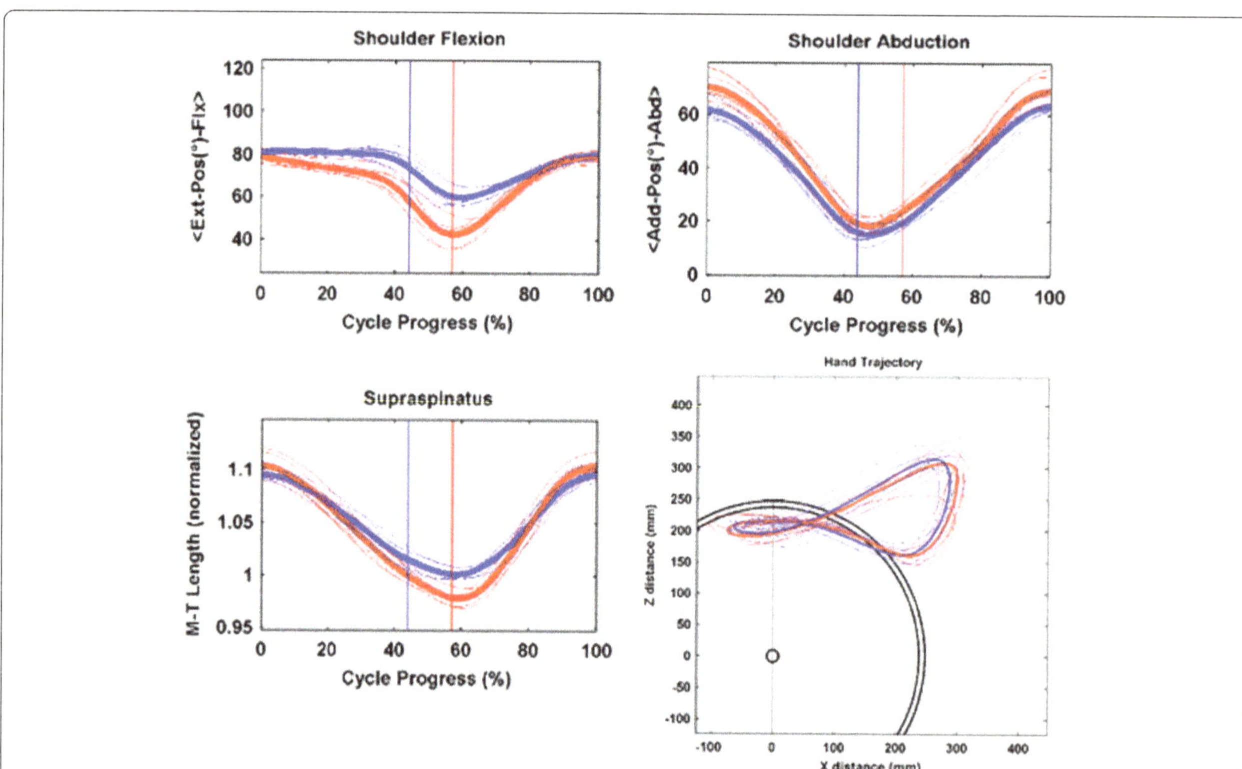


**Fig. 4** Example Clinical Outputs. Joint Kinematics and Spatiotemporal Parameters for exemplar subject, age 15, with spina bifida – Joint kinematics (top), musculotendon excursion (bottom, left) and propulsion pattern (bottom, right). The subject propelled using the same wheelchair and settings used for everyday mobility, at a self-selected speed and propulsion pattern. The thin lines represent individual trials, and thick lines are average of all trials for left and right extremities

pushrim at each time point, and is then able to estimate which data points have hand contact with the pushrim and which points are recovery phase. In the lower left corner of the first page, values are tabulated for range of motion, peak angular velocity, and peak angular acceleration of key joints. The values are averages across all trials, with left and right extremities presented separately.

To analyze the sensitivity of the model for the wheelchair propulsion task, sensitivity analysis was used to relate shoulder and elbow joint motion to musculotendon excursions for the muscles which cross the respective joints. The purpose of this analysis is to mathematically determine the effect of changing joint angle on musculotendon motion. This was performed by perturbing the model throughout the range of shoulder and elbow mobility expected in wheelchair propulsion and recording the musculotendon response length response for each measurement. Then, a linear regression model was used to determine the musculotendon sensitivity to joint motion. This method of using sensitivity analysis to assess the model performance is adapted from the formulas of Rump et al. [30], and while it has not yet been widely applied in the field of biomechanics, it has been a useful procedure in several other scientific fields, such as automotive injury prediction [31–35].

To more specifically address the sensitivity of the model at key points of interest in manual wheelchair propulsion, the shoulder joint is assessed at the start (hand contact) and end (hand release) of a typical wheelchair propulsion cycle, since these transition points represent the most significant potential for injury, as suggested by Rankin et al. [36], who found peak power output at the transition points. To set up the analysis, the model is fixed to the start and end points (based on typical values collected from subjects), and the other joints not being perturbed are fixed at those values. Thus, only the joint of interest is being perturbed for the sensitivity analysis. Analysis is performed using a dimensionless sensitivity coefficient. This is computed as:

((MTL + 5%)-(MTL -5%))/(Initial MTL)]/[((JA + 5%)-(JA-5%))/(Initial JA)

Where MTL = musculotendon length, and JA = joint angle.

Inter-trial measurement repeatability was analyzed using correlation analysis to provide scatterplots and Pearson correlation coefficients describing the repeatability of the measurement system, to test the hypothesis that the system reliably measures parameters between trials. Thirty wheelchair propulsion assessments of pediatric manual wheelchair users were conducted as part of an institutional review board-approved study. The subjects ranged from 6 to 17 years of age with 6 females and 24 males, 12 with spina bifida, 3 with Charcot-Marie-Tooth disorder, and 15 with cerebral palsy, all of whom use a manual wheelchair as a primary means of daily mobility.

The assessment included spatiotemporal parameters, such as push time, cycle time, frequency, contact angle, and proportion of push and recovery, similar to those used in Vegter et al. [37]. Additionally, joint range of motion and musculotendon excursion were analyzed, which differs from the kinetic analysis of Vegter. Each assessment was performed twice, and the two assessments are statistically compared – each subject's data from the first assessment was compared to the same subject's data from the second assessment. The subjects were instructed to perform the same manual propulsion task during each assessment. Speed and power output were results of the subject's standard, self-selected pattern. A Lilliefors test of normality was performed on the differences between the data sets to ensure that the normality assumption of the parametric Pearson correlation analysis was satisfied.

## Results

### Application and kinematic results

The system developed during this project has been rigorously assessed in both laboratory and real-world applications. The cost-effectiveness, at approximately $8000 for the complete system, and simplified assessment protocol make the system viable in several key environments. Our pilot experience in field-testing the system suggested that assessments typically can be completed in under 15 min and training clinicians to use the system is feasible and can be accomplished in 30–45 min.

### Sensitivity analysis

Sensitivity analysis was performed on the musculoskeletal model to determine the relationship between joint motion and musculotendon excursion. For shoulder elevation, the anterior and posterior deltoid musculotendon complexes (– 0.08%/degree and 0.15%/degree, respectively), along with the coracobrachialis (– 0.09%/degree), have high sensitivity to shoulder elevation and rotation when compared to the other musculotendon complexes studied, mostly in the range of 0.01–0.03%/degree. Sensitivity analysis describing musculotendon response to shoulder elevation shows a clear transition to a point where the joint is not moving very much, but the muscles are quickly changing in length. The high linear velocity of the musculotendon complexes strongly suggests a high potential for injury during the wheelchair propulsion cycle.

Thus, each coefficient presented in Table 2 below is dimensionless, and the higher the coefficient, the more sensitive the muscle is to joint angle changes within the specified propulsion area. These coefficients are then

**Table 2** Sensitivity of Musculotendon Complexes to Shoulder Motion at Start and End Points of Propulsion

| Muscle | Shoulder Elevation (Start Point) | Shoulder Elevation (End Point) | Shoulder Rotation (Start Point) | Shoulder Rotation (End Point) |
|---|---|---|---|---|
| Ant Deltoid | 0.435* | −0.185 | 0.243 | 0.810** |
| Lat Deltoid | −1.409** | 0.039 | −0.779** | −0.175 |
| Post Deltoid | −2.038** | 0.288 | −1.137** | −1.268** |
| Supraspinatus | 0.118 | −0.041 | 0.068 | 0.183 |
| Infraspinatus | 0.466* | −0.025 | 0.257 | 0.112 |
| Subscapularis | −0.494* | 0.028 | −0.273 | −0.124 |
| Teres Minor | 0.828** | 0.038 | 0.456* | −0.169 |
| Teres Major | 1.493** | 0.191 | 0.819** | −0.840** |
| Pectoralis Major | 0.701* | −0.038 | 0.385 | 0.169 |
| Latissimus Dorsi | 1.369** | 0.103 | 0.751** | −0.453* |
| Coracobrachialis | 1.719** | −0.155 | 0.952** | 0.683* |
| Triceps-Long | 0.372 | 0.088 | 0.203 | −0.385 |
| Triceps-Medial | −0.287 | −0.103 | − 0.156 | 0.454* |
| Biceps-Long | 0.778** | 0.029 | 0.406* | −0.131 |
| Biceps-Short | 1.874** | 0.008 | 1.010** | −0.037 |
| Brachialis | 0.318 | 0.077 | 0.172 | −0.341 |

Values presented as dimensionless sensitivity coefficients with +/− 5% perturbation at the start and end points of propulsion; Shoulder thoracohumeral angles describe the arm position – consistent with the coordinate system used in the musculoskeletal model. The start point represents initial contact of the hand with the pushrim, and end point is the instant when the hand leaves the pushrim
* = Sensitive (coefficient magnitudes > 0.40); ** = Highly sensitive (coefficient magnitudes > 0.75)

categorized as moderately sensitive ($0.40 < s < 0.75$) or highly sensitive ($s > 0.75$).

The results in Table 2 show several key points. Musculotendon complexes are most sensitive at the beginning of the propulsion cycle (hand contact), with fewer musculotendon complexes showing high sensitivity at the end of the propulsion cycle (hand release). Further, several muscles exhibit significantly higher sensitivity than others, including the posterior and lateral deltoid, teres major, latissimus dorsi, coracobrachialis, and biceps brachii. These results suggest that there is a higher risk of injury during initial hand contact over hand release, and that at the hand contact these key muscles are most sensitive to the angular changes, and thus at risk for injury. The longer muscles overall appear to have lower sensitivity, and hypersensitivity in the shorter musculotendons suggests a higher risk of injury. These results may be limited due to the use of only data from a representative subject for this model analysis. Future work is suggested to confirm these results in a larger population.

### Repeatability analysis

For each pediatric manual wheelchair propulsion assessment, two separate kinematic trials were recorded for each subject. Statistical correlation analysis was performed to determine inter-trial measurement repeatability of the system, with the results summarized in Table 3. Figure 5 shows that the Pearson correlation coefficients for spatiotemporal parameters, joint range of motion, and musculotendon excursion were high and correlations were significant for all parameters, demonstrating inter-trial measurement repeatability of the system. An additional finding of note is that the metrics with higher Pearson correlation coefficients are the metrics with the least standard deviation in the data, and vice versa. This may suggest that within-subject variability is inversely related to the repeatability of inter-trial measurements.

## Discussion

The system developed in this project uses a combination of consumer-grade hardware and open-source musculoskeletal modeling software to create a unique, cost-effective, efficient, and appropriate analysis technique for clinical research in manual wheelchair biomechanics. The Microsoft Kinect was chosen because of its low cost and

**Table 3** Inter-Trial Measurement Repeatability

| Metric Type | Pearson Correlation Coefficient | Significance (p) |
|---|---|---|
| Spatiotemporal Parameters | 0.792 | 0.001* |
| Joint Range of Motion | 0.853 | 0.001* |
| Musculotendon Excursion | 0.931 | 0.001* |

Results of correlation analysis. Each individual subject was tested twice under self-selected conditions with no control of speed or power output, and the two measures for each subject are compared
* $p$-value significant at $\alpha = 0.05$

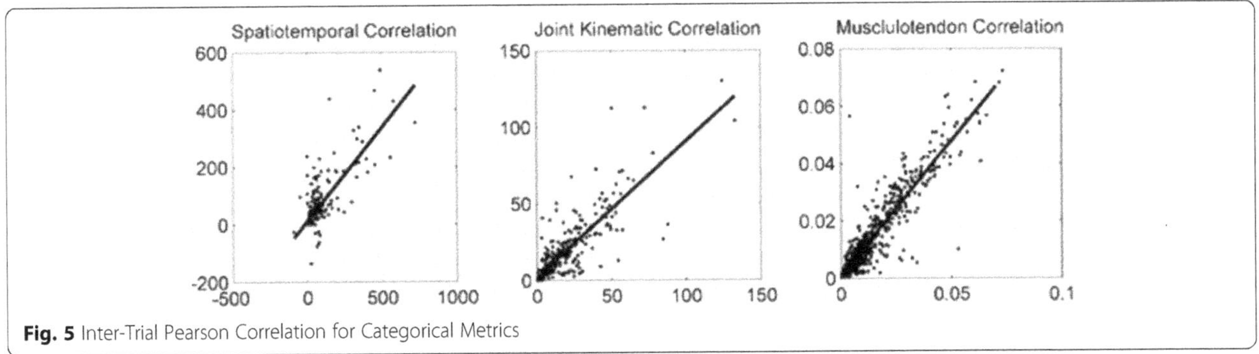


**Fig. 5** Inter-Trial Pearson Correlation for Categorical Metrics

ease of use. The OpenSim upper extremity model brings significant computational power to the system, and the interface allowing its use with the Kinect and automating the protocol is the key development of this work. Characterization of the system in several settings has demonstrated its effectiveness for its intended applications. The system adds value to clinical assessments by extracting metrics that other methods, such as standardized outcome tools, cannot.

Comparison of the markerless wheelchair propulsion assessment system against other common outcome measurement protocols reveals several key differences. When compared to laboratory marker-based motion analysis techniques [1] the markerless system requires less space (due to the stationary wheelchair platform), reduced training requirements, and allows faster assessment. However, the marker-based systems have higher precision, and include kinetic assessment and EMG data. Inertial measurement units [4] and instrumented wheels ([11], and [10]) have similar ease of use when compared to the markerless system, and require less time and training to implement than marker-based systems. However, inertial measurement units and instrumented wheels do not provide complete kinematic outputs, but only partial or supplemental data. Inertial measurement units and instrumented wheels are possible future expansion options for the markerless system to permit kinetics to be included in the model. Standardized outcome measures [12] have fewer equipment and technological requirements, but do require trained observation. It is these evaluations that the markerless system is intended to supplement, by adding objective, quantitative outcomes, while adding minimal time and expense.

Manual wheelchair skills capacity is a key therapeutic outcome and can affect quality of life [38]. Assessments using standardized outcome tools in a community setting have documented improvements in response to therapy [39], and shown that levels of physical activity can impact the risk of shoulder pain [40]. Additionally, clinical guidelines suggest monitoring propulsion pattern and technique employed, daily activities, and exercise or therapeutic activities, as a means to assess risk of upper limb pain or injury [41]. This monitoring is important to continuing research in the field, given the paucity of research into therapeutic outcomes of manual wheelchair users [42]. The system proposed here has several key benefits and limitations relative to both standardized outcome tools and complex laboratory motion capture. Clinically, a physical therapist can use the system to assess the UE kinematics and propulsion pattern of wheelchair users as part of routine therapy visits, to track progress. The system provides reliable quantitative data to track patient progress, which is rarely available in current physiotherapeutic clinical assessments. Propulsion pattern, for instance, impacts upper extremity muscle power and stress in manual wheelchair users [43], and is produced by the system. The system enhances the capability of therapists to obtain quantitative data without requiring overly complex and detailed analysis, such as laboratory motion capture [1]. Basic data includes speed, cadence, and propulsion pattern, which are difficult to measure accurately by video or other techniques. The system is effective in this role because it requires minimal training – the software is largely automated and therapists can be trained to use it in a short time.

Several limitations of this work exist and should be addressed. The present system does not include power output, which would require kinetic detection hardware, and is a common metric presented in wheelchair biomechanics literature [37, 44]. The present system also does not measure wheel speed directly, but an algorithm has been implemented to estimate ground speed based on a known wheel diameter and hand motion. Cadence is measured by the system and the resistance level is adjustable based on wheelchair wheel diameter. The lack of speed and power output data reduces the ability to control testing conditions for these metrics, which would be desirable for consistency, such as tracking changes among preferred cadence, wheelchair specifications, and personal differences over time. The addition of speed and power output would increase usability of the system, allowing tracking of changes in preferred cadence, wheelchair styles, and personal performance.

Inter-trial repeatability was significant for spatiotemporal parameters, joint kinematics, and musculotendon excursions. This suggests that the markerless wheelchair propulsion kinematic assessment system is a repeatable measurement tool for pediatric manual wheelchair users, and detects changes that are greater than the inherent normal variability in the population. Given inter-trial repeatability with significant correlation, the system is recommended for further quantitative assessment use in pediatric manual wheelchair users. However, it should be noted that the markerless technology has limitations in precision of kinematic detection, and more advanced technologies may be required to obtain higher precision.

## Conclusions

There is a significant deficit in current literature on manual wheelchair propulsion biomechanics and physiotherapeutic treatment for this population. The system is suggested for immediate implementation in novel research to resolve these key deficiencies in current literature, leading to more effective point-of-care clinical outcome assessments for manual wheelchair users, provided the limitations of markerless technology are taken into account. In the future, home use and telerehabilitation development are suggested as possible directions for the project.

**Acknowledgments**
The authors acknowledge the support of Bay Cliff Health Camp Children's Therapy and Wellness Center and the Orthopaedic and Rehabilitation Engineering Center (OREC) at Marquette University. The contents of this paper were developed under a grant from the National Institute on Disability, Independent Living, and Rehabilitation Research (NIDILRR Grant number 90RE5006-01-00, Formerly H133E100007). NIDILRR is a Center within the Administration for Community Living (ACL), Department of Health and Human Services (HHS). The contents of this paper do not necessarily represent the policy of NIDILRR, ACL, HHS, and you should not assume endorsement by the Federal Government.

**Funding**
The contents of this paper were developed under a grant from the National Institute on Disability, Independent Living, and Rehabilitation Research (NIDILRR Grant number 90RE5006-01-00, Formerly H133E100007). NIDILRR is a Center within the Administration for Community Living (ACL), Department of Health and Human Services (HHS). The contents of this paper do not necessarily represent the policy of NIDILRR, ACL, HHS, and you should not assume endorsement by the Federal Government.

**Authors' contributions**
JR and SR performed the data collection and analysis. GH, BS, JW, and JK contributed to the interpretation of results and writing of the manuscript. All authors read and approved the final manuscript.

**Competing interests**
The authors declare that they have no competing interests.

**Author details**
[1]Orthopaedic and Rehabilitation Engineering Center (OREC), Marquette University, Olin Engineering Suite 323, Milwaukee, WI 53201-1881, USA. [2]Department of Biomedical Engineering, Marquette University, Olin Engineering Suite 323, Milwaukee, WI 53201-1881, USA. [3]Department of Orthopaedic Surgery, Medical College of Wisconsin, Milwaukee, WI 53201-1881, USA. [4]Shriners Hospitals for Children, Chicago, IL, USA. [5]Midwestern University, Physical Therapy Program, 555 31st St., Alumni Hall 340C, Downers Grove, IL 60515, USA. [6]University of Wisconsin-Milwaukee, 2400 E Hartford Ave, Rm. 983, Milwaukee, WI 53211, USA. [7]Marquette University, Biomedical Engineering, Milwaukee, WI 53201-1881, USA.

**References**

1. Schnorenberg AJ, Slavens BA, Wang M, Vogel LC, Smith PA, Harris GF. Biomechanical model for evaluation of pediatric upper extremity joint dynamics during wheelchair mobility. J Biomech. 2014;47(1):269. https://doi.org/10.1016/j.jbiomech.2013.11.014.
2. van der Woude LH, Veeger D, Dallmeijer AJ, Janssen TW, Rozendaal LA. Biomechanics and physiology in active manual wheelchair propulsion. Med Eng Phys. 2001;23(10):713–33. https://doi.org/10.1016/S1350-4533(01)00083-2.
3. Vegter RJK, Hartog J, de Groot S, Lamoth CJ, Bekker MJ, van der Scheer Jan W, Veeger Dirkjan HEJ. Early motor learning changes in upper-limb dynamics and shoulder complex loading during handrim wheelchair propulsion. J Neuroeng Rehabil. 2015;12. Retrieved from https://doi.org/10.1186/s12984-015-0017-5.
4. Bergamini E, Morelli F, Marchetti F, Vannozzi G, Polidori L, Paradisi F, Delussu, AS. Wheelchair propulsion biomechanics in junior basketball players: a method for the evaluation of the efficacy of a specific training program. BioMed Res Int. 2015;275965. Retrieved from https://doi.org/10.1155/2015/275965.
5. Leving M, Horemans H, Vegter R, de Groot S, Bussmann J, van der Woude L. Validity of consumer-grade activity monitor to identify manual wheelchair propulsion in standardized activities of daily living. PLoS One. 2018;13(4):e0194864.
6. van der Slikke R, Berger M, Bregman D, Lagerberg A, Veeger H. Opportunities for measuring wheelchair kinematics in match settings: reliability of a three inertial sensor configuration. J Biomech. 2015;48:3398–405.
7. van der Slikke R, Berger M, Bregman D, Veeger H. From big data to rich data: the key features of athlete wheelchair mobility performance. J Biomech. 2016;49:3340–6.
8. van der Slikke R, Mason B, Berger M, Goosey-Tolfrey V. Speed profiles in wheelchair court sports: comparison of two methods for measuring wheelchair mobility performance. J Biomech. 2017;65:221–5.
9. van der Slikke R, Bregman D, Berger M, de Witte A, Veeger D. The future of classification in wheelchair sports: can data science and technological advancement offer an alternative point of view? Int J Sports Physiol Perform. 2018;13:742–9.
10. Dellabiancia F, Porcellini G, Merolla G. Instruments and techniques for the analysis of wheelchair propulsion and upper extremity involvement in patients with spinal cord injuries: current concept review. Muscles Ligaments Tendons J. 2013;3(3):150–6.
11. Conger SA, Scott SN, Fitzhugh EC, Thompson DL, Bassett DR. Validity of physical activity monitors for estimating energy expenditure during wheelchair propulsion. J Phys Act Health. 2015;12(11):1520. Retrieved from https://doi.org/10.1123/jpah.2014-0376.
12. Kenny S, Gowran RJ. Outcome measures for wheelchair and seating provision: a critical appraisal. Br J Occup Ther. 2014;77(2):67–77. https://doi.org/10.4276/030802214X13916969447119.
13. Boninger M, Koontz A, Sisto S, Dyson-Hudson T, Chang M, Price R, Cooper R. Pushrim biomechanics and injury prevention in spinal cord injury: recommendations based on CULP-SCI investigations. J Rehabil Res Dev. 2005;42(3):9–20.

14. Bonnechère, B., Jansen, B., Salvia, P., Bouzahouene, H., Omelina, L., Moiseev, F., . . . Van Sint Jan, S. (2014). Validity and reliability of the kinect within functional assessment activities: comparison with standard stereophotogrammetry. Gait Posture, 39(1), 593–598. doi:https://doi.org/10.1016/j.gaitpost.2013.09.018.
15. Dutta T. Evaluation of the Kinect sensor for 3-D kinematic measurement in the workplace. Appl Ergon. 2012;43:645–9.
16. Galna B, Barry G, Jackson D, Mhiripiri D, Olivier P, Rochester L. Accuracy of the microsoft kinect sensor for measuring movement in people with parkinson's disease. Gait Posture. 2014;39(4):1062–8. https://doi.org/10.1016/j.gaitpost.2014.01.008.
17. Otte K, Kayser B, Mansow-Model S, Verrel J, Paul F, Brandt AU, Schmitz-Hübsch T. Accuracy and reliability of the kinect version 2 for clinical measurement of motor function. PLoS One. 2016;11(11):e0166532. https://doi.org/10.1371/journal.pone.0166532.
18. van Diest M, Stegenga J, Wortche HJ, Postema K, Verkerke GJ, Lamoth CJC. Suitability of kinect for measuring whole body movement patterns during exergaming. J Biomech. 2014;47(12):2925–32. https://doi.org/10.1016/j.jbiomech.2014.07.017.
19. Huber ME, Seitz AL, Leeser M, Sternad D. Validity and reliability of kinect skeleton for measuring shoulder joint angles: a feasibility study. Physiotherapy. 2015;101(4):389–93. https://doi.org/10.1016/j.physio.2015.02.002.
20. Lee SH, Yoon C, Chung SG, Kim HC, Kwak Y, Park H, Kim K. Measurement of shoulder range of motion in patients with adhesive capsulitis using a kinect. PLoS One. 2015;10(6):e0129398. https://doi.org/10.1371/journal.pone.0129398.
21. Mobini A, Behzadipour S, Saadat M. Test-retest reliability of kinect's measurements for the evaluation of upper body recovery of stroke patients. Biomed Eng Online. 2015;14:75 Retrieved from https://doi.org/10.1186/s12938-015-0070-0.
22. Rammer J, Krzak J, Riedel S, Harris G. Evaluation of upper extremity movement characteristics during standardized pediatric functional assessment with a Kinect®-based Markerless motion analysis system. IEEE Eng Med Biol Conf. 2014:2525–8. https://doi.org/10.1109/EMBC.2014.6944136.
23. Delp SL, Anderson FC, Arnold AS, Loan P, Habib A, John CT, Thelen DG. OpenSim: open-source software to create and analyze dynamic simulations of movement. IEEE Trans Biomed Eng. 2007;54(11):1940–50. https://doi.org/10.1109/TBME.2007.901024.
24. Holzbaur K, Murray W, Delp S. A model of the upper extremity for simulating musculoskeletal surgery and analyzing neuromuscular control. Ann Biomed Eng. 2005;33(6):829–40. https://doi.org/10.1007/s10439-005-3320-7.
25. Saul K, Hu X, Goehler C, Vidt M, Daly M, Velisar A, Murray W. Benchmarking of dynamic simulation predictions in two software platforms using an upper limb musculoskeletal model. Comput Methods Biomech Biomed Engin. 2014;18(13):1–14.
26. DiGiovine CP, Cooper RA, Boninger ML. Dynamic calibration of a wheelchair dynamometer. J Rehabil Res Dev. 2001;38(1):41 Retrieved from http://www.ncbi.nlm.nih.gov/pubmed/11322470.
27. Rammer J, Riedel S, Harris G. A portable, low-cost wheelchair ergometer design based on a mathematical model of pediatric wheelchair dynamics. In: Proceedings of 2015 Annual Meeting of the Rehabilitation Engineering and Assistive Technology Society of North America; 2015. p. 4.
28. de Groot S, Vegter R, Vuijik C, van Dijk F, et al. WHEEL-I: development of a wheelchair propulsion laboratory for rehabilitation. J Rehabil Med. 2014;46: 493–503.
29. Kwarciak A, Turner J, Guo L, Richter W. Comparing handrim biomechanics for treadmill and overground wheelchair propulsion. Spinal Cord. 2011;49:457–62.
30. Rump S. Rigorous sensitivity analysis for systems of linear and nonlinear equations. Math Comput. 1990;54(190):721–36.
31. Autrique L, Lormel C. Numerical design of experiment for sensitivity analysis—application to skin burn injury prediction. IEEE Trans Biomed Eng. 2008;55(4):1279–90.
32. Fang X, Yang T. Regression methodology for sensitivity analysis of solar heating walls. Appl Therm Eng. 2008;28:2289–94.
33. Small D. Sensitivity analysis for instrumental variables regression with overidentifying restrictions. J Am Stat Assoc. 2007;102(49):1049–58.
34. Suchao X, Hongqi T. Influencing factors and sensitivity analysis of occupant impact injury in passenger compartment. Traffic Injury Prevention. 2013;14: 816–22.
35. Tondel K, Vik J, Martens J, Indahl U, Smith N, Omholt S. Hierarchical multivariate regression-based sensitivity analysis reveals complex parameter interaction patterns in dynamic models. Chemom Intell Lab Syst. 2013;120:25–41.
36. Rankin J, Richter W, Neptune R. Individual muscle contributions to push and recovery subtasks during wheelchair propulsion. J Biomech. 2011; 44:1246–52.
37. Vegter R, de Groot S, Lamoth C, Veeger D, van der Woude L. Initial skill acquisition of handrim wheelchair propulsion: a new perspective. IEEE Trans Neural Syst Rehabil Eng. 2014;22(1):104–13.
38. Hosseini SM, Oyster ML, Kirby RL, Harrington AL, Boninger ML. Manual wheelchair skills capacity predicts quality of life and community integration in persons with spinal cord injury. Arch Phys Med Rehabil. 2012;93(12):2237. https://doi.org/10.1016/j.apmr.2012.05.021.
39. Best K, Kirby R, Smith C, MacLeod D. Wheelchair skills training for community-based manual wheelchair users: a randomized controlled trial. Arch Phys Med Rehabil. 2005;86:2316–24.
40. Mulroy SJ, Hatchett P, Eberly VJ, Haubert LL, Conners S, Requejo PS. Shoulder strength and physical activity predictors of shoulder pain in people with paraplegia from spinal injury: prospective cohort study. Phys Ther. 2015;95(7):1027–38. https://doi.org/10.2522/ptj.20130606.
41. Consortium for Spinal Cord Medicine. Preservation of upper limb function following spinal cord injury: a clinical practice guideline for health-care professionals; 2005. p. 48.
42. Zwinkels, M. G. J., Verschuren, O. W., Janssen, T., Ketelaar, M., Takken, T., Backx, F. J. G., . . . Volman, M. (2014). Exercise training programs to improve hand rim wheelchair propulsion capacity: a systematic review. Clin Rehabil, 28(9), 847–861. doi:https://doi.org/10.1177/0269215514525181.
43. Slowik J, Requejo P, Mulroy S, Neptune R. The influence of wheelchair propulsion hand pattern on upper extremity muscle power and stress. J Biomech. 2016;49:1554–61.
44. Dallmeijer A, Kilkens O, Post M, de Groot S, Angenot E, van Asbeck F, van der Woude L. Hand-rim wheelchair propulsion capacity during rehabilitation of persons with spinal cord injury. J Rehabil Res Dev. 2005;42(3 Suppl 1):55 Retrieved from http://www.ncbi.nlm.nih.gov/pubmed/16195963.
45. Clark RA, Bower KJ, Mentiplay BF, Paterson K, Pua Y. Concurrent validity of the microsoft kinect for assessment of spatiotemporal gait variables. J Biomech. 2013a;46(15):2722–5. https://doi.org/10.1016/j.jbiomech.2013.08.011.
46. Clark RA, Pua Y, Fortin K, Ritchie C, Webster KE, Denehy L, Bryant AL. Validity of the microsoft kinect for assessment of postural control. Gait Posture. 2012;36(3):372–7. https://doi.org/10.1016/j.gaitpost.2012.03.033.
47. Clark, R. A., Pua, Y., Oliveira, C. C., Bower, K. J., Thilarajah, S., McGaw, R., . . . Mentiplay, B. F. (2015a). Reliability and concurrent validity of the microsoft xbox one kinect for assessment of standing balance and postural control. Gait Posture, 42(2), 210–213. doi:https://doi.org/10.1016/j.gaitpost.2015.03.005.
48. Clark, R. A., Vernon, S., Mentiplay, B. F., Miller, K. J., McGinley, J. L., Pua, Y. H., . . . Bower, K. J. (2015b). Instrumenting gait assessment using the kinect in people living with stroke: reliability and association with balance tests. J Neuroeng Rehabil, 12(1), 15. doi:https://doi.org/10.1186/s12984-015-0006-8.
49. Boninger M, Souza A, Cooper R, Fitzgerald S, Koontz A, Fay B. Propulsion patterns and Pushrim biomechanics in manual wheelchair propulsion. Arch Phys Med Rehabil. 2002;83:718–23.
50. Clark RA, Pua Y, Bryant AL, Hunt MA. Validity of the microsoft kinect for providing lateral trunk lean feedback during gait retraining. Gait Posture. 2013b;38(4):1064–6. https://doi.org/10.1016/j.gaitpost.2013.03.029.
51. de Groot S, Veeger H, Hollander A, van der Woude L. Consequence of feedback-based learning of effective hand rim wheelchair force production on mechanical efficiency. Clin Biomech. 2002;17:219–26.
52. de Groot S, de Bruin M, Noomen S, van der Woude L. Mechanical efficiency and propulsion technique after 7 weeks of low-intensity wheelchair training. Clin Biomech. 2008;23:434–41.
53. Finley M, Rodgers M. Prevalence and identification of shoulder pathology in athletic and nonathletic wheelchair users with shoulder pain: a pilot study. J Rehabil Res Dev. 2004;38:395–402.
54. Gonzalez-Jorge H, Rodríguez-Gonzálvez P, Martínez-Sánchez J, González-Aguilera D, Arias P, Gesto M, Díaz-Vilariño L. Metrological comparison between kinect I and kinect II sensors. Measurement. 2015;70:21–6. https://doi.org/10.1016/j.measurement.2015.03.042.
55. Guo L, Kwarciak A, Rodriguez R, Sarkar N, Richter W. Validation of a biofeedback system for wheelchair propulsion training. Rehabil Res Pract. 2011;2011:590780 7p.
56. Kertis JD, Fritz JM, Long JT, Harris GF. Static and dynamic calibration of an eight-camera optical system for human motion analysis. Crit Rev Phys Rehabil Med. 2010;22(1–4):49–60.

57. Knippenberg E, Verbrugghe J, Lamers I, Palmaers S, Timmermans A, Spooren A. Markerless motion capture systems as training device in neurological rehabilitation: a systematic review of their use, application, target population, and efficacy. J NeuroEng Rehabil. 2017;14:61 11p.
58. Morrow M, Van Straaten M, Murthy N, Braman J, Zanella E, Zhao K. Detailed shoulder MRI findings in manual wheelchair users with shoulder pain. Biomed Res Int. 2014;2014:1–7.
59. Rankin J, Kwarciak A, Richter W, Neptune R. The influence of wheelchair propulsion technique on upper extremity muscle demand: a simulation study. Clin Biomech. 2012;27:879–86.
60. Rice I, Gagnon D, Gallagher J, Boninger M. Hand rim wheelchair propulsion training using biomechanical real-time visual feedback based on motor learning theory principles. J Spinal Cord Med. 2010;33:33–42.
61. Rice I, Pohlig R, Gallagher J, Boninger M. Handrim wheelchair propulsion training effect on Overground propulsion using biomechanical real-time visual feedback. Arch Phys Med Rehabil. 2013;94:256–63.
62. Richter W, Kwarciak A, Guo L, Turner J. Effects of single-variable biofeedback on wheelchair Handrim biomechanics. Arch Phys Med Rehabil. 2011;92:572–7.
63. Rodgers M, Keyser R, Gardner E, Russell P, Gorman P. Influence of trunk flexion on biomechanics of wheelchair propulsion. J Rehabil Res Dev. 2000; 37(3):283–95.
64. Sabick M, Korajarvi B, An K. A new method to quantify demand on the upper extremity during manual wheelchair propulsion. Arch Phys Med Rehabil. 2004;85:1151–9.
65. Veeger H, Rozendaal L, van der Helm F. Load on the shoulder in low-intensity wheelchair propulsion. Clin Biomech. 2002;17:211–8.
66. Wu G, van der Helm F, Veeger H, Makhsous M, Van Roy P, Anglin C, Nagels J, Karduna A, McQuade K, Wang X, Werner F, Buchholz B. ISB recommendation on definitions of joint coordinate systems of various joints for the reporting of human joint motion—part II: shoulder, elbow, wrist, and hand. J Biomech. 2005;38:981–92.
67. Xu X, McGorry RW. The validity of the first and second generation microsoft kinect™ for identifying joint center locations during static postures. Appl Ergon. 2015;49:47–54. https://doi.org/10.1016/j.apergo.2015.01.005.
68. Xu X, McGorry RW, Chou L, Lin J, Chang C. Accuracy of the microsoft kinect for measuring gait parameters during treadmill walking. Gait Posture. 2015; 42(2):145–51. https://doi.org/10.1016/j.gaitpost.2015.05.002.

# 15

# Speed-adaptive control of functional electrical stimulation for dropfoot correction

Guangtao Chen[1], Le Ma[1], Rong Song[1*], Le Li[2], Xiaoyun Wang[3] and Kaiyu Tong[4]

**Abstract**

**Background:** Functional electrical stimulation is an important therapy technique for dropfoot correction. In order to achieve natural control, the parameter setting of FES should be associated with the activation of the tibialis anterior.

**Methods:** This study recruited nine healthy subjects and investigated the relations of walking speed with the onset timing and duration of tibialis anterior activation. Linear models were built for the walking speed with respect to these two parameters. Based on these models, the speed-adaptive onset timing and duration were applied in FES-assisted walking for nine healthy subjects and ten subjects with dropfoot. The kinematic performance of FES-assisted walking triggered by speed-adaptive stimulation were compared with those triggered by the heel-off event, and no-stimulation walking at different walking speeds.

**Results:** Higher ankle dorsiflexion angle was observed in heel-off stimulation and speed-adaptive stimulation conditions than that in no-stimulation walking condition at all the speeds. For subjects with stroke, the ankle plantarflexion angle in speed-adaptive stimulation condition was similar to that in no-stimulation walking condition, and it was significant larger than that in heel-off stimulation condition at all speeds.

**Conclusions:** The improvement in ankle dorsiflexion without worsening ankle plantarflexion in speed-adaptive stimulation condition could be attributed to the appropriate stimulation timing and duration. These results provide evidence that the proposed stimulation system with speed-related parameters is more physiologically appropriate in dropfoot correction, and it may have great potential value in future clinical applications.

**Keywords:** Electromyography, Walking speed, Dropfoot, Functional electrical stimulation

## Background

About three quarters of stroke survivors experience different levels of brain dysfunction and movement disorder [1], which result in lower living quality and limited ability in social activities [2]. Of these subjects, 20% suffer from impaired motor function in the lower extremities. One of such impairments is dropfoot, which is characterized by poor ankle dorsiflexion during the swing phase and an inability to achieve heel strike at the initial contact [3, 4]. Abnormal gaits such as circumduction gait and abnormal foot clearance on the affected side are often found as a method of compensating for excessive hip abduction and pelvis elevation on the unaffected side [5]. This results in gait asymmetry and slow walking speed [6].

Functional electrical stimulation was a representative intervention to correct dropfoot and Liberson et al. first introduced functional electrical stimulation (FES) to correct dropfoot for chronic hemiplegic subjects in the 1960s [7]. An electrical charge is delivered via a pair of electrodes to activate the tibialis anterior (TA), which results in ankle dorsiflexion. Yan et al. applied two dual-channel

* Correspondence: songrong@mail.sysu.edu.cn
[1]Key Laboratory of Sensing Technology and Biomedical Instrument of Guangdong Province, School of Engineering, Sun Yat-sen University, Guangzhou, Guang Dong, China
Full list of author information is available at the end of the article

stimulators to the quadriceps, hamstring, gastrocnemius, and TA to recover motor function of the lower extremities in an early stage after stroke [8]. The stimulation was followed by a predetermined sequence of muscle activations that mimic a healthy gait cycle [9]. The duration of stimulation was five seconds in Yan et al.'s study. However, subjects with different severities of impairment might have different walking speeds [10], which means that a fixed stimulation duration might not be able to account for different walking patterns.

Liberson et al. used the heel-off event detected by a footswitch to trigger the stimulation [7]. However, the reliability of the footswitch controller was significantly reduced when subjects who dragged their feet during walking encountered a slope or an obstacle [11]. Bhadra et al. proposed a manual switch to trigger stimulation as a walking aid for subjects with spinal cord injury (SCI) [12]. However, manual control may distract subjects from maintaining balance and lead to an increased risk of falls [13, 14]. Furthermore, the cable between the control sensor and stimulator was inconvenient for walking [15].

Instead of a footswitch, Mansfield et al. [16] and Monaghan et al. [17] detected the heel event of the gait cycle in FES-assisted walking using an accelerometer and a uniaxial gyroscope, respectively. The commercially available product WalkAide also uses an accelerometer for this purpose [18]. Electromyography (EMG) signal is also applied as a control source in FES-assisted walking for the detection of volitional intent of muscle [19]. Yeom et al. amplified the EMG signal of the TA and modulated the stimulation intensity in proportion to the integrated EMG envelope. The electrical pulses are then sent to the common peroneal nerve for dropfoot correction [20].

In these studies, FES applied to the TA was mainly triggered by the heel-off event. However, this event occurs during the push-off phase and before TA activation [17]. An earlier start of TA stimulation results in reduced ankle plantarflexion [21]. Spaich et al. suggested implementing a constant time interval before the onset timing of TA stimulation to extend the push-off phase before the ankle dorsiflexion [21]. Some studies have found that walking speed can affect the activation of TA [22, 23]. Shiavi et al. found that the duration of EMG activity decreased as speed increased [22]. In Winter et al.'s study, the shape of the EMG patterns generally remained similar at the different walking speeds and the duration of EMG activity was closely related to the normalized stride time [23]. Although the duration of TA activation changes with the walking speeds has been reported [24], the selection of speed-adaptive FES parameters for TA has not been investigated.

The objective of this study is to find a more physiologically appropriate FES design for dropfoot correction. Firstly, speed-related changes in onset timing and the duration of TA activation were examined. Next, linear models were built for the walking speed and time interval from the heel-off event to the onset timing of TA activation, as well as for the walking speed and the duration of the TA activation. The speed-adaptive stimulation (SAS) timing and duration were then calculated based on the two models and applied for FES-assisted walking. Finally, the performance of stimulation triggered by SAS, heel-off event (HOS) and no stimulation (NS) were compared during FES-assisted walking on both subjects with stroke and healthy subjects at different walking speeds.

## Methods

### Modeling of stimulation parameters based on TA

Nine healthy subjects (five male and four female) were recruited, and none of them had musculoskeletal impairment or injury that affected walking. The mean (±standard deviation (SD)) age of the volunteers was 22.7 (±1.3) years. Two different sensors were attached to the healthy subjects. A footswitch (Tekscan Inc., Boston, MA, USA) was placed on the hindfoot using medical tape to record heel-off and heel-strike events. Circular silver-silver chloride (Ag-AgCl) electrodes (M2223, 3 M Inc., USA) with a diameter of 5 mm and inter-electrode distance of 20 mm were attached to the skin surface of the TA with one reference electrode near the lateral epicondyle of the femur [25].

The surface electromyography signal from the TA was recorded and amplified at a gain of 4000 using a tele-EMG system (MyoSystem2400, Noraxon, USA). Labview software (Labview 2010, National Instruments Corporation, Austin, Texas, USA) stored the footswitch and EMG signals synchronously on a hard disk for offline processing. The signals were obtained using an A-D converter (PCIe-6341, National Instrument, Texas, USA) with 16-bit resolution.

The sampling rate of the footswitch and EMG signals were 100 Hz and 1000 Hz, respectively. The raw EMG signal was first filtered with a bandwidth of 10–150 Hz to remove the low-frequency drift and high-frequency noise [26]. The time interval of TA stimulation was defined as the time span from the heel-off event to the onset timing of TA activation. The timing was defined when the full-wave rectified EMG signal exceeded three times the SD of baseline activity [27]. The baseline activity was the average amplitude of the full-wave rectified EMG signal at rest. The EMG duration was the length of the TA contraction time from the onset timing to the offset timing. The offset timing of TA activation was detected when the rectified signal was less than three times SD of baseline activity [27].

During the experiment, the healthy subjects were instructed to walk on a treadmill (BH, G6425-F3, Spain).

A total of eleven speeds were selected for each subject, which ranged from 0.5 to 1.5 m/s with intervals of 0.1 m/s [24]. In each trial, subjects were asked to walk at a certain speed for at least two minutes. The order of the gait speeds was randomly arranged, and a 2-min rest was given between each two trials to avoid muscle fatigue.

Subjects used their preferred stride length during the experiment. To exclude the influence of acceleration and deceleration during treadmill walking, more than 20 steps were averaged for each subject at each speed [28]. After correlation analysis and least square curve fitting, the linear models were built for the walking speed with respect to the time interval from the heel-off event to the onset timing of TA activation, as well as for the EMG duration.

### Speed-adaptive stimulation

The speed-adaptive stimulation is presented in Fig. 1a. A total of five spherical 12-mm reflective markers were attached to the subject's lower limb (Fig. 1c). Five markers from top to bottom were placed on the following anatomical reference locations: the mid thigh sufficiently distal to the hip, lateral knee joint, the mid shank sufficiently distal to the knee joint, lateral malleolus, and the space between the second and third metatarsal heads [29]. A motion capture system (OptiTrack, NaturalPoint, USA) consisting of six infrared cameras was used to detect the markers at a sampling rate of 100 Hz. The coordinates of each marker were recorded using Tracking Tools software (NaturalPoint, USA) to calculate the walking speed and for further kinematic analysis. The footswitch signal was recorded by the A-D converter and stored in a PC for heel event detection. When the heel-strike event was detected, the walking speed was calculated from the kinematic data of the lower limb marker. The average walking speed of the previous five steps was considered as the current walking speed [25]. After the heel-off event, the time interval before onset timing and duration of TA stimulation were defined from the linear models according to the walking speed. After the speed-adaptive time interval, TA stimulation was triggered. After the speed-adaptive duration, TA stimulation was terminated. The time interval and duration of TA stimlation can be changed according to the different walking speeds. If the walking speeds were lower, the time interval and duration would be longer and if the walking speeds were higher, the time interval and duration would be shorter. A flowchart of the SAS control system is shown in Fig. 2.

### Participants

Nine healthy subjects from the previous experiment and ten subjects with stroke (nine male and one female) with

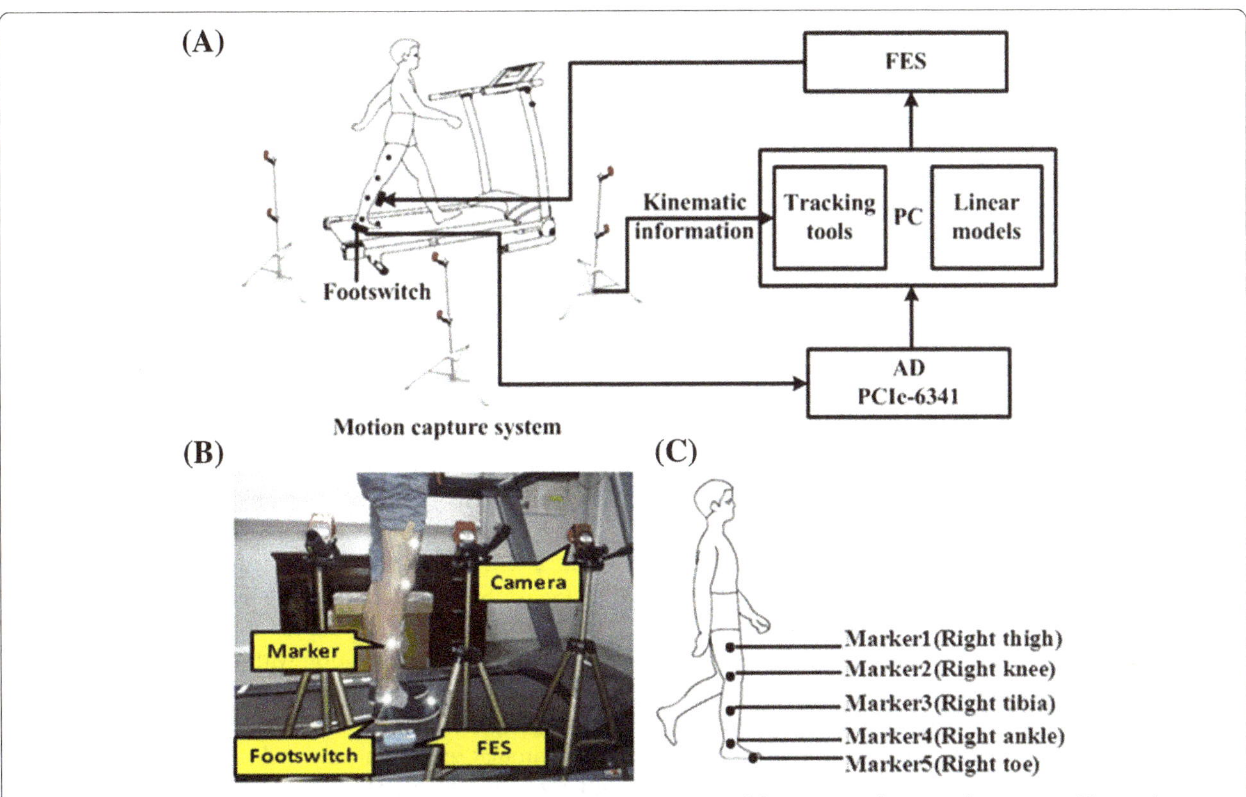


**Fig. 1** **a** The experiment setup of SAS condition; **b** one healthy subject on the treadmill for system evaluation; **c** the position of five markers on the right leg

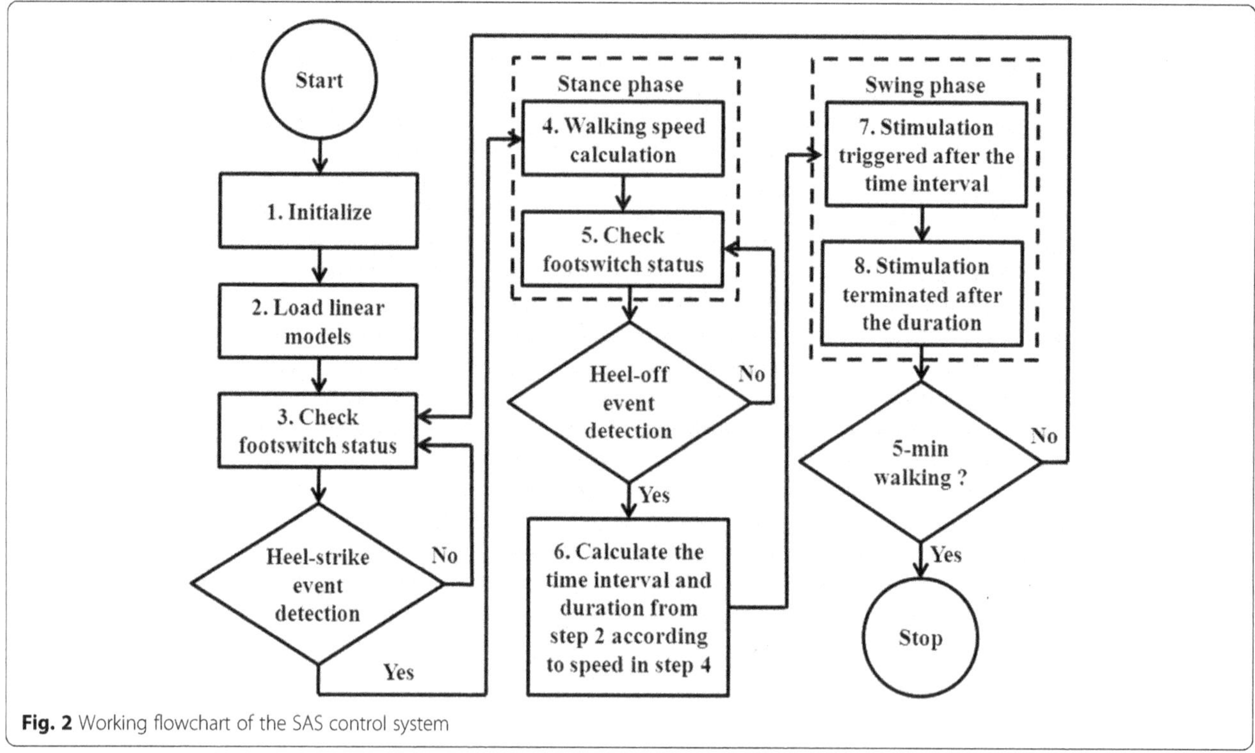


**Fig. 2** Working flowchart of the SAS control system

dropfoot (Table 1) took part in the study for system evaluation. The subjects with stroke had sustained a single stroke within at least 6 months prior to study participation and were able to walk on a treadmill independently at multiple speeds without any help. For safety, the subjects with stroke held on to a front handrail during walking [6]. Before participating in the experiment, written informed consent was collected from all the subjects. This study was approved by the ethics committee of the Guangdong Work Injury Rehabilitation Center.

### Experiment setup

Healthy subjects were instructed to walk on a treadmill at four speeds: 0.3, 0.6, 0.9, and 1.2 m/s. The subjects walked in four conditions at each speed: FES-assisted walking triggered by the heel-off event (HOS), FES-assisted walking triggered by SAS, and no-stimulation walking (NS). Subjects with stroke were instructed to walk at three speeds: slow, free, and fast. The free speed was the comfortable walking speed on the treadmill, while the fast speed was the maximal tolerable speed (approximately 25 to 30% larger than the free walking speed) [30]. The slow speed was smaller than the free speed with the same proportion. The healthy and subjects with stroke walked in the three conditions at each speed: FES-assisted walking triggered by the heel-off event (HOS), FES-assisted walking triggered by SAS, and no-stimulation walking (NS). Each subject with stroke should finish nine trials and each healthy subject should finish twelve trials. The order of the trials was randomly arranged for both healthy subjects and subjects with stroke. In each trial, the subjects were required to walk for not more than five minutes at a certain speed. A 2-min rest was given between each two trials to avoid muscle fatigue. During the experiment, a footswitch placed on the hindfoot and five reflective makers attached to the lower limb (Fig. 1b) were used for heel event detection and kinematics acquisition.

In the HOS condition, TA stimulation was triggered and terminated by the heel-off and heel-strike events, respectively [31], the TA stimulation was ahead of the TA activation, and the stimulation duration was shorter than the TA duration in healthy subjects (Fig. 3). In the SAS condition, the onset timing and duration of TA stimulation were defined from the linear models according to the walking speed (Fig. 4), the onset timing and duration of the TA stimulation were in agreement with the onset timing of the TA activation and the TA duration. In the HOS and SAS conditions, the stimulation intensity was set when the subjects achieved a neutral ankle angle (0 degrees) in a seated position with the foot hanging freely in a plantar-flexed position [32]. A functional electrical stimulator was selected (P2–9632, Faisco, China), and the stimulation frequency and the pulse width were 40 Hz and 400 μs, respectively [33]. The shape of the stimulation burst is rectangular, and the stimulation amplitude of the stimulator ranged from 0 to 120 mA.

**Table 1** Demographic and clinical information about subjects with stroke

| No. | Sex | Age | Lesion side | Months after stroke | FMA-L | Slow speed (km/h) | Free speed (km/h) | Fast speed (km/h) |
|---|---|---|---|---|---|---|---|---|
| 1 | M | 44 | L | 20 | 20/34 | 0.9 | 1.2 | 1.5 |
| 2 | M | 38 | L | 8 | 25/34 | 1.0 | 1.3 | 1.6 |
| 3 | F | 47 | R | 15 | 24/34 | 1.3 | 1.7 | 2.1 |
| 4 | M | 52 | R | 12 | 22/34 | 1.0 | 1.2 | 1.4 |
| 5 | M | 51 | L | 24 | 20/34 | 1.0 | 1.3 | 1.6 |
| 6 | M | 45 | R | 12 | 23/34 | 1.0 | 1.2 | 1.4 |
| 7 | M | 65 | L | 7 | 31/34 | 1.0 | 1.3 | 1.6 |
| 8 | M | 41 | L | 12 | 23/34 | 1.0 | 1.3 | 1.6 |
| 9 | M | 53 | R | 12 | 26/34 | 1.0 | 1.3 | 1.6 |
| 10 | M | 48 | L | 12 | 25/34 | 1.0 | 1.3 | 1.6 |

*M* male, *F* female, *L* left, *R* right, *FMA-L* Fugl-meyer motor assessment for lower limb

## Data analysis

The peak knee flexion angle, maximum ankle dorsiflexion angle during the swing phase, and ankle angle at the toe-off event were extracted from kinematic data of the lower limb. A second-order low-pass Butterworth filter was used to filter the kinematic signal with a cutoff frequency of 15 Hz, since a majority of the power was less than 15 Hz according to power spectral analysis [34].

The toe-off event was determined by marker5 on the toe as in a previous study [35], and the heel-off and heel-strike events were detected by a footswitch under the heel. Kolmogorov-Smirnov test was used to assess all the variables for the normality of distribution. Over six strides were averaged for the subjects with stroke in this study [6]. To analyze the influence of the stimulation condition (NS, HOS, and SAS) on ankle dorsiflexion, ankle plantarflexion, and knee flexion angles, one-way analysis of variance with repeated measures was applied to compare the angle values among different conditions. $P > 0.05$ corresponded to the null hypothesis of no significant effect. If there was a significant difference, post-hoc analysis was then carried out using the Bonferroni between different conditions [32]. In each condition, one-way analysis of variance was applied to compare the angle values between the healthy subjects and the subjects with stroke. All statistical analyses were performed using SPSS 19 (SPSS Inc., Chicago, USA), and the level of significance was set at 0.05.

## Results

The Kolmogorov-Smirnov test was applied to the variables, and the results indicated that all the variables followed Gaussian distribution ($P > 0.05$).

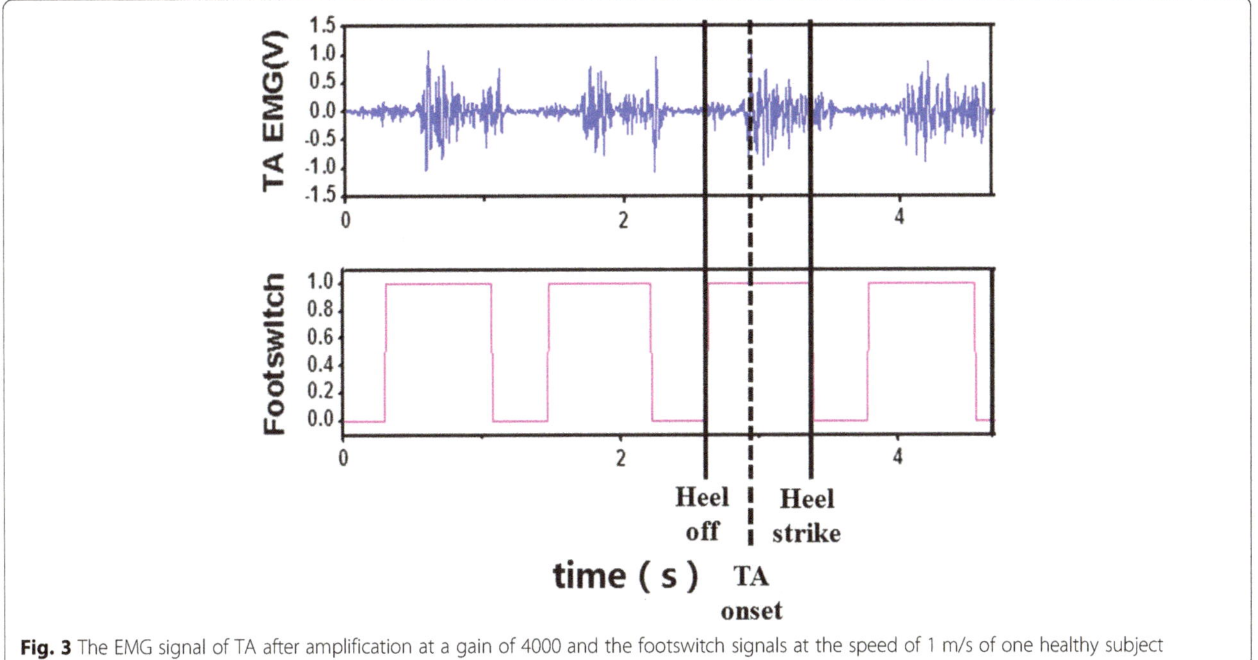


**Fig. 3** The EMG signal of TA after amplification at a gain of 4000 and the footswitch signals at the speed of 1 m/s of one healthy subject

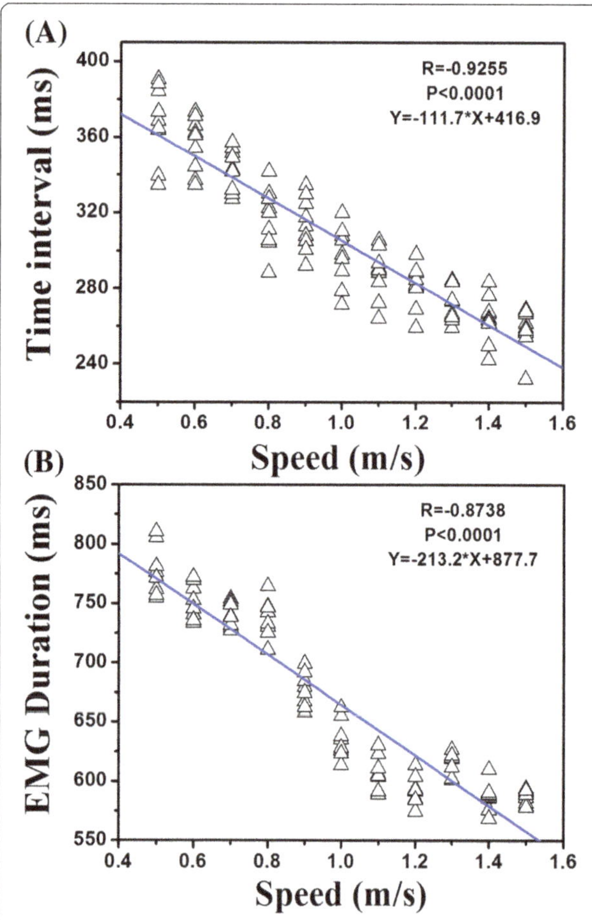


**Fig. 4** Linear models of walking speed with time interval and EMG duration. R: negative linear correlation; P: significantly linear correlation; The equation is the negative linear model established

### Effect of walking speed on time interval and EMG duration

A graphical representation of the TA activation and footswitch signals is shown in Fig. 3. The first vertical solid line indicates a heel-off event when the footswitch signal is converted from zero to one. The dashed line is the onset timing of TA activation. The time interval is between the first two lines. The second solid line indicates a heel-strike event when the footswitch signals converted from one to zero.

The mean time interval was 367.6 ms in a gait cycle at the slowest speed, which decreased to 258.4 ms at the highest speed. The relationships between the walking speed and time interval are presented in Fig. 4a, and significantly negative linear correlation was found between the walking speed and time interval ($R = -0.93$, $P < 0.0001$). There was a significant effect of the walking speed on the EMG duration ($P < 0.01$). The mean EMG duration at the speed of 0.5 m/s was 780.0 ms, which decreased by 23.6% to 595.7 ms at the highest speed of 1.5 m/s (Table 2).

For most pairs of successive speeds, the change in EMG duration was statistically significant ($P < 0.05$). As

**Table 2** The mean (±S.D.) of time interval and EMG duration at each speed

| Speed (m/s) | Time interval (ms) | EMG duration (ms) |
|---|---|---|
| 0.5 | 367.63 ± 19.89 | 780.02 ± 21.23 |
| 0.6 | 356.11 ± 14.41 | 753.83 ± 15.71[a] |
| 0.7 | 343.01 ± 11.06[a] | 742.82 ± 12.67 |
| 0.8 | 316.90 ± 16.14[a] | 741.84 ± 19.75 |
| 0.9 | 314.01 ± 14.03 | 684.78 ± 25.73[a] |
| 1.0 | 297.49 ± 15.61[a] | 640.95 ± 22.42[a] |
| 1.1 | 287.96 ± 13.27 | 611.67 ± 18.06[a] |
| 1.2 | 281.40 ± 11.48 | 596.69 ± 16.55 |
| 1.3 | 270.56 ± 8.93 | 620.92 ± 17.76[a] |
| 1.4 | 263.89 ± 12.43 | 593.67 ± 22.95[a] |
| 1.5 | 258.40 ± 11.18 | 595.71 ± 27.91 |

[a]indicated significant difference between the current speed and the previous speed

shown in Fig. 4b, there was a significantly negative linear correlation between walking speed and EMG duration ($R = -0.87$, $P < 0.0001$). After the correlation analysis and least square curve fitting, the linear models of walking speed for the time interval and EMG duration were built, as shown in Fig. 4a and b.

### Ankle and knee angle in the three stimulation conditions

Figure 5a and b present the ankle and knee angles during the gait cycle for nine healthy subjects at 0.9 m/s, respectively. The solid line and the shadow indicate the average angles and the standard deviation during the gait cycle, respectively. Figure 6 presents the maximum ankle dorsiflexion angles during the swing phase, ankle plantarflexion angles at toe-off events, and peak knee flexion angles during the swing phase for the nine healthy subjects. Compared to the NS condition, only the SAS condition achieved larger ankle dorsiflexion angle (2.8 degrees) during the swing phase at 1.2 m/s ($P < 0.05$), but the value SAS condition were significantly smaller compared to that in the HOS condition at 0.3 m/s and 0.6 m/s ($P < 0.05$)

The plantarflexion angle at toe-off events in the HOS condition had the lowest value, which was significantly smaller than the values in the SAS conditions at all speeds ($P < 0.05$) and the value in the NS condition at 1.2 m/s (13.8 degrees versus 18.1 degrees). The plantarflexion angle in the SAS condition was not statistically different from that in the NS condition at all speeds. The peak knee flexion angle in the NS condition was similar to that in the SAS condition at most of speeds, and the peak knee flexion angle in the SAS condition (44.8 degrees) was significantly larger than that in the NS condition (43.7 degrees) at 0.9 m/s ($P < 0.05$). The peak knee flexion angle in the HOS condition had the

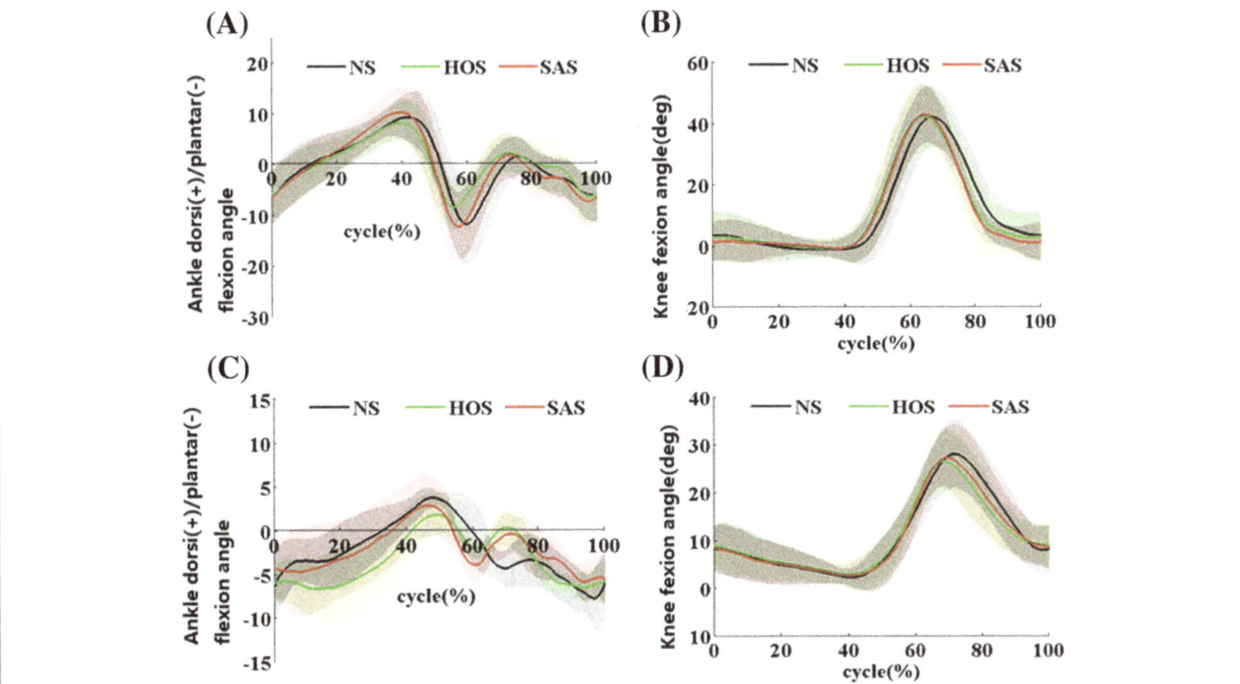


**Fig. 5 a** Ankle angles (mean ± std) during the gait cycle for nine healthy subjects at 0.9 m/s; **b** knee angles (mean ± std) during the gait cycle for nine healthy subjects at 0.9 m/s; **c** ankle angles (mean ± std) during the gait cycle for ten post-stroke subjects at free speed; **d** knee angles (mean ± std) during the gait cycle for ten post-stroke subjects at free speed

lowest values at 0.3 m/s and 0.6 m/s, which were significantly lower than that in the SAS condition at 0.3 m/s.

Figure 5c and d present the ankle and the knee angles during the gait cycle for ten subjects with stroke at free speed, respectively. The solid line and the shadow indicate the average angles and the standard deviation during the gait cycle, respectively. Figure 7 presents the maximum ankle dorsiflexion angles during the swing phase, ankle plantarflexion angles at toe-off events, and peak knee flexion angles during the swing phase for ten subjects with stroke. The HOS and SAS conditions achieved larger ankle dorsiflexion angles during the swing phase compared to that in the NS condition, and the values in the SAS conditions were significantly smaller than that in the HOS condition at free speed ($P < 0.05$). At fast speed, the value in the SAS condition (0.23 degrees) was smaller than that in the HOS condition (1.0 degrees). The plantarflexion angles during toe-off events in the HOS condition had the minimum values at all the speeds, which were significantly smaller than that in the NS and SAS conditions. And the plantarflexion angles in the SAS condition were similar with that in the NS condition at all the speeds. The peak knee flexion angles in the SAS condition were significantly larger than that in the HOS condition at slow and free speeds. There were smaller knee flexion angles at free and fast speeds in the HOS condition than that in the NS condition. No significant differences were found in the peak knee flexion angle between the NS and SAS conditions ($P > 0.05$)

Figure 8 presents the maximum ankle dorsiflexion angles during the swing phase, ankle plantarflexion angles at toe-off events, and peak knee flexion angles during the swing phase for the nine healthy subjects at 0.9 m/s and ten stoke subjects at free speed. In all conditions, healthy subjects achieved larger ankle and knee angles than stoke patients ($P < 0.05$).

## Discussion

FES-assisted ankle dorsiflexion has mainly been triggered by heel-off events in previous studies [7, 33, 36–38]. However, the heel-off event occurred ahead of the actual onset timing of TA activation in healthy subjects (shown in Fig. 3). Therefore, FES-assisted ankle dorsiflexion triggered by the heel-off event may lead to adverse effects, such as reduced ankle plantar flexion during the push-off phase and decreased knee flexion in the swing phase resulted from the decreased forward propulsive force [12]. The stimulation duration to the TA was set to a fixed value for different situations in a previous study, but the physiological relevance of the fixed value was not investigated [8]. If the stimulation duration is shorter than what is needed by the subjects with stroke, a heel-strike event may not emerge due to insufficient ankle dorsiflexion, which might still cause a risk of falls. However, if the stimulation duration is longer than what is needed, the gait may be too unstable for the forefoot to touch the ground immediately after the heel strike, and

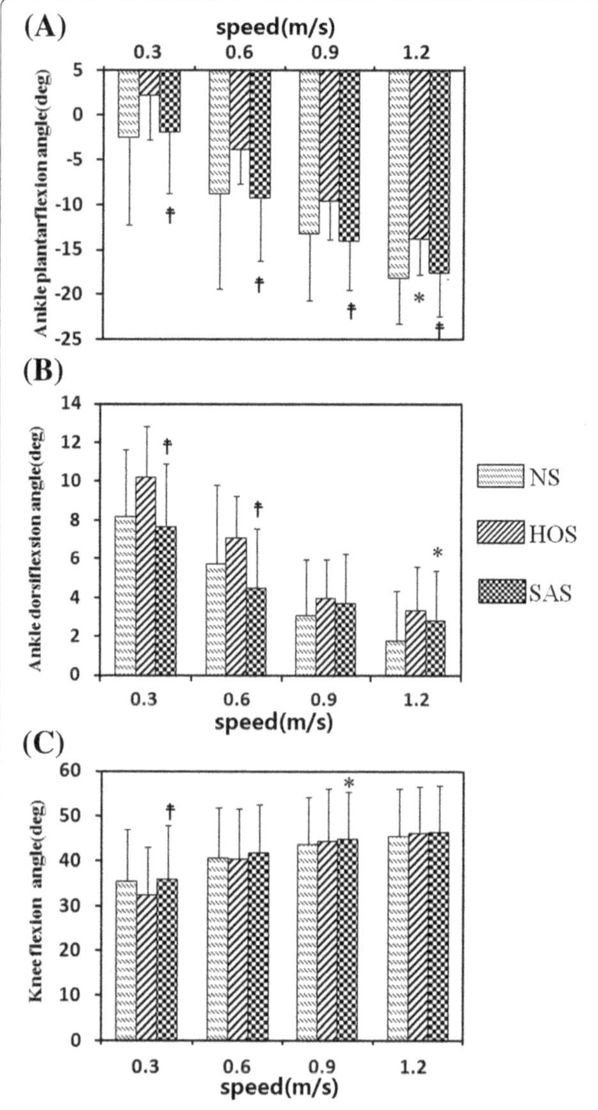


**Fig. 6** Nine healthy subjects' results for: **a** Ankle plantarflexion angles at toe-off event; **b** maximum ankle dorsiflexion angles during swing phase; **c** peak knee flexion angles during swing phase; *: significant difference from NS ($P < 0.05$);  : significant difference from HOS ($P < 0.05$)

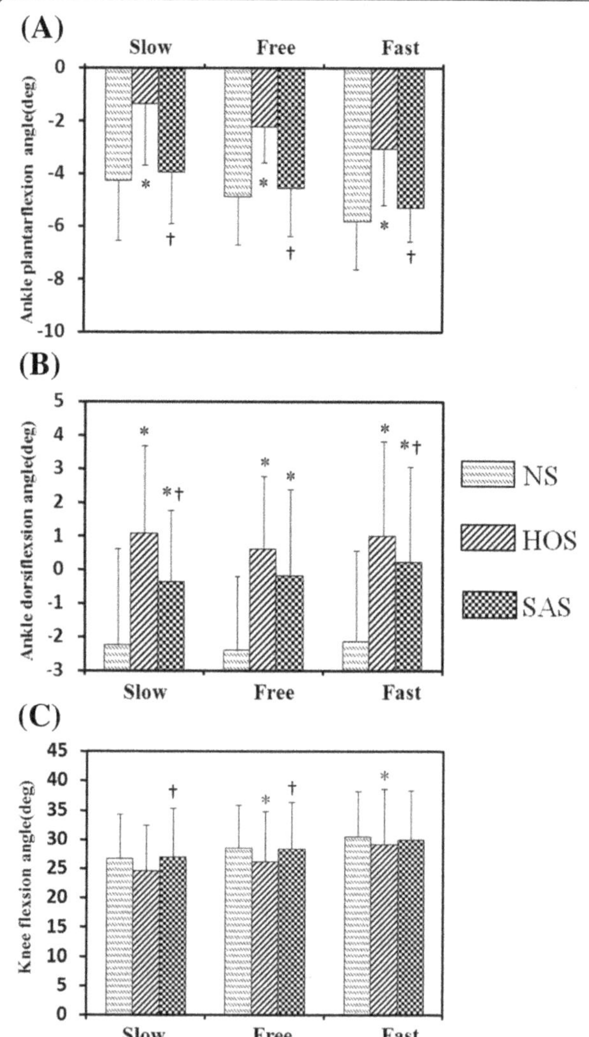


**Fig. 7** The results of ten subjects with stroke: **a** ankle plantarflexion angles at toe-off event, **b** maximum ankle dorsiflexion angles during swing phase, and **c** peak knee flexion angles during swing phase, *: significant difference from NS ($P < 0.05$), †: significant difference from HOS ($P < 0.05$)

the subjects would have to decrease their speed to make an adjustment. This might also be adverse for long-term rehabilitation.

Kesar et al. and Springer et al. terminated ankle dorsiflexion stimulation using heel-strike events [32, 37]. As shown in Fig. 3, the TA was still activated at heel-strike events, which suggests that it is not the optimal timing for terminating the dorsiflexion stimulation [32, 37]. Burridge et al. used the Odstock drop foot stimulator (ODFS) for chronic hemiplegic subjects with adjustable extension time after heel strike [39], but they did not mention the selection criteria of stimulation duration [39].

A higher ankle dorsiflexion angle could be observed at all speeds in the subjects with stroke in the HOS and SAS conditions than that in the NS condition. This implies that ankle dorsiflexion can be improved by using FES. An ideal FES intervention for dropfoot correction would be involved increasing ankle dorsiflexion during the swing phase without affecting ankle plantarflexion at toe-off events. However, although the FES produced greater ankle dorsiflexion during the swing phase in previous studies, the ankle plantarflexion at toe-off events was worsened [6, 21, 32], and a similar result could also be found in subjects in the HOS conditions. One of the reasons was that the stimulation timing in the HOS condition was triggered by heel-off events, as in most previous studies [7, 33, 36–38]. However, as shown in Fig. 3, the heel-off event was ahead of the actual onset timing of TA activation.

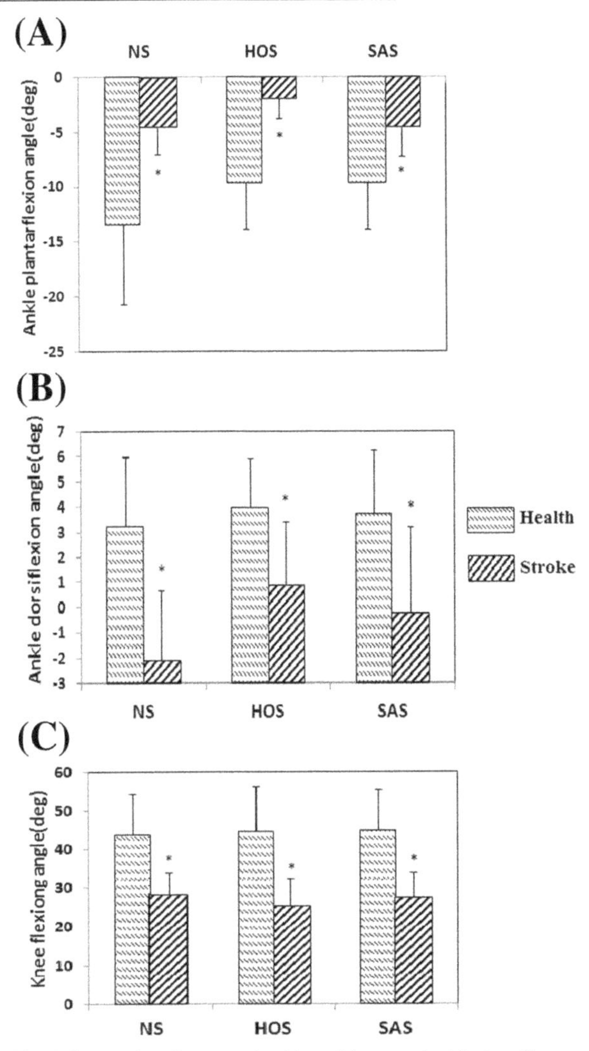


**Fig. 8** Comparison between healthy subjects and subjects with stroke: **a** Ankle plantarflexion angles at toe-off event; **b** maximum ankle dorsiflexion angles during swing phase; **c** peak knee flexion angles during swing phase; *: $P < 0.05$. The error bars represented the standard deviations

In the SAS condition, the FES was triggered by heel-off events after a time interval, which was associated with the linear model. Moreover, the duration of the time interval in the SAS condition was speed-adaptive rather than a constant value, as in previous study [8]. This design could account for the effect of walking speed. In Table 2 the standard deviation of time interval and EMG duration was very small, so the model was accurate enough to account for the individual differences. The FES parameter settings of the patients were referred to the EMG data of healthy subjects in previous studies, because FES that mimicked normal TA activation mode can help subjects with stroke recover in a natural way [8, 9, 25]. Therefore, data from healthy young subjects were collected to develop the linear model offline, and the model was applied to control the speed-adaptive onset timing and duration for the FES system in real time. In our study, the ankle plantarflexion angle in the SAS condition was similar to that in no-stimulation walking, and it was significantly larger than that in the HOS condition. This indicates that the speed-adaptive onset timing and duration are physiologically appropriate for dropfoot correction. According to the previous studies, the knee angle plays a critical role in generating the forward propulsion [40, 41]. An improper take-off during the push-off phase is attributed to reduced ankle plantar flexion during the push-off phase and decreased knee flexion in the swing phase resulted from the decreased forward propulsive force [6, 12]. Therefore, the larger peak knee flexion angle might be the consequence of the larger reserved forward propulsive force in the SAS condition than that in the HOS condition in this study. We implemented an adaptive time interval before the onset timing of TA stimulation to extend the push-off phase, so the knee angle in SAS condition was similar with that in NS condition and different from that in HOS condition. The plantarflexion angle in the SAS condition more closely approached that in the NS condition and was larger than that in the HOS condition at all the speeds.

## Conclusions

Although FES could correct dropfoot, it may come at the cost of impairing ankle plantarflexion if the ankle dorsiflexion stimulation is not well timed. In this study, the footswitch sensor and the kinematic signal combined with the linear model contributed a correction in stimulation timing in order to improve ankle plantarflexion. It should be noted that the phases of stimulation between HOS and SAS was not evaluated in detail in this study. In future work, it will be further investigated to improve the performance of FES-assisted walking together with many other factors that may also account for the change of ankle dorsiflexion and plantarflexion angles, such as stimulation intensity, step length, swing/stance time. Although the model was developed by the data from healthy young subjects within limited range of walking speeds (0.5 m/s - 1.5 m/s), linear relationships can be extended to a wider range of speeds. Moreover, the speed-adaptive stimulation system could be transferred from treadmill walking to walking over ground. An inertial sensor could be used to replace the motion capture system for the walking speed calculation and the FES controller. The SAS system with the motion tracking system is useful for indoor treadmill-based rehabilitation mainly in hospital. If the inertial sensor is used to estimate the walking speeds in outdoor daily-life walking for subjects with stroke, where acceleration or deceleration exits, the linear model can also apply to calculate the speed adaptive time interval and duration of stimulation.

**Acknowledgements**
We thank Lin Chen for conducting the clinical assessments.

**Funding**
This work was supported by the Guangdong Science and Technology Plan Project Grant No. 2015B020214003, 2017B010110015, and 2017B020210011; Science and Technology Program of Guangzhou Grant No. 201604020108.

**Authors' contributions**
GTC and RS designed the study, and GTC carried out the experiments. LL and XYW helped to carry out the experiments. GTC, KYT, LM and RS analyzed the data, interpreted the results, and drafted and revised the manuscript. All the authors approved the final version of the manuscript.

**Competing interests**
The authors declare that they have no competing interests.

**Author details**
[1]Key Laboratory of Sensing Technology and Biomedical Instrument of Guangdong Province, School of Engineering, Sun Yat-sen University, Guangzhou, Guang Dong, China. [2]Department of Rehabilitation Medicine, Sun Yat-sen First Affiliated Hospital, Sun Yat-sen University, Guangzhou, China. [3]Guangdong Work Injury Rehabilitation Center, Guangzhou, China. [4]Department of Biomedical Engineering, the Chinese University of Hong Kong, Hong Kong, China.

**References**

1. Lawrence ES, Coshall C, Dundas R, Stewart J, Rudd AG, Howard R, et al. Estimates of the prevalence of acute stroke impairments and disability in a multiethnic population. Stroke. 2001;32:1279–84.
2. Summers D, Leonard A, Wentworth D, Saver JL, Simpson J, Spilker JA, et al. Comprehensive overview of nursing and interdisciplinary care of the acute ischemic stroke patient: a scientific statement from the American Heart Association. Stroke. 2911;2009:40.
3. RW T, SK B, NC F, MR S. Gait retraining post stroke. Top Stroke Rehabil. 2003; 10:34–65.
4. JH B, DE W, PN T, DL M. Indices to describe different muscle activation patterns, identified during treadmill walking, in people with spastic drop-foot. Med Eng Phys. 2001;23:427.
5. Olney SJ, Richards C. Hemiparetic gait following stroke. Part I: characteristics. Gait Posture. 1996;4:136–48.
6. Kesar TM, Perumal R, Jancosko A, Reisman DS, Rudolph KS, Higginson JS, et al. Novel patterns of functional electrical stimulation have an immediate effect on dorsiflexor muscle function during gait for people poststroke. Phys Ther. 2010;90:55–66.
7. Liberson WT, Holmquest HJ, Scot D, Dow M. Functional electrotherapy: stimulation of the peroneal nerve synchronized with the swing phase of the gait of hemiplegic patients. Arch Phys Med Rehabilit. 1961;42:101.
8. Yan T, HuiChan CWY, Li LSW. Functional electrical stimulation improves motor recovery of the lower extremity and walking ability of subjects with first acute stroke a randomized placebo-controlled trial. Stroke. 2005;36:80–5.
9. Perry J, ST K, Davids JR. Gait analysis: normal and pathological function. JAMA J Am Med Assoc. 1992;304:907.
10. Perry J, Garrett M, Gronley JK, Mulroy SJ. Classification of walking handicap in the stroke population. Stroke. 1995;26:982.
11. Aminian K, Rezakhanlou K, De AE, Fritsch C, Leyvraz PF, Robert P. Temporal feature estimation during walking using miniature accelerometers: an analysis of gait improvement after hip arthroplasty. Med Biol Eng Comput. 1999;37:686–91.
12. Bhadra N, Kilgore KL, Peckham PH. Implanted stimulators for restoration of function in spinal cord injury. Med Eng Phys. 2001;23:19–28.
13. Sabut SK, Kumar R, Mahadevappa M. Design of an insole embedded foot pressure sensor controlled FES system for foot drop in stroke patients. Int Conf Sys Med Biol. 2010;237–41. https://doi.org/10.1109/ICSMB.2010.5735379.
14. Lundin-Olsson L, Nyberg L, Gustafson Y. "Stops walking when talking" as a predictor of falls in elderly people. Lancet. 1997;349:617.
15. Willemsen AT, Bloemhof F, Boom HB. Automatic stance-swing phase detection from accelerometer data for peroneal nerve stimulation. Biomed Eng IEEE Trans. 1990;37:1201–8.
16. Mansfield A, Lyons GM. The use of accelerometry to detect heel contact events for use as a sensor in FES assisted walking. Med Eng Phys. 2003;25:879–85.
17. Monaghan CC, WJBMv R, Veltink PH. Control of triceps surae stimulation based on shank orientation using a uniaxial gyroscope during gait. Med Biol Eng Comput. 2009;47:1181–8.
18. Weber DJ, Stein RB, Chan KM, Loeb G, Richmond F, Rolf R, et al. BIONic WalkAide for correcting foot drop. IEEE Trans Neural Syst Rehabil Eng. 2005; 13:242–6.
19. Dutta A, Kobetic R, Triolo RJ. Gait initiation with electromyographically triggered electrical stimulation in people with partial paralysis. J Biomech Eng. 2009;131:081002.
20. Yeom H, Chang YH. Autogenic EMG-controlled functional electrical stimulation for ankle dorsiflexion control. J Neurosci Methods. 2010;193:118–25.
21. Spaich EG, Bøg MF, Erkocevic E, Smidstrup A, Andersen OK, Nielsen JF. Gait orthosis Lokomat combined with functional electrical stimulation for foot drop correction: a feasibility study. Int Conf NeuroRehabilitation. 2014;7:751–7.
22. Shiavi R, Bugle HJ, Limbird T. Electromyographic gait assessment, part 1: adult EMG profiles and walking speed. J Rehabil Res Dev. 1987;24:13.
23. Yang JF, Winter DA. Surface EMG profiles during different walking cadences in humans. Electroencephalogr Clin Neurophysiol. 1985;60:485–91.
24. Byrne CA, O'Keeffe DT, Donnelly AE, Lyons GM. Effect of walking speed changes on tibialis anterior EMG during healthy gait for FES envelope design in drop foot correction. J Electromyogr Kinesiol. 2007;17:605–16.
25. Chen M, Wu B, Lou X, Zhao T, Li J, Xu Z, et al. A self-adaptive foot-drop corrector using functional electrical stimulation (FES) modulated by tibialis anterior electromyography (EMG) dataset. Med Eng Phys. 2013;35:195.
26. Sartori M, Lloyd DG, Reggiani M, Pagello E. A stiff tendon neuromusculoskeletal model of the knee. Advanced Robotics and ITS Social Impacts. 2009. p. 132–8.
27. Di FR. Reliability of computerized surface electromyography for determining the onset of muscle activity. Phys Ther. 1987;67:43.
28. Gabel RH, Brand RA. The effects of signal conditioning on the statistical analyses of gait EMG. Electroencephalogr Clin Neurophysiol. 1994;93:188–201.
29. Kadaba MP, Ramakrishnan HK, Wootten ME. Measurement of lower extremity kinematics during level walking. J Orthop Res. 1990;8:383–92.
30. Tyrell CM, Roos MA, Rudolph KS, Reisman DS. Influence of systematic increases in treadmill walking speed on gait kinematics after stroke. Phys Ther. 2011;91:392.
31. Pappas IPI, Keller T, Mangold S, Popovic MR, Dietz V, Morari M. A reliable gyroscope-based gait-phase detection sensor embedded in a shoe insole. IEEE Sensors J. 2004;4:268–74.
32. Kesar TM, Perumal R, Reisman DS, Jancosko A, Rudolph KS, Higginson JS, et al. Functional electrical stimulation of ankle plantarflexor and dorsiflexor muscles: effects on poststroke gait. Stroke. 2009;40:3821.
33. RK T, MF N, LS L, EF S. Gait training of patients after stroke using an electromechanical gait trainer combined with simultaneous functional electrical stimulation. Phys Ther. 2006;86:1282–94.

34. Bogert AJVD, Koning JJD. On optimal filtering for inverse dynamics analysis. Proceedings of the IXth biennial conference of the Canadian society for Biomechanics. Vancouver: Simon Fraser University; 1996. 214–5.
35. Jr ZJ, Richards JG, Higginson JS. Two simple methods for determining gait events during treadmill and overground walking using kinematic data. Gait Posture. 2008;27:710–4.
36. Sabut SK, Lenka PK, Kumar R, Mahadevappa M. Effect of functional electrical stimulation on the effort and walking speed, surface electromyography activity, and metabolic responses in stroke subjects. J Electromyogr Kinesiol. 2010;20:1170–7.
37. Springer S, Vatine JJ, Lipson R, Wolf A, Laufer Y. Effects of dual-channel functional electrical stimulation on gait performance in patients with hemiparesis. Sci World J. 2012;2012:530906–13.
38. Heller BW, Clarke AJ, Good TR, Healey TJ, Nair S, Pratt EJ, et al. Automated setup of functional electrical stimulation for drop foot using a novel 64 channel prototype stimulator and electrode array: results from a gait-lab based study. Med Eng Phys. 2013;35:74–81.
39. Burridge J, Taylor P, Hagan S, Wood D, Swain I. The effects of common peroneal stimulation on the effort and speed of walking: a randomized controlled trial with chronic hemiplegic patients. Clin Rehabil. 1997;11:201.
40. Neptune RR, Kautz SA, Zajac FE. Contributions of the individual ankle plantar flexors to support, forward progression and swing initiation during walking [J]. J Biomech. 2001;34(11):1387–98.
41. Anderson FC, Goldberg SR, Pandy MG, Delp SL. Contributions of muscle forces and toe-off kinematics to peak knee flexion during the swing phase of normal gait: an induced position analysis. J Biomech. 2004;37:731–7.

# 16

# Manual wheelchair downhill stability: an analysis of factors affecting tip probability

Louise Thomas[1,3], Jaimie Borisoff[2,3] and Carolyn J. Sparrey[1,3*]

**Abstract**

**Background:** For people who use manual wheelchairs, tips and falls can result in serious injuries including bone fractures, concussions, and traumatic brain injury. We aimed to characterize how wheelchair configuration changes (including on-the-fly adjustments), user variables, and usage conditions affected dynamic tip probability while rolling down a slope and contacting a small block.

**Methods:** Rigid body dynamic models of a manual wheelchair and test dummy were created using multi-body software (Madymo, TASS International, Livonia, MI), and validated with 189 experiments. Dynamic stability was assessed for a range of seat angles (0 to 20° below horizontal), backrest angles (0 to 20°), rear axle positions (0 to 20 cm from base of backrest), ground slopes (0 to 15°), bump heights (0 to 4 cm), wheelchair speeds (0 to 20 km/hr), user masses (50 to 115 kg), and user positions (0 to 10 cm from base of backrest). The tip classifications (forward tip, backward tip, rolled over bump, or stopped by bump) were investigated using a nominal logistic regression analysis.

**Results:** Faster wheelchair speeds significantly increased the probability of tipping either forward or backward rather than stopping, but also increased the probability of rolling over the bump ($p < 0.001$). When the rear axle was positioned forward, this increased the risk of a backward tip compared to all other outcomes ($p < 0.001$), but also reduced the probability of being stopped by the bump ($p < 0.001$ compared to forward tip, $p < 0.02$ compared to rolling over). Reclining the backrest reduced the probability of a forward tip compared to all other outcomes ($p < 0.001$), and lowering the seat increased the probability of either rolling over the bump or tipping backwards rather than tipping forward ($p < 0.001$). In general, the wheelchair rolled over bumps < 1.5 cm, and forwards tipping was avoided by reducing the speed to 1 km/hr.

**Conclusions:** The probability of forward tipping, corresponding to the greatest risk of injury, was significantly reduced for decreased speeds, smaller bumps, a reclined backrest, and a lower rear seat height. For wheelchairs with dynamic seating adjustability, when travelling downhill, on-the-fly adjustments to the seat or backrest can increase the likelihood of safely rolling over a bump.

**Keywords:** Wheelchair stability, Mobility devices, Rigid body dynamics, Simulation, Motion capture

## Background

It is estimated that approximately 1% of the population in developed countries require the use of a wheelchair [1, 2]. Each year, 3.3% of people who use wheelchairs in the United States are involved in serious accidents [3], sometimes resulting in traumatic brain injury, bone fractures, and concussions [4]. For active manual wheelchair users, the risk is even higher. Over a three year period from January 2006 to December 2008, 60.7% of people using manual wheelchairs ($n = 56$) reported tipping and falling at least once [5]. In the developed world, that equates to over 1.5 million manual wheelchair tips and falls every year.

The risk of a wheelchair tipping is related to its stability. Manual wheelchair static stability is defined by ISO 7176-1: 2014 as the angle at which a wheelchair and user tip over at rest [6]. However, there are currently no standards for determining manual wheelchair dynamic stability, that is, the risk of tipping while moving. Previous

* Correspondence: csparrey@sfu.ca
[1]School of Mechatronic Systems Engineering, Simon Fraser University, SFU Surrey Campus, 250-13450 102 Ave, Surrey, BC, Canada
[3]International Collaboration on Repair Discoveries (ICORD), Blusson Spinal Cord Centre, 818 West 10th Avenue, Vancouver, BC V5Z 1M9, Canada
Full list of author information is available at the end of the article

studies have considered manual wheelchair dynamic stability as the maximum speed that causes the wheelchair to stop rather than tip when rolling down a slope with a 5 cm bump at the end (while varying seat position and caster diameter) [7, 8]. Yet this fails to consider a range of obstacles that wheelchair users encounter, some of which they would be able to safely roll over. The lack of more comprehensive dynamic stability studies is likely due to the difficulties of experimentally controlling variables such as wheeling speed in a safe environment, and the considerable number of variables that affect the stability of a wheelchair in use. Such difficulties can be minimized by integrating computer simulations, validated with controlled experiments.

Rigid body dynamics are commonly used for biomechanical analyses of injuries [9] and falls [10], and are characterized by equations relating the kinematics of a system to the corresponding kinetic forces [11]. A key simplifying assumption, as suggested by the name, is the absence of deformation. This reduces the degrees of freedom, enabling problems to be solved without needing to calculate the stresses and strains in each segment. Compared to finite element analysis, rigid body dynamic simulations are therefore much more efficient and computationally inexpensive for analyzing large motions of bodies, making it an ideal method of studying wheelchair dynamics [12].

Our aim was to determine how fixed and spontaneous changes to a manual wheelchair configuration can affect the dynamic stability of the wheelchair rolling down a slope with a small bump at the end; a wheelchair skill that poses well-known safety concerns [13, 14]. Currently most manual wheelchairs are designed with a fixed frame [15], but more recent innovative designs allow users to adjust the seat angle and backrest angle 'on-the-fly' to suit their purposes [16]. These changes affect static stability by changing the centre of gravity of the system [17]. However, these changes are also likely to affect the inertia of the system and the resulting dynamic stability. The purpose of this study was to determine the effects of on-the-fly wheelchair configuration adjustments (seat angle and backrest angle), fixed wheelchair configuration changes (rear axle position), user variables (user mass, user positioning), and usage conditions (wheelchair velocity, slope of the ground, and bump height), on the dynamic tip probability of a wheelchair when moving down a slope.

## Methods

This study was comprised of a combination of simulations, experiments, and statistical analyses. First a rigid body simulation of the adjustable wheelchair was created and a sensitivity analysis was performed on that simulation. Simulations were then constructed and validated using matched experiments. Multinomial logistic parameter estimations were determined from the simulation results. Finally, the multinomial logistic model was explored to determine the effects of on-the-fly adjustability on downhill manual wheelchair stability (Fig. 1).

### Simulation

To quantify the effects of on-the-fly wheelchair configuration adjustments, fixed wheelchair configuration changes, user variables (Fig. 2a), and usage conditions on downhill stability, a rigid body dynamic model of a wheelchair (Fig. 2b, 16× 16, first generation Elevation™ model with 24 rear wheels and 5 casters, PDG Mobility, Vancouver, BC) and ISO standard test dummy were developed using MADYMO software (TASS International, Livonia, MI) and placed on a sloped ramp with a small obstacle at the end. The model was used to simulate a manual wheelchair and user rolling down a slope, over a small bump. The initial velocity of the wheelchair was assigned to the chair center of mass when the wheelchair front axles were 10 cm from the bump. The chair was then released to freely roll down the incline and impact the bump.

The wheelchair model was defined by seven components: the seat, backrest, front wheels (× 2), rear wheels (× 2), and frame. The point mass and inertia of each of these components were taken from a CAD model provided by PDG Mobility (Table 1). The mass distribution of the CAD model had been previously validated by comparing the tipping angle to that of the physical wheelchair for both forward and backward static stability [17]. The initial dummy measurements were taken from a CAD model of a 250 lb. test dummy, the same one used for the experimental validation. When varying the user mass, segment masses and centre of gravities (CoGs) were calculated from ISO 7176-11 [18]. The dummy was rigidly attached to the chair in the simulations to prevent relative motion between the dummy and the chair.

The loading characteristics of the rear wheels and casters, which define the compression response during contact, were calculated by measuring the static deflection of each wheel under masses ranging from 0 to 40 kg, and fitting a curve to the results. The unloading curve was defined as a percentage of the loading curve. For the rear wheels, this was calculated by measuring the reduction in bounce height of the wheels when they were dropped from heights of 15–30 cm, which was recorded and analyzed using motion capture. The mean unloading/loading ratio for the rear wheels was 0.810 ($\sigma = 0.027$). For the casters, the assembly was measured as a whole since the housing also has a significant effect on contact characteristics [19]. For the cases where the wheelchair was stopped by the bump during the

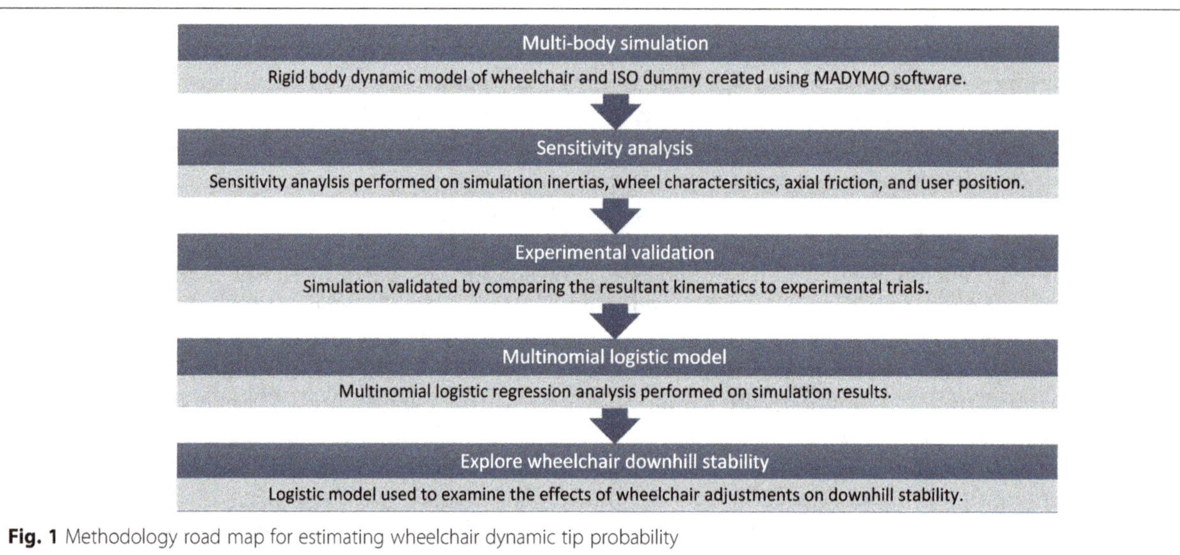


**Fig. 1** Methodology road map for estimating wheelchair dynamic tip probability

experimental testing, the unloading percentage was calculated using the average distance the wheelchair rolled back up the slope after impact with the bump. Using this method, the mean unloading/loading ratio for the caster housing was 0.294 ($\sigma = 0.145$). The axial friction in the wheels were found experimentally by rotating each of the wheels and recording the deceleration using motion capture. The process was repeated three times for each wheel, with the frictional torque calculated from the wheels' inertias and the resulting angular decelerations. The front wheels had a mean frictional torque of 0.000918 N/m, and the rear wheels 0.00263 N/m.

A sensitivity analysis was performed to determine the accuracy and sensitivity of various model inputs, including the inertia of each segment, wheel loading and unloading characteristics, axial frictions, and offsets between the user and the wheelchair backrest. Each parameter was altered independently at least 5 times for a set of simulations (66 trials), and evaluated by the number of simulation outcomes matching the experimental results. Additional simulations were run with variations to caster diameter (4″, 5″ and 6″). These were separate from the rest of the sensitivity analysis as the wheelchair caster diameter was known, but changes to that diameter (if different casters were used) would likely have a significant impact on the probability of rolling over. For these simulations, all other wheelchair configuration variables were held constant (seat angle 10°, backrest angle 10°, rear axle position 10 cm from the base of the backrest, a slope of 4.8°, user mass of 75 kg, and no offset between the user position and backrest), and the bump height was increased in increments of 1 mm until the wheelchair no longer rolled over the bump. This procedure was followed for 3 different speeds (1, 3, and 5 km/h).

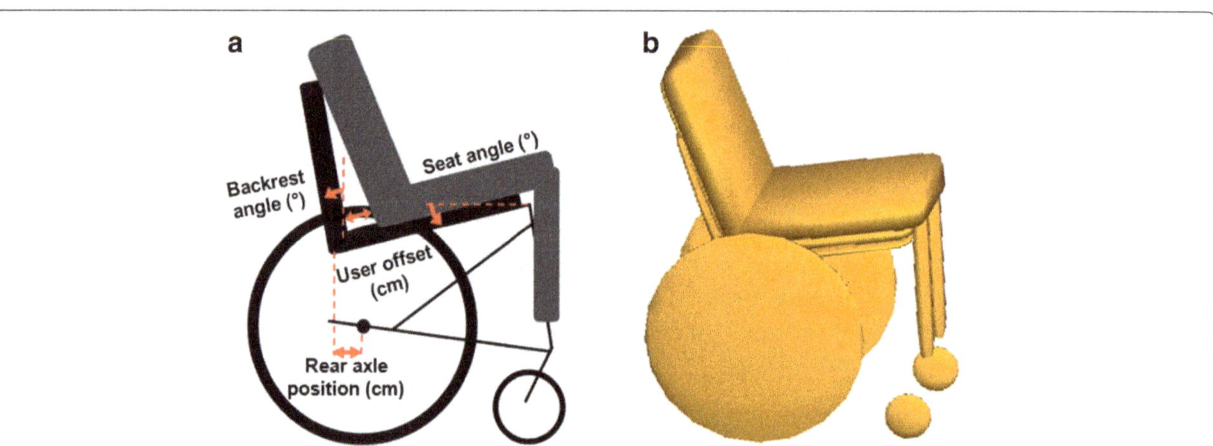


**Fig. 2** Diagram of wheelchair model. Variations were made to the wheelchair seat angle, backrest angle, rear axle position and user position (**a**), as well as user mass, wheelchair speed, ground slope, and bump height in the simulations. The Madymo model is shown on the right (**b**) Fig. 3 Experimental setup for testing wheelchair downhill stability

**Table 1** Mass and inertia for all wheelchair and dummy components included in model

| | Mass (kg) | Inertia: $I_{xx}$, $I_{yy}$, $I_{zz}$, $I_{xy}$, $I_{yz}$, $I_{xz}$ (kg.m$^2$) |
|---|---|---|
| Component | | |
| Front wheels (×2) | 0.38 | 0.0005, 0.0009, 0.0005, 0, 0, 0 |
| Rear wheels (×2) | 1.80 | 0.0670, 0.1323, 0.0670, 0.0023, 0, 0 |
| Seat (inc. gas springs) | 3.21 | 0.0645, 0.0529, 0.0892, 0, 0, – 0.0044 |
| Backrest | 1.24 | 0.0435, 0.0253, 0.0242, 0 0–0.0016 |
| Wheelchair frame | 3.19 | 0.1328, 0.1187, 0.2024, 0 0–0.0117 |
| Total wheelchair mass | 12.00 | |
| Torso | 62.80 | 0.9439, 0.6674, 1.3138, 0, 0, 0.0730 |
| Thigh | 42.16 | 0.9682, 0.5219, 1.2659, 0, 0, 0.0702 |
| Legs (×2) | 4.16 | 0.0182, 0.1022, 0.0871, 0, 0, 0.0163 |
| Total dummy mass | 113.28 | |

**Table 2** Wheelchair seat and backrest configurations used for validation tests

| Configuration type | Seat angle | Backrest angle |
|---|---|---|
| 1 | 16.1° below horizontal | −1.0° |
| 2 | | 17.4° |
| 3 | | 34.7° |
| 4 | 1.3° below horizontal | −1.0° |
| 5 | | 14.6° |
| 6 | | 29.0° |
| 7 | 13.6° above horizontal | −1.0° |
| 8 | | 6.1° |
| 9 | | 17.6° |

Seat angles ranged from 16.1° below horizontal to 13.6° above horizontal, and back angles ranged from vertical to a recline of 34.7°

### Experimental validation

The model was validated by comparing simulations of the user and wheelchair rolling down a slope and into a bump to the kinematics of the physical wheelchair and test dummy, which was captured using 3D motion capture (Vicon, Oxford, UK). The dummy was strapped to the chair during testing to minimize relative motion between the dummy and the wheelchair, and padding was placed at the end of the ramp to minimise damage when forward tipping (Fig. 3).

The wheelchair was tested for a full-factorial combination of nine seat and backrest configurations (Table 2), two ramp angles (4.8° and 7.8°), three bump heights (1.3 cm, 1.9 cm, and 3.2 cm), and at least four speeds (up to 5.3 km/hr). This resulted in a total of 189 trials. The lower ramp angle was a 1:12 slope as recommended by the Americans with Disabilities Act [20], and the bumps were created from standard timber to represent small obstacles typical of activities of daily living. Speed was varied by changing the release distance from the bump to the front wheels. Wheelchair kinematic behaviour was classified into four categories; rolled over bump, stopped by bump, tipped forwards, or tipped backwards. These classifications were used to compare the simulations to the physical experimental results.

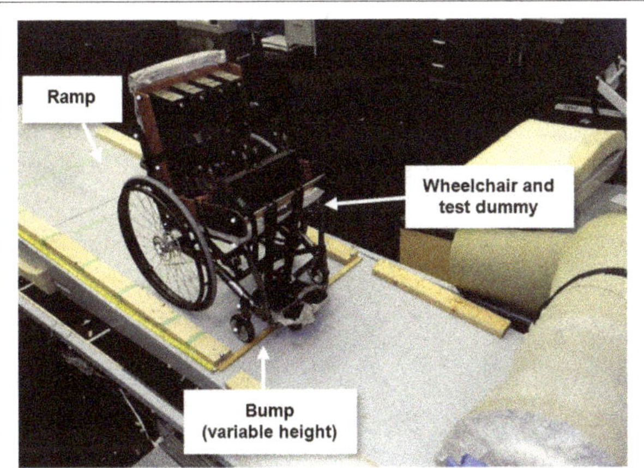


**Fig. 3** Experimental setup for testing wheelchair downhill stability

### Analysis

Due to the number of variables, a Latin Hypercube experimental design [21, 22] was used to run 2000 variations of the validated model. The independent variables were the seat angle (0 to 20° below horizontal), backrest angle (0 to 20° from vertical), rear axle position (0 to 20 cm from base of backrest), slope of the ground (0 to 15°), bump height (0 to 4 cm), and speed of the wheelchair (0 to 20 km/hr), user mass (50 to 115 kg), and user position from base of backrest (0 to 10 cm). The geometry of the dummy model was constant for all user masses, and the CoG of the torso, thigh, and leg sections changed according to the wheelchair dummy standards [18]. The inertia values were scaled by the change in mass of each segment, and transformed using parallel axis theorem for changes in CoG locations. The observed dependent variable was the tip condition of the chair after impact with the bump. The final position of the wheelchair after impact with the bump was characterized as tipped forward, tipped backward, rolled over or stopped. A nominal logistic regression analysis was performed on the tip classifications using JMP software to determine the effects of the independent wheelchair configuration and user variables on the resulting tip behaviour (v13, SAS Institute, NC, USA). *P*-values less than 0.05 were considered significant, with results grouped by $p < 0.001$, $p < 0.02$ and $p < 0.05$.

## Results

### Simulation sensitivity analyses

The wheel unloading curve for the front casters had the greatest impact on model accuracy (Table 3). Rear wheel friction had an increased effect because, for the sensitivity

**Table 3** Sensitivity of wheelchair model to set parameter changes

| | Parameter variation | Percentage change in correct simulations |
|---|---|---|
| Torso inertia | 50–150% of original | 4.5% |
| Thigh inertia | 50–150% of original | 7.6% |
| Wheel unloading characteristics | 50–150% of original | 21.2% |
| Wheel loading characteristics | 50–150% of original | 6.1% |
| Rear wheel friction | 50–150% of original | 10.6% |
| Caster wheel friction | 50–150% of original | 1.5% |
| Position offset between user and base of backrest | ±1.5 cm from original | 6.1% |
| Position offset between user and top of backrest | ±1.5 cm from original | 10.6% |

analysis, speed was controlled by releasing the wheelchair from varied distances up the slope (the same as the experiment) and so axial friction affected impact speed. However, for the final simulations, an initial velocity was assigned to the wheelchair directly before hitting the bump, thus mitigating the effect of axial friction. User positioning also had a considerable effect on model sensitivity, highlighting the need to consider posture and user movement when configuring manual wheelchairs. For each inch increase in caster diameter, the maximum bump height that the wheelchair could successfully roll over increased by 2–3 mm (Table 4). For situations where the wheelchair could not roll over the bump, results differed depending on speed: for higher bumps, the wheelchair stopped when travelling at slower speeds (≤ 3 km/h), tipped forward when travelling at higher speeds (≥ 5 km/h). The effect of caster diameter on dynamic stability had been previously studied [7], and was not included in the main model as it is well known that larger diameter casters assist in rolling over higher bumps.

### Validation with experiments

Of the 189 validation simulations performed, 168 (89%) achieved the same tip classification as the experimental results (Tables 5, 6 and 7). The most common occurrence was rolling over the bump (84 out of 189 experimental trials), while tipping backwards was least likely to occur (Table 5). Backwards tipping was also the least accurately modelled case, with a positive predictive value (PPV) of 64.3% (Table 6). The simulations were most accurate for low bumps (1.27 cm) and least accurate when the bump height was 1.91 cm (Table 7). The majority of trials rolled over the low bump, and were stopped or tipped forward for the high (3.18 cm) bump. The tip outcomes were more variable for the mid-sized bump.

At higher speeds, the front of the wheelchair often became airborne on impact with the bump (Fig. 4). In some cases, this assisted in rolling over the bump, but also increased the probability of a backwards tip. Backwards tipping generally occurred when the wheelchair launched over the bump and the casters did not come down after clearing the bump. With the large test dummy, flex was observed in the wheelchair frame on impact with the bump, particularly to the backrest. For higher bumps, the wheelchair rolled over the bump using a rocking motion that popped the castors up (Fig. 5).

**Table 4** Maximum bump height that the wheelchair rolled over for different caster diameters and speeds

| Speed | Caster diameter | | |
|---|---|---|---|
| | 4 in | 5 in | 6 in |
| 1 km/h | 1.2 cm | 1.5 cm | 1.7 cm |
| 3 km/h | 1.7 cm | 1.9 cm | 2.2 cm |
| 5 km/h | 2.1 cm | 2.4 cm | 2.7 cm |

### Multinomial logistic model

The multinomial logistic parameter estimations (Table 8) showed bump height and speed were the most influential parameters on tip outcomes; rear axle position and backrest angle had the greatest effect of the wheelchair configuration variables. Speed had a significant effect on all tip classifications, and the backrest angle had a significant effect ($p < 0.001$) on all comparisons apart from 'rolled vs stop'. Lowering the seat made the wheelchair significantly more likely to roll over the bump or tip backwards rather than tipping forwards.

The results of the logistic analysis, considering only linear terms, had a generalized $R^2$ value of 0.908 and a

**Table 5** Experimental vs. simulation confusion matrix

| Experimental result | Simulation result | | | | |
|---|---|---|---|---|---|
| | Forward tip | Backward tip | Rolled over | Stopped | Total |
| Forward tip | 28 | 2 | – | 6 | 36 |
| Backward tip | – | 9 | 1 | 1 | 11 |
| Rolled over | 3 | 2 | 78 | 1 | 84 |
| Stopped | – | 1 | 4 | 53 | 58 |
| Total | 31 | 14 | 83 | 61 | 189 |

**Table 6** Classification statistics for simulations compared to experimental results

| | Tip category | | | |
|---|---|---|---|---|
| | Forward tip | Backward tip | Rolled over | Stopped |
| Prevalence | 0.190 | 0.058 | 0.444 | 0.307 |
| Sensitivity | 0.778 | 0.818 | 0.929 | 0.914 |
| Specificity | 0.980 | 0.972 | 0.952 | 0.939 |
| PPV | 0.903 | 0.643 | 0.940 | 0.869 |
| NPV | 0.949 | 0.989 | 0.943 | 0.961 |
| $F_1$ score | 0.836 | 0.720 | 0.934 | 0.891 |

Rolling over the bump was the most common scenario, followed by being stopped by the bump. The $F_1$ score was greatest for rolling over the bump, and least accurate for backward tips

misclassification rate of 10.2% (Table 9). The majority of simulations (1093 of 2000) rolled over the bump, and rolling over was accurately predicted by the logistic model 94.9% of the time (Table 10). Backwards tips were the most likely behaviour to be misclassified, and 37.9% of the simulations that tipped backwards were misclassified as rolling over. Being stopped by the bump was the least likely scenario, occurring for 7.55% of simulations with a model sensitivity of 92.1% (Table 10). With interaction terms included in the analysis, the generalized $R^2$ value increased to 0.942 and the misclassification rate was reduced to 7.6%. Significant interaction effects with $p < 0.001$ were found for [speed]*[bump height], [rear axle position]*[bump height], [speed]*[rear axle position], [speed]*[slope], [backrest angle]*[rear axle position], [slope]*[bump height], and [user position]*[speed]. At the $p < 0.02$ level, interaction effects were also seen for [slope]*[backrest angle] and [user mass]*[user position].

To explore the effects of on-the-fly adjustability on downhill stability, the expected wheelchair tip classifications from the logit model were plotted for different backrest angles, seat angles, speeds, and bump heights (Fig. 6). Rear axle position was held constant at 10 cm, slope was set to 4.8 degrees (equivalent to 1:12, a wheelchair standard for maximum ramp inclines), user mass set to 75 kg, and the user was positioned with no offset to the backrest. The plots show that bumps of 1.5 cm or less are unlikely to be an issue for manual wheelchairs to roll over, and forwards tipping over higher bumps can be avoided by reducing speed to 1 km/hr. Bumps of 2.5 cm and greater could generally not be rolled over regardless of variable configurations (except at higher speeds). For speeds of 1 km/h and 3 km/h, lowering the seat angle moved the expected outcomes of forward tipping or stopping to the safer results of stopping or rolling over. Similar results are shown for backrest recline, where a reclined backrest increases the likelihood of stopping rather than tipping forward and, for bumps < 2 cm, increases the probability of rolling over the bump instead of stopping. However, under greater backrest angle conditions, backwards tips are also possible.

## Discussion

Manual wheelchairs are an invaluable mobility aid for those that require them, but can pose a risk of tipping when traveling on sloped and uneven surfaces. Of manual wheelchair users that have experienced a fall, it is reported that 46.3% of falls were in the forward direction [21], which is also the tip direction most likely to result in a serious injury [22]. The top three self-reported causes of wheelchair related accidents are inexperience, uneven surfaces, and obstacles [5]. This study explored the stability of a manual wheelchair when wheeling down a slope and into a small bump using a combination of experiments and simulations. A comprehensive map of the effects of on-the-fly manual wheelchair configuration adjustments (seat angle and backrest

**Table 7** Comparison of simulation and experimental results, grouped by slope and bump height

| Slope angle | Bump height | Sims correct | Sims incorrect | Discrepancies | | Percentage correct |
|---|---|---|---|---|---|---|
| | | | | Simulations | Experiments | |
| 7.8° | 1.3 cm | 24 | 1 | Rolled | Backwards tip | 96.0 |
| 7.8° | 1.9 cm | 22 | 5 | 3x forward tip<br>2x backward tip | Rolled<br>Rolled | 81.5 |
| 7.8° | 3.2 cm | 24 | 3 | Stopped<br>Stopped<br>Backward tip | Backward tip<br>Forward tip<br>Forward tip | 88.9 |
| 4.8° | 1.3 cm | 33 | 1 | Stopped | Rolled | 97.1 |
| 4.8° | 1.9 cm | 33 | 6 | 4x rolled<br>2x stopped | Stopped<br>Forward tip | 84.6 |
| 4.8° | 3.2 cm | 32 | 5 | 3x stopped<br>Backward tip<br>Backward tip | Forward tip<br>Forward tip<br>Stopped | 86.5 |

For 189 trials, 88.9% of the simulations gave the same results as the experiment

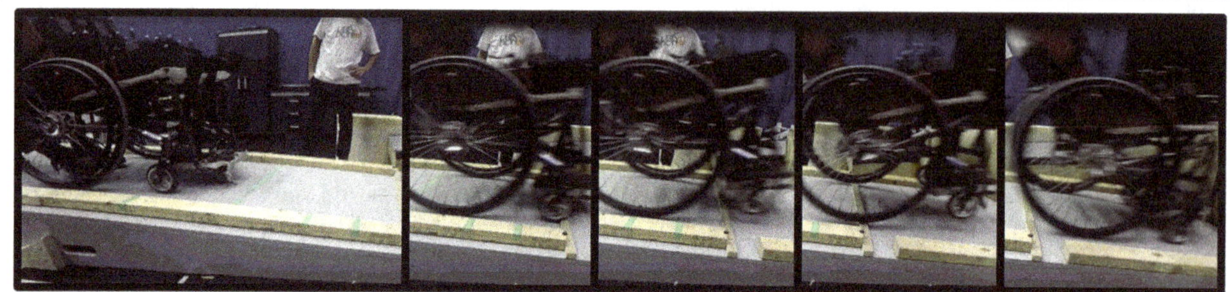

**Fig. 4** Experimental sequence of events for wheelchair rolling over a medium bump (1.91 cm) at 3.92 km/h.(1) wheelchair released on slope, (2) casters impact bump, (3) the momentum of the wheelchair causes the casters to launch over bump, (4) rear wheels impact bump while casters are still in the air, (5) wheelchair continues rolling down slope

angle), fixed wheelchair configuration changes (rear axle position), user variables (user mass, user positioning), and usage conditions (wheelchair velocity, slope of the ground, and bump height) on tip risk when wheeling downhill was determined. Bump height, wheeling speed and rear axle position were the most significant determinants of tipping probability, while on-the-fly adjustments to the seat angle and backrest angle could also change the outcome.

While standards exist for static stability [6], there are currently no standards for manual wheelchair dynamic stability. Previous studies considered dynamic stability rolling down a slope with a large (5 cm) bump at the bottom [7, 8, 23], where the outcome was either a stop or forwards tip. One such study showed that by moving the horizontal position of the seat (and therefore CoG) forward, the speed required to cause a forward tip decreases [8]. This agrees with our results, which show that forward movement of the CoG (by reducing the backrest angle or increasing user position offset from the backrest) increases the risk of a forward tip (Table 8).

A forward tip is the worst case scenario, and most likely to result in injuries requiring medical attention [22]. The parameters that had the greatest effect on forward tip probability were bump height, speed, and rear axle position. As the bump height increased, the speed required to roll over (assuming no torso movement) also increased. However, increasing speed also increased the risk of tipping rather than stopping. For lower bumps (≤2 cm), speed could be used to assist in overcoming obstacles, but this increases the risk of causing greater injury if a tip does occur. These results agree with prior work and highlight the importance of training wheelchair users to effectively navigate obstacles during downhill wheeling, including by adjusting their wheeling speed for different obstacles [13]. Lowering the seat significantly increased the probability of rolling over the bump and reduced the risk of a forward tip. When considering functional mobility, reclining the seat is also commonly used to improve balance and reach [24]. It is therefore recommended to lower the seat as far as possible, if the wheelchair includes this function, for downhill wheeling.

When wheeling downhill, the ideal outcome is for the wheelchair to roll over the bump. This occurred for 95% of simulations with a bump lower than 1 cm and backrest angle less than 20 degrees. However, if rolling over is not possible, it is much better for the wheelchair to be stopped by the bump rather than tip. In general, encountering a bump at 1 km/h (slow speed) allowed

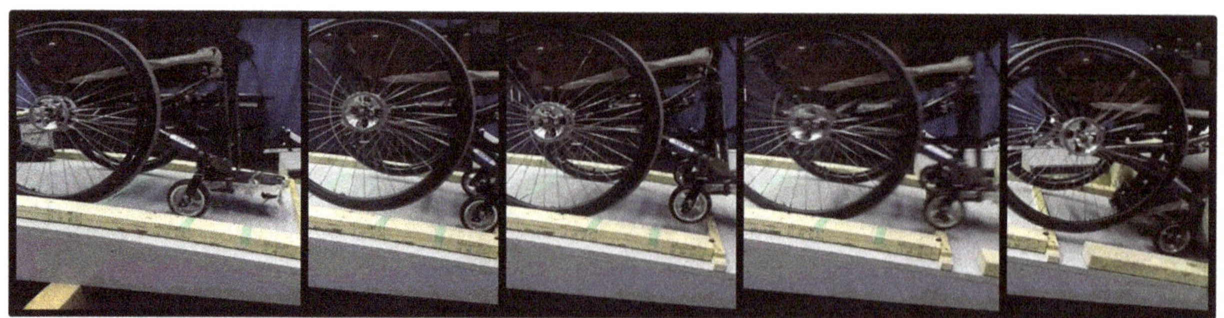

**Fig. 5** Experimental sequence of events for wheelchair rolling over a high bump (3.18 cm) at 2.59 km/h.(1) wheelchair released to roll down slope, (2) casters impact bump and rear wheels lift, (3) the rear wheels return to the ground, but the momentum causes the casters to lift, (4) casters clear bump, (5) the rear wheels follow, also clearing the bump

**Table 8** Multinomial logistic parameter estimations, with standard errors in brackets

| | Forward tip vs Stop | Backward tip vs Stop | Rolled vs Stop | Backward vs Forward tip | Rolled vs Forward tip | Rolled vs Backward tip |
|---|---|---|---|---|---|---|
| Bump height (cm) | −0.127 (0.292) | −6.088*** (0.467) | −7.612*** (0.473) | −5.962*** (0.439) | −7.486*** (0.448) | −1.524*** (0.148) |
| Speed (km/hr) | 2.311*** (0.235) | 2.684*** (0.244) | 2.851*** (0.244) | 0.373*** (0.041) | 0.540*** (0.040) | 0.167*** (0.022) |
| Rear axle position (cm) | 0.170*** (0.038) | 0.547*** (0.049) | 0.128** (0.042) | 0.377*** (0.035) | −0.042 (0.024) | −0.419*** (0.032) |
| Backrest angle (°) | −0.119*** (0.035) | 0.258*** (0.042) | 0.042 (0.038) | 0.377*** (0.031) | 0.160*** (0.026) | −0.216*** (0.022) |
| Slope (°) | 0.532*** (0.065) | 0.439*** (0.068) | 0.493*** (0.067) | −0.094** (0.034) | −0.039 (0.030) | 0.054* (0.025) |
| User position (cm) | 0.006 (0.073) | −0.375*** (0.084) | −0.170* (0.080) | − 0.382*** (0.054) | −0.176*** (0.047) | 0.205*** (0.039) |
| Seat angle (°) | −0.059 (0.035) | 0.086* (0.041) | 0.029 (0.039) | 0.145*** (0.026) | 0.087*** (0.023) | −0.058** (0.019) |
| User mass (kg) | 0.025** (0.011) | −0.010 (0.012) | 0.005 (0.012) | −0.035*** (0.008) | −0.020** (0.007) | 0.015** (0.006) |

The first three columns use the 'stop' condition as the reference category, the next two use 'forward tip' as the reference category, and the final column uses 'backward tip' as the reference. In that way comparisons were made between all categories. Bump height and wheelchair speed were the most influential parameters, with the rear axle position and backrest angle having the greatest effect of the parameters directly relating to wheelchair configuration (*$p < 0.05$, **$p < 0.02$, ***$p < 0.001$)

the user to safely stop without tipping. On level ground, comfortable propulsion speeds range from 3.7 km/h [25] to 4.6 km/h [26], with downhill wheeling sometimes faster. For terrain with bumps these speeds may become unsafe, thus for controlled wheeling the user may be required to slow down. Common obstacles encountered when wheeling downhill include potholes, rocks, and differences in pavement height, most of which are unlikely to be more than 2 cm in height. Wheelchair users can overcome higher obstacles such as curbs using torso rotation and controlled wheelies [27]. A similar type of movement was shown in Fig. 5, where the wheelchair pitched back and forth over the high bump. User movements (such as balancing in a wheelie) could be used in addition to configuration changes and speed to further improve downhill stability over bumps.

Reclining the backrest increased the probability of rolling over the bump or stopping rather than tipping forward. This did increase the risk of a backwards tip, but this was the least common outcome (5.8% of experiments and 10.6% of final simulations), was only an issue at very high backrest angles typically not used during active wheeling, and has been shown to be less dangerous than a tip forward [22]. The angle of the backrest can be the difference between a forward tip, being stopped by the bump, rolling over, or tipping backward (Fig. 6). A reclined backrest assists in maneuvering over bumps, but once the angle is more than 20 degrees there becomes a risk of tipping backward. This is similar to the static stability of the wheelchair, where a more reclined backrest enables the wheelchair to be more maneuverable, but less stable [17]. For wheelchairs without adjustable backrests, the user will usually have to perform a wheelie to go down steep inclines [13], which many users find unsafe or are unable to perform [28]; reclining the backrest may negate the need to do this. However, users with fixed framed wheelchairs may also benefit from knowing the quantified effects of backrest and seat angle on dynamic downhill stability, as it could assist in selecting the correct configuration for daily usage conditions. Depending on individual stability requirements, adjusted results from this study could be used to create guidelines to inform users and therapists of customized stability limits and maneuverability changes resulting from different wheelchair configurations.

User positioning has been previously shown to have a significant effect on stability [29]. When the user's pelvis was positioned at an offset from the backrest, the probability of tipping backward was significantly reduced in comparison to all other behaviours. However, the

**Table 9** Confusion matrix for the logit model

| Simulation result | Predicted logit model result | | | | |
|---|---|---|---|---|---|
| | Forward tip | Backward tip | Rolled over | Stopped | Total |
| Forward tip | 506 | 17 | 17 | 5 | 545 |
| Backward tip | 11 | 114 | 80 | 6 | 211 |
| Rolled over | 14 | 40 | 1037 | 2 | 1093 |
| Stopped | 8 | 1 | 3 | 139 | 151 |
| Total | 539 | 172 | 1137 | 152 | 2000 |

**Table 10** Classification statistics for logit model compared to simulations

| | Tip category | | | |
|---|---|---|---|---|
| | Forward tip | Backward tip | Rolled over | Stopped |
| Prevalence | 0.270 | 0.106 | 0.547 | 0.076 |
| Sensitivity | 0.928 | 0.540 | 0.949 | 0.921 |
| Specificity | 0.977 | 0.954 | 0.869 | 0.993 |
| PPV | 0.939 | 0.663 | 0.912 | 0.914 |
| NPV | 0.973 | 0.933 | 0.913 | 0.994 |
| $F_1$ score | 0.934 | 0.595 | 0.930 | 0.917 |

Overall there was a 10.2% misclassification rate when comparing the predicted result from the multinomial logistic analysis to the simulation results. Categories 'forward tip', 'rolled over', and 'stopped' all had F1 scores over 0.9, and 'backward tip' was the least accurate category with an F1 score of 0.595

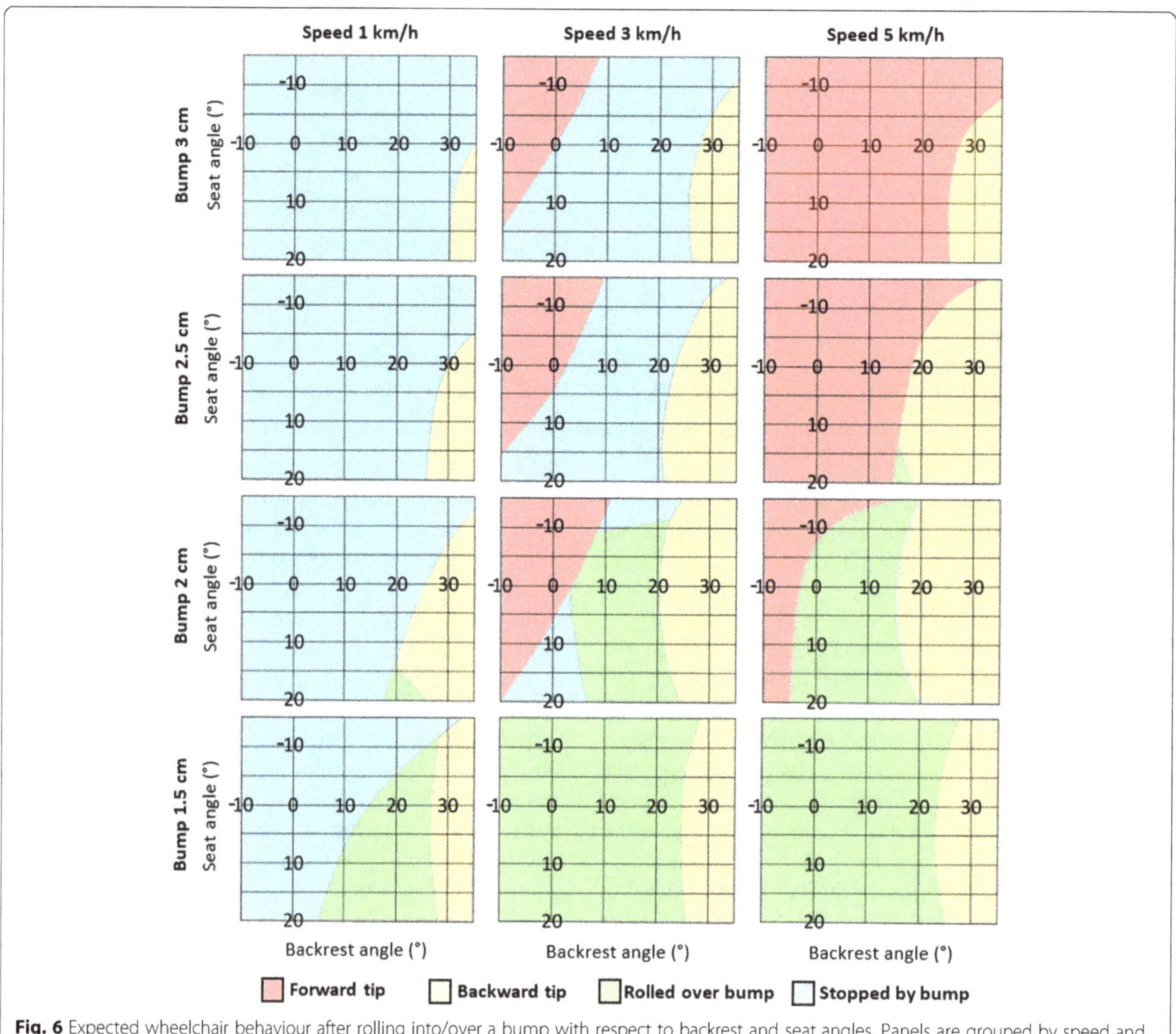


**Fig. 6** Expected wheelchair behaviour after rolling into/over a bump with respect to backrest and seat angles. Panels are grouped by speed and bump height

probability of tipping forward rather than rolling over was also increased. For users that sit with their hips forward from the base of the seat, configuring the wheelchair with the rear axle further forward can permanently reverse the ensuing stability effects, or a reclined backrest could be used to temporarily adjust the stability as needed. As suggested by the Wheelchair Skills Training Program Manual, users should therefore be encouraged to reposition themselves as far back in the wheelchair as possible during downhill wheeling [13] to reduce the risk of a forward tip.

In general, configuration changes that made the wheelchair more likely to roll over the bump (lowering the seat, reclining the backrest, moving the rear axle forward) did so by shifting the system CoG towards the rear axles. On level ground, backward shifts in the CoG position also increase maneuverability [30]. The position of the rear axle had the greatest effect on tip response at slower speeds and when the bump was between 1.5 and 2.5 cm. For these cases, the outcome was less predictable and the position of the rear axle could be the deciding factor of whether the wheelchair tipped or rolled over. Moving the rear axle further forward made the chair more likely to tip backwards; interestingly, it also slightly increased the probability of rolling over the bump or tipping forwards rather than being stopped. Rolling over probability was likely increased due to shifting the CoG towards the rear axle, which reduced the load on the front wheels, making it easier for them to clear the bump. The increase in forward tipping probability may be owing to the weight of the rear wheels shifting the CoG forwards in relation to the front wheels. The effect of wheel position on dynamic rolling stability highlights the need for therapists and industry professionals to properly configure the wheelchair for each particular user. These results relate to previous research on manual wheelchair static stability, which showed that forward movements of the rear axle reduced stability, but increased maneuverability for a straight trajectory (defined as minimizing rolling resistance) [17]. It also suggests an opportunity for future designs offering a rear axle (or CoG) 'on-the-fly' adjustment capability that could significantly improve wheeling stability on slopes.

Changes in wheelchair configuration that affect downhill stability will also affect maneuverability and biomechanical demand during manual wheelchair propulsion [24, 31, 32]. The mobility of a manual wheelchair is a function of both the biomechanics of the user and the dynamics of the wheelchair itself. For situations where the user is pushing the chair (i.e. most dynamic cases apart from wheeling downhill), reducing rolling resistance and improving push biomechanics are important for minimizing the risk of upper limb overuse injuries [31, 33–35]. Increasing the load on the rear wheels reduces rolling resistance for straight trajectories [32], such as the modelled case of wheeling downhill, but does so at the cost of reducing rear stability [17, 30]. In addition to mechanical advantages due to reduced rolling resistance, shifting the rear axle forward increases the biomechanical push angle and shoulder ROM [24], and decreases needed muscle activity for the triceps, anterior deltoids and biceps [36]. The optimal seat angle for propulsion efficiency is still unknown [24], but a horizontal seat has been linked to the development of shoulder pain [37]. However, small changes in system tilt and seat to backrest angle (up to 10°) did not show any effect on joint angles or shoulder moments in manual wheeling [38]. Though a lower seat may be biomechanically superior for wheeling, an elevated seat can improve daily tasks such as transferring and reaching, and provide psychosocial benefits such as reducing eye to eye level discrepancies with others [39]. In daily life, wheelchair users perform a variety of maneuvers including movements forward, backward, turning, and accelerating. During straight motion the majority of propulsion energy is converted to translational energy, with some rotational kinetic energy for the wheels and casters, but during turning up to 71% of the system energy is converted to turning kinetic energy [38]. Therefore it is also important to consider multi-directional wheelchair maneuverability when evaluating complete wheelchair performance, where an increase in rear wheel loading corresponds to an increase in resistive forces due to turning [40]. Better dynamic wheelchair performance is likely a balance between stability, rolling resistance, and turning resistance, with the optimal configuration dependent on task specific requirements. Thus, the ability to change wheelchair configurations on-the-fly to emphasize different performance advantages may be beneficial to wheelchair users.

Our analysis demonstrated that on-the-fly adjustments to wheelchair configurations can improve downhill wheeling stability; however, the dominant factors in determining tip risk were bump height, wheeling speed and rear axle position which are not affected by on-the-fly alterations. Furthermore, an incorrectly positioned adjustable wheelchair can decrease stability. Therefore, training users to effectively use on-the-fly adjustments and defining the limits of operation will be important for optimizing the potential stability benefits of the technology. The results of the analysis also show that backrest angle had a greater effect on downhill rolling stability than seat angle. As a result, a chair with an adjustable backrest alone [42] could provide most of the potential downhill wheeling stability benefits observed in this study.

### Strengths and limitations

Computational models are an efficient method for studying wheelchair dynamics, however they are limited by model input accuracy [12]. The use of passive dummy models is a particular limitation, as it disregards any active movements of the user. For the case of rolling down a slope this is not a major issue as users are advised to maintain their weight towards the rear of the wheelchair when descending [13]. However, when navigating obstacles and for other situations where the user actively changes their position, future models will need to be modified to simulate user activity. Since the mass of the user represents the majority of the system mass, dummy stature is another limitation. The ISO dummies used represent the average stature of a wheelchair user [18], but individual variations may affect model accuracy by changing the mass distribution and therefore the inertial characteristics and centre of mass of the user.

Discrepancies between the simulation and experimental results were likely due to the model being highly sensitive to the material properties of the wheels, and limitations in the method of measuring axial friction of the wheels. This is demonstrated by the increased sensitivity of the model to the wheel unloading characteristics (Table 3). Rigid body models are unable to fully capture the dynamics of collisions [41]. Since some deformation occurs on impact with the bump, finite-element methods could improve the accuracy of the tire contact calculations. Including tire deformation would also allow the rolling resistance of the wheelchair to be more accurately modelled. However, using finite element analysis in the model would greatly increase computational time and limit the number of simulations that could feasibly be run. The measured physical properties of the wheelchair were another possible source of error in the model.

In particular, the accuracy of the wheel contact characteristics and the axial friction were limited by the methods used to measure them. Since the loading of the wheels were measured statically, they would not precisely match the dynamic loading characteristics during a collision. Measuring the dynamic loading of the wheels was outside the scope of this study. Using an unloaded axial friction load was also a limitation, but provided a reasonable approximation. Estimating friction coefficients from the deceleration of the wheels resulted in a less accurate model than using the friction loads from the unloaded wheel.

## Conclusion

A combination of skills training and dynamic wheelchair adjustability could greatly improve user safety when wheeling over obstacles. The most significant factors for downhill wheeling stability were bump height, speed, and rear axle position. Increased speed can be used to overcome smaller bumps where the user is confident they will not tip, but for larger bumps a more controlled method should be employed. Generally, this will involve skilled user movements such as balancing in a wheelie. However, the need for this maneuver could be negated by changing the CoG of the wheelchair system using on-the-fly wheelchair seat and backrest adjustability. The significant impact of rear axle position on both stability and wheelchair maneuverability also suggests an opportunity for the future development of wheelchairs with dynamically adjustable rear axles. This would enable users to optimize the balance between wheelchair stability and maneuverability as required throughout the day. Having developed a validated computer simulation of wheelchair tip dynamics, future research could include the effects of user movements on manual wheelchair stability and maneuverability. Since the weight of user constitutes such a large proportion of the total system, small movements in user position may have a large effect on the dynamics as a whole. This can be comprehensively explored using the developed simulation methods. This also underlines the importance of effective wheelchair skills training [13] in combination with good wheelchair design for safe and reliable wheelchair use. Consequently, manual wheelchair dynamics need to be analyzed as a complex system with interactions between the wheelchair itself, the user, and the environmental conditions.

On-the-fly adjustments to the seat and backrest could be used in certain situations to reduce the probability of tipping and/or increase the probability of rolling over a bump. The quantified general downhill rolling stability results could also be used to guide the configuration of fixed-frame wheelchairs or those with adjustable backrests only [42] to define more optimal operating limits. For wheelchairs with dynamic seat and backrest adjustability, when travelling downhill the seat should be lowered as far as possible to increase the likelihood of safely rolling over a bump. Reclining the backrest may also help in overcoming obstacles, but should be adjusted with caution as reclining will also increase the probability of a backwards tip.

**Acknowledgements**

The authors would like to acknowledge the support of Garrett Kryt and Rory Dougall in conducting the experiments and Tanuj Singla in constructing portions of the wheelchair model.

**Funding**

This work was funded by the Natural Sciences and Engineering Research Council of Canada (NSERC), the Canada Foundation for Innovation (CFI) and Mitacs with support from the Rick Hansen Institute.

**Authors' contributions**

LT, JB and CJS devised and designed the study. LT completed the experiments and simulations with assistance from CJS. LT analyzed the data with input from JB and CJS. LT drafted the manuscript and JB and CJS critically revised it. All authors approved the final manuscript.

**Competing interests**

JB is a product consultant to PDG Mobility, the manufacturer of the ElevationTM wheelchair. In addition, JB has a financial interest in the sale of the ElevationTM wheelchair product and is named on the following patents related to the ElevationTM wheelchair: US 7,950,684, US 7,845,665, US 8,042,824, US 8,801,020. LT and CJS declare that they have no competing interests.

**Author details**

[1]School of Mechatronic Systems Engineering, Simon Fraser University, SFU Surrey Campus, 250-13450 102 Ave, Surrey, BC, Canada. [2]British Columbia Institute of Technology, BCIT Centre for Applied Research & Innovation, 4355 Mathissi Pl, Burnaby, BC, Canada. [3]International Collaboration on Repair Discoveries (ICORD), Blusson Spinal Cord Centre, 818 West 10th Avenue, Vancouver, BC V5Z 1M9, Canada.

**References**

1. Sapey B, Stewart J, Donaldson G. The social implications of increases in wheelchair use. Dep Appl Soc Sci Lanc Univ. 2004. https://disability-studies.leeds.ac.uk/wp-content/uploads/sites/40/library/sapey-BOBs-REPORT.pdf.
2. Van Drongelen A, Roszek B, Hilbers-Modderman E, Kallewaard M, Wassenaar C. Wheelchair incidents. In: RIVM repos; 2002.
3. Calder CJ, Kirby RL. Fatal wheelchair-related accidents in the United States. Am J Phys Med & Rehabil/Assoc Acad Physiatr. 1990;69:184–90. https://doi.org/10.1097/00002060-199008000-00003.
4. Opalek JM, Graymire VL, Redd D. Wheelchair falls: 5 years of data from a level I trauma center. J Trauma Nurs : Off J Soc Trauma Nurses. 2009;16:98–102. https://doi.org/10.1097/JTN.0b013e3181ac920e.

5. Chen W-Y, Jang Y, Wang J-D, Huang W-N, Chang C-C, Mao H-F, et al. Wheelchair-related accidents: relationship with wheelchair-using behavior in active community wheelchair users. Arch Phys Med Rehabil. 2011;92:892–8. https://doi.org/10.1016/j.apmr.2011.01.008.
6. ISO - International Organization for Standardization. ISO 7176-1: 2014 Wheelchairs - part 1 determination of static stability. 2014.
7. Kirby RL, McLean AD, Eastwood BJ. Influence of caster diameter on the static and dynamic forward stability of occupied wheelchairs. Arch Phys Med Rehabil. 1992;73:73–7.
8. Majaess GG, Kirby RL, Ackroyd-Stolarz SA, Charlebois PB. Influence of seat position on the static and dynamic forward and rear stability of occupied wheelchairs. Arch Phys Med Rehabil. 1993;74:977–82.
9. Silva MPT, Ambrosio JAC, Pereira MS. Biomechanical model with joint resistance for impact simulation. Multibody Syst Dyn. 1997;1:65–84.
10. Romick-Allen R, Schultz AB. Biomechanics of reactions to impending falls. J Biomech. 1987;21:591–600. https://doi.org/10.1016/0021-9290(88)90222-9.
11. Schiehlen W. Multibody system dynamics: roots and perspectives. Multibody Syst Dyn. 1997;1:149–88.
12. Erickson B, Hosseini MA, Mudhar PS, Soleimani M, Aboonabi A, Arzanpour S, et al. The dynamics of electric powered wheelchair sideways tips and falls: experimental and computational analysis of impact forces and injury. J Neuroengineering Rehabil. 2016;13:20. https://doi.org/10.1186/s12984-016-0128-7.
13. Kirby RL, Smith C, Parker K, MacLeod D, McAllister M. Rushton P., editors. In: Wheelchair skills training program (WSTP) manual; 2016.
14. Edlich RF, Kelley AR, Morton K, Gellman RE, Berkey R, Greene JA, et al. A case report of a severe musculoskeletal injury in a wheelchair user caused by an incorrect wheelchair ramp design. J Emerg Med. 2010;38:150–4. https://doi.org/10.1016/j.jemermed.2007.07.067.
15. Brubaker CE. Wheelchair prescription: an analysis of factors that affect mobility and performance. J Rehabil Res Dev. 1986;23:19–26.
16. Mattie J, Borisoff J, Miller WC, Noureddin B. Characterizing the community use of an ultralight wheelchair with "on the fly" adjustable seating functions: a pilot study. PLoS One. 2017;12:e0173662. https://doi.org/10.1371/journal.pone.0173662.
17. Thomas L, Borisoff J, Sparrey C. Quantifying the effects of on-the-Fly changes of seating configuration on the stability of a manual wheelchair. In: The 39th annual international conference of the IEEE engineering in medicine and biology society; 2017.
18. ISO - International Organization for Standardization. ISO 7176-11: 2012 Wheelchairs - part 11 test dummies. 2012.
19. Bertocci GE, Esteireiro J, Cooper RA, Young TM, Thomas C. Testing and evaluation of wheelchair caster assemblies subjected to dynamic crash loading. J Rehabil Res Dev. 1999;36:32–41.
20. US Department of Justice. ADA Standards for Accessible Design. 2010.
21. Kirby RL, Ackroyd-Stolarz SA, Brown MG, Kirkland SA, MacLeod DA. Wheelchair-related accidents caused by tips and falls among noninstitutionalized users of manually propelled wheelchairs in Nova Scotia. Am J Phys Med & Rehabil/Assoc Acad Physiatr. 1994;73:319–30. https://doi.org/10.1097/00002060-199409000-00004.
22. Gaal RP, Rebholtz N, Hotchkiss RD, Pfaelzer PF. Wheelchair rider injuries: causes and consequences for wheelchair design and selection. J Rehabil Res Dev. 1996;34:58–71.
23. Kirby RL, Atkinson SM, MacKay EA. Static and dynamic forward stability of occupied wheelchairs: influence of elevated footrests and forward stabilizers. Arch Phys Med Rehabil. 1988;70:681–6.
24. Medola FO, Elui VMC, Santana C d S, Fortulan CA. Aspects of manual wheelchair configuration affecting mobility: a review. J Phys Ther Sci. 2014;26:313–8.
25. Slowik JS, Requejo PS, Mulroy SJ, Neptune RR. The influence of speed and grade on wheelchair propulsion hand pattern. Clin Biomech. 2015;30:927–32. https://doi.org/10.1016/j.clinbiomech.2015.07.007.
26. Chow JW, Millikan TA, Carlton LG, Chae W, Lim Y, Morse MI. Kinematic and electromyographic analysis of wheelchair propulsion on ramps of different slopes for young men with paraplegia. Arch Phys Med Rehabil. 2009;90: 271–8. https://doi.org/10.1016/j.apmr.2008.07.019.
27. Kirby RL, Bennett S, Smith C, Parker K, Thompson K. Wheelchair curb climbing: randomized controlled comparison of highly structured and conventional training methods. Arch Phys Med Rehabil. 2008;89:2342–8. https://doi.org/10.1016/j.apmr.2008.04.028.
28. MacPhee AH, Kirby RL, Coolen AL, Smith C, MacLeod DA, Dupuis DJ. Wheelchair skills training program: a randomized clinical trial of wheelchair users undergoing initial rehabilitation. Arch Phys Med Rehabil. 2004;85:41–50. https://doi.org/10.1016/S0003-9993(03)00364-2.
29. Kirby RL, Sampson MT, Thoren FA, MacLeod DA. Wheelchair stability: effect of body position. J Rehabil Res Dev. 1995;32:367–72.
30. Thomas L, Borisoff J, Sparrey C. Defining the stability limits of a manual wheelchair with adjustable seat and backrest. In: Rehabilitation Engineering and Assistive Technology Society of North America Conference. New Orleans, LA; 2017.
31. Sauret C, Vaslin P, Lavaste F, de Saint Remy N, Cid M. Effects of user's actions on rolling resistance and wheelchair stability during handrim wheelchair propulsion in the field. Med Eng Phys. 2013;35:289–97.
32. Tomlinson JD. Managing maneuverability and rear stability of adjustable manual wheelchairs: an update. Phys Ther. 2000;80:904–11. https://doi.org/10.1093/ptj/80.9.904.
33. Guo L-Y, Zhao KD, Su F-C, An K-N. Modeling the effect of seat height and fore-aft position on wheelchair propulsion. J Biomech. 2006;39. https://doi.org/10.1016/S0021-9290(06)85019-0.
34. Masse LC, Lamontagne M, O'Riain MD. Biomechanical analysis of wheelchair propulsion for various seating positions. J Rehabil Res Dev. 1991;29. https://doi.org/10.1682/JRRD.1992.07.0012.
35. van der Woude LH, Veeger HE, Dallmeijer AJ, Janssen TW, Rozendaal LA. Biomechanics and physiology in active manual wheelchair propulsion. Med Eng & Phys. 2001;23:713–33. https://doi.org/10.1016/S1350-4533(01)00083-2.
36. Bertolaccini G d S, Carvalho Filho IFP, Christofoletti G, Paschoarelli LC, Medola FO. The influence of axle position and the use of accessories on the activity of upper limb muscles during manual wheelchair propulsion. Int J Occup Saf Ergon. 2018;24:311–5. https://doi.org/10.1080/10803548.2017.1294369.
37. Giner-Pascual M, Alcanyis-Alberola M, Millan González L, Aguilar-Rodríguez M, Querol F. Shoulder pain in cases of spinal injury: influence of the position of the wheelchair seat. Int J Rehabil Res Int Z fur Rehabil Rev Int de Rech de readaptation. 2011;34:282–9. https://doi.org/10.1097/MRR.0b013e32834a8fd9.
38. Medola FO, Dao PV, Caspall JJ, Sprigle S. Partitioning kinetic energy during freewheeling wheelchair maneuvers. IEEE transactions on neural systems and rehabilitation engineering. 2014;22:326–33. https://doi.org/10.1109/TNSRE.2013.2289378.
39. Arva J, Schmeler M, Lange M, Lipka D, Rosen L. RESNA position on the application of seat-elevating devices for wheelchair users. Assist Technol. 2009;21:69–72.
40. Lin J-T, Huang M, Sprigle S. Evaluation of wheelchair resistive forces during straight and turning trajectories across different wheelchair configurations using free-wheeling coast-down test. J Rehabil Res Dev. 2015;52:763–74. https://doi.org/10.1682/JRRD.2014.10.0235.
41. Eichberger A, Schittenhelm M. Implementations, applications and limits of Tyre models in multibody simulation. Veh Syst Dyn. 2005;43:18–29. https://doi.org/10.1080/00423110500109349.

# 17

# Kinect-based assessment of proximal arm non-use after a stroke

K. K. A. Bakhti[1,2,5*], I. Laffont[1,2,5], M. Muthalib[1,3], J. Froger[1,4,5] and D. Mottet[1,5]

**Abstract**

**Background:** After a stroke, during seated reaching with their paretic upper limb, many patients spontaneously replace the use of their arm by trunk compensation movements, even though they are able to use their arm when forced to do so. We previously quantified this proximal arm non-use (PANU) with a motion capture system (Zebris, CMS20s). The aim of this study was to validate a low-cost Microsoft Kinect-based system against the CMS20s reference system to diagnose PANU.

**Methods:** In 19 hemiparetic stroke individuals, the PANU score, reach length, trunk length, and proximal arm use (PAU) were measured during seated reaching simultaneously by the Kinect (v2) and the CMS20s over two testing sessions separated by two hours.

**Results:** Intraclass correlation coefficients (ICC) and linear regression analysis showed that the PANU score (ICC = 0.96, $r^2 = 0.92$), reach length (ICC = 0.81, $r^2 = 0.68$), trunk length (ICC = 0.97, $r^2 = 0.94$) and PAU (ICC = 0.97, $r^2 = 0.94$) measured using the Kinect were strongly related to those measured using the CMS20s. The PANU scores showed good test-retest reliability for both the Kinect (ICC = 0.76) and CMS20s (ICC = 0.72). Bland and Altman plots showed slightly reduced PANU scores in the re-test session for both systems (Kinect: $-4.25 \pm 6.76$; CMS20s: $-4.71 \pm 7.88$), which suggests a practice effect.

**Conclusion:** We showed that the Kinect could accurately and reliably assess PANU, reach length, trunk length and PAU during seated reaching in post stroke individuals. We conclude that the Kinect can offer a low-cost and widely available solution to clinically assess PANU for individualised rehabilitation and to monitor the progress of paretic arm recovery.

**Keywords:** Arm non-use, Stroke, Rehabilitation, Kinect v2, Movement analysis

## Background

A large proportion of post-stroke individuals develop upper limb (UL) disabilities that result in difficulties in performing daily activities such as reaching for objects, which impacts their quality of life [1]. Stroke survivors often compensate for these impairments by adapting their movement patterns to incorporate additional degrees of freedom at new joints and body segments. One of the most common compensatory behaviours following a stroke is when patients replace the use of the paretic UL with the use of the less impaired UL. When such compensation persists even if the paretic UL recovered enough to be used, this is called learned non-use. However, the learned non-use phenomenon is more general than this specific case: it applies to any cases when individuals "forget" how to use their affected body parts and instead overuse compensations [2]. Here, our reasoning is that the learned non-use phenomenon explains why some patients "forget" how to use the proximal joints of their paretic UL and instead overuse trunk compensation during reaching [3]. Long-term non-use of shoulder-elbow movements is suspected to be detrimental to optimal upper limb functional recovery, due the "Use it or lose it" principle of use-dependent neural plasticity [4, 5]. Long-term overuse of trunk compensation may lead to suboptimal motor

* Correspondence: k-bakhti@chu-montpellier.fr
[1]Euromov, University of Montpellier, Montpellier, France
[2]Physical Medicine and Rehabilitation, Montpellier University Hospital, Montpellier, France
Full list of author information is available at the end of the article

recovery of the paretic UL and secondary complications such as muscle contractures [6–8].

In our previous study [3], the non-use of shoulder-elbow joints during reaching has been termed "proximal arm non-use" and was assessed with the PANU score. The PANU score was derived by the substraction of the spontaneous proximal arm-use (SPAU) from the maximal proximal arm-use (MPAU). MPAU was recorded when patients were asked to voluntarily maximise proximal arm-use (or constrained not to use trunk movement). SPAU was recorded when patients were free to spontaneously balance their use of trunk and their use of proximal arm joints during hand reaching while seated. We showed that the PANU score had very good test-retest reliability. The PANU score did not depend on time since stroke. Higher PANU scores were moderately related to higher UL impairment (Fugl Meyer Assessment of the upper extremity-FM-UE) and to lower UL function (Box & Block Test-BBT). Moreover, 61% of patients with lower impairment (FM-UE proximal > 28/42) exhibited proximal arm non-use (PANU score > 6.5%) [3]. The PANU score provides information about the remaining functional motor reserve of the paretic arm (at the level of shoulder and elbow movements), which is not used by the post-stroke individual. Consequently, the PANU score potentially enables a clinician to select patients for specific rehabilitation programs focusing on the "maximal-use" of shoulder and elbow movements as well as to monitor the paretic arm recovery/compensation post-stroke. Therefore, PANU assessment is complementary with routine clinical measures of arm impairment (e.g., FM-UE) and function (e.g., BBT).

In our previous study [3], the PANU measurement using the ultrasound 3D motion capture system (CMS20s, Zebris) required to position markers on the hands and on the trunk of the patient during the seated reaching task. Although most clinical motion analysis devices use a marker-based system [9], recent developments in computer gaming technology brought forward marker-less infrared sensors such as the Microsoft's Kinect (v2 for Xbox One) that captures the users' body movement and allows them to interact within video games [10–14]. The Kinect sensor is low cost and reliably tracks body joints in real-time without requiring markers attached to the body [15]. Moreover, the official Microsoft software development kit (SDK) release permitted the Kinect sensor to be used not only as a gaming device, but also as a measurement system. The Kinect sensor showed comparable results in measuring shoulder movements during validity assessment with an established highly accurate 3D motion capture (6-camera Vicon) system [16, 17]. However, very few studies are available on the validity of the Kinect to accurately track trunk and hand motions in a reaching task [18], which is essential for quantifying the PANU score in post-stroke individuals.

Therefore, the purpose of this study was to validate a Kinect-based system to accurately and reliably quantify PANU scores and related kinematic parameters in individuals who sustained a stroke. Our clinical question was whether the Kinect sensor could replace the existing CMS20s sensor for PANU assessment. We hypothesised that the Kinect will provide similar measurements of PANU score and related kinematic variables as those determined by the reference CMS20s system.

## Methods

### Participants

A total of 19 people who had suffered a stroke (59 ± 3 years; 9 women) participated in this study. Participants were recruited during their in-patient hospitalization at the department of Physical and Rehabilitation Medicine of Nimes University Hospital Le Grau du Roi and Montpellier University Hospital Lapeyronie. We chose patients with no restriction of time since their stroke, which resulted in a wide range from 13 to 4027 days after stroke with a median of 96 days after a stroke. Inclusion criteria were: single supratentorial cerebral vascular accident; either haemorrhagic or ischemic; any time after the stroke; either left or right affected hemisphere; aged ≥18 and ≤ 90; able to carry out a seated hand reaching task with the paretic arm (see details of reaching task in Fig. 1). Participants were excluded if they had shoulder pain or perceptual-cognitive deficits (hemi-negligence, ataxia, receptive aphasia) [19]. The study was approved by The Ethics Committee of Montpellier, France (N° ID-RCB: 2014-A00395–42) and registered in Clinical Trials (N° NCT02326688). Written informed consent was obtained from all participants prior to their inclusion.

### Clinical assessments

The patient's UL sensorimotor impairment was assessed with the FM-UE, allowing a maximal score of 66 [20]. The patient's UL functional capacity was measured using the BBT, with scores expressed as the number of blocks transferred using the paretic arm [21]. These two clinical tests were used in our previous study [3].

### Experimental protocol

The participants seated comfortably on a chair with armrests and were invited to reach a target placed in front of them, with each of their hands. The start position was standardised with the pronated forearm resting on the armrest such that the hand was hanging without support. The target was aligned to the midline of the body at a distance determined in terms of the length of the participant's active stretched arm measured from the medial axilla to the distal wrist crease [3, 22]. The reaching movement speed was self-selected and was repeated 5 times with the paretic hand, then 5 times with the less-impaired hand. This reaching sequence was first performed in a spontaneous proximal arm-use condition (SPAU), and then in a maximal proximal arm-use

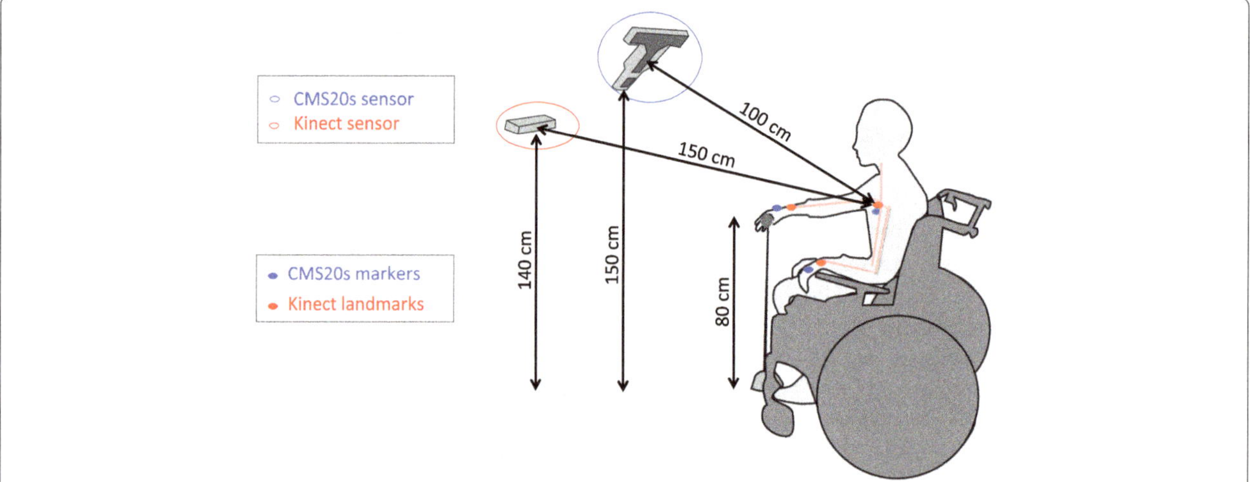


**Fig. 1** Experimental setup. The quantification of the proximal arm non-use (PANU) score was simultaneously determined by the Kinect (blue encircled) and CMS20s (red encircled) movement recording systems. The CMS20s recorded the position of 3 markers placed on the manubrium, right dorsal hand, left dorsal hand (blue spots). The Kinect provided a skeleton of the person (orange) out of which we retained 3 "joints" corresponding best to the position of CMS20s markers on the body: Spine-Shoulder, WristRight, WristLeft

condition (MPAU). In the SPAU condition, after a "go" signal, the participant was free of any constraint and reached the target in a spontaneous manner. In the MPAU condition, the therapist provided manual (Therapists' hands on the patient's sternum) and verbal feedback to ensure that the patient minimized trunk use as much as possible and, consequently, maximised the use of the elbow-shoulder joints [3, 23, 24]. The sequence of reaching tasks to assess the PANU score was repeated 2 h later to determine test-retest reliability and to know the magnitude of the difference in PANU score due to the patient repeating the tasks (i.e., practice effect). The same therapist evaluated each participant in the two assessments.

### Experimental setup

The experimental setup of the Kinect and CMS20s sensors allowed simultaneous capturing of the 3D motion of the trunk and hand during seated reaching by the two systems (see details in Fig. 1).

#### *CMS20s sensor*

The CMS20s (Zebris, Isny, Germany) is a marker-based system that measures body kinematics based on travel-time measurement of ultrasound impulses. Ultrasound impulses emitted by miniature markers placed on the body are triangulated by three microphones built into the CMS20s receiver at a sampling rate of 50 Hz. The CMS20s was chosen as the gold standard motion analysis system (accuracy lower than 1/10 mm). Three markers were placed so as to be clearly visible by the CMS20s during the whole reach sequence: on the right and left head of the second metacarpal (dorsal hand) and on the manubrium (blue spots in Fig. 1). The CMS20s system was connected to a PC running the "WinData" software (Zebris, Isny, Germany) to record the position time-series of the markers.

#### *Kinect sensor*

The Kinect (v2, Microsoft, USA) is a marker-less motion capture system combining 3 sensors (a RGB colour camera, a depth sensor, an infrared sensor) to provide the 3D position (X, Y, Z) of 25 landmarks on a skeleton with a sampling rate of 30 Hz [25]. The Kinect processes the distance data by a time-of-flight camera system with proprietary algorithms. The time-of-flight system is able to reconstruct the 3D scene through the measurement of the time elapsed between the emission of a light ray and its collection after reflection on the target [26]. The Kinect was connected to a PC running the "MaCoKi" software (NaturalPad, Montpellier, France) developed from the Kinect SDK (v2.0_1409, Microsoft, USA) to record the position time-series of the hands and trunk.

#### *Position of landmarks for the CMS20s and for the Kinect*

The Kinect records the movements of 25 predefined body "joints" that approximately correspond to the centre of the anatomical joint or body part. For the comparison of the position of landmarks between the two systems, we had to select the Kinect joints best corresponding to the CMS20s markers. Preliminary tests revealed that the "wrist" in the Kinect-based records corresponded best to the positioning of the CMS20s marker on the dorsal face of the hand. Similarly, "Spine-Shoulder" in the Kinect-based records corresponded best to a marker on the manubrium (red spots in Fig. 1), which confirms the Kinect joints chosen by Ozturk et al., (2016) and Valdes et al., (2017) to determine trunk compensation during a reaching task [18, 27]. Consequently, the Kinect joints chosen to be corresponding to CMS20s markers were the right wrist (for right head of the second metacarpal with CMS20s), left wrist (for left head of the second metacarpal with CMS20s) and spine-shoulder (for manubrium with CMS20s).

### *Materials/ setup*

To prevent the Kinect from mistakenly identifying furniture as part of the body [28], we omitted the table, used a chair with small arm rests and positioned the target on a narrow stand in front of the participant (see details in Fig. 1). In addition, while the Kinect detects human bodies, it is not able to determine the position of the target. As a consequence, we designed a calibration procedure in which we asked the participants to show where the target was. By asking the participant to reach and stay at the target, with the paretic hand, then with the non-paretic hand, the therapist verified visually that the hand was in place, and recorded for 2 s. The average position of both hands gave the target position. This calibration procedure also included the positioning of the patient in relation to the target.

### Data processing

For the computation of the PANU score, we reasoned that, because the very nature of a reaching task is to reduce the distance to the target [29], we could summarize the relevant information into a 1D space that is, the Euclidean distance from the hand or the trunk to the target. One important consequence of working with distance is to increase the robustness of the assessment procedure in a clinical context: clinician can avoid a long and complex calibration procedure to ensure the orientation of the X, Y and Z axes and the position of the origin. From the Euclidean distance time-series, we first determined the start and the end of each reaching movement. The start position corresponded to the moment when the Euclidean distance from the hand to the target began to decrease (i.e., the hand was no longer resting on the armrest). The end position corresponded to the shortest Euclidean distance from the hand to the target (i.e., the target was reached). Second, we determined the change in Euclidean distance to the target during the reaching movement, for the trunk and for the hand (i.e., ΔTrunk and ΔHand) so to obtain a value of proximal arm use (i.e., PAU = (ΔHand - ΔTrunk) / ΔHand) for each reaching movement. From the PAU values, we could compute a PANU score as PANU = MPAU – SPAU, where SPAU is the median PAU value in the spontaneous proximal arm use condition and MPAU is the median PAU value in the maximal proximal arm use condition [3]. In the present study, the PANU scores were calculated for the paretic arm only.

We also extended the comparison of the Kinect and CMS20s towards classical kinematic variables often used to describe upper limb movements after a stroke: Movement Time (MT) and Number of Velocity Peaks (NVP) [30]. The MT was defined as the duration from reach start to reach end. The NVP was defined as the number of peaks in the tangential velocity profile of the hand over the MT.

Finally, we explored one potential added value of the Kinect over the marker-based CMS20s, the fact that it provides the time series of 25 body landmarks that can be used to assess joint movements. Joint angles of interest in relation to the PANU score were measured only with the Kinect system. Since PANU is related to elbow-shoulder non-use [23], we calculated the shoulder and elbow angles using the Kinect data. The elbow flexion/extension was computed from the angle between two vectors in 3D: the forearm (from Elbow to Wrist Kinect joint) and the arm (from Elbow to Shoulder). In order to define the shoulder flexion/extension angle, a frontal plane was created by using 3D coordinates of the spine-mid, spine-shoulder and shoulder right/left. Then, the shoulder flexion/extension was defined as the angle between the vector directed from the shoulder left/right to elbow left/right and the frontal plane [27]. From the shoulder flexion/extension angle and elbow flexion-extension angle, we computed the non-use of shoulder flexion (nuSF) and the non-use of elbow extension (nuEE) by subtracting the angular amplitude in the SPAU condition from the angular amplitude in the MPAU condition.

### Statistical analysis

To quantify the degree to which CMS20s and Kinect measurements are related, we used intra-class correlation coefficient (ICC) and linear regression analysis [31], complemented with Bland and Altman plots with 95% limits of agreement [32]. In these analyses, we pooled the data from the test and retest sessions and compared the values obtained with the Kinect and CMS20s.

To quantify the degree to which the test and retest measurements are related, we again used ICC, linear regression and Bland and Altman plots. In these analyses, we did not pool the data from the Kinect and CMS20s, such that we compared the values obtained over the test and retest sessions separately for the two devices.

In addition, the non-parametric Spearman correlation coefficients were used to assess the relationship between the PANU score, arm impairment (FM-UE score) and arm function (BBT) on the paretic side.

Statistical analyses were performed using R (version 3.5.0). The level of significance for all tests was set at $p < 0.05$.

## Results

### Validity of the Kinect against the CMS20s

The Bland and Altman analysis is a powerful, yet simple, graphical method to assess the agreement of measurements by 2 methods/sessions: a scatter plot visualises the difference between two measures as a function of the average of the two measures. In a Bland and Altman plot, the central horizontal line indicates the mean of the differences (systematic bias), which is 0 for a perfect agreement. The horizontal lines above and below represent the 95% limits of agreement (average difference ± 1.96 standard deviation of the difference), which tells how far apart measurements by the 2 methods/sessions were more likely to be for most individuals. We complemented the Bland and Altman plots with linear regression analysis, which also provides the equation of the linear regression of one method/session against the other. In Figs. 2, 3, 4, 5 the left panels represent the Bland and Altman plots, and the right panels indicate ICC and regression plots, including the regression equation. Table 1 presents the descriptive statistics of PANU, MT and NVP per recording device.

The agreement between the Kinect and CMS20s measurements of PANU score, PANU components (PAU, ΔHand, and ΔTrunk) and kinematic variables (MT and NVP) is assessed using ICC, linear regression and Bland and Altman plots (Figs. 2 and 3).

### *PANU scores*

PANU scores obtained from Kinect and CMS20s were strongly correlated with ICC = 0.96. The slope of the linear regression (0.91) indicated that Kinect underestimated PANU (Fig. 2, right panel). The Bland and Altman plots indicated that PANU scores obtained from Kinect were on average 0.37 ± 3.40 shorter than the PANU scores obtained from CMS20s (Fig. 2, left panel).

### *PANU components*

**Pau** For the comparison of PAU, the measurements from the Kinect and CMS20s were strongly correlated with ICC = 0.97. The slope of the linear regression (0.93) indicated that Kinect underestimated PAU (Fig. 3 top panel, right column). The Bland and Altman plots indicated that PAU scores obtained from Kinect were on average 0.06 ± 2.84 shorter than the PAU scores obtained from CMS20s (Fig. 3, top panel, left column).

**Δtrunk** Change in distance to the target due to trunk movement (ΔTrunk) from the Kinect and CMS20s were correlated with ICC = 0.97. The slope of the linear regression (0.90) indicated that Kinect underestimated ΔTrunk (Fig. 3, middle panel, right column). The Bland and Altman plots indicated that ΔTrunk obtained from Kinect was on average 0.79 ± 8.79 mm shorter (Fig. 3, middle panel, left column).

**Δhand** Change in distance to the target due to the hand movement (ΔHand) from the Kinect and CMS20s were also highly correlated with an ICC = 0.81. The slope of the linear regression (0.78) indicated that Kinect underestimated ΔHand (Fig. 3, bottom panel, right column). The Bland and Altman plots indicated that ΔHand obtained from Kinect was on average 11.41 ± 33.50 mm shorter (Fig. 3, bottom panel, left column).

### *Related kinematic variables*

**Mt** For the movement time (MT), the measurements from the Kinect and CMS20s were correlated with ICC = 0.82. The slope of the linear regression (0.81) indicated that Kinect underestimated MT (Fig. 4, top panel, right column). The Bland and Altman plots indicated that, on average, Kinect measured MT 0.06 s less than CMS20s with a standard deviation of 0.37 s (Fig. 4, top panel, left column).

**NVP** For the number of velocity peaks (NVP), the measurements from the Kinect and CMS20s were poorly correlated with ICC = 0.34. The large intercept (4.70) did not

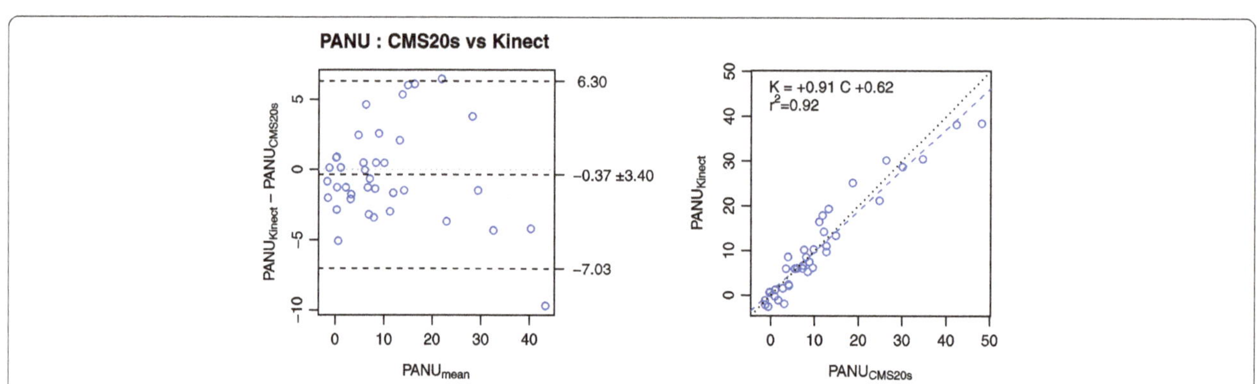


**Fig. 2** Comparison of PANU scores obtained with the Kinect and CMS20s systems. The left panel presents the Bland and Altman plot and the right panel presents the linear regression plot. PANU scores obtained with the Kinect and CMS20s were strongly correlated, yet with a small underestimate with the Kinect

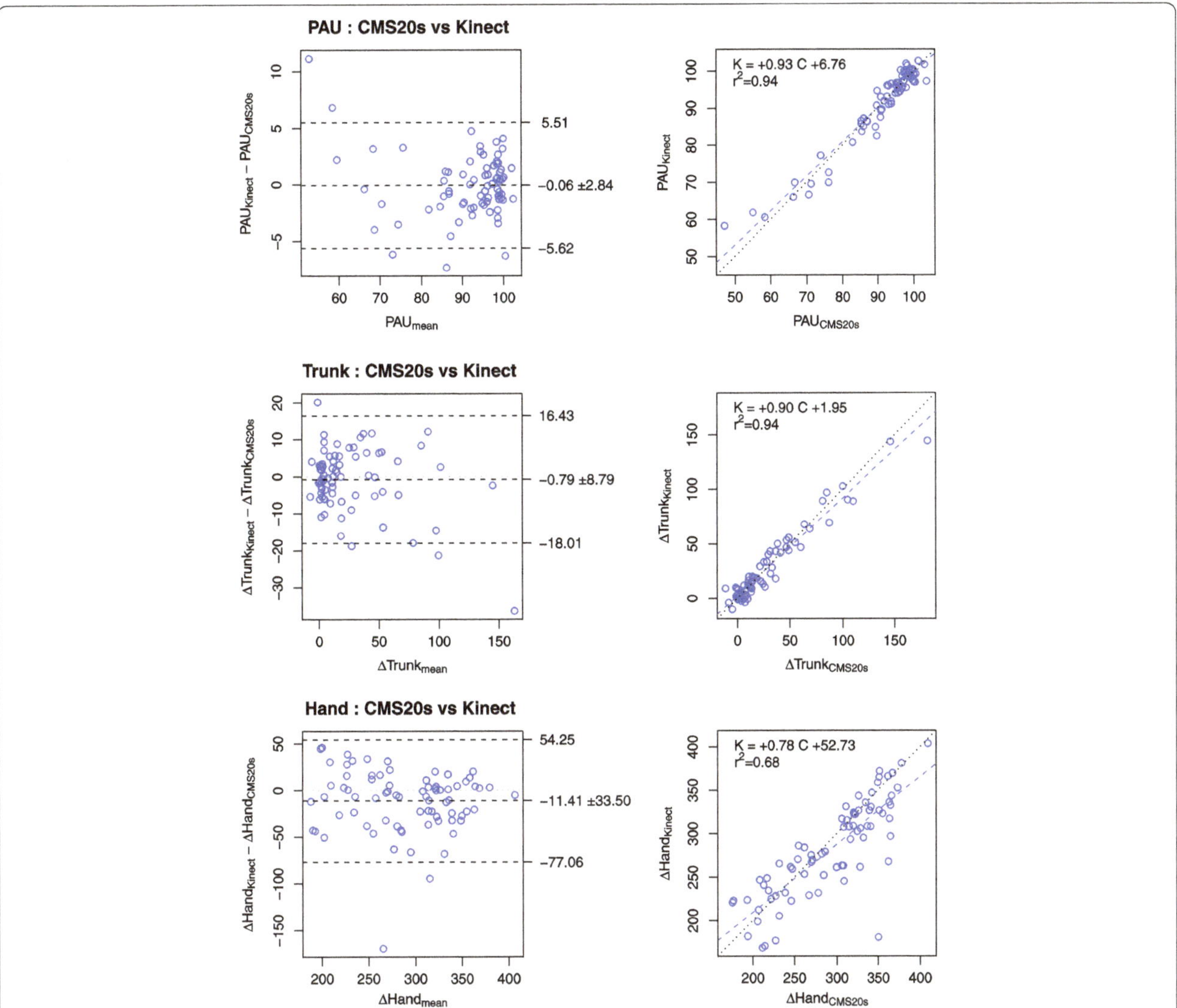


**Fig. 3** Comparison of PANU components obtained with the Kinect and CMS20s systems. Panels in the first row illustrate proximal arm-use (PAU). Panels in the second row illustrate trunk movement amplitude (ΔTrunk). Panels in the third row illustrate reach length (ΔHand). For each row, the left panel represents the Bland and Altman plot and the right panel represents the linear regression plot. The three components are adequately determined by the Kinect, yet with a small underestimate for ΔHand (11 mm)

allow the simple interpretation that the slope of the linear regression (0.80) indicates an underestimate (Fig. 4, bottom panel, right column). The Bland and Altman plots indicated that, on average, Kinect measured NVP with 3.39 peaks more than CMS20s and a standard deviation of 2.89 peaks (Fig. 4, bottom panel, left column).

#### Additional kinematic variables that have only been measured by the Kinect

From the time series of 25 body landmarks provided by Kinect data, we examined the non-use of joints determining the PANU score, namely the non-use of shoulders and elbows. We found that PANU was significantly related to nuEE (PANU = + 0.55 nuEE + 7.17 with $r^2 = 0.35$, $F(1,36) = 19.63$, $p = 0.0001$) but not to nuSF (PANU = – 0.27 nuSF + 10.77 with $r^2 = 0.02$, $F(1,36) = 0.85$, $p = 0.3631$).

#### Relation of PANU scoring with the clinical assessments

Spearman's correlation indicated that the PANU score was not significantly linked to the FM-UE (CMS20s: $r = -0.32$, $p = 0.24$ and Kinect: $r = -0.26$, $p = 0.13$) and not significantly linked to the BBT (CMS20s: $r = -0.10$, $p = 0.57$ and Kinect: r = – 0.10, $p = 0.42$). The FM-UE and BBT scores were linked ($r = 0.60$, p = 0.00). The PANU scores measured by the CMS20s and by the Kinect were very strongly linked ($r = 0.95$, p = 0.00) as expected from Fig. 2, right panel.

In addition, among the 11 post-stroke individuals with a high proximal FM-UE score (> 28/42: low impairment),

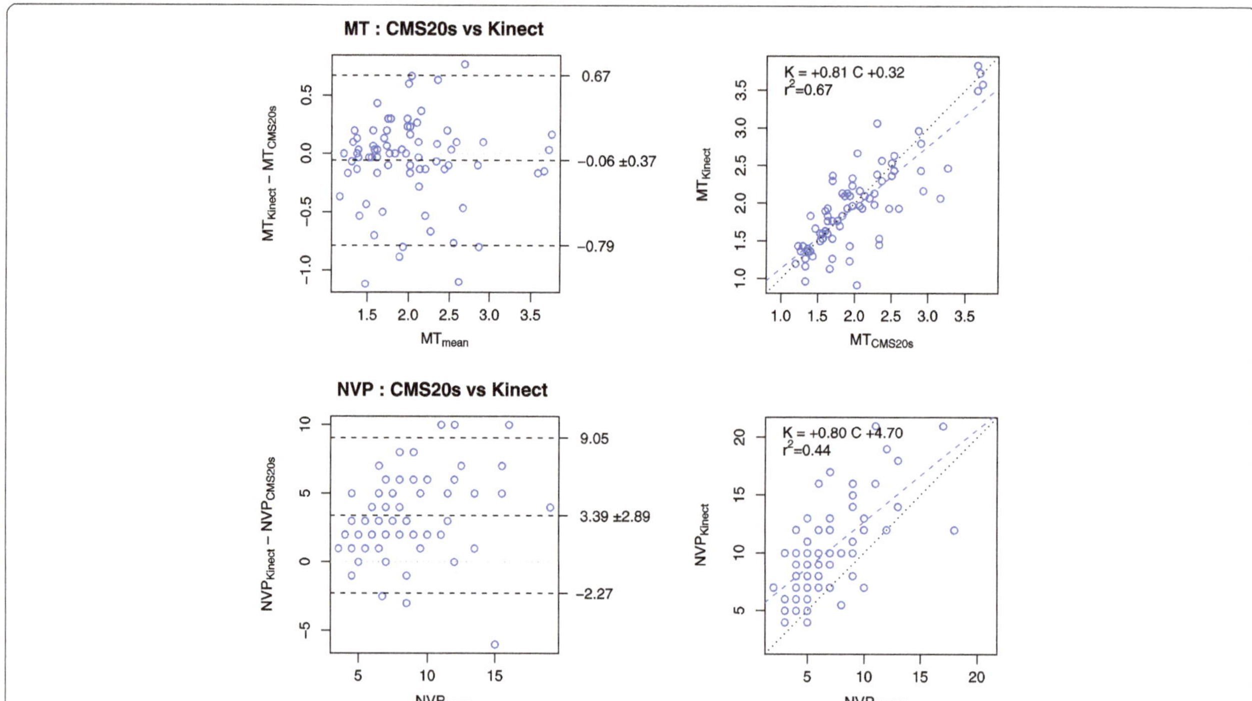


**Fig. 4** Comparison of movement kinematics obtained with the Kinect and CMS20s. Panels in the first row illustrate the movement time (MT). Panels in the second row illustrate the number of velocity peaks (NVP). For each row, the left panel represents the Bland and Altman plot and the right panel represents the linear regression plot. The movement time is adequately determined by the Kinect, but not the number of velocity peaks

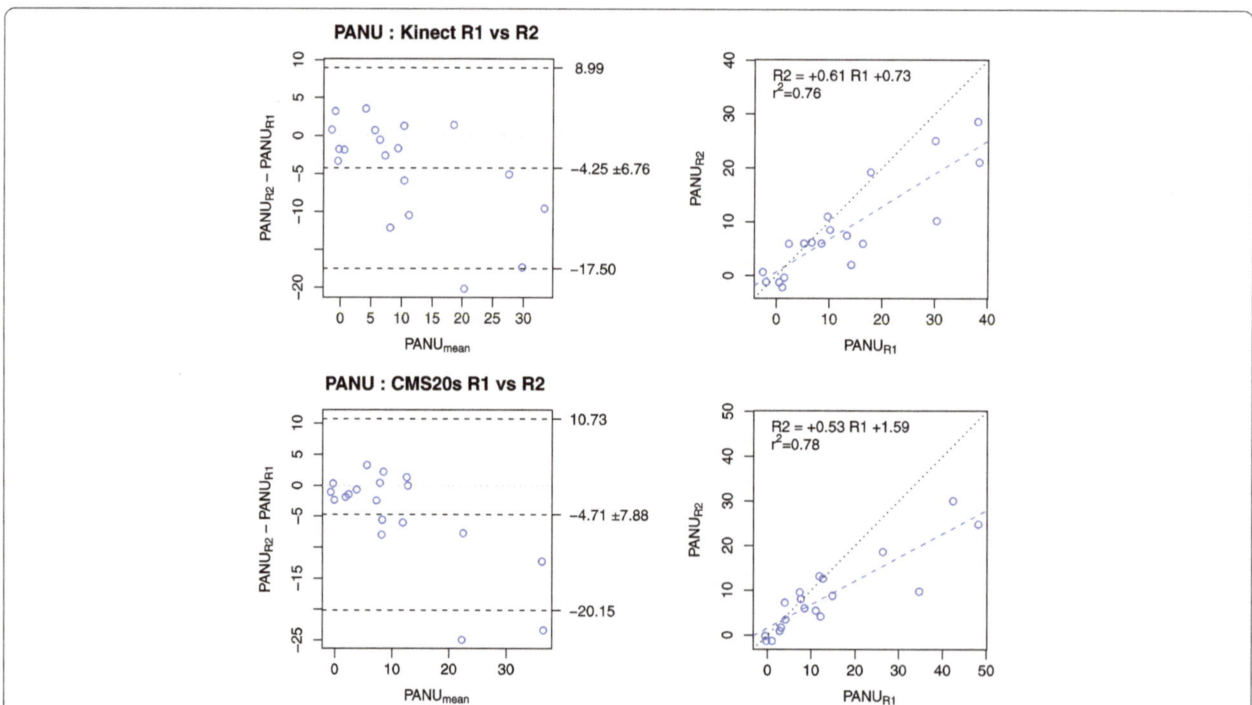


**Fig. 5** Test-retest of PANU scores with the Kinect and CMS20s. Each panel compares the PANU scores in the test (R1) and retest (R2) sessions. Panels in the first row illustrate repeatability with the Kinect. Panels in the second row illustrate repeatability with the CMS20s. For each row, the left panel represents the Bland and Altman plot and the right panel represents the linear regression plot. The constant bias in the Bland and Altman plots (− 4.25 for Kinect; − 4.71 for CMS20s) indicates that the PANU scores decrease over repetitions, which was accurately determined by the Kinect and the CMS20s

**Table 1** Descriptive statistics of the main measures

| | $MT_{[CMS20s]}$ | $NVP_{[CMS20s]}$ | $PANU_{[CMS20s]}$ | $MT_{[Kinect]}$ | $NVP_{[Kinect]}$ | $PANU_{[Kinect]}$ |
|---|---|---|---|---|---|---|
| Min | 1.20 | 2.00 | −1.27 | 0.92 | 4.00 | −2.53 |
| Max | 3.73 | 18.00 | 48.15 | 3.83 | 21.00 | 38.46 |
| Median | 1.90 | 6.00 | 7.88 | 1.93 | 9.00 | 7.11 |
| IQR | 0.74 | 4.00 | 9.56 | 0.77 | 5.00 | 14.87 |
| Mean | 2.03 | 6.51 | 10.92 | 1.97 | 9.90 | 10.55 |
| SD | 0.62 | 3.15 | 11.75 | 0.61 | 3.77 | 11.16 |

The three leftmost columns summarize the distributions of MT, NVP and PANU obtained with the CMS20s. The three rightmost columns summarize the distributions of MT, NVP and PANU obtained with the Kinect. The rows in the table indicate the range (Min and Max), the central tendency (Mean and Median) and the variability (SD: standard deviation and IQR: inter quartile range) of the corresponding distribution

greater than 50% were found with a PANU score > 6.5% (Kinect: 6/11 and CMS20s: 7/11).

### Repeatability of PANU scoring using Kinect and CMS20s

#### *Kinect*

With the Kinect, PANU scores between the two testing sessions (R1 and R2) were highly correlated (ICC = 0.76) (Fig. 5, top panels). The slope of the linear regression (0.61) indicated that PANU scores at the second session were systematically lower than at the first session. Bland and Altman plots revealed that PANU score at the second session was 4.25 ± 6.76 lower than at the first session.

#### *CMS20s*

With the CMS20s, PANU scores between the two testing sessions were highly correlated (ICC = 0.72) (Fig. 5, bottom panels). The slope of the linear regression (0.53) indicated that PANU scores at the second session were systematically lower than at the first session. Bland and Altman plots revealed that PANU score at the second session was 4.71 ± 7.88 lower than at the first session.

## Discussion

This study validated that the Kinect sensor can adequately measure proximal arm non-use (PANU score) in patients after a stroke. PANU scores determined by the Kinect system and the reference CMS20s system were highly correlated and repeatable. The components of the PANU score (PAU, ΔTrunk, ΔHand) were similarly highly correlated. However, other kinematic variables such as movement time (MT) and number of velocity peaks (NVP) obtained from the Kinect were less correlated to that obtained from the CMS20s. Thus, more precise systems than the Kinect are likely necessary to properly detail arm movement organisation. Nevertheless, the marker-less Kinect affords easier access to joint angles than the marker based CMS20s. Furthermore, the Kinect results showed that the majority of patients with lower impairment (FM-UE proximal > 28/42) exhibited proximal arm non-use (PANU score > 6.5%), as already found in our previous paper [3].

### Validation of the PANU scores obtained by Kinect

Our clinical question was whether the Kinect, which is a low cost and marker-less motion capture system, could replace the reference CMS20s system in the assessment of PANU score in patients after a stroke. An ideal model would claim that the measurements obtained by the Kinect and CMS20s would give exactly the same results. So, ideally, this would result in an ICC = 1, linear regression line with slope = 1 and intercept = 0 and coefficient of determination $r = 1$. In the same way, all the differences between paired measurements in the Bland and Altman plots would be equal to 0 (systematic bias) and limits of agreements = 0 as well. Our results indicate that PANU scores determined by the Kinect were similar to those determined by the CMS20s, yet with small discrepancies compared to the ideal expectations.

The origin of the small discrepancies in PANU scores between CMS20s and Kinect are likely twofold: due to measurement inaccuracies of the Kinect and due to the different positioning of the "markers" with the two systems. Inaccuracy in Kinect's joints is likely due to higher measurement noise (because marker-less tracking is not as stable as marker-based tracking) and also probably due to joint-tracking errors depending on the orientation of body segments relative to the Kinect sensor [33]. For example, it is likely that the Kinect can more easily lose track of the wrist when the hand and elbow are aligned with the sensor (e.g., when moving back after reaching a target).

The different positioning of the "markers" with the two systems can also account for the discrepancies, especially when assessing the trunk movement amplitude (ΔTrunk) and the reach length (ΔHand). For ΔTrunk, the Kinect joint was close to the anatomical seventh cervical vertebra while the CMS20s marker was at the body surface on the manubrium, hence with a distance of about 100 mm between the two. However, the error in ΔTrunk measured by the Kinect was about 1 mm on average. Hence, this distance of 100 mm did not impair much the assessment of trunk movements, probably because the trunk moved mainly along the Z axis and remained at a large distance from the centre of the space (i.e., the target). For ΔHand, the Kinect joint was close to the anatomical centre of the wrist joint while the CMS20s marker was at the body surface, hence about 30 mm to the anatomical centre of the wrist joint. The error in ΔHand determined by the Kinect was about 11 mm on average, which is small in comparison to the total reach length (280 mm). This might be because the "wrist" was not very well tracked by the Kinect. However, another cause of error was due to (i) the wrist joint, which is close to but not exactly at the position of the CMS20s marker and (ii) the difference generates an error that depends on upper limb orientation. The combination of these causes likely explains the small differences in ΔHand measurements between the Kinect and CMS20s.

#### Test-retest reliability of the Kinect based PANU assessment compared to CMS20s

PANU scores between the two testing sessions were highly correlated with the Kinect (ICC = 0.76) as well as with the CMS20s (ICC = 0.72). Moreover, Bland and Altman plots revealed that PANU scores at the second session were lower than at the first session (4.25 ± 6.76 for the Kinect and 4.71 ± 7.88 for the CMS20s). Therefore, the size of the error in PANU score with the Kinect (– 0.37 ± 3.40) was smaller than the size of the error between two repeated measures on the same person – 4.71 ± 7.88.

The slope of the linear regression (0.61 and 0.53) indicated that the second session PANU scores were systematically lower than the first session with both Kinect and CMS20s. The differences in PANU scores over repetitions might be due to practice effect, since patients may remember to not move their trunk in the spontaneous arm condition of the PANU assessment, as already found with a between day test-retest [3]. This could be explained by patients remembering the task requirements of minimising their trunk movements in the maximal arm use condition of the PANU assessment. However, this provides evidence that specific training reduces trunk compensation that overtime will reduce PANU to acceptable levels.

#### Application of Kinect for clinical assessment of PANU

Compared to our previous study [3], the results of this study did not show a significant correlation between FM-UE or BBT and PANU (Kinect and CMS20s). The lower number of patients used in the correlation analysis in the present ($n$ = 19) vs previous ($n$ = 45) study is a factor for not being able to get a significant relationship between PANU and FM-UE/BBT. However, the non-significant correlation is not fully unexpected, and this is because PANU measures the non-use while FM-UE/BBT measures the deficit/function. On the one hand, PANU measures the non-use as a difference: the difference between spontaneous arm use and maximum arm use. On the other hand, FM-EU/BBT measures the deficit/function as single measure: the ability to make movements on request, which is in a maximal arm use condition. Hence, it is expected that the FM-UE/BBT and the PANU are largely independent measures, because they assess two different domains of arm movement. In addition, the fact that we consistently found in the present and previous [3] study that the majority of patients with lower impairment exhibited proximal arm non-use indicates that these patients do not make full use of their arm movement ability in everyday life. Thus, we think that PANU represents novel information on arm use, which is useful to complement clinical measures of arm impairment and function.

To accomplish arm-reaching movements after damage to the central nervous system, post-stroke individuals generally use an abnormal coupling of the shoulder abduction and elbow flexion generally termed "flexion synergy" [34–36] which is due to a difficulty to isolate degrees of freedom especially in the elbow [23, 37]. Some post-stroke individuals presenting a pathological flexion synergy show a smaller range of motion of the shoulder and elbow joints which is compensated by excessive trunk movements in order to achieve the forward reaching task [6, 38]. Moreover, the amount of compensatory trunk movements classically increases with the severity of the motor deficit [23]. However, two cases of compensatory trunk movements need to be distinguished. On the one hand, if the motor deficit is very severe, trunk compensation might be mandatory to ensure the success of the arm reaching: the patient will not succeed in the same reaching task if trunk movements are restricted. Typically, those patients do not have PANU because they use the same trunk compensation in both the spontaneous and the maximum arm-use conditions. On the other hand, if the motor deficit is not too severe, trunk compensation might not be mandatory: the patient will succeed in the same reaching task whether trunk movements are restricted or not. In the latter case, the amount of trunk compensation to perform the reaching task reveals the *bad-use* of the existing synergies [4, 23]. Here, the compensatory bad-use of the trunk and elbow-shoulder during reaching is diagnosed by the PANU score. This phenomenon can be observed whatever the level of UL recovery.

With the Kinect, the PANU score is assessed with a small error of – 0.37 ± 3.40 compared to the gold standard CMS20s. We think that this small error in PANU score is not clinically important when compared to the typical range of PANU scores of Kinect and CMS20s (– 2.53 to + 48.15, with a median value of 7.58). As a consequence, our results indicate that the Kinect sensor can accurately and reliably determine the PANU score in clinical routine. This result is important because the Kinect has many advantages that are important for practical reasons in clinical routine. The main asset of the Kinect is that it is a marker-less system. Using markers presents several potential problems, including soft tissue artefacts, lost markers, training of therapist to position markers and potentially uncomfortable exposure of areas of the body such as the thorax region that could be a problem for female patients [39]. The other advantages of a Kinect-based system are: 1) easy to set up - no further physical equipment is needed, 2) safe - no additional trip hazards with wires, and 3) inexpensive - the Kinect is a widely available low cost consumer device [12]. These characteristics enable its use in a clinical routine, but also for rehabilitation following-up in conditions where expensive marker-based motion capture methods are difficult to use (e.g., at the patient's home).

Although the use of the Kinect sensor for virtual rehabilitation of the UL is well demonstrated [12, 40–45], the use of the Kinect for clinical assessment is less well developed. Nevertheless, several studies have already shown that the Kinect sensor can be used for clinical assessments of the UL [13, 46–55], and trunk kinematics [18, 56]. In the present study, the reliability of a Kinect-based assessment of PANU score, which allows differentiating reaching strategies following a stroke, provides a strong case for the routine clinical application of the Kinect for this purpose. Our work reveals that a Kinect based assessment of PANU score is fast, easy to undertake and accurate, hence potentially clinically useful in classifying targeted rehabilitation and follow-up monitoring of UL recovery of post-stroke individuals.

In the present study, in addition to accurately and reliably measuring PANU with the Kinect, we were able to estimate shoulder-elbow angles and derive nuSF and nuEE, with the same setup. In fact, we could measure these angles with CMS20s but this would have required to add more markers on the post-stroke individual. Consequently, we do not have a referent device, which hinders the possibility of a precise assessment of the reliability of the Kinect in the measure of angles (that would be analysed in future work). Our findings with the Kinect suggested that proximal arm non-use (PANU) was mainly due to the inability to extend the elbow joint (nuEE). Yet, Massie et al. found that elbow extension predicts motor impairment and performance after stroke [57]. In our study, the PANU score seems to be more associated with the non-use of elbow extension (nuEE). This additional information can provide useful global UL functional measure to help therapists in focusing the therapy.

To our knowledge, the only clinical scale measuring excessive trunk movements during reach-to-grasp tasks is the Reaching Performance Scale [58]. Unlike the objective PANU assessment, the Reaching Performance Scale focuses on direct assessor's observation of compensatory movement patterns performed during a reaching task in post-stroke individuals including trunk compensation [58]. Although the Reaching Performance Scale evaluates reach-to-grasp performance (motor impairment) and trunk compensation, it does not provide an indication of arm non-use [2, 59–61]. Recently, Valdés et al. measured changes in anterior trunk displacement using the Kinect in a bimanual reaching task [18]. The PANU method also evaluates trunk displacement using the ΔTrunk variable. Yet, because PANU compares ΔTrunk when reaching in a spontaneous condition to ΔTrunk in a maximal proximal arm use condition, the PANU assessment method provides new information on maladaptive trunk compensation (i.e., compensations that are not mandatory to succeed at the task).

## Limitations

This study faces several limitations. First, the sample size used in the clinical evaluation was small. Second, the comparison of the Kinect system was done on only one reference (CMS20s), but we could also compare to others systems such as optoelectronic devices. Moreover, the generalization of these results has to be confirmed since the participants in our sample do not reflect all types of deficits after a stroke. Finally, the chosen task is a horizontal forward reaching task and other tasks may yield different results.

## Future work

Motion capture systems such as the Kinect are mostly conducted in upper limb stroke rehabilitation to increase the motivation during training and may assist improvement on one or more International Classification of Functioning, Disability and Health (ICF) levels.

Future work could focus on developing a video game that reduces the PANU. We believe that PANU score is important for rehabilitation programs that aim to reduce the arm non-use in the paretic arm itself and to promote previous movement patterns. Future research might test the hypothesis that post-stroke individuals with a high PANU score might improve with rehabilitation focused on the use of their paretic shoulder flexion and elbow extension.

Therefore, the PANU score could be implemented in virtual/robotic rehabilitation to monitor arm-use over time. Ideally, the new game would be able to measure the proximal arm non-use (assessment part). The game would also be able to meaningfully use arm movements and PANU score as inputs and use incentives and disincentives to reduce PANU (treatment part). The selected training programmes would incorporate different forms of forced-use feedback of the paretic arm provided by the virtual or robotic devices [18] and games to promote the use of the paretic arm [62].

In addition, motion-based games show promise for motivating patients to perform stroke rehabilitation exercises at home by themselves [12]. In the near future and with further validation, Kinect based video games may prove useful as a home-based assessment tool (e.g., monitoring recovery of proximal arm-use [3], monitoring changes in motor synergy patterns over time [63]) and as home-based self-rehabilitation tool [64].

## Conclusion

Our goal was to develop a valid and reliable methodology, easy to administer, time-efficient, and cost-effective to capture PANU in individuals post-stroke. The present study showed that the low cost and marker-less Kinect based motion capture system could accomplish these goals adequately. PANU assessment using the Kinect motion capture sensor could be recommended for rehabilitation (both in hospitals and private practices), such that we envision that PANU could help therapists to 1) classify patients as a function of their unused motor reserve, so to guide patients towards specific arm-use rehabilitation, 2) monitor recovery over time, to assess the effectiveness of interventions.

**Acknowledgements**

This research was funded by the French Ministère des Affaires Sociales et de la Santé (KINE-INDEX, PHRIP: 13-0433). The funders had no role in study design, data collection and analysis, decision to publish, or preparation of the manuscript.

**Funding**

French Ministère des Affaires Sociales et de la Santé (KINE-INDEX, PHRIP: 13–0433).

**Authors' contributions**

All authors contributed to the design of the experiment. K.B. and D.M. performed the experiments. The manuscript was written by K.B. and D.M. with input from all authors. All authors read and approved the final manuscript.

**Competing interests**

The authors declare that they have no competing interests.

**Author details**

[1]Euromov, University of Montpellier, Montpellier, France. [2]Physical Medicine and Rehabilitation, Montpellier University Hospital, Montpellier, France. [3]Silverline Research, Brisbane, Australia. [4]Physical Medicine and Rehabilitation, Nîmes University Hospital, Le Grau du Roi, France. [5]Federative Institute for Research on Handicap, Paris, France.

**References**

1. van Kordelaar J, van Wegen EEH, Nijland RHM, de Groot JH, Meskers CGM, Harlaar J, et al. Assessing longitudinal change in coordination of the paretic upper limb using on-site 3-dimensional kinematic measurements. Phys Ther. 2012;92:142–51.
2. Taub E, Uswatte G, Mark VW, Morris DMM. The learned nonuse phenomenon: implications for rehabilitation. Eura Medicophys. 2006;42:241–56.
3. Bakhti KKA, Mottet D, Schweighofer N, Froger J, Laffont I. Proximal arm non-use when reaching after a stroke. Neurosci Lett. 2017;657:91–6.
4. Kleim JA, Jones TA. Principles of experience-dependent neural plasticity: implications for rehabilitation after brain damage. J Speech Lang Hear Res. 2008;51:S225–39.
5. Jones TA. Motor compensation and its effects on neural reorganization after stroke. Nat Rev Neurosci. 2017;18:267–80.
6. Wee SK, Hughes A-M, Warner M, Burridge JH. Trunk restraint to promote upper extremity recovery in stroke patients: a systematic review and meta-analysis. Neurorehabil Neural Repair. 2014;28:660–77.
7. Hatem SM, Saussez G, Della Faille M, Prist V, Zhang X, Dispa D, et al. Rehabilitation of motor function after stroke: a multiple systematic review focused on techniques to stimulate upper extremity recovery. Front Hum Neurosci. 2016;10:442.
8. Spasticity LS, Recovery M. Neural plasticity after stroke. Front Neurol. 2017;8:120.
9. Bonnechère B, Jansen B, Salvia P, Bouzahouene H, Omelina L, Moiseev F, et al. Validity and reliability of the Kinect within functional assessment activities: comparison with standard stereophotogrammetry. Gait Posture. 2014;39:593–8.
10. Butkiewicz T. Low-cost coastal mapping using Kinect v2 time-of-flight cameras. IEEE. 2014:1–9.
11. Eltoukhy M, Oh J, Kuenze C, Signorile J. Improved kinect-based spatiotemporal and kinematic treadmill gait assessment. Gait Posture. 2017; 51:77–83.
12. Knippenberg E, Verbrugghe J, Lamers I, Palmaers S, Timmermans A, Spooren A. Markerless motion capture systems as training device in neurological rehabilitation: a systematic review of their use, application. target population and efficacy J Neuroeng Rehabil. 2017;14:61.
13. Kurillo G, Han JJ, Obdržálek S, Yan P, Abresch RT, Nicorici A, et al. Upper extremity reachable workspace evaluation with Kinect. Stud Health Technol Inform. 2013;184:247–53.
14. Pagliari D, Pinto L. Calibration of Kinect for Xbox one and comparison between the two generations of Microsoft sensors. Sensors (Basel). 2015;15:27569–89.
15. Galna B, Barry G, Jackson D, Mhiripiri D, Olivier P, Rochester L. Accuracy of the Microsoft Kinect sensor for measuring movement in people with Parkinson's disease. Gait Posture. 2014;39:1062–8.
16. Kuster RP, Heinlein B, Bauer CM, Graf ES. Accuracy of KinectOne to quantify kinematics of the upper body. Gait & Posture. 2016;47:80–5.
17. Zulkarnain RF, Kim G-Y, Adikrishna A, Hong HP, Kim YJ, Jeon I-H. Digital data acquisition of shoulder range of motion and arm motion smoothness using Kinect v2. J Shoulder Elb Surg. 2017;26:895–901.
18. Valdés BA, Schneider AN, Van der Loos HFM. Reducing trunk compensation in stroke survivors: a randomized crossover trial comparing visual and force feedback modalities. Arch Phys Med Rehabil 2017;98:1932–1940.
19. Michaelsen SM, Levin MF. Short-term effects of practice with trunk restraint on reaching movements in patients with chronic stroke: a controlled trial. Stroke. 2004;35:1914–9.
20. Fugl-Meyer AR, Jääskö L, Leyman I, Olsson S, Steglind S. The post-stroke hemiplegic patient. 1. A method for evaluation of physical performance. Scand J Rehabil Med. 1975;7:13–31.
21. Mathiowetz V, Volland G, Kashman N, Weber K. Adult norms for the box and block test of manual dexterity. Am J Occup Ther. 1985;39:386–91.
22. Levin MF, Michaelsen SM, Cirstea CM, Roby-Brami A. Use of the trunk for reaching targets placed within and beyond the reach in adult hemiparesis. Exp Brain Res. 2002;143:171–80.
23. Levin MF, Liebermann DG, Parmet Y, Berman S. Compensatory versus noncompensatory shoulder movements used for reaching in stroke. Neurorehabil Neural Repair. 2016;30:635–46.
24. Subramanian SK, Lourenço CB, Chilingaryan G, Sveistrup H, Levin MF. Arm motor recovery using a virtual reality intervention in chronic stroke: randomized control trial. Neurorehabil Neural Repair. 2013;27:13–23.

25. Zhang Z. Microsoft Kinect sensor and its effect. IEEE Multimedia. 2012;19:4–10.
26. Corti A, Giancola S, Mainetti G. Sala R. a metrological characterization of the Kinect V2 time-of-flight camera. Robot Auton Syst. 2016;75:584–94.
27. Ozturk A, Tartar A, Ersoz Huseyinsinoglu B. Ertas AH. A clinically feasible kinematic assessment method of upper extremity motor function impairment after stroke. Measurement. 2016;80:207–16.
28. Obdrzálek S, Kurillo G, Ofli F, Bajcsy R, Seto E, Jimison H, et al. Accuracy and robustness of Kinect pose estimation in the context of coaching of elderly population. Conf Proc IEEE Eng Med Biol Soc. 2012;2012:1188–93.
29. Mottet D, Guiard Y, Ferrand T, Bootsma RJ. Two-handed performance of a rhythmical fitts task by individuals and dyads. J Exp Psychol Hum Percept Perform. 2001;27:1275–86.
30. van Dokkum L, Hauret I, Mottet D, Froger J, Métrot J, Laffont I. The contribution of kinematics in the assessment of upper limb motor recovery early after stroke. Neurorehabil Neural Repair. 2014;28:4–12.
31. Shrout PE, Fleiss JL. Intraclass correlations: uses in assessing rater reliability. Psychol Bull. 1979;86:420–8.
32. Bland JM, Altman DG. Statistical methods for assessing agreement between two methods of clinical measurement. Lancet. 1986;1:307–10.
33. Gonzalez A, Hayashibe M, Fraisse P. Estimation of the center of mass with Kinect and Wii balance board. IEEE. 2012:1023–8.
34. Bourbonnais D, Vanden Noven S, Carey KM, Rymer WZ. Abnormal spatial patterns of elbow muscle activation in hemiparetic human subjects. Brain. 1989;112(Pt 1):85–102.
35. Cirstea MC, Levin MF. Compensatory strategies for reaching in stroke. Brain. 2000;123(Pt 5):940–53.
36. Twitchell TE. The restoration of motor function following hemiplegia in man. Brain. 1951;74:443–80.
37. Levin MF. Interjoint coordination during pointing movements is disrupted in spastic hemiparesis. Brain. 1996;119(Pt 1):281–93.
38. Roby-Brami A, Feydy A, Combeaud M, Biryukova EV, Bussel B, Levin MF. Motor compensation and recovery for reaching in stroke patients. Acta Neurol Scand. 2003;107:369–81.
39. Mündermann L, Corazza S, Andriacchi TP. The evolution of methods for the capture of human movement leading to markerless motion capture for biomechanical applications. J Neuroeng Rehabil. 2006;3:6.
40. Da Gama A, Fallavollita P, Teichrieb V, Navab N. Motor rehabilitation using Kinect: a systematic review. Games for Health Journal. 2015;4:123–35.
41. Pastor I, Hayes HA. Bamberg SJM. A feasibility study of an upper limb rehabilitation system using Kinect and computer games. Conf Proc IEEE Eng Med Biol Soc. 2012;2012:1286–9.
42. Scano A, Caimmi M, Malosio M, Tosatti LM. Using Kinect for upper-limb functional evaluation in home rehabilitation: a comparison with a 3D stereoscopic passive marker system. IEEE. 2014:561–6.
43. Scano A, Caimmi M, Chiavenna A, Malosio M, Tosatti LM. Kinect one-based biomechanical assessment of upper-limb performance compared to clinical scales in post-stroke patients. Conf Proc IEEE Eng Med Biol Soc. 2015;2015: 5720–3.
44. Yates M, Kelemen A, Sik Lanyi C. Virtual reality gaming in the rehabilitation of the upper extremities post-stroke. Brain Inj. 2016;30:855–63.
45. Zheng H, Black ND, Harris ND. Position-sensing technologies for movement analysis in stroke rehabilitation. Med Biol Eng Comput. 2005;43:413–20.
46. Gregorij K, Alic C, Ruzena B. J HJ. Evaluation of upper extremity reachable workspace using Kinect camera. Technol Health Care. 2013:641–56.
47. Han JJ, de Bie E, Nicorici A, Abresch RT, Anthonisen C, Bajcsy R, et al. Reachable workspace and performance of upper limb (PUL) in duchenne muscular dystrophy. Muscle Nerve. 2016;53:545–54.
48. Huber ME, Seitz AL, Leeser M, Sternad D. Validity and reliability of Kinect skeleton for measuring shoulder joint angles: a feasibility study. Physiotherapy. 2015;101:389–93.
49. Kim W-S, Cho S, Baek D, Bang H, Paik N-J. Upper extremity functional evaluation by Fugl-Meyer assessment scoring using depth-sensing camera in hemiplegic stroke patients. PLoS One. 2016;11:e0158640.
50. Lee SH, Yoon C, Chung SG, Kim HC, Kwak Y, Park H-W, et al. Measurement of shoulder range of motion in patients with adhesive capsulitis using a Kinect. PLoS One. 2015;10:e0129398.
51. Matsen FA, Lauder A, Rector K, Keeling P, Cherones AL. Measurement of active shoulder motion using the Kinect, a commercially available infrared position detection system. J Shoulder Elb Surg. 2016;25:216–23.
52. Metcalf CD, Robinson R, Malpass AJ, Bogle TP, Dell TA, Harris C, et al. Markerless motion capture and measurement of hand kinematics: validation and application to home-based upper limb rehabilitation. IEEE Trans Biomed Eng. 2013;60:2184–92.
53. Rammer JR, Krzak JJ, Riedel SA, Harris GF. Evaluation of upper extremity movement characteristics during standardized pediatric functional assessment with a Kinect®-based markerless motion analysis system. Conf Proc IEEE Eng Med Biol Soc. 2014;2014:2525–8.
54. Seo NJ, Fathi MF, Hur P, Crocher V. Modifying Kinect placement to improve upper limb joint angle measurement accuracy. J Hand Ther. 2016;29:465–73.
55. Sevick M, Eklund E, Mensch A, Foreman M, Standeven J, Engsberg J. Using free internet videogames in upper extremity motor training for children with cerebral palsy. Behavioral Sciences. 2016;6:10.
56. Macpherson TW, Taylor J, McBain T, Weston M, Spears IR. Real-time measurement of pelvis and trunk kinematics during treadmill locomotion using a low-cost depth-sensing camera: a concurrent validity study. J Biomech. 2016;49:474–8.
57. Massie CL, Fritz S, Malcolm MP. Elbow extension predicts motor impairment and performance after stroke. Rehabilitation Research and Practice. 2011; 2011:1–7.
58. Levin MF, Desrosiers J, Beauchemin D, Bergeron N, Rochette A. Development and validation of a scale for rating motor compensations used for reaching in patients with hemiparesis: the reaching performance scale. Phys Ther. 2004;84:8–22.
59. Andrews K, Stewart J. Stroke recovery: he can but does he? Rheumatol Rehabil. 1979;18:43–8.
60. Han CE, Kim S, Chen S, Lai Y-H, Lee J-Y, Osu R, et al. Quantifying arm nonuse in individuals poststroke. Neurorehabil Neural Repair. 2013;27:439–47.
61. Sterr A, Freivogel S, Schmalohr D. Neurobehavioral aspects of recovery: assessment of the learned nonuse phenomenon in hemiparetic adolescents. Arch Phys Med Rehabil. 2002;83:1726–31.
62. Brokaw EB, Eckel E, Brewer BR. Usability evaluation of a kinematics focused Kinect therapy program for individuals with stroke. Technol Health Care. 2015:143–51.
63. Dipietro L, Krebs HI, Fasoli SE, Volpe BT, Stein J, Bever C, et al. Changing motor synergies in chronic stroke. J Neurophysiol. 2007;98:757–68.
64. Johnson L, Bird M-L, Muthalib M, Teo W-P. Innovative STRoke interactive virtual thErapy (STRIVE) online platform for community-dwelling stroke survivors: a randomised controlled trial protocol. BMJ Open. 2018;8:e018388.

18

# Effects of a powered ankle-foot orthosis on perturbed standing balance

Amber R. Emmens*, Edwin H. F. van Asseldonk and Herman van der Kooij

**Abstract**

**Background:** Lower extremity exoskeletons are mainly used to provide stepping support, while balancing is left to the user. Designing balance controllers is one of the biggest challenges in the development of exoskeletons. The goal of this study was to design and evaluate a balance controller for a powered ankle-foot orthosis and assess its effect on the standing balance of healthy subjects.

**Methods:** We designed and implemented a balance controller based on the subject's body sway. This controller was compared to a simple virtual-ankle stiffness and a zero impedance controller. Ten healthy subjects wearing a powered ankle-foot orthosis had to maintain standing balance without stepping while receiving anteroposterior pushes. Center of mass kinematics, ankle torques and muscle activity of the lower legs were analyzed to assess the balance performance of the user and exoskeleton.

**Results:** The different controllers did not significantly affect the center of mass responses. However, the body sway based controller resulted in a decrease of 29% in the biological ankle torque compared to the zero impedance controller and a decrease of 32% compared to the virtual-ankle stiffness. Furthermore, the soleus muscle activity of the left and right leg decreased on average with 8%, while the tibialis anterior muscle activity increased with 47% compared to zero impedance.

**Conclusion:** The body sway based controller generated human-like torque profiles, whereas the virtual-ankle stiffness did not. As a result, the powered ankle-foot orthosis with the body sway based controller was effective in assisting the healthy subjects in maintaining balance, although the improvements were not seen in the body sway response, but in the subjects' decreased biological ankle torques to counteract the perturbations. This decrease was a combined effect of decreased soleus muscle activity and increased tibialis anterior muscle activity.

**Keywords:** Standing balance, Ankle-foot orthosis, Balance control, Exoskeleton

## Background

Exoskeletons have been successfully used to provide assistance during walking. For people with lower extremity impairments the focus is generally on rehabilitation [1–5] and gait restoration [6–8], while exoskeletons for healthy people aim to augment strength and endurance [9–13]. A particular goal of exoskeleton studies is to reduce the effort of walking. A key challenge is to overcome the negative effects of the added weight of the exoskeleton. It was shown that when using passive or powered ankle-foot orthoses, it is possible to reduce the muscle activity of ankle plantar flexors [14, 15].

While healthy users are able to walk in wearable exoskeletons without assistive devices, for people suffering from neurological disorders such as a spinal cord injury this is difficult, if not impossible, due to affected balance. Except for the wearable exoskeleton "REX" (REX Bionics Ltd, New Zealand) [16], commercially available devices are used in combination with crutches for maintaining balance [3, 7, 17]. This implies that the users are supporting themselves and the exoskeleton in staying upright, although it would be desirable if the exoskeleton supported the user's balance. Still, only a few studies have addressed the topic of exoskeleton balance control during walking [6, 18] or standing [19]. These studies mainly focused on controller implementation and not on its effect

*Correspondence: a.r.emmens@utwente.nl
Department of Biomechanical Engineering, University of Twente, Drienerlolaan 5, 7522 Enschede, NB, the Netherlands

on the user. In maintaining balance, the user and the exoskeleton should work in harmony and therefore it is essential to incorporate knowledge about human balance control in the controller.

Several models have been suggested to describe the dynamics of human standing balance control, based on full-state feedback [20–24] or center of mass (CoM) motion [25–27]. For full-state feedback, the human is modeled as a multi-link system and the state vector generally consists of the angles and velocities of all the joints. How much each state contributes to a certain joint torque can be acquired by fitting experimental data to a model simulation. Through forward simulations it was shown that full-state feedback models are able to predict ankle and hip torques well in a linearized double-inverted pendulum model [22]. For CoM feedback, the whole body CoM of a human is generally assumed to sway like a single inverted pendulum. This model has proven to be sufficient for describing the balance strategy known as the "ankle strategy," which is dominant in quiet stance and other simple balancing tasks [28]. Furthermore, previous studies have shown that CoM kinematics can be used to reproduce muscle activity patterns during both ankle and hip postural responses [25, 26].

Our goal is to use an exoskeleton for balancing during standing and walking and to assess the effect of the exoskeleton on the user. As a first step we tested balance controllers on a powered Ankle-Foot Orthosis (pAFO) and investigated standing balance in the sagittal plane. Since task-space goals, such as controlling CoM excursion, play an important role in standing balance, the proposed balance controller was based on CoM kinematics. The assistive torque that the pAFO had to deliver was calculated from the CoM kinematics according to the single inverted pendulum model. We compared this control law to a simple stiffness applied around the ankle, which is equivalent to controlling a single inverted pendulum model in joint-space. Furthermore, ankle stiffness corresponded to the largest feedback gain of the full-state feedback law for the ankle torque [21]. As such, the stiffness controller had the potential to positively affect balance performance, without introducing task-space complexity. The goal of this research is twofold: 1. to assess the effects of the controlled pAFO on the balance performance of the user; and 2. to assess the effects of the controlled pAFO on human effort, based on ankle joint kinetics and muscle activity of the ankle plantar and dorsiflexors. We hypothesized that, with the application of an assistive torque that is similar to the ankle torque humans generate when being perturbed, balance performance would improve, while the human contribution to the balance response would decrease.

## Methods

In this study, the assistive effect of the pAFO was tested by means of experiments in which subjects had to maintain balance, without stepping, while receiving external perturbations. Ethical approval for the experiments was given by the Human Research Ethics Committee of Delft University of Technology, the Netherlands.

### Subjects

Ten healthy subjects (three female, mass 71 $\pm$6 kg, height 1.81 $\pm$0.07 m, mean $\pm$ SD) participated in the experiment after giving written informed consent. Based on the size and stiffness characteristics of the pAFO, participants were selected who were taller than 1.65 m, weighed less than 84 kg, and had an EU shoe size between 36 and 44.

### Experimental setup

Figure 1 shows a schematic overview of the experimental setup.

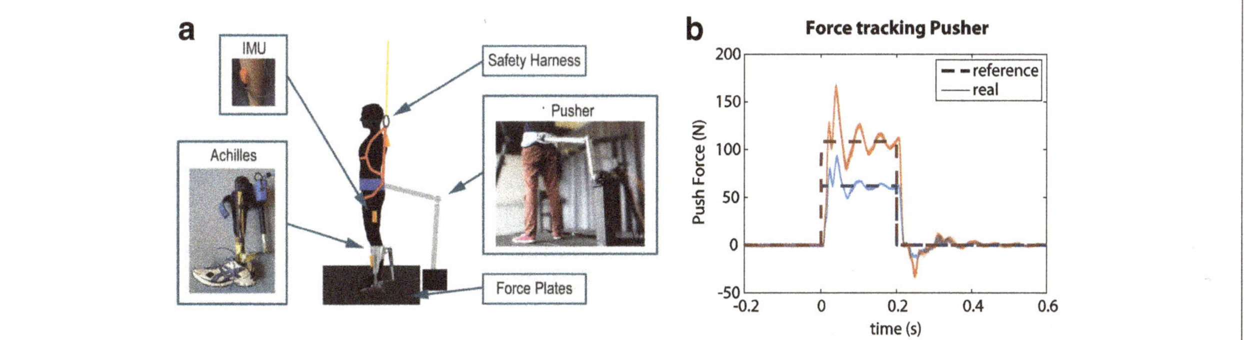


**Fig. 1** Experimental setup and perturbation profile. **a** Overview of the setup. In addition to the shown hardware, EMG sensors were placed on the lower legs. **b** The reference perturbation force profile of the Pusher and all generated perturbation forces (real) for a representative subject. There is little variation in the generated perturbation forces

***The Achilles Ankle-Foot Orthosis***

The Achilles [29] is a pAFO that can provide dorsi/plantar flexion torques to the ankle, while inversion and eversion are constrained. As shown in Fig. 2, the complete device consists of two orthoses (1.5 kg each) and a backpack. Each orthosis is actuated by a motor and ball-screw spindle. The spindle is connected to the rotational joint of the orthosis by a leaf spring, such that a translation of the spindle results in a spring deflection and a joint torque. To measure the motor stroke and the joint rotation, the device is equipped with an incremental encoder on the motor axis and an absolute encoder at the rotational joint. Since the spring stiffness is known (410 Nm/rad) and the spring deflection depends on the motor stroke and the joint rotation, the torque that is delivered by the Achilles can be estimated [29]. In the Zero Impedance (ZI) condition the spring deflection, and therefore the joint torque, is minimized. There is no compensation for the inertia of the device, but the inertia effects are small.

The Achilles backpack contains a PC (EtherCAT Master) that communicates with the orthoses and other hardware modules (EtherCAT slaves). EtherLab (IgH, Essen, Germany) is used to compile MATLAB/Simulink (2014b, The Mathworks, Natick, MA, USA) models into executables that can run on the real-time Linux EtherCAT master. During the experiments subjects did not wear the backpack, but it was hung behind them on parallel bars.

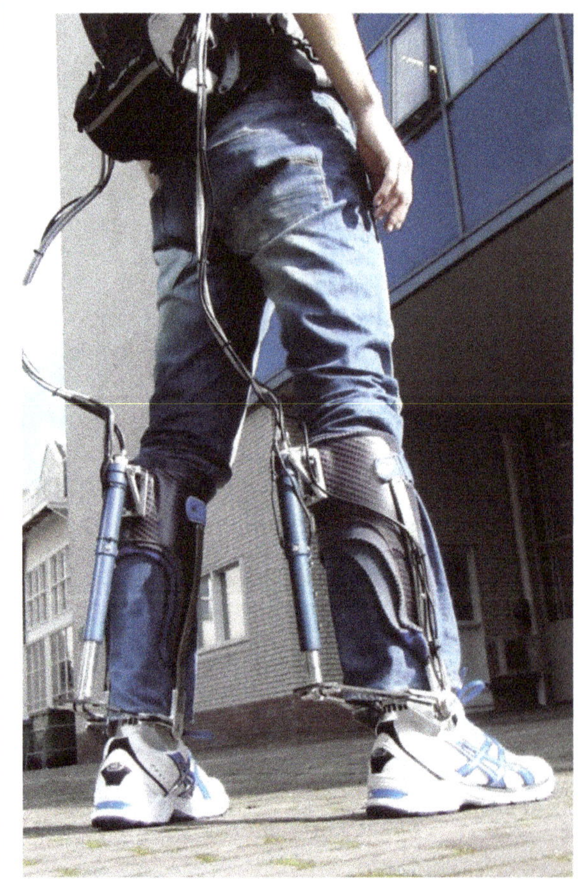

**Fig. 2** The Achilles ankle-foot orthosis

***Perturbation device***

An actuated perturbation device, referred to as "Pusher," was used to provide forward perturbations at the pelvis. It consists of a motor (SMH60, Moog, Nieuw-Vennep, The Netherlands) located at the rear of the treadmill, connected to a horizontal push rod (0.8 m) through a lever arm (0.3 m), as shown on the right of Fig. 1a. At the end of the push rod a modified universal hip abduction brace (Distrac Wellcare, Hoegaarden, Belgium) was attached, which was tightly worn by the subjects. A load cell (model QLA131, FUTEK, Los Angeles, CA, USA) was mounted on the lever arm for torque sensing [30]. The rod was approximately horizontal, such that a motor torque would result in forward pushing or backward pulling of the subject. The motor was admittance controlled over Ethernet (User Datagram Protocol) at 1000 Hz, using xPC-target (The Mathworks, Natick, MA, USA).

***Measurement equipment***

To measure the orientation of the shank, thigh and trunk of the subject, three Xsens (Xsens Technologies B.V., Enschede, the Netherlands) MTx Inertial Measurement Units (IMUs) with integrated magnetometers were used. Each IMU was calibrated by defining its axis of rotation to correspond to the mediolateral axis of the subject. This calibration was done by letting each subject make a squat movement at the start of the first trial. The signals of the IMU's were made available real-time on the EtherCAT Master through a RS232-EtherCAT converter box.

Besides motion data, ground reaction forces and moments were collected using a custom made split-belt instrumented treadmill (Y- Mill, Motek, Amsterdam, The Netherlands) that has force plates beneath each belt. Furthermore, the Delsys Trigno Lab Wireless EMG System (Delsys Inc., Natick, USA) was used to record EMG signals on the gastrocnemius medialis (GM), soleus (Sol) and tibialis anterior (TA). The EMG sensors were equipped with a fourth order Butterworth filter that rejected frequencies above 450 Hz.

**Control strategies**

***Body sway controller***

The body sway controller, $PD_{com}$, was a proportional-derivative controller that uses the CoM motion as an input. The control law for the $PD_{com}$ is as follows:

$$\tau_{ach} = K_P\left(\theta_{sway_d} - \theta_{sway}\right) + K_D\left(\dot{\theta}_{sway_d} - \dot{\theta}_{sway}\right) \quad (1)$$

where $\tau_{ach}$ is the desired support torque that the Achilles should deliver, $K_P$ and $K_D$ are proportional and derivative gains respectively, $\theta_{sway}$ is the body sway angle that is dependent on the CoM location, and subscript $d$ indicates a desired value. A complete list of symbols is provided in Table 1.

The body sway angle was estimated using a human model with four rigid segments: a foot, shank, thigh and head-arms-trunk segment (Fig. 3). The segments' lengths, inertias, masses and CoM were estimated using the method described by Winter (2009) [31]. The orientation of the segments was obtained using the IMUs and expressed with respect to the baseline pose. For this baseline pose, subjects were requested to stand comfortably upright. Then all segment angles and the body sway angle were set to zero.

The desired body sway $\theta_{sway_d}$ was initially set to zero, matching the zero value of the body sway angle when a subject was in the baseline pose. During the experiments the desired body sway angle was adapted to match the body sway angle of the steady pose that subjects maintained between perturbations, allowing for natural variation in forward lean. This was done by updating the desired body sway when the subject's body sway velocity remained close to zero for over 3 s. In a pilot test this duration had shown to be long enough to be sure that the subject had adopted a new steady pose. The updated desired body sway angle was then set to the average of the body sway angle during those 3 s. To prevent instantaneous changes in the desired body sway angle, the signal was passed through a second order low-pass filter with a cut-off frequency of 0.5 Hz Fig. 4 shows an example of how the desired body sway was adapted to a new steady pose.

To obtain the body sway velocity the body sway was differentiated and low-pass filtered using a second order Butterworth filter with a cut-off frequency of 10 Hz. The desired body sway velocity was set to zero.

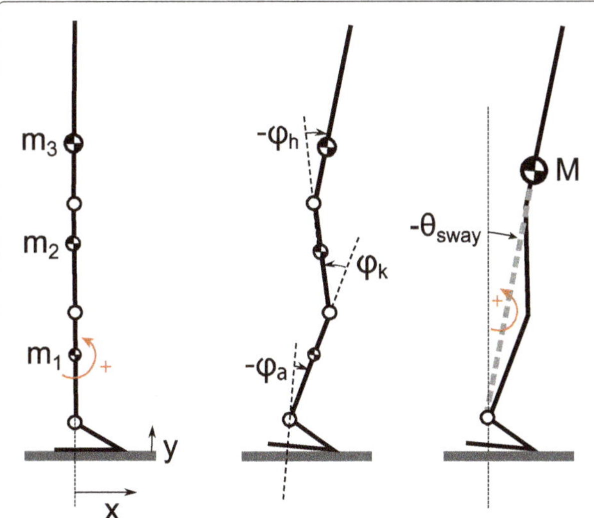


**Fig. 3** Computation of the body sway angle $\theta_{sway}$ based on a rigid link model. The variables $\phi$ are joint angles, $m_n$ are segment masses and $M$ is the total body mass located at the CoM. Left: model in baseline pose. The model is placed in a right-handed coordinate system. Segment rotations and moments are positive in counter-clockwise direction. Middle: the joint angles are defined as the angle of segment $n$ minus the angle of segment $n-1$. Segment angles are zero in the baseline pose. Right: Based on segment angles and masses the whole body CoM can be computed. Similar to the segment angles, the body sway angle is positive in counter-clockwise direction

The control gains were normalized to the subject mass $M$ and CoM height $l_{com}$:

$$K_P = Mgl_{com}$$

$$K_D = 0.3\sqrt{Ml_{com}^2 K_P}$$

where $g$ is the gravitational acceleration. The gain $K_P$ was chosen such that the gravitational stiffness was compensated. As a result the subject was able to stabilize the system with a minimal amount of effort. The subject was assumed to provide the feed-forward torque to maintain the desired body sway angle himself. Gain $K_D$ consisted of a normalization part $\sqrt{Ml_{com}^2 K_P}$ multiplied by a factor 0.3. This multiplication factor was tuned empirically to maximize damping without causing unstable oscillations.

**Table 1** List of symbols

| Symbol | Meaning |
|---|---|
| $F_{Cx}$ | Sum of the horizontal ground reaction forces |
| $F_{Cy}$ | Sum of the vertical ground reaction forces |
| $g$ | Gravitational acceleration |
| $h_A$ | Ankle height w.r.t. the ground |
| $K$ | Proportional gain of the virtual-ankle stiffness controller |
| $K_D$ | Derivative gain of the body sway controller |
| $K_P$ | Proportional gain of the body sway controller |
| $l_{com}$ | Height of the center of mass in upright stance w.r.t. the ankle |
| $M$ | Subject mass |
| $p_x$ | Center of pressure in anteroposterior direction w.r.t. the ankle |
| $\theta_{sway}$ | Body sway angle |
| $\tau_{ach}$ | Desired support torque |
| $\tau_{tot}$ | Total torque applied to the ankle |
| $\phi_A$ | Ankle joint angle |

### *Virtual-ankle stiffness controller*

The $P_{ankle}$ is a proportional controller that takes the ankle angle $\phi_A$ as an input:

$$\tau_{ach} = K(\phi_{A_d} - \phi_A) \quad (2)$$

where $\tau_{ach}$ is the output torque from the Achilles and $K$ the proportional gain. The controller is similar to a simple ankle stiffness that tries to keep the ankle angle at its desired value. The desired ankle angle was initialized

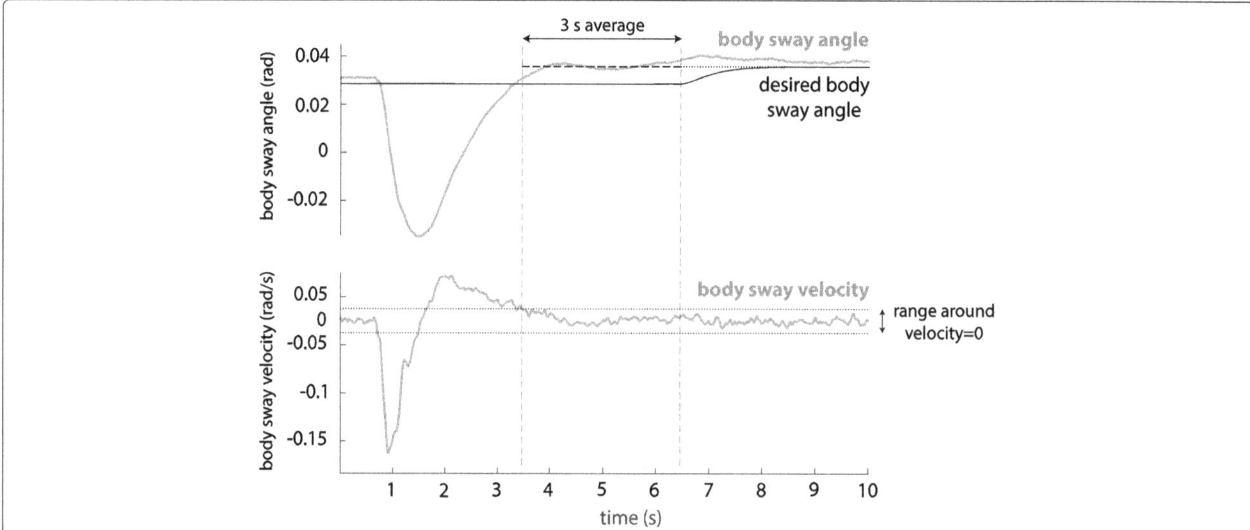


**Fig. 4** Adaptation of the desired body sway angle to a new steady pose. Top: body sway angle and desired body sway angle, bottom: scaled body sway velocity. The vertical dashed lines indicate the time frame in which the body sway velocity is approximately zero and the average body sway angle is computed. The desired body sway angle moves slowly to this average value

as zero, corresponding to the ankle angle in the baseline pose. Damping was provided by the inherent properties of the ankle of the subject wearing the Achilles.

The control gain $K$ was chosen to be:

$$K = Mgl_{com}$$

which is the same as the gain $K_P$ of the body sway controller. Therefore, if subjects adjusted only their ankle joint angle, while keeping their foot flat on the floor, the stiffness provided by the virtual-ankle stiffness controller was equal to the stiffness provided by the body sway controller.

### Protocol

Each subject was tested under three control conditions: ZI, $PD_{com}$ and $P_{ankle}$ in a total of 10 trials. The order of the $P_{ankle}$ controller (three trials) and the $PD_{com}$ controller (three trials) was reversed after each subject. To prevent familiarization with the first controller influencing the second, a ZI trial was done in between. Furthermore, the $PD_{com}$ and $P_{ankle}$ controlled trials were preceded and succeeded by ZI trials to let subjects get used to the perturbation, collect baseline data and to check for fatiguing effects. The trial order for each subject is shown in Table 2.

Before the first $P_{ankle}$ and $PD_{com}$ trial, subjects had one minute of practice time to feel the effect of the controller.

Before the experiments, subjects were instructed to try their best to maintain balance without stepping. They were instructed to cross their arms in front of their chests to prevent arm movement while they balanced. Furthermore, subjects were asked to stand still between perturbations with their weight evenly distributed on both feet and to return to their initial pose after a perturbation was applied.

During the experiments, the Pusher provided forward perturbations at the pelvis with two different magnitudes. The reference force was a rectangular pulse with a duration of 0.2 s and a magnitude of $0.08\,Mg$ (N) (small perturbation) or $0.14\,Mg$ (N) (large perturbation), where $M$ is the mass of the subject and $g$ the gravitational acceleration. Although the Pusher motors could not track the desired reference force perfectly, the provided pushes were consistent (Fig. 1b). In each trial subjects received 15 perturbations of each magnitude in a random order. The time between perturbations varied between 8 and 12 s.

After five trials subjects had a five minute break. Additionaly, subjects were given a rest when needed.

### Data collection and analysis

IMU data and joint encoder data were collected on the PC in the Achilles backpack at a sampling frequency of 100 Hz. Force plate, Pusher and EMG sensor data were collected on the XPC target PC at 1000 Hz. A signal composed of pseudo-random numbers and a time interval of

**Table 2** Experiment trial order. Subjects were coded S02 to S11

| Subject Code | Controller Order |
|---|---|
| $Sxx_{odd}$ | ZI(0), ZI(1), $P_{ankle}$(1), $P_{ankle}$(2), $P_{ankle}$(3), ZI(2), $PD_{com}$(1), $PD_{com}$(2), $PD_{com}$(3), ZI(3) |
| $Sxx_{even}$ | ZI(0), ZI(1), $PD_{com}$(1), $PD_{com}$(2), $PD_{com}$(3), ZI(2), $P_{ankle}$(1), $P_{ankle}$(2), $P_{ankle}$(3), ZI(3) |

Suffixes (0) to (3) refer to the trial number, where (0) indicates a practice trial

0.1 s between two samples was sent from the XPC target to the Achilles PC and logged on both systems. By computing the cross-correlation between the (resampled) signals, the delay between the systems was established and used for data synchronization.

The force plate and Pusher data were resampled to 100 Hz. To remove movement artifacts, the EMG data were first high-pass filtered in both forward and reverse direction using a third order Butterworth filter, resulting in a sixth order zero-lag filter with a cut-off frequency of 10 Hz. Then they were rectified by taking the absolute value, and low-pass filtered using a sixth order zero-lag filter with a cut-off frequency of 5 Hz to create an envelope.

To investigate the effect of the controllers on the standing balance of the subjects, the body sway, body sway velocity, ankle torques and muscle activity of the lower legs were analyzed. The body sway velocity was used to obtain a measure of the recovery time, that is, the subject was assumed to have recovered from a perturbation when the absolute body sway velocity was lower than 0.015 rad/s. This threshold was based on the computed variation in body sway velocity during quiet stance. The ankle torques can provide insight on how much the subject and device are contributing to maintaining balance. Therefore, the total ankle torque was derived using the center of pressure (CoP) in the anteroposterior (AP) direction. The CoP was computed from the measured moments and normal forces of the force plates. Since the location of the subject on the force plates was known, the CoP could be computed with respect to the time varying location of the ankle of the subject. For the estimation of the total dorsi/plantar flexion torque applied to the ankle, the mass of the foot was neglected. Then the total ankle torque

$$\tau_{tot} = F_{Cy}p_x - F_{Cx}h_A \quad (3)$$

where $F_{Cy}$ is the vertical ground reaction force, $p_x$ the CoP in the AP direction with respect to the ankle, $F_{Cx}$ the horizontal ground reaction force and $h_A$ the time varying ankle height. The torque applied by the Achilles was estimated from the leaf spring deflection, which was measured by differences between the motor and joint encoder readings. Subtracting this torque from the total torque resulted in the torque applied by the subject (the biological ankle torque). The total ankle torque, the torque applied by Achilles and the biological ankle torque were all normalized by a factor $\frac{1}{Mgl_{com}}$.

Data from the ZI, $P_{ankle}$ and $PD_{com}$ trials 1:3 were used; ZI(0) was excluded from the analysis to reduce the effect of subjects not being used to the perturbations. All data were split to extract the perturbation responses, and the first two perturbations of each trial were removed.

We computed the root-mean-square (RMS) of the body sway response from the start of a perturbation until the average time at which subjects had recovered. Therefore, the average recovery time was calculated over all subjects and all controllers. Additionally, we computed the RMS of the normalized total ankle torque, normalized biological ankle torque and the EMG envelopes, but then from the start of a perturbation until the average time at which the normalized total ankle torque had returned to a steady-state value. We assumed that the steady-state was reached when the absolute value of the derivative of the normalized total ankle torque was smaller than 0.02. This threshold was based on the variation in the derivative of the normalized total ankle torque during quiet stance. To compare the muscle activities of all the subjects, the means of the RMS values of the EMG envelopes in the $P_{ankle}$ and $PD_{com}$ trials were normalized by dividing them by the mean from the corresponding ZI trials.

#### *Statistical procedures*

Since not all data were normally distributed, we analyzed the data using non-parametric tests. Furthermore, both perturbation levels were analyzed separately. We used the Wilcoxon signed rank test to compare $PD_{com}$ to the two other control cases by testing for differences in RMS body sway, recovery time, RMS torque and RMS EMG.

The overall significance level was set at $p < 0.05$. To correct for two comparisons ($PD_{com}$ compared to ZI and $P_{ankle}$), for each test the significance level was set at $p < 0.05/2$. To compute effect sizes, $z$-values were transferred to $r$-values, where $r$=0.1 corresponded to a small effect, $r$=0.3 a medium effect and $r$=0.5 a large effect [32]. All statistical procedures were carried out with SPSS Statistics 24 (IBM, New York, USA).

## Results

### Controller behavior

We investigated the general behavior of the virtual-ankle stiffness and body sway controllers by analyzing their inputs and outputs.

Due to heel lift, the ankle showed plantar flexion in response to a perturbation, represented by an increasing ankle angle in Fig. 5a. However, the body's CoM sways forward, which corresponds to the decreasing body sway angle in Fig. 5b. Therefore, the output torques of the $P_{ankle}$ controller and $PD_{com}$ controller are in opposite directions: a dorsiflexion torque for $P_{ankle}$ (Fig. 5c) and a plantar-flexion torque for $PD_{com}$ (Fig. 5d).

A comparison of the normalized controller output torques and the normalized biological ankle torque in the ZI trial shows that for both perturbation sizes the torques generated by the $PD_{com}$ follow a similar trend to the biological ankle torque in the ZI trial, while the torques generated by the $P_{ankle}$ do not (Fig. 6). Despite

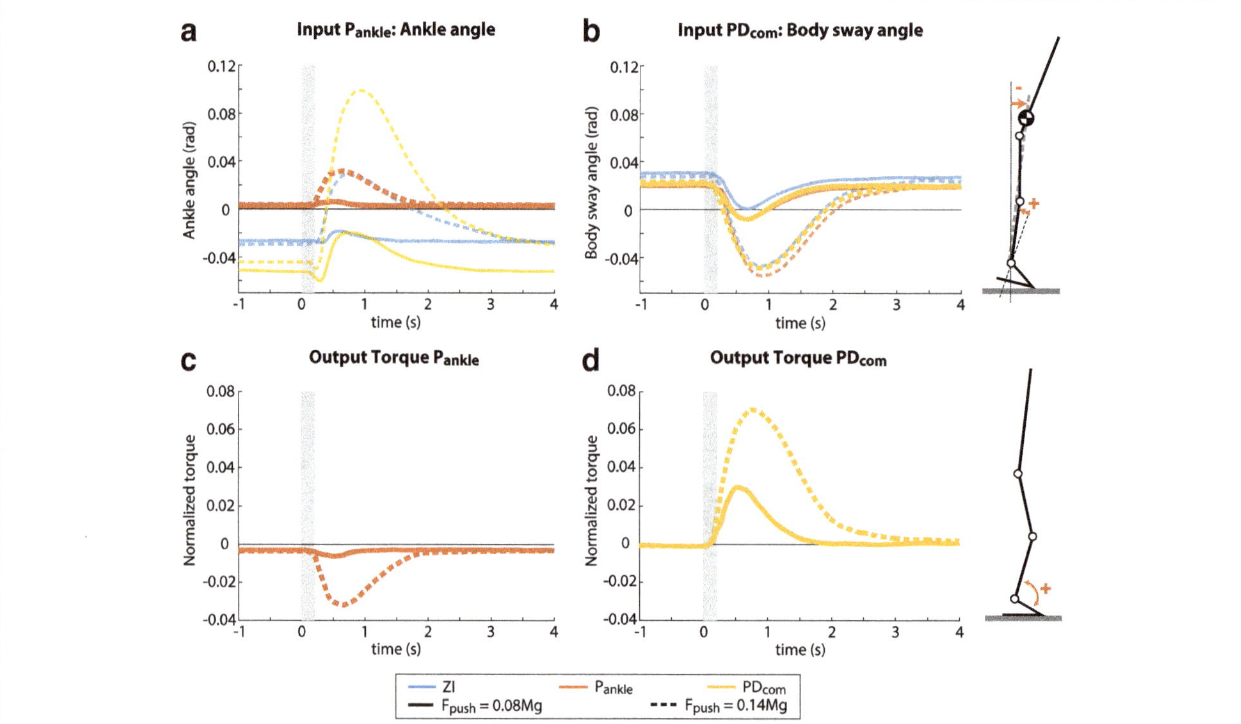


**Fig. 5** Controller inputs and outputs of a representative subject. **a** The ankle angle measured in the $P_{ankle}$ trial, compared to the ankle angle measured in the ZI and $PD_{com}$ trials. **b** The body sway angle measured in the $PD_{com}$ trial, compared to the body sway angle measured in the ZI and $P_{ankle}$ trials. **c** Normalized output torque of the $P_{ankle}$ controller. **d** Normalized output torque of the $PD_{com}$ controller. The grey boxes indicate the perturbation period. Similar results were obtained at a group level

these differences in support torques, subjects were able to maintain balance using each of the controllers.

### CoM kinematics

The amplitude of the body sway angle hardly varied with the controller used, for small and large perturbations (Fig. 7a, b). The results of the Wilcoxon signed rank tests showed no significant differences between the RMS body sway with the $PD_{com}$ and the RMS body sway with the other two controllers (Table 3). The recovery time, derived from the body sway velocity, was also similar for all controllers for small perturbations and large perturbations (Fig. 7c, d, Table 3).

### Ankle joint kinetics

For both the small and the large perturbations, there was a decrease in normalized biological ankle torque for the $PD_{com}$ compared to ZI and $P_{ankle}$, while the normalized

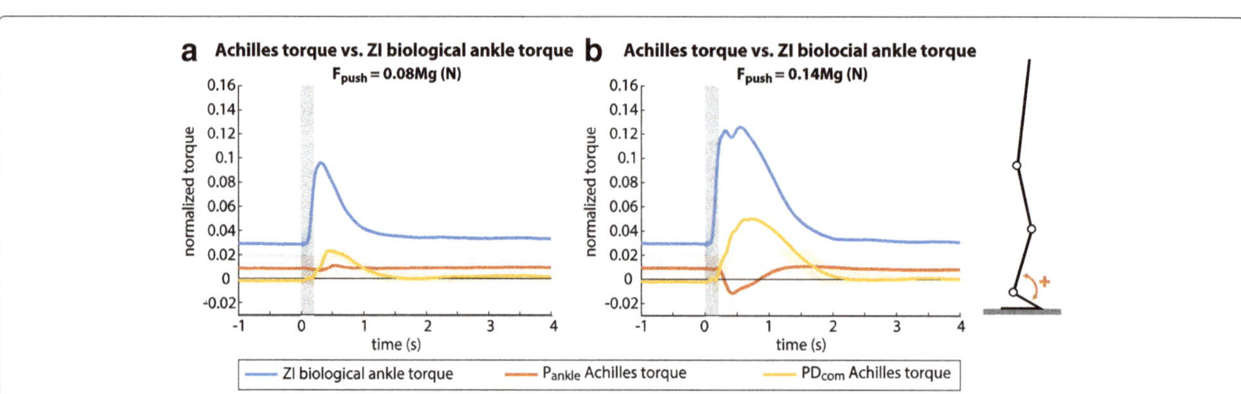


**Fig. 6** Normalized biological ankle torque in ZI compared to normalized torques provided by Achilles, mean over all subjects. **a** Torques in response to small perturbations. **b** Torques in response to large perturbations. The grey boxes indicate the perturbation period

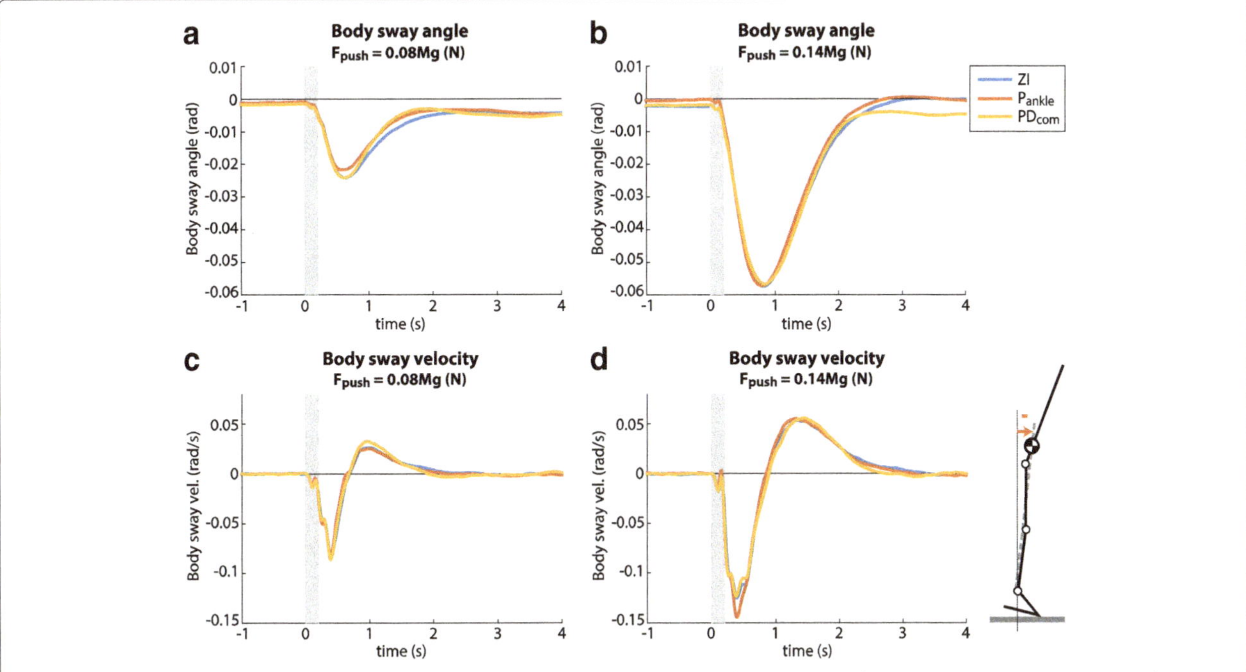


**Fig. 7** CoM kinematics, mean over all subjects. **a** Body sway angle in response to a small perturbation. **b** Body sway angle in response to a large perturbation. **c** Body sway velocity in response to a small perturbation. **d** Body sway velocity in response to a large perturbation. The grey boxes indicate the perturbation period

total ankle torque stayed approximately the same (Fig. 8, Table 4). The decrease in median RMS normalized biological ankle torque over all subjects was particularly large in response to large perturbations: a 29% decrease for the $PD_{com}$ compared to ZI, and a 32% decrease compared to $P_{ankle}$. The results of the Wilcoxon signed rank tests showed that the normalized biological ankle torque response to a large perturbation was significantly lower for the $PD_{com}$ ($Mdn = 0.063$) compared to ZI ($Mdn = 0.089$), $z = -2,701$, $p = 0.004$, $r = 0.60$ and $P_{ankle}$ ($Mdn = 0.093$), $z = -2,803$, $p = 0.002$, $r = 0.63$. For the small perturbations the normalized biological ankle torque was only significantly lower for the $PD_{com}$ compared to ZI, but not compared to $P_{ankle}$ (Table 4).

### EMG

Generally, when the $PD_{com}$ was applied, subjects showed a decrease in soleus activity and an increase in tibialis anterior activity with respect to ZI, as shown for a representative subject in Fig. 9. These results were also observed at a group level in response to large perturbations (Fig. 10b). In that case the median of the RMS soleus muscle activity of the left and right leg decreased on average with 8%, while the median of the RMS tibialis anterior muscle activity increased with 47% compared to zero impedance. The results of the Wilcoxon signed rank tests showed that the SolL activity significantly decreased using the $PD_{com}$ ($Mdn = 0.88$) compared to ZI, $z = -2,293$, $p = 0.02$, $r = 0.50$, while the TAR activity significantly increased ($Mdn = 1.56$), $z = -2,803$, $p = 0.002$, $r = 0.63$. Test results for the opposite leg (SolR and TAL) were nearly significant (Table 5). In comparison with $P_{ankle}$, $PD_{com}$ also resulted in a consistent decrease in plantar flexor muscle activity and an increase in dorsiflexor muscle activity (Fig. 10), however this effect was only significant for TAR (Table 5).

## Discussion

The goal of this study was to develop a controller that provides balance support for an exoskeleton and its user. Therefore, we tested a body sway controller and a virtual-ankle stiffness controller on a pAFO, and assessed their

**Table 3** Statistical test results from the Wilcoxon signed rank test for RMS body sway and recovery time

| | Small perturbations | | Large perturbations | |
|---|---|---|---|---|
| Variable | $PD_{com}$ vs. ZI | $PD_{com}$ vs. $P_{ankle}$ | $PD_{com}$ vs. ZI | $PD_{com}$ vs. $P_{ankle}$ |
| body sway | $z = -0.255, p = 0.846$ | $z = -0.153, p = 0.922$ | $z = -0.357, p = 0.770$ | $z = -0.051, p = 1.000$ |
| recovery time | $z = -0.204, p = 0.865$ | $z = -0.510, p = 0.643$ | $z = -1.989, p = 0.047$ | $z = -0.306, p = 0.785$ |

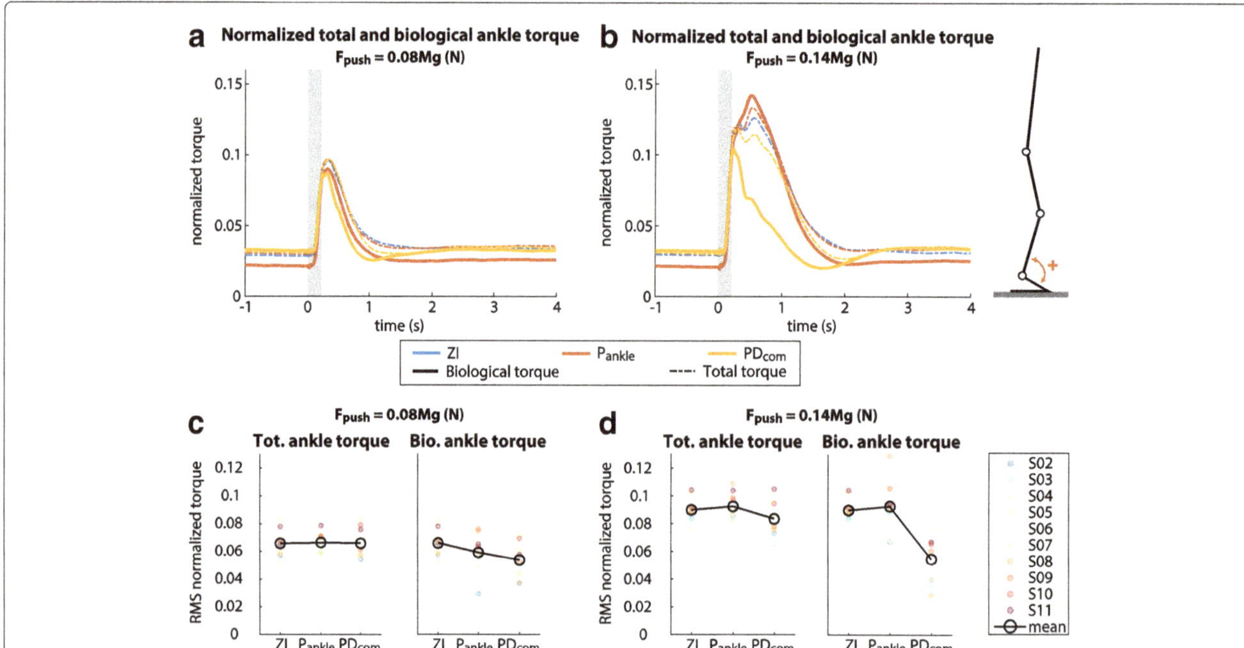


**Fig. 8** Normalized total ankle torque and normalized biological ankle torque. **a** Torques in response to a small perturbation, mean over all subjects. **b** Torques in response to a large perturbation, mean over all subjects. **c** Mean of the RMS of the normalized torques computed over the small perturbations for each subject individually. **d** Mean of the RMS of the normalized torques computed over the large perturbations for each subject individually. The grey boxes indicate the perturbation period

effects on standing balance. The body sway controller showed a human-like torque profile—that is, a torque profile that has the same shape as the biological ankle torque profile with ZI—whereas the virtual-ankle stiffness did not (Fig. 6). Firstly, we hypothesized that when a human-like assistive torque was provided, the balance performance would improve. However, subjects did not use the provided support torques to respond to a perturbation quicker or to reduce the body sway. Secondly, we expected that the human contribution to the balance response would decrease. This was indeed the case, since subjects reduced their own contribution to ankle torque.

In the scarce literature on balance control of exoskeletons, no results have been reported on the effects on the biological ankle torque. Therefore, we compared our results with pAFO locomotion studies. For torque assisted walking, decreased biological ankle torques were reported [14, 15]. In these studies, the support torque profiles that were provided to the subjects had a similar shape as the biological torque profiles during unassisted walking. These findings are similar to our results on pAFO balance control.

In our work, the decrease in biological ankle torque was a combined effect of decreased soleus activity (8%) and increased tibialis anterior activity (47%). This is similar to the results of Collins et al. (2015) [14], who also reported decreased soleus activity and increased tibialis anterior activity (although in combination with affected gastrocnemicus activity), but different from the results of Kao et al. (2010), who only showed a decrease in soleus activity, without an increase of tibialis anterior activity [15]. For balance recovery from a forward perturbation the plantar flexor muscles are typically active. However, the increased tibialis anterior activity indicates co-contraction of the lower leg muscles. A possible reason for this co-contraction is that it could be an effective way to decrease

**Table 4** Statistical test results from the Wilcoxon signed rank test for RMS normalized total ankle torque and RMS normalized biological ankle torque

| | Small perturbations | | Large perturbations | |
|---|---|---|---|---|
| Variable | $PD_{com}$ vs. ZI | $PD_{com}$ vs. $P_{ankle}$ | $PD_{com}$ vs. ZI | $PD_{com}$ vs. $P_{ankle}$ |
| total torque | $z = -0.866, p = 0.432$ | $z = -0.051, p = 1.000$ | $z = -1.886, p = 0.064$ | $z = -1.886, p = 0.064$ |
| bio. torque | $z = -2.497, p = 0.010$[a] | $z = -0.866, p = 0.432$ | $z = -2.701, p = 0.004$[a] | $z = -2.803, p = 0.002$[c] |

[a] RMS $PD_{com}$ < RMS ZI
[c] RMS $PD_{com}$ < RMS $P_{ankle}$

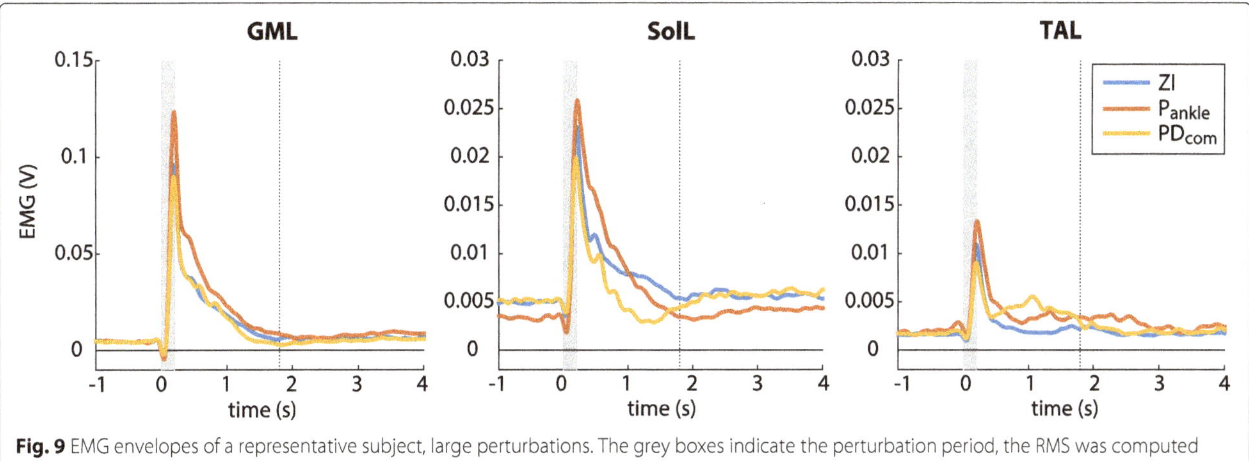


**Fig. 9** EMG envelopes of a representative subject, large perturbations. The grey boxes indicate the perturbation period, the RMS was computed from the start of the perturbation period until the dotted vertical line. Similar results were obtained at a group level

the biological ankle torque, while limiting the decrease in joint stiffness. Since the EMG signal is not an absolute measure of muscle activity, it is not immediately clear whether the total muscle activity has increased or decreased. We expressed the RMS EMG with respect to ZI to be able to analyze the effects of the controllers on the muscle activities across subjects. This normalization impacts the interpretation of the 8% decrease in soleus activity and 47% increase in tibialis anterior activity, since the absolute changes in muscle activity are dependent on the muscle activity with ZI. Although in unassisted standing balance mainly plantar flexor muscles are active during recovery from a forward perturbation, Maki and Ostrovski (1993) reported a substantial amount of co-

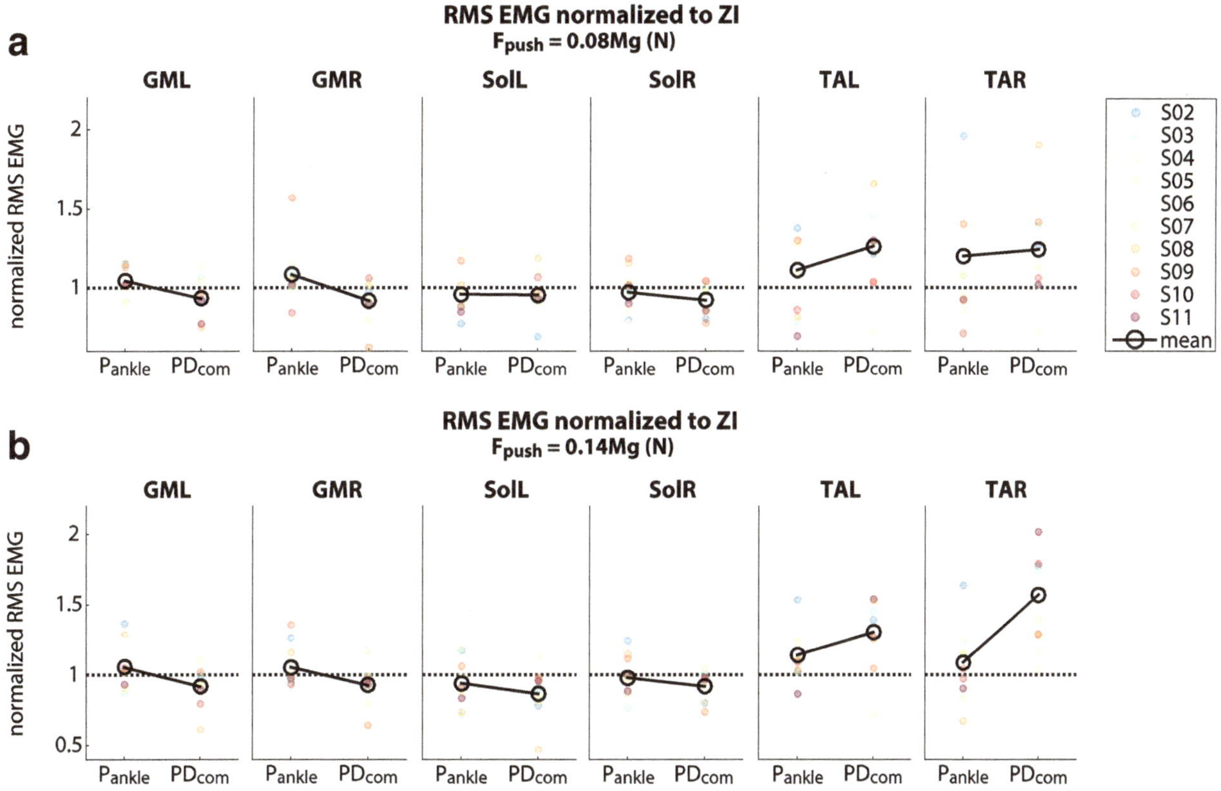

**Fig. 10** RMS EMG normalized to the RMS EMG for ZI condition. The dashed lines at 1 represent ZI. **a** Mean of the normalized RMS EMG over the small perturbations for each subject individually. **b** Mean of the normalized RMS EMG over the large perturbations for each subject individually

**Table 5** Statistical test results from the Wilcoxon signed rank test for normalized RMS EMG. RMS EMG was normalized to ZI

| | Small perturbations | | Large perturbations | |
|---|---|---|---|---|
| Muscle | $PD_{com}$ vs. 1 (ZI) | $PD_{com}$ vs. $P_{ankle}$ | $PD_{com}$ vs. 1 (ZI) | $PD_{com}$ vs. $P_{ankle}$ |
| GML | $z = -1.362, p = 0.203$ | $z = -1.599, p = 0.129$ | $z = -1.599, p = 0.129$ | $z = -1.244, p = 0.250$ |
| GMR | $z = -1.988, p = 0.049$ | $z = -1.988, p = 0.049$ | $z = -1.988, p = 0.049$ | $z = -1.580, p = 0.131$ |
| SolL | $z = -0.968, p = 0.375$ | $z = -0.255, p = 0.846$ | $z = -2.293, p = 0.02$ [a] | $z = -1.274, p = 0.232$ |
| SolR | $z = -1.784, p = 0.084$ | $z = -0.633, p = 0.557$ | $z = -2.090, p = 0.037$ | $z = -0.561, p = 0.625$ |
| TAL | $z = -2.073, p = 0.039$ | $z = -0.889, p = 0.426$ | $z = -2.192, p = 0.027$ | $z = -1.362, p = 0.203$ |
| TAR | $z = -2.191, p = 0.027$ | $z = -0.561, p = 0.625$ | $z = -2.803, p = 0.002$ [b] | $z = -2.701, p = 0.004$ [d] |

[a] RMS $PD_{com}$ < RMS ZI
[b] RMS $PD_{com}$ > RMS ZI
[d] RMS $PD_{com}$ > RMS $P_{ankle}$

contraction for forward sway induced by backward platform perturbations, which increased with increasing perturbation magnitude [33]. Since the perturbations in our experiment were also large, we would expect substantial tibialis anterior activity with ZI. Therefore, it is not likely that the relatively large increase in tibialis anterior activity is a result of low muscle activity with ZI. Consequently, we cannot be conclusive about whether the total muscle activity increased or decreased through the concurrent changes in soleus and tibialis anterior activity.

Modifications in muscle activity can be feedback-driven or a result of feed-forward adaptation over time. In this study the reduction of the biological ankle torque when using the body sway controller could already be observed at the beginning of a trial and remained constant over that trial. This suggests that the changes in muscle activity are feedback-driven adjustments. Similar feedback responses were found for walking pattern adaptation in the presence of velocity-dependent resistance, but in combination with feed-forward adaptive strategies [34]. It is possible that feed-forward adaptation could also occur in standing balance with a controlled pAFO, but that the exposure to the controllers in this study was too short to demonstrate this effect.

Besides a decreased human contribution to ankle torque, we also found that the body sway in response to a perturbation was similar for all controllers (Fig. 7). It is not directly clear why the body sway stayed approximately the same, while the provided support torques were different. For example, in Functional Electrical Stimulation controlled standing, healthy subjects do change their body sway when different control strategies are applied [35]. There are several explanations for the unchanged body sway. Firstly, according to the inverted-pendulum model, the total torque must increase to reduce the body sway. In case of a large perturbation, it is not always possible to increase the total ankle torque considering that the CoP is already at the edge of the base of support. However, this does not hold for the small perturbations for which we saw the same effects. Secondly, subjects may prefer to reduce effort when they are confidently able to maintain balance. This preference of minimizing effort instead of body sway was also reported in standing balance identification studies, through comparisons between optimal and identified neural feedback gains [23, 36]. Lastly, subjects may need more training time to adapt their body sway response.

The body sway controller and virtual-ankle stiffness controller are not only different in the sense that they require different angles as an input. For one thing, for the sake of simplicity the virtual-ankle stiffness controller did not have a velocity dependent component. For another thing, for the body sway controller we allowed the desired body sway angle to adapt to a new steady pose, while for the virtual-ankle stiffness controller the desired ankle angle was kept constant. As a result of keeping the desired ankle angle constant, there was an offset from zero in the torque provided by the pAFO when the virtual-ankle stiffness controller was used (Fig. 6). Yet, we do not think that the virtual-ankle stiffness controller would perform better when a velocity dependent component was added to the control law, or when the desired ankle angle was updated to a new steady pose, since the support torques generated in response to a perturbation would still be small, or even in opposite direction (Fig. 6).

A practical implication of using a body sway controller is that the CoM location should be estimated on-line. Generally, exoskeletons are equipped with joint encoders. While this is sufficient for simpler controllers, such as the virtual-ankle stiffness, for a body sway estimation additional sensors are needed, which adds complexity to the setup. We used three IMUs and anthropometric data to make a CoM estimation, but simpler methods based on a single IMU have already been shown to give good results during walking [37].

Ultimately, our goal is to improve the standing balance of people with a spinal cord injury. Based on the results of the tests with healthy subjects, it is difficult to predict how paraplegics would respond to the provided support torques. Healthy subjects reduced their own ankle torques, but paraplegics, depending on the level of injury, typically cannot provide large ankle torques themselves. Therefore, a future direction is to test these balance controllers on an exoskeleton for people with a spinal cord injury who are not able to balance themselves.

## Conclusion

In this study we investigated the effects of a pAFO on the standing balance of healthy subjects. We found that when using a controller that uses body sway motion as an input, effective supportive torques could be provided. These support torques followed a similar trend to the ankle torques humans would provide in response to a perturbation. Although the CoM kinematics remained unaffected with the body sway controller, the biological ankle torque decreased when the body sway controller was implemented. This was a result of the combined effects of decreased soleus muscle activity and increased tibialis anterior muscle activity. Subjects were able to quickly adapt their ankle torques to the torques provided by the ankle exoskeleton. These results suggest a preference for reducing effort instead of further improving body sway.

**Funding**

The work presented here was performed in the SYMBITRON project which is supported by EU research program FP7, FET-Proactive initiative "Symbiotic human-machine interaction" (ICT-2013-10) under project contract #611626. The SYMBITRON project is coordinated by the University of Twente.

**Authors' contributions**

AE, EA and HK designed the study and interpreted the data, AE collected and analyzed the data and wrote the manuscript, EA and HK revised the manuscript. All authors read and approved the final manuscript.

**Competing interests**

The authors declare that they have no competing interests.

**References**

1. Veneman JF, Kruidhof R, Hekman EEG, Ekkelenkamp R, van Asseldonk EHF, van der Kooij H. Design and evaluation of the lopes exoskeleton robot for interactive gait rehabilitation. IEEE Trans Neural Syst Rehabil Eng. 2007;15(3):379–86.
2. Jezernik S, Colombo G, Keller T, Frueh H, Morari M. Robotic orthosis lokomat: A rehabilitation and research tool. Neuromodulation. 2003;6(2): 108–15.
3. Strickland E. Good-bye, wheelchair. IEEE Spectrum. 2012;49(1):30–2.
4. Bortole M, Venkatakrishnan A, Zhu F, Moreno JC, Francisco GE, Pons JL, Contreras-Vidal JL. The h2 robotic exoskeleton for gait rehabilitation after stroke: Early findings from a clinical study. J NeuroEngineering Rehabil. 2015;12(1):1–14.
5. Aach M, Cruciger O, Sczesny-Kaiser M, Höffken O, Meindl RC, Tegenthoff M, Schwenkreis P, Sankai Y, Schildhauer TA. Voluntary driven exoskeleton as a new tool for rehabilitation in chronic spinal cord injury: A pilot study. Spine J. 2014;14(12):2847–53.
6. Wang S, Wang L, Meijneke C, van Asseldonk E, Hoellinger T, Cheron G, Ivanenko Y, La Scaleia V, Sylos-Labini F, Molinari M, Tamburella F, Pisotta I, Thorsteinsson F, Ilzkovitz M, Gancet J, Nevatia Y, Hauffe R, Zanow F, van der Kooij H. Design and control of the mindwalker exoskeleton. IEEE Trans Neural Syst Rehabil Eng. 2015;23(2):277–86.
7. Esquenazi A, Talaty M, Packel A, Saulino M. The rewalk powered exoskeleton to restore ambulatory function to individuals with thoracic-level motor-complete spinal cord injury. Am J Phys Med Rehabil. 2012;91(11):911–21.
8. Farris RJ, Quintero HA, Murray SA, Ha KH, Hartigan C, Goldfarb M. A preliminary assessment of legged mobility provided by a lower limb exoskeleton for persons with paraplegia. IEEE Trans Neural Syst Rehabil Eng. 2014;22(3):482–90.
9. Zoss AB, Kazerooni H, Chu A. Biomechanical design of the berkeley lower extremity exoskeleton (bleex). IEEE/ASME Trans Mechatron. 2006;11(2):128–38.
10. Walsh CJ, Endo K, Herr H. A quasi-passive leg exoskeleton for load-carrying augmentation. Int J Humanoid Robot. 2007;4(3):487–506.
11. Zhang J, Fiers P, Witte KA, Jackson RW, Poggensee KL, Atkeson CG, Collins SH. Human-in-the-loop optimization of exoskeleton assistance during walking. Science. 2017;356(6344):1280–3.
12. Panizzolo FA, Galiana I, Asbeck AT, Siviy C, Schmidt K, Holt KG, Walsh CJ. A biologically-inspired multi-joint soft exosuit that can reduce the energy cost of loaded walking,. J Neuroengineering Rehabil. 2016;13(1):43.
13. Ding Y, Panizzolo FA, Siviy C, Malcolm P, Galiana I, Holt KG, Walsh CJ. Effect of timing of hip extension assistance during loaded walking with a soft exosuit. J Neuroengineering Rehabil. 2016;13(1):87.
14. Collins SH, Bruce Wiggin M, Sawicki GS. Reducing the energy cost of human walking using an unpowered exoskeleton. Nature. 2015;522(7555):212–5.
15. Kao P-C, Lewis CL, Ferris DP. Invariant ankle moment patterns when walking with and without a robotic ankle exoskeleton. J Biomech. 2010;43(2):203–9.
16. Rex Bionics Ltd. Product Information - Rex Bionics. https://www.rexbionics.com/product-information/. Accessed 13 Nov 2017.
17. Parker Haniffin Corp. Indego-Powering People Forward. http://www.indego.com/indego/en/home. Accessed 14 Dec 2016.
18. Wang L, Wang S, van Asseldonk EHF, van der Kooij H. Actively controlled lateral gait assistance in a lower limb exoskeleton. In: 2013 IEEE/RSJ International Conference on Intelligent Robots and Systems. 445 Hoes Lane Piscataway: IEEE; 2013. p. 965–970.
19. Rajasekaran V, Aranda J, Casals A, Pons JL. An adaptive control strategy for postural stability using a wearable robot. Robot Auton Syst. 2015;73:16–23.
20. Kuo AD. An optimal control model for analyzing human postural balance. IEEE Trans Biomed Eng. 1995;42(1):87–101.
21. Park S, Horak FB, Kuo AD. Postural feedback responses scale with biomechanical constraints in human standing. Exp Brain Res. 2004;154(4):417–27.
22. Afschrift M, Jonkers I, de Schutter J, de Groote F. Mechanical effort predicts the selection of ankle over hip strategies in nonstepping postural responses. J Neurophys. 2016;116(4):1937–45.
23. Kiemel T, Zhang Y, Jeka JJ. Identification of neural feedback for upright stance in humans: Stabilization rather than sway minimization. J Neurosci. 2011;31(42):15144–53.
24. Engelhart D, Pasma JH, Schouten AC, Aarts RGKM, Meskers CGM, Maier AB, van der Kooij H. Adaptation of multijoint coordination during standing balance in healthy young and healthy old individuals. J Neurophys. 2016;115(3):1422–35.
25. Welch TDJ, Ting LH. A feedback model reproduces muscle activity during human postural responses to support-surface translations. J Neurophys. 2008;99(2):1032–8.

26. Welch TDJ, Ting LH. A feedback model explains the differential scaling of human postural responses to perturbation acceleration and velocity. J Neurophys. 2009;101(6):3294–309.
27. Peterka RJ. Sensorimotor integration in human postural control. J Neurophys. 2002;88(3):1097–118.
28. Horak FB, Nashner LM. Central programming of postural movements: Adaptation to altered support-surface configurations. J Neurophys. 1986;55(6):1369–81.
29. van Dijk W, Meijneke C, van der Kooij H. Evaluation of the achilles ankle exoskeleton. IEEE Trans Neural Syst Rehabil Eng. 2017;25(2):151–60.
30. Vlutters M, van Asseldonk EHF, van der Kooij H. Center of mass velocity-based predictions in balance recovery following pelvis perturbations during human walking. J Exp Biol. 2016;219(10):1514–23.
31. Winter DA. Anthropometry. In: Biomechanics and Motor Control of Human Movement. New York: John Wiley & Sons, Inc.; 2009. p. 82–106.
32. Cohen J. A power primer. Psychol Bull. 1992;112(1):155–9.
33. Maki B, Ostrovski G. Scaling of postural responses to transient and continuous perturbations. Gait Posture. 1993;1(2):93–104.
34. Lam T, Anderschitz M, Dietz V. Contribution of feedback and feedforward strategies to locomotor adaptations. J Neurophys. 2006;95(2):766–73.
35. Huryn TP, Blouin JS, Croft EA, Koehle MS, der Loos HFMV. Experimental performance evaluation of human balance control models. IEEE Trans Neural Syst Rehabil Eng. 2014;22(6):1115–27.
36. van der Kooij H, Peterka RJ. Non-linear stimulus-response behavior of the human stance control system is predicted by optimization of a system with sensory and motor noise. J Comput Neurosci. 2011;30(3):759–78.
37. Floor-Westerdijk MJ, Schepers HM, Veltink PH, van Asseldonk EHF, Buurke JH. Use of inertial sensors for ambulatoryassessment of center-of-mass displacements during walking. IEEE Trans Biomed Eng. 2012;59(7):2080–4.

# Mechanisms of electrical vasoconstriction

Mark Brinton[1*], Yossi Mandel[2], Ira Schachar[3] and Daniel Palanker[3,4]

## Abstract

**Background:** Electrical vasoconstriction is a promising approach to control blood pressure or restrict bleeding in non-compressible wounds. We explore the neural and non-neural pathways of electrical vasoconstriction in-vivo.

**Methods:** Charge-balanced, asymmetric pulses were delivered through a pair of metal disc electrodes. Vasoconstriction was assessed by measuring the diameter of rat saphenous vessels stimulated with low-voltage (20 V, 1 ms) and high-voltage (150 V, 10 μs) stimuli at 10 Hz for 5 min. Activation pathways were explored by topical application of a specific neural agonist (phenylephrine, alpha-1 receptor), a non-specific agonist (KCl) and neural inhibitors (phenoxybenzamine, 25 mg/ml; guanethidine, 1 mg/ml). Acute tissue damage was assessed with a membrane permeability (live-dead) fluorescent assay. The Joule heating in tissue was estimated using COMSOL Multiphysics modeling.

**Results:** During stimulation, arteries constricted to 41 ± 8% and 37 ± 6% of their pre-stimulus diameter with low- and high-voltage stimuli, while veins constricted to 80 ± 18% and 40 ± 11%, respectively. In arteries, despite similar extent of constriction, the recovery time was very different: about 30 s for low-voltage and 10 min for high-voltage stimuli. Neural inhibitors significantly reduced low-voltage arterial constriction, but did not affect high-voltage arterial or venous constriction, indicating that high-voltage stimuli activate non-neural vasoconstriction pathways. Adrenergic pathways predominantly controlled low-voltage arterial but not venous constriction, which may involve a purinergic pathway. Viability staining confirmed that stimuli were below the electroporation threshold. Modeling indicates that heating of the blood vessels during stimulation (< 0.2 °C) is too low to cause vasoconstriction.

**Conclusions:** We demonstrate that low-voltage stimuli induce reversible vasoconstriction through neural pathways, while high-voltage stimuli activate non-neural pathways, likely in addition to neural stimulation. Different stimuli providing precise control over the extent of arterial and venous constriction as well as relaxation rate could be used to control bleeding, perfusion or blood pressure.

**Keywords:** Electrical stimulation, Electroceuticals, Vasoconstriction

## Background

For decades, electrical stimulation of cardiac striated muscle has been successfully utilized in pacemakers and defibrillators. Recently, electrical control of vascular smooth muscle has been proposed to treat bleeding in non-compressible wounds [1–3]. Understanding the vasoconstriction pathways activated by electrical stimuli will help create safe and effective devices for electrical control of blood vessels.

Constriction of blood vessel involves both neural and non-neural pathways. Non-neural vasoconstriction mechanisms include mechanical stretching (myogenic) [4, 5], release of endothelin-1 [6, 7] and uridine adenosine tetraphosphate [8] from endothelial cells, circulating hormones (i.e. angiotensin II) [9, 10], and damaged platelets [11]. The dominant neural mechanisms include a fast ionotropic (P2X receptor) pathway activated by adenosine triphosphate (ATP) [12–15]; slower metabotropic (alpha-1 and -2 adrenoreceptor) pathways with the release of norepinephrine [15–21]; and the release of neuropeptide Y, which potentiates constriction from norepinephrine and ATP [22–25].

Electrical stimulation of blood vessels has been used to study the neural pathways in constriction [15–18, 26–29] and dilation [30, 31], including identification of norepinephrine and ATP involvement by in-vitro stimulation of the rat saphenous artery [13]. Direct electric current induces vasoconstriction and thrombosis [32–34] but also causes tissue damage. In-vitro studies have demonstrated both neural and non-neural electrically induced

* Correspondence: brintonmr@gmail.com
[1]Department of Bioengineering, University of Utah, 20 S. 2030 E., Salt Lake City, UT 84112, USA
Full list of author information is available at the end of the article

vasoconstriction using a variety of vessel types (pulmonary, somatic and umbilical); however, they used direct or sinusoidal alternating current, which can damage tissue, and could not directly compare the arterial and venous responses to the same stimulation because the vessels were harvested from different locations [28, 29].

Based on our previous studies of vasoconstriction thresholds as a function of pulse duration, frequency, and amplitude [1, 2], we hypothesized that two distinct stimuli (20 V, 1 ms and 150 V, 10 μs pulses at 10 Hz) could constrict the rat saphenous artery to a similar extent. Using these stimuli, we sought to compare the extent and recovery time of arterial and venous constriction in-vivo, describe the underlying pathways, and determine whether the stimulation required for vasoconstriction damages the rat saphenous vessels.

## Methods

### Animals

Male, Long Evans rats (Charles River), aged 50–60 days, with average weight of 309 g (range: 220-380 g) were used in this study with approval by the Stanford Administrative Panel on Laboratory Animal Care. Fourteen animals were used to confirm the maximum constriction using the proposed electrical parameters. In the vasoconstriction and neural inhibition study 7 animals were used for control, 6 for phenoxybenzamine and 5 for guanethidine blockade. Additional five animals were used to assess vessel damage with the live-dead assay. Before surgery, animals were anesthetized using ketamine HCl (75 mg/kg) and xylazine (5 mg/kg), with an additional half dose given every 45 min thereafter.

For surgery, the animal was placed in the supine position and the rectal temperature was kept at 37 ± 1 °C. The saphenous artery and vein were exposed by removing the skin. Hartman's Lactated Ringer solution (~ 37 °C) dripped at about 1 Hz onto the surgical site during the surgery and stimulation.

### Neural inhibitors

Neural inhibitors were applied topically once the skin was removed and vessels exposed. Phenoxybenzamine HCl (Sigma Aldrich) was dissolved in DMSO (25 mg/ml, Sigma Aldrich). Guanethidine monosulfate (MyBioSource) was first dissolved in de-ionized water (10 mg/ml) and then diluted with DMSO to 1 mg/ml. Control animals ($n$ = 7) received DMSO without inhibitors. The solutions were applied liberally (~ 200 μl) to the exposed vessels and covered with a thin piece of plastic to prevent desiccation ($n$ = 6 for phenoxybenzamine and $n$ = 5 for guanethidine). Fresh solution was added (~ 200 μl) about every 5 min for a total of 25 min. The superficial fascia was removed to improve visualization of the vessel diameters prior to electrical stimulation. Following electrical stimulation, some animals from each group—control (DMSO), phenoxybenzamine and guanethidine—received about 200 μl of potassium chloride (25 mg/ml, Sigma Aldrich; $n$ = 5) and, after several minutes of washout, one drop of phenylephrine HCl (25 mg/ml, Akorn; n = 5 for the Guanethidine block and $n$ = 7 for the phenoxybenzamine block) as positive controls of vasoconstriction. Potassium chloride acts directly to depolarize nerves and smooth muscle while phenylephrine selectively activates the alpha-1 adrenergic receptors on smooth muscle cells. Because maximal constriction occurred within several seconds, only the maximal constriction from phenylephrine and potassium chloride are reported.

### Vessel stimulation and data collection

Electrical stimulation and video monitoring of the vessels' width were performed as previously described [2]. Briefly, stainless steel disc electrodes, 1.6 mm diameter, were placed 3.5 mm apart, with the saphenous vessels between them. An anodic square pulse from a customized pulse generator was delivered through an 11 μF capacitor to the electrodes to assure charge balance. Electrical parameters were selected based on previous studies [1, 2], which demonstrated that 150 V, 10 μs stimulation at 1 Hz induced a maximum constriction (about 30% of the original diameter), and 20 V, 1 ms stimulation at 1 Hz produced constriction to 40–45% of the original diameter. Since constriction also increased with pulse frequency, we hypothesized that 20 V, 1 ms (referred to as low-voltage) and 150 V, 10 μs (referred to as high-voltage) stimulation, pulsed at 10 Hz, would both reach the state of maximum constriction in the arteries. We first tested these stimulation parameters in 14 animals (7 animals with both 20 V and 150 V, 5 with only 150 V and 2 with only 20 V stimuli) without pharmacological treatment (Fig. 1). When multiple stimulations were delivered on the same animal, the second occurred at least 15 min later and about 1 cm proximal to first stimulation. Stimulations lasted for 5 min, and the waveforms were monitored using an oscilloscope (Tektronix, TDS 210). The low-voltage stimulus delivered 12.5 mA (250 μJ/pulse) and the high-voltage stimulus 120 mA (180 μJ/pulse), as measured with a 100 Ω series resistor. The inner diameter of blood vessels was measured in ImageJ (NIH) from video captured with a digital camera (Sentech Inc., TC202USB-A).

Data were normalized by the vessel diameter prior to stimulation and presented as mean ± stdev. Statistical significance ($p$ < 0.01) was determined using one-way ANOVA, and, where appropriate, the post hoc Tukey-Kramer test to determine statistical significance between study groups. To compare vessel constriction between study groups, the vessel diameters were averaged over the 5-min stimulation period.

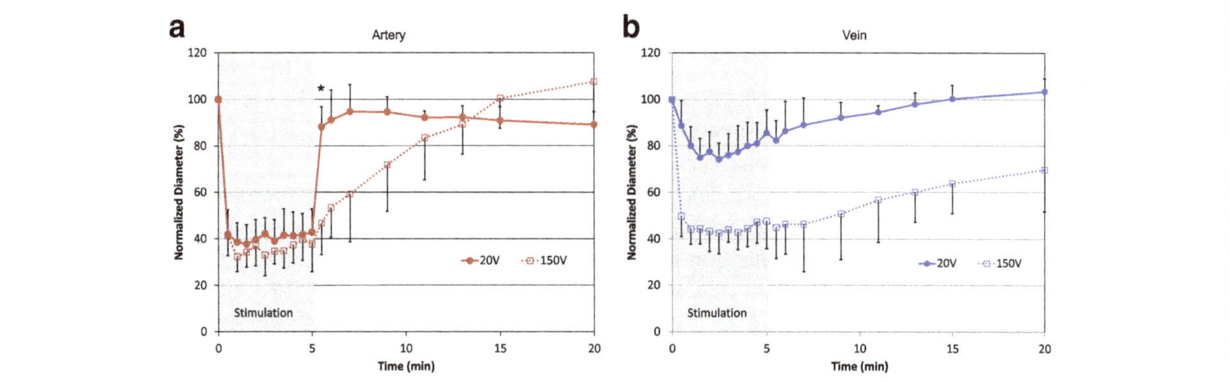


**Fig. 1** Arterial and venous constriction in response to 10 Hz stimulation at 20 V, 1 ms and 150 V, 10 µs pulses. **a** Both stimuli constrict arteries to similar extent, but arteries dilate faster after low-voltage stimulation (*$p < 0.01$, $F(3,34) = 23.24$; one-way ANOVA with Tukey-Kramer multi-comparison test). **b** Veins constrict similarly to arteries at high voltage, but significantly less at low-voltage stimulation ($p < 0.01$, $F(3,34) = 27.31$; one-way ANOVA with Tukey-Kramer multi-comparison test on the average vessel diameter during stimulation); $N = 7$ for 20 V and $N = 12$ for 150 V stimuli

### Viability staining and vessel damage

To determine whether the electrical stimulation damaged cell membranes, the saphenous vessels were surgically exposed, as described above, and the femoral artery of anesthetized rats was cannulated with a micro-renethane catheter (Braintree Scientific, MRE-025) [35]. The catheter was advanced to the sapheneous branch, at which time the animal was euthanized. Immediately, the distal end of the saphenous vessel was cut and the vessel flushed with Ringers solution. A viability/cytotoxicity assay (BIOTUM, 30002-T) was pumped through the vessel at a rate of 0.02 ml/min (New Era Pump Systems, NE-300). While the staining solution flowed through the vessel, electrical pulses (20 V, 1 ms; 150 V, 10 µs; and 300 V, 10 µs) were applied at 10 Hz for 2 min to the exposed vessel in ascending order of voltage, moving the electrodes by 5 mm for each voltage setting, along the saphenous vessels. Stimulation at 300 V, 10 µs was included as a positive control to validate the viability/cytotoxicity assay. After the last stimulation, the assay continued to flow through the vessel for another 15 min. The vessel was then again flushed with Ringers solution, excised and mounted on a glass slide for imaging. The cytotoxicity component of the assay (ethidium homodimer III) only enters the cells with damaged membranes to label the nuclei red, while the viability component (calcein AM) crosses uncompromised cell membranes to label the entire viable cell green. The number of damaged cells (red) for each case were counted using ImageJ and normalized by the control.

### Multiphysics modeling

Based on the electric current measured in-vivo, we modeled the electric field and tissue heating using COMSOL Multiphysics. The model assumed symmetry with respect to the plane passing through the middle of the disk electrodes to reduce the modeling volume to $1.5 \times 0.5 \times 0.7$ cm. The tissue was assumed to have electrical and thermal properties of a muscle, and a 0.1 cm thick saline layer covered the 0.6 cm thick muscle layer. Blood vessels in the muscle layer were modeled as 0.6 cm long cylinders of blood. Because constricted vessel diameters (especially the vein) depend on the electrical stimulus applied, we modeled the vessel diameters and flow rates according to our previous work (Table 1) [2]. The vessels passed between two 0.16 cm diameter electrodes, separated by 0.35 cm, center-to-center. Electrodes were placed in direct contact with the muscle tissue, covered with saline.

To model the electric field in the tissue, we applied voltage pulses and chose tissue conductivity so that the total injected current matched the current measured in-vivo (12.5 mA and 120 mA for the 20 V and 150 V stimuli). The in-vivo currents delivered through the electrodes were measured using a 100-Ohm resistor in series with the electrode and tissue. The current was measured at pulse onset, before the capacitive interface had charged.

The electrode surfaces were defined as equipotential, and all other model boundaries were insulating. To calculate the average Joule heating, the power density from one electrical pulse was multiplied by the duty cycle of the pulsed stimulus (0.01 for 1 ms pulses and 0.0001 for 10 µs pulses). To account for the recharge

**Table 1** In-vivo vessel diameters and flow rates used in the COMSOL modeling

| Applied Stimulus | Artery | Vein |
|---|---|---|
| | *Constricted Diameter (µm)* | |
| 20 V, 1 ms | 200 | 430 |
| 150 V, 10 µs | 160 | 180 |
| | *Flow Rate* [2] *(m/s)* | |
| 20 V, 1 ms | 0.15 | 0.12 |
| 150 V, 10 µs | 0.09 | 0.07 |

phase of the pulse, we took the conservative approach and doubled this average power density. In muscle, a perfusion term was included according the bioheat equation [36]. Blood flow rates through the vessels were estimated using the in-vivo measured diameters and published bleeding rates of electrically constricted saphenous vessels in rats of similar weight and electrical parameters (Table 1) [2]. Blood flowed into the model at 37 °C from the rear in both vessels, a simplification justified by the minimal blood heating due to high flow rates. The boundaries of the modeled volume were set to 37 ° C to reflect body temperature and the drip of the warm saline onto the vessels. The muscle and saline boundaries in the plane of symmetry were insulating. The electrical and thermal material properties are detailed in Table 2. The tissue temperature reached a steady state within about 25 s of stimulation. For comparison, the tissue heating without blood flow through the vessels was also calculated.

## Results

### Electrical stimulation of blood vessels

Upon electrical stimulation, arteries constricted to 41 ± 8% and 37 ± 6% of the initial vessel diameter with 20 V, 1 ms and 150 V, 10 μs pulses repeated at 10 Hz (Fig. 1(a)). While arteries constricted to a similar extent with both stimuli, vessels treated with low-voltage dilated back to 90% of the initial diameter within 30 s, while vessels treated with high-voltage stimulation recovered after 10 min.

With high-voltage stimulation, veins constricted to a similar extent as arteries (40 ± 11% of the initial vessel diameter), but to only 80 ± 18% of the initial diameter with low-voltage stimuli (Fig. 1(b), $p < 0.01$). Veins also recovered slower than arteries—10 min with low-voltage and 15 min with high-voltage stimuli.

### Assessment of tissue damage by electrical stimulation

To evaluate tissue damage by electrical stimulation (electroporation), we applied a fluorescent cell viability assay to stimulated blood vessels: cells with permeabilized membranes fluoresce in red, while intact cells are stained with green. As shown in Fig. 2, a limited number of damaged cells (red) can be seen in the vessel walls of the control (no stimulation), as well as in the 20 V and 150 V samples. However, with the 300 V stimuli (10 μs pulses at 10 Hz), the number of damaged cells increased 4.7-fold ($p < 0.05$) and the arterial wall dilated beyond its non-stimulated diameter.

**Table 2** Electrical and thermal parameters for COMSOL modeling

| | Muscle | Blood | Saline |
|---|---|---|---|
| Electrical Conductivity (S/m) | 0.376 (20 V)[51] 0.475(150 V)[51] | 0.76 [52, 53] | 1.0 |
| Thermal Conductivity (W/m·K) | 0.52 [53] | 0.5 [54] | 0.59 |
| Specific Heat Capacity (J/kg·K) | 3550 [54] | 3840 [54] | 4173 [55] |
| Density (kg/m$^3$) | 1041[53] | 1055[56] | 1000[56] |
| Perfusion Parameter (s$^{-1}$) | 0.00067 [53] | NA | NA |

### Tissue heating by electrical stimulation

To assess the extent of tissue heating during stimulation, we modeled the system using COMSOL Multiphysics. At steady-state, about 25 s of stimulation, the temperature on electrode surface increases by about 0.96 °C with 20 V and by 0.61 °C with 150 V stimuli (Fig. 3). Near blood vessels, the steady-state temperature rise was only 0.2 °C and 0.15 °C from the 20 V and 150 V stimuli, respectively. As a worst-case scenario, we modeled tissue heating without blood flow in the vessels, which yielded temperature rise of 0.6 °C and 0.35 °C at the vessel walls. Both temperatures are well within the range of physiological variations, and less than the temperature change from each drip of warmed saline on the vessel surface.

### Neural pathways

To determine whether neural and non-neural pathways were involved in electrically-induced vasoconstriction, we applied a selective neural agonist (phenylephrine), neural inhibitors (phenoxybenzamine and guanethidine), and a non-specific depolarizing agent (potassium chloride). Phenylephrine is a synthetic analog of norepinephrine that binds and activates the alpha-1 adrenergic receptor. Phenoxybenzamine (PBZ) prevents binding to the alpha-1 and alpha-2 adrenergic receptors on smooth muscle cells. Guanethidine prevents the release of adrenergic (norepinephrine) and purinergic (adenosine triphosphate) neurotransmitters from sympathetic nerves. Potassium chloride (KCl) depolarizes both nerves and smooth muscle cells directly. Since maximal constriction occurred immediately following application, we plot these levels of constrictions as horizontal lines in Fig. 4, for comparison.

#### *Neural inhibition: Arteries*

As expected, phenylephrine induced a strong arterial constriction to 45 ± 16% of the initial vessel diameter (Fig. 4, green dash). Pretreatment with PBZ completely blocked arterial constriction with phenylephrine—the vessel constricted to only 96 ± 5% of the initial diameter ($p < 0.01$; Fig. 4(a) and Fig. 4(c), green solid). Potassium Chloride (KCl) produced the largest arterial constriction, down to 25 ± 2% of the initial diameter (Fig. 4, black dash). PBZ slightly reduced the constrictive effect of KCl, presumably by preventing the endogenous norepinephrine, released from depolarized nerves, from binding to the adrenergic receptors (alpha-1 and alpha-2) on the smooth muscle cells; however, KCl can also act on smooth muscle directly

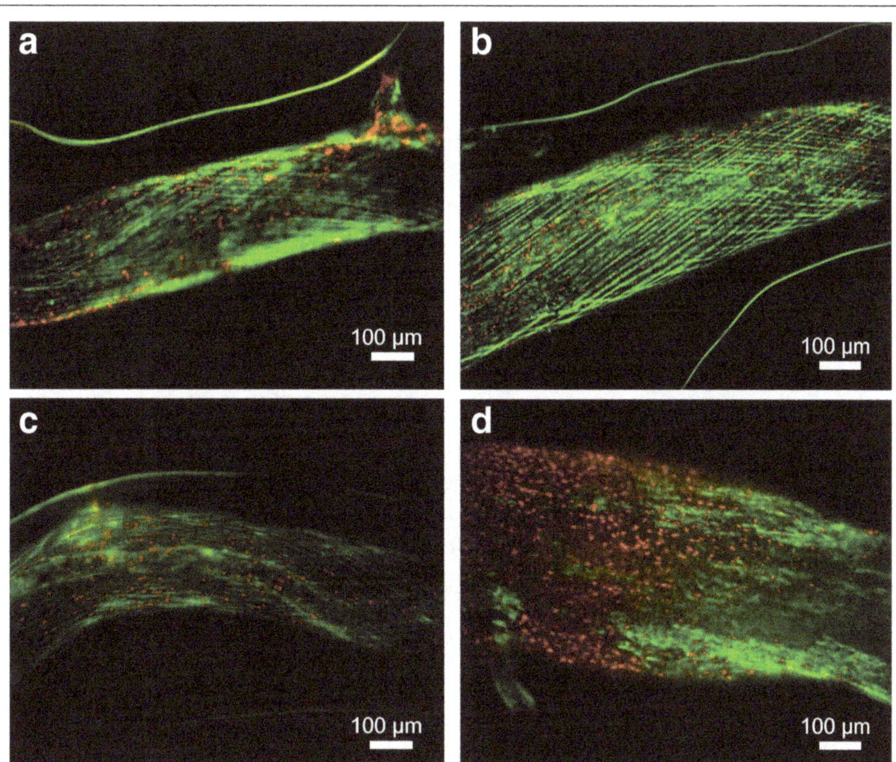


**Fig. 2** Cell viability assay of the stimulated artery. Green color indicates intact cells, while red shows cells with compromised cell membrane. **a** Control with no stimulation (*N* = 5); **b** 20 V, 1 ms pulses (N = 5); **c** 150 V, 10 μs pulses (N = 5); or **d** 300 V, 10 μs pulses (*N* = 4). All treatments were applied at 10 Hz for 2 min. The number of compromised cells was normalized by the control. Vessels stimulated by 20 V and 150 V pulses exhibited damage similar to the control (1.1 ± 0.8, 1.2 ± 0.3, and 1.0 ± 0.6 a.u., respectively), while the 300 V stimulation damaged 4.7-fold more cells (4.7 ± 1.6 a.u.; $p < 0.05$ and $F(3,15) = 18.15$, using one-way ANOVA and Tukey-Kramer multi-comparison tests)

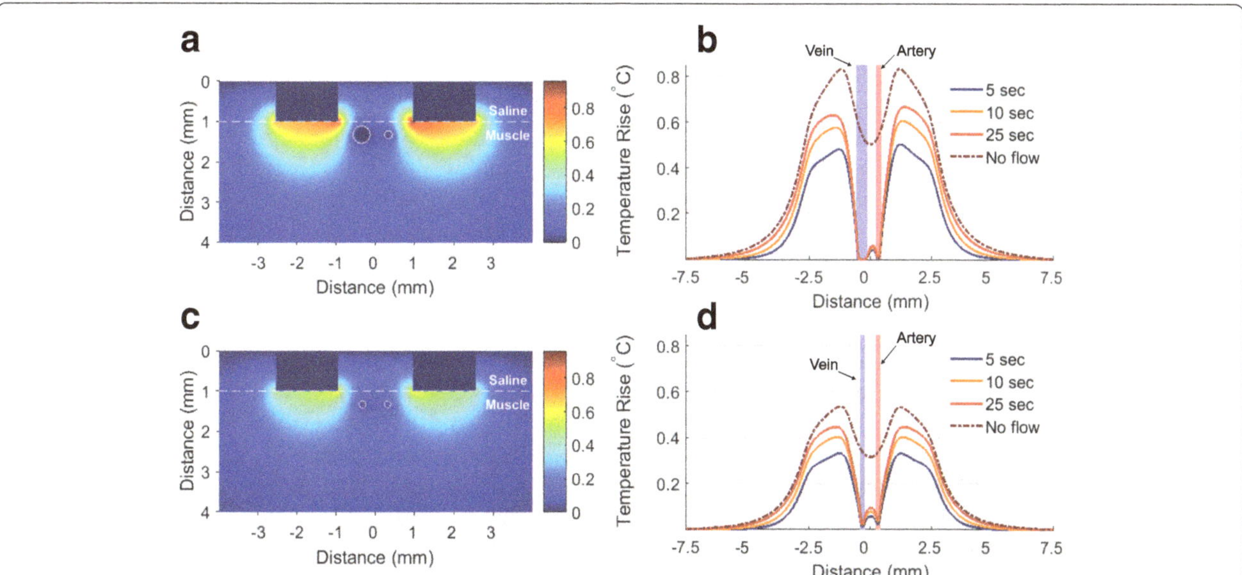


**Fig. 3** Temperature rise from **a** 20 V, 1 ms and **c** 150 V, 10 μs stimulations, pulsed at 10 Hz. The temperature rise along the line passing through the center of blood vessels for **b** 20 V and **d** 150 V stimulation

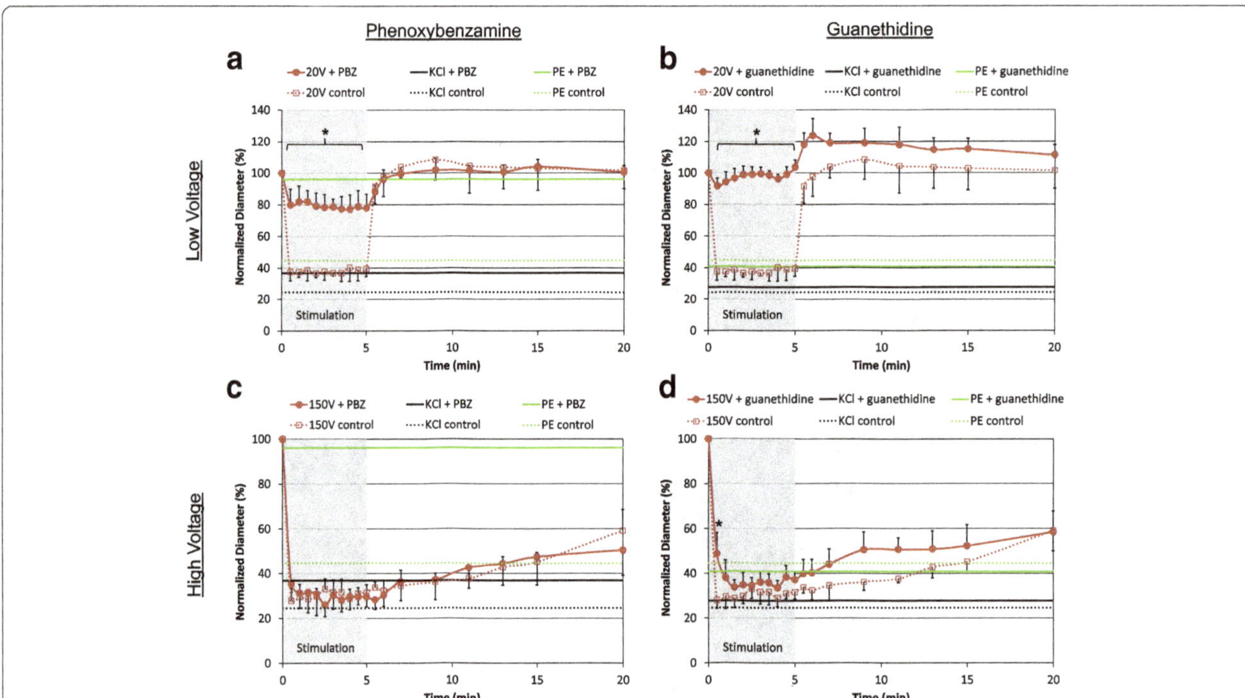


**Fig. 4** Neural inhibition of arterial vasoconstriction. **a** Phenoxybenzamine (PBZ) completely blocked constriction induced by phenylephrine (PE, $p < 0.01$), slightly inhibited constriction induced by potassium chloride (KCl, $p = 0.02$), and significantly decreased constriction induced by 20 V stimulation (*$p < 0.01$, $F(11,58) = 98.75$; one-way ANOVA and Tukey-Kramer multi-comparison test). **b** Guanethidine had no effect on vasoconstriction induced by PE or by KCl, but it blocked the 20 V-induced constriction (*$p < 0.01$). **c** PBZ did not inhibit constriction induced by 150 V stimulation, demonstrating that vasoconstriction by high-voltage bypassed the adrenergic neural pathway. **d** On average, guanethidine failed to block constriction induced by 150 V stimuli. However, the onset of constriction was slowed down (*$p < 0.01$; one-way ANOVA, $F(1,9) = 25.14$). Unless specified, significance was determined using one-way ANOVA and Tukey-Kramer multi-comparison test, $F(11,58) = 98.75$; N = 5 for guanethidine, $N = 6$ for PBZ and N = 7 control. Horizontal lines indicate the maximum constriction achieved with phenylephrine (green) and potassium chloride (black)

so the arterial diameter increased only slightly from 25 ± 2% to 37 ± 4% of the initial ($p < 0.01$; Fig. 4(a) and Fig. 4(c), black). As expected, Guanethidine failed to inhibit constriction induced by phenylephrine because phenylephrine acts downstream of the guanethidine blockade: 45 ± 16% (without guanethidine) compared with 40 ± 3% (with guanethidine) of the initial arterial diameter (Fig. 4(b) and Fig. 4(d), green). Guanethidine also had a limited effect on KCl, which can act directly on smooth muscle cells and constricted the vessel to 25 ± 2% of the initial vessel diameter without guanethidine, compared with 30 ± 4% with guanethidine (Fig. 4(b) and Fig. 4(d), black).

Pretreatment with PBZ reduced arterial constriction with low-voltage (20 V) stimuli and increased the vessel diameter from 38 ± 4% (without PBZ) to 79 ± 7% (with PBZ) of the original ($p < 0.01$; Fig. 4(a), red). Pretreatment with guanethidine completely eliminated arterial constriction with low-voltage (20 V) stimulation—the stimulated vessel diameter increased from 38 ± 4% (without guanethidine) to 98 ± 2% (with guanethidine) of the pre-stimulus value ($p < 0.01$; Fig. 4(b), red). Guanethidine pretreatment also revealed a post-stimulus dilation up to 127% of the initial diameter, which trends back to normal over 15 min.

With high-voltage stimulation, PBZ and guanethidine failed to block arterial vasoconstriction: the vessel diameters were 30 ± 3% of the initial without PBZ or guanethidine, 30 ± 6% with PBZ, and 37 ± 2% with guanethidine (Fig. 4(c) and Fig. 4(d), red). However, in the presence of guanethidine, arterial constriction occurred slower: at 30 s the artery constricted to 28 ± 4% of the initial diameter in the control but just 51 ± 11% of the initial diameter with guanethidine ($p < 0.01$; Fig. 4(d), red).

#### *Neural inhibition: Veins*

The neural agonists and antagonists affected veins quite differently. Phenylephrine (PE) failed to elicit a venous contraction and was not affected by the neural inhibitors: the vein diameters were 97 ± 13% (PE alone), 102 ± 7% (with PBZ) and 92 ± 10% (with guanethidine). PBZ reduced venous constriction with KCl and increased the vessel diameter from 51 ± 21% of the initial to 95 ± 6% with PBZ ($p = 0.02$). The increase to 74 ± 19% with guanethidine was not significant (Fig. 5, black).

PBZ did not affect the venous response to low-voltage stimulation (78 ± 9% of the pre-stimulus diameter compared with 73 ± 15% with PBZ), while guanethidine

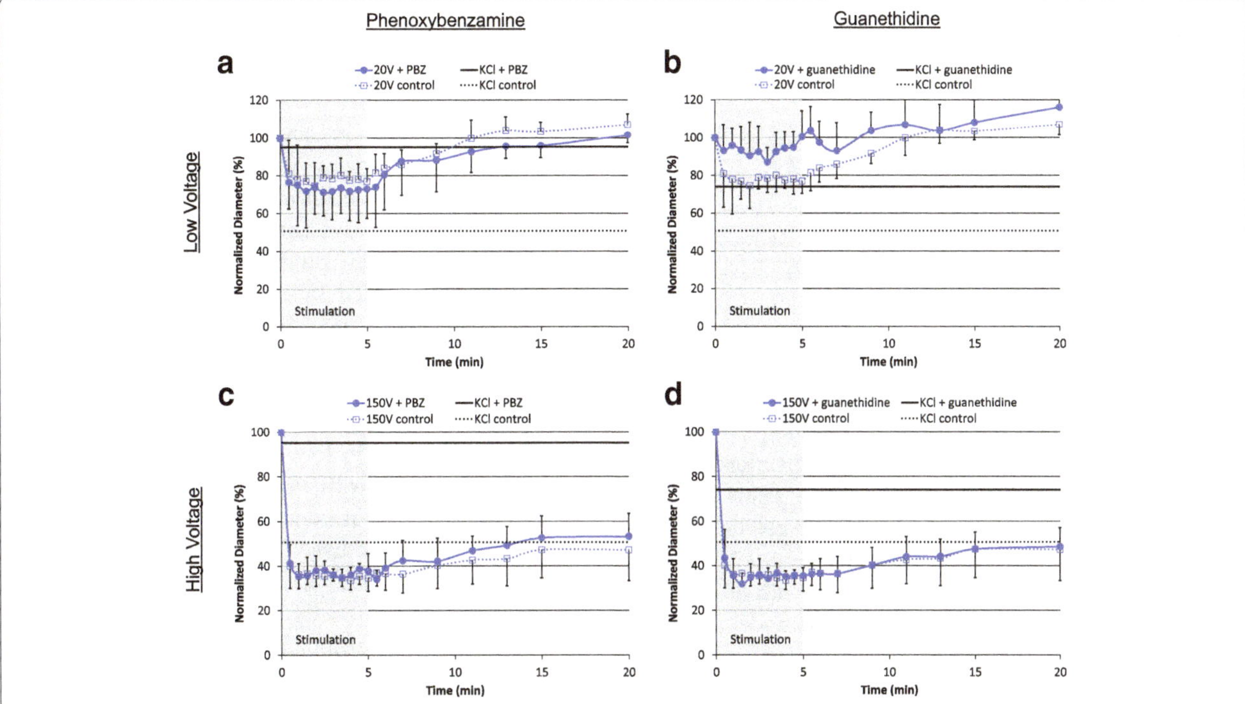


**Fig. 5** Neural inhibition of venous vasoconstriction. **a** PBZ did not block venous constriction induced by 20 V stimulation, but it reduced constriction induced by KCl (p = 0.02). **b** Although guanethidine reduced constriction induced by 20 V stimulation and by KCl, the differences were not significant. **c** PBZ did not affect vasoconstriction induced by 150 V stimulation. **d** Guanethidine did not affect constriction induced by 150 V stimulation. Horizontal lines indicate the maximum constriction achieved with potassium chloride (black). Statistical comparisons were performed using the one-way ANOVA and the Tukey-Kramer multi-comparison test; N = 5 for guanethidine, N = 6 for PBZ and N = 7 for control

reduced constriction, but not significantly (94 ± 10% of the pre-stimulus diameter) (Fig. 5(a) and Fig. 5(b), blue). Neither PBZ nor guanethidine affected venous constriction by high-voltage stimulation: vein diameters were 36 ± 4% of the pre-stimulus diameter (control), compared with 37 ± 4% (PBZ) and 36 ± 5% (guanethidine) (Fig. 5(c) and Fig. 5(d), blue).

## Discussion

### Electrical stimulation of blood vessels

We found that both, high- and low-voltage stimuli constrict saphenous arteries to a similar extent in-vivo, but low-voltage engages a neural pathway that recovers quickly (within 30 s), while high-voltage activates a non-neural pathway that recovers slowly (over several minutes). We also show that high-voltage stimulation constricts veins as much as arteries, but low-voltage constricts only half that amount. These observations suggest that different vasoconstriction pathways could be activated by electrical stimulation.

### Electrical stimulation below damage threshold

Strong electric field can permeabilize and damage cell membranes; however, our cellular viability assay showed no damage to arteries with the 20 and 150 V stimuli. Extremely high electric field (tens of kV/cm) with pulse durations shorter than the typical cell polarization time (~ 50 ns) can selectively polarize intracellular organelles [37]. While damage of intracellular organelles would not be detected with our viability assay, it is unlikely that our stimulus induced any direct action on them because electric field in our case is about 2 orders of magnitude lower than that required for activation of intracellular organelles. Extremely high electric fields can also activate platelets in the blood, which may constrict vessels by releasing thromboxane or serotonin [11, 38, 39], but this pathway is also unlikely due to the substantially lower electric fields in our study.

With 150 V, 0.01 ms stimuli, charge density at the electrode surface ($60\mu C/cm^2$) is close to the capacitive coupling limit for stainless steel-electrolyte interface ($40\text{-}50\mu C/cm^2$), and may be delivered without electrolysis due to surface roughness. For the 20 V, 1 ms stimulus, charge density ($625\mu C/cm^2$) exceeds the capacitive coupling limit, so the current was sustained via electrolysis of water [40]. However, even with electrolysis, it is unlikely that gas byproducts or changes in pH affected vasoconstriction since the electrodes were located several millimeters away from the vessels and warm saline was continuously washing the tissue surface. To avoid hydrolysis in clinical applications, electrodes should have sufficiently high capacitance, such as sputtered iridium oxide films (SIROF), which can safely supply charge densities exceeding $1mC/cm^2$ [40].

### Heating by electrical stimulation

For electric field modeling, we selected the muscle conductivity, so that the total current matched the in-vivo measured current (12.5 mA or 120 mA for the 20 or 150 V electrodes). This approach resulted in a slightly lower value of muscle conductivity for the 20 V stimulus (likely due to gas formation at the electrode-electrolyte interface). Our thermal modeling demonstrated a temperature rise below 1 °C on electrodes, and only 0.15–0.2 °C on the vessel walls, even without considering cooling from convection at the exposed saline surface. Such a minimal heating is very unlikely to induce vasoconstriction since temperature pulsation by a few degrees from a drip of warm saline (37 °C) did not affect the vessel diameter. The vessel heating is similar to our previous reported values (about 2.5 °C with 150 V, 100 μs pulses at 10 Hz and 0.02 °C with 80 V, 1 μs pulses at 10 Hz) [1]; however, our current model predicts even less heating with 150 V pulses because of about 10-fold less charge per pulse and luminal blood flow based on in-vivo measurements. While variations in blood flow affect the modeling results, even without blood flow, the vessels will heat no more than 0.6 and 0.35 °C for 20 V and 150 V stimuli—again less temperature variation than that produced by the dripping 37 °C saline.

### Neural pathways

The thermal modeling and cell viability assay suggest that vasoconstriction was not induced by electroporation or vessel heating. To understand the mechanisms of electrical vasoconstriction we applied the pharmacological inhibitors PBZ and guanethidine. PBZ partially blocks neuro-mediated constriction by preventing the neurotransmitter (norepinephrine) from binding to alpha-1 and alpha-2 receptors on the smooth muscle cells [13], while guanethidine provides a complete neural block by preventing the release of adrenergic (norepinephrine) and purinergic (adenosine triphosphate) neurotransmitters from sympathetic nerves [15].

### Neural inhibition during low-voltage stimulation

In-vivo, low-voltage constriction in arteries was neuro-mediated, with about 65% of the effect due to the adrenergic pathway and additional 30–35% from the purinergic pathway, as evidenced by the partial and complete inhibition with PBZ and guanethidine, respectively. Neural vasoconstriction through adrenergic (dominant) and purinergic pathways was also observed ex-vivo in rat saphenous arteries using similar electrical parameters [13]. Low-voltage stimulation did not depolarize the arterial smooth muscle directly in-vivo (evidenced by complete blockage of vasoconstriction with guanethidine) which confirmed ex-vivo observations [13].

PBZ completely eliminated venous constriction by KCl, while it was only slightly reduced in case of arterial constriction by KCl. This implies that KCl induces venous constriction by depolarizing neurons that release norepinephrine. Because phenylephrine, a pure alpha-1 agonist, did not affect the vein, we conclude that saphenous vein constriction occurs primarily through the alpha-2 receptors, which are activated by norepinephrine, blocked by PBZ and unaffected by phenylephrine. The alpha-2 receptor pathway was also shown to be the dominant venous constriction pathway in dogs [19, 20]. Interestingly, the adrenergic pathway (alpha-1 and -2 receptors) does not appear to be involved in low-voltage venous constriction because pretreatment with PBZ failed to block constriction. Low-voltage venous constriction may involve activation of a purinergic pathway because veins treated with guanethidine constricted less than without purinergic blockage (Fig. 5(b)).

Low-voltage, neural stimulation primarily affects arterial constriction and flow, which could be useful to control hemorrhage [2], blood perfusion or blood pressure in a localized tissue or organ. The neural pathway provides rapid constriction and dilation and can safely constrict vessels for hours [2]. However, chronic stimulation will require electrode materials capable of safely injecting $625\mu C/cm^2$, such as SIROF or TiN [40, 41].

Arterial dilation following low-voltage stimulation was observed most clearly in guanethidine treated vessels (Fig. 4(b)), and it may be mediated by release of nitric oxide or prostaglandins [42, 43]. Because the dilation presented only when the neurotransmitters were blocked, the dilatory effect appears to be overpowered under normal stimulation conditions (no pharmacological blockade). Further studies could determine whether this effect could be exploited to increase blood flow in tissue with poor circulation.

### Neural inhibition during high-voltage stimulation

In-vivo, high-voltage vasoconstriction was not dependent on a neural pathway, since it was not affected by neurotransmitter blockers and confirms previous in-vitro studies showing both arterial and venous constriction in the presence of neural inhibitors [28, 29]. Direct depolarization of smooth muscle with high-voltage stimuli is unlikely because high-voltage constriction persists for several minutes after stimulation, unlike KCl-induced constriction which directly depolarizes smooth muscle and reverses within a minute of rinsing the solution. Furthermore, it has been shown that contractility of smooth muscle decreased rapidly below $165\mu C/cm^2$ per pulse at 20 Hz [44]. Our high-voltage stimulation generates 8-fold less charge density per pulse ($20\mu C/cm^2$) at the arterial wall with half the pulse frequency (10 Hz), further indicating that a direct effect on smooth muscle is unlikely in our case.

High-voltage electrical vasoconstriction may result from release of endothelin-1 by endothelial cells in the lumen of arteries and veins: endothelin-1 constricts vessels to a similar extent as KCl, and does not readily wash-out (vessels remain constricted for more than 10 min) [6, 7, 45, 46]. Endothelial cells under mechanical stress can also release uridine adensosine tetraphosphate and induce potent vasoconstriction [8]. Since vasoconstriction is localized between the electrodes, circulating agents (such as angiotensin) are unlikely to play a role because they would diffuse downstream rather than constrict the vessel only locally.

For some applications, high-voltage, non-neural vasoconstriction has the advantage of constricting veins nearly as much as arteries. This could help control traumatic bleeding in highly perfused tissue, where the major arterial blood supply may be difficult to locate or reach, or in sacral and pelvic cavities where venous hemorrhage can be significant [47–49].

Since high-voltage stimulation uses 40% less energy per pulse, achieves maximum constriction with 10-fold lower pulse frequency [1], and could be applied intermittently because constriction lasts several minutes, it could enable smaller, more power efficient devices for long lasting vessel control. At 1 Hz, high-voltage delivers 14-fold less power than the low-voltage stimulation.

### Limitations

One limitation of this study is that we have not shown safety for clinically relevant durations of stimulation (i.e. greater than 30 min). However, histological examination of the rat saphenous vessels showed no vessel damage one week after a 60-min-long stimulation with identical electrodes at low voltage (20 V, 1 ms pulses at 10 Hz) [2]. In addition, a previous study demonstrated that the threshold of cellular damage by electroporation does not decrease beyond about 50 pulses, suggesting that longer stimulations should also be safe [50].

The DMSO used in the inhibitor experiments extended the arterial recovery time after high-voltage constriction (comparing Fig. 1(a) and Fig. 4(c)). However, it did not affect the extent of constriction, so comparisons between neural inhibitors and their controls are accurate. Even without DMSO (Fig. 1(b)), the vein did not fully dilate 15 min after high-voltage stimulation, perhaps due to lower blood pressure compared with the artery.

Electrical stimulation capable of inducing vasoconstriction also activates the nearby muscles and sensory nerves. Due to depletion of acetylcholine at the neuromuscular junction, muscle contraction with each stimulus pulse decreased over time, and was almost gone after about a minute. Unwanted activation of the muscle and sensory nerves could be reduced by using sensors to identify and stimulate only near the source of hemorrhage, by using neuromuscular blocking agents available during general anesthesia or by intermixing the vasoconstriction stimuli with high pulse frequency stimuli, capable of blocking the pain or completely exhausting the neuromuscular junction.

## Conclusions

Pulsed electrical stimulation provides a reversible and non-damaging approach to blood vessel control in-vivo. Low-voltage stimuli engage neural vasoconstriction pathways, while high-voltage also activates non-neural pathways to induce maximum arterial constriction. The low- and high-voltage stimuli provide different extent of constriction and rates of dilation, which could be useful in a variety of applications for control of bleeding, perfusion, or blood pressure.

**Funding**
This work was funded by the U.S. Department of Defense, Air Force Office of Scientific Research (FA9550–14–1-0074).

**Authors' contributions**
All authors conceived experiments and helped to draft the manuscript. MB performed the experiments and analyzed the data. All authors read and approve the final manuscript

**Competing interest**
The authors declare that they have no competing interests.

**Author details**
[1]Department of Bioengineering, University of Utah, 20 S. 2030 E., Salt Lake City, UT 84112, USA. [2]Faculty of Life Sciences, Bar Ilan University, 5290002 Ramat-Gan, Israel. [3]Department of Ophthalmology, Stanford University, 2452 Watson Court Palo Alto, Stanford, CA 94303, USA. [4]Hansen Experimental Physics Laboratory, Stanford University, 452 Lomita Mall, Stanford, CA 94305, USA.

**References**
1. Mandel Y, et al. "Vasoconstriction by electrical stimulation: new approach to control of non-compressible hemorrhage," (in eng). Sci Rep. 2013;3:2111.
2. Brinton MR, Mandel Y, Dalal R, Palanker D. "Miniature electrical stimulator for hemorrhage control," (in eng). IEEE Trans Biomed Eng. 2014;61(6):1765–71.
3. Mandel Y, et al. Hemorrhage control of liver injury by short electrical pulses. PLoS One. 2013;8(1):e49852.

4. Davis MJ, Hill MA. "Signaling mechanisms underlying the vascular myogenic response," (in eng). *Physiol Rev*. 1999;79(2):387–423.
5. Schubert R, Lidington D, Bolz SS. "The emerging role of Ca2+ sensitivity regulation in promoting myogenic vasoconstriction," (in eng). Cardiovasc Res. 2008;77(1):8–18.
6. Furchgott RF, Vanhoutte PM. "Endothelium-derived relaxing and contracting factors," (in eng). *FASEB J*. 1989;3(9):2007–18.
7. Tykocki NR, Watts SW. "The interdependence of endothelin-1 and calcium: a review," (in eng). Clin Sci (Lond). 2010;119(9):361–72.
8. Jankowski V, et al. "Uridine adenosine tetraphosphate: a novel endothelium-derived vasoconstrictive factor," (in eng). Nat Med. 2005;11(2):223–7.
9. Majumder K, Wu J. "Molecular targets of antihypertensive peptides: understanding the mechanisms of action based on the pathophysiology of hypertension," (in eng). Int J Mol Sci. 2014;16(1):256–83.
10. Zhou Y, Chen Y, Dirksen WP, Morris M, Periasamy M. "AT1b receptor predominantly mediates contractions in major mouse blood vessels," (in eng). Circ Res. 2003;93(11):1089–94.
11. Golino P, et al. "Local platelet activation causes vasoconstriction of large epicardial canine coronary arteries in vivo. Thromboxane A2 and serotonin are possible mediators," (in eng). Circulation. 1989;79(1):154–66.
12. Kur J, Newman EA. "Purinergic control of vascular tone in the retina," (in eng). *J Physiol*. 2014;592(3):491–504.
13. Tarasova O, Sjöblom-Widfeldt N, Nilsson H. "Transmitter characteristics of cutaneous, renal and skeletal muscle small arteries in the rat," (in eng). Acta Physiol Scand. 2003;177(2):157–66.
14. Burnstock G, Ralevic V. "Purinergic signaling and blood vessels in health and disease," (in eng). Pharmacol Rev. 2014;66(1):102–92.
15. Burnstock G, Warland JJ. "A pharmacological study of the rabbit saphenous artery in vitro: a vessel with a large purinergic contractile response to sympathetic nerve stimulation," (in eng). Br J Pharmacol. 1987;90(1):111–20.
16. Sjöblom-Widfeldt N, Nilsson H. "Sympathetic transmission in small mesenteric arteries from the rat: influence of impulse pattern," (in eng). Acta Physiol Scand. 1990;138(4):523–8.
17. Ferrell WR, Khoshbaten A. Responses of blood-vessels in the rabbit knee to electrical-stimulation of the joint capsule. J Physiol. 1990;423:569–78.
18. Khoshbaten A, Ferrell WR. Nerve-mediated responses of blood-vessels in the rabbit knee-joint. J Vasc Res. 1993;30(2):102–7.
19. De Mey J, Vanhoutte PM. "Uneven distribution of postjunctional alpha 1- and alpha 2-like adrenoceptors in canine arterial and venous smooth muscle," (in eng). *Circ Res*. 1981;48(6 Pt 1):875–84.
20. MacLennan SJ, Luong LA, Jasper JR, To ZP, Eglen RM. "Characterization of alpha 2-adrenoceptors mediating contraction of dog saphenous vein: identity with the human alpha 2A subtype," (in eng). *Br J Pharmacol*. 1997;121(8):1721–9.
21. Dunn WR, McGrath JC, Wilson VG. "Postjunctional alpha-adrenoceptors in the rabbit isolated distal saphenous artery: indirect sensitivity to prazosin of responses to noradrenaline mediated via postjunctional alpha 2-adrenoceptors," (in eng). *Br J Pharmacol*. 1991;103(2):1484–92.
22. Donoso MV, Steiner M, Huidobro-Toro JP. "BIBP 3226, suramin and prazosin identify neuropeptide Y, adenosine 5'-triphosphate and noradrenaline as sympathetic cotransmitters in the rat arterial mesenteric bed," (in eng). J Pharmacol Exp Ther. 1997;282(2):691–8.
23. Cortés V, et al. "Synergism between neuropeptide Y and norepinephrine highlights sympathetic cotransmission: studies in rat arterial mesenteric bed with neuropeptide Y, analogs, and BIBP 3226," (in eng). J Pharmacol Exp Ther. 1999;289(3):1313–22.
24. Cheung DW. "Neuropeptide Y Potentiates specifically the purinergic component of the neural responses in the Guinea pig saphenous artery," (in eng). Circ Res. 1991;68(5):1401–7.
25. Fabi F, Argiolas L, Ruvolo G, del Basso P. "Neuropeptide Y-Induced potentiation of noradrenergic vasoconstriction in the human saphenous vein: involvement of endothelium generated thromboxane," (in eng). Br J Pharmacol. 1998;124(1):101–10.
26. Paterson G. "The response to transmural stimulation of isolated arterial strips and its modification by drugs," (in eng). J Pharm Pharmacol. 1965;17:341–9.
27. Yates CM, Gillis CN. "The response of rabbit vascular tissue to electrical and drug stimulation," (in eng). J Pharmacol Exp Ther. 1963;140:52–9.
28. Furchgott RF. "The pharmacology of vascular smooth muscle," (in eng). Pharmacol Rev. 1955;7(2):183–265.
29. Somlyo AV, Somlyo AP. "Electromechanical and pharmacomechanical coupling in vascular smooth muscle," (in eng). J Pharmacol Exp Ther. 1968; 159(1):129–45.
30. Hughes J, Vane JR. Relaxations of isolated portal vein of rabbit induced by nicotine and electrical stimulation. Br J Pharmacol. 1970;39(3):476–89.
31. Torok J, Zemancikova A. Relaxation of rat main pulmonary artery to electrical stimulation: role of nitric oxide. Act Nerv Super Rediviva. 2010;52(2):151–6.
32. Guarini S. A highly reproducible model of arterial thrombosis in rats. J Pharmacol Toxicol Methods. 1996;35(2):101–5.
33. Hladovec J. "Experimental arterial thrombosis in rats with continuous registration," (in eng). Thromb Diath Haemorrh. Oct 1971;26(2):407–10.
34. Bourgain RH, F S. A continuous registration method in experimental arterial thrombosis in the rat. Thromb Res. 1974;4:599–607.
35. Jespersen B, Knupp L, Northcott CA. "Femoral arterial and venous catheterization for blood sampling, drug administration and conscious blood pressure and heart rate measurements," (in eng). Vis Exp. 2012;(59). doi: https://doi.org/10.3791/3496.
36. Pennes HH. "Analysis of tissue and arterial blood temperatures in the resting human forearm," (in eng). *J Appl Physiol*. 1948;1(2):93–122.
37. Vernier PT, Sun Y, Marcu L, Salemi S, Craft CM, Gundersen MA. "Calcium bursts induced by nanosecond electric pulses," (in eng). Biochem Biophys Res Commun. 2003;310(2):286–95.
38. Torres AS, et al. "Platelet activation using electric pulse stimulation: growth factor profile and clinical implications," (in eng). *J Trauma Acute Care Surg*. 2014;77(3 Suppl 2):S94–S100.
39. Frelinger AL, et al. "Modification of pulsed electric field conditions results in distinct activation profiles of platelet-rich plasma," (in eng). *PLoS One*. 2016; 11(8):e0160933.
40. Merrill DR, Marom B, GR JJ. Electrical stimulation of excitable tissue: design of efficacious and safe protocols. J Neurosci Methods. 2005;141(2):171–98.
41. Cogan SF. Neural stimulation and recording electrodes. Annu Rev Biomed Eng. 2008;10:275–309.
42. Zhang W, Edvinsson L, Lee TJ. "Mechanism of nicotine-induced relaxation in the porcine basilar artery," (in eng). J Pharmacol Exp Ther. 1998;284(2):790–7.
43. Tartas M, et al. Cathodal current-induced vasodilation to single application and the amplified response to repeated application in humans rely on aspirin-sensitive mechanisms. J Appl Physiol (1985). 2005;99(4):1538–44.
44. van Mastrigt R, Glerum JJ. "Electrical stimulation of smooth muscle strips from the urinary bladder of the pig," (in eng). *J Biomed Eng*. 1985;7(1):2–8.
45. Burnstock G. "Endothelium-derived vasoconstriction by purines and pyrimidines," (in eng). Circ Res. 2008;103(10):1056–7.
46. Yanagisawa M, et al. "A novel potent vasoconstrictor peptide produced by vascular endothelial cells," (in eng). Nature. 1988;332(6163):411–5.
47. Lallemand MS, et al. "Resuscitative endovascular balloon occlusion of the aorta for major abdominal venous injury in a porcine hemorrhagic shock model," (in eng). *J Trauma Acute Care Surg*. 2017;83(2):230–6.
48. Weber DG, Bendinelli C, Balogh ZJ. "Damage control surgery for abdominal emergencies," (in eng). *Br J Surg*. 2014;101(1):e109–18.
49. Costantini TW, et al. "Current management of hemorrhage from severe pelvic fractures: results of an American Association for the Surgery of Trauma multi-institutional trial," (in eng). *J Trauma Acute Care Surg*. 2016; 80(5):717–23; discussion 723–5.
50. Butterwick A, Vankov A, Huie P, Freyvert Y, Palanker D. Tissue damage by pulsed electrical stimulation. IEEE Trans Biomed Eng. 2007;54(12):2261–7.
51. Bronzino JD. The biomedical engineering handbook, 2nd ed. (electrical engineering handbook series). Boca Raton, FL: CRC Press; 2000. p. 2. volumes
52. Duck F. Physical Properties of Tissues: A Comprehensive Reference Book. London: Academic Press; 1990.
53. Duck FA. Physical properties of tissue : a comprehensive reference book. London: Academic Press; 1990.
54. K. Giering, I. Lamprecht, and O. Minet, "Specific heat capacities of human and animal tissues," Proc SPIE 2624, Laser-Tissue Interaction and Tissue Optics, vol. 188, 1996.
55. Fegler G. "Measurement of cardiac output in anaesthetized animals by a thermodilution method," (in eng). Q J Exp Physiol Cogn Med Sci. 1954;39(3):153–64.
56. Christensen DA. Ultrasonic bioinstrumentation. New York: Wiley; 1988. pp. xiii, 235 pages

# Permissions

All chapters in this book were first published in JNER, by BioMed Central; hereby published with permission under the Creative Commons Attribution License or equivalent. Every chapter published in this book has been scrutinized by our experts. Their significance has been extensively debated. The topics covered herein carry significant findings which will fuel the growth of the discipline. They may even be implemented as practical applications or may be referred to as a beginning point for another development.

The contributors of this book come from diverse backgrounds, making this book a truly international effort. This book will bring forth new frontiers with its revolutionizing research information and detailed analysis of the nascent developments around the world.

We would like to thank all the contributing authors for lending their expertise to make the book truly unique. They have played a crucial role in the development of this book. Without their invaluable contributions this book wouldn't have been possible. They have made vital efforts to compile up to date information on the varied aspects of this subject to make this book a valuable addition to the collection of many professionals and students.

This book was conceptualized with the vision of imparting up-to-date information and advanced data in this field. To ensure the same, a matchless editorial board was set up. Every individual on the board went through rigorous rounds of assessment to prove their worth. After which they invested a large part of their time researching and compiling the most relevant data for our readers.

The editorial board has been involved in producing this book since its inception. They have spent rigorous hours researching and exploring the diverse topics which have resulted in the successful publishing of this book. They have passed on their knowledge of decades through this book. To expedite this challenging task, the publisher supported the team at every step. A small team of assistant editors was also appointed to further simplify the editing procedure and attain best results for the readers.

Apart from the editorial board, the designing team has also invested a significant amount of their time in understanding the subject and creating the most relevant covers. They scrutinized every image to scout for the most suitable representation of the subject and create an appropriate cover for the book.

The publishing team has been an ardent support to the editorial, designing and production team. Their endless efforts to recruit the best for this project, has resulted in the accomplishment of this book. They are a veteran in the field of academics and their pool of knowledge is as vast as their experience in printing. Their expertise and guidance has proved useful at every step. Their uncompromising quality standards have made this book an exceptional effort. Their encouragement from time to time has been an inspiration for everyone.

The publisher and the editorial board hope that this book will prove to be a valuable piece of knowledge for researchers, students, practitioners and scholars across the globe.

# List of Contributors

**Philippe Malcolm, Samuel Galle, Pieter Van den Berghe and Dirk De Clercq**
Department of Biomechanics and Center for Research in Human Movement Variability, University of Nebraska Omaha, Omaha, NE 68182, USA
Department of Movement and Sports Sciences, Ghent University, B-9000 Ghent, Belgium

**Taeyong Sim, Hakje Yoo, Dongjun Lee, Hyunggun Kim and Joung Hwan Mun**
Department of Bio-Mechatronic Engineering, College of Biotechnology and Bioengineering, Sungkyunkwan University, Natural Sciences Campus, 2066 Seobu-ro, Jangan-gu, Suwon-si, Gyeonggi-do 16419, South Korea

**Seung-Woo Suh and Jae Hyuk Yang**
Department of Orthopedics, Scoliosis Research Institute, Korea University Medical College, Guro Hospital, 148 Gurodong-ro, Guro-gu, Seoul 08308, South Korea

**Iason Batzianoulis and Aude Billard**
Learning Algorithms and Systems Laboratory (LASA), School of Engineering, École Polytechnique Fédérale de Lausanne (EPFL), Route Cantonale, CH-1015 Lausanne, Switzerland

**Nili E. Krausz, Ann M. Simon and Levi Hargrove**
Center for Bionic Medicine, Shirley Ryan AbilityLab, E Erie St., 60611 Chicago, IL, USA

**Nili E. Krausz, Ann M. Simon and Levi Hargrove**
Dept. of Physical Medicine and Rehabilitation, Northwestern University, N Lake Shore, 60611 Chicago, IL, USA

**Levi Hargrove**
Dept. of Biomedical Engineering, Northwestern University, 60208 Evanston, IL, USA

**Gyrd Thrane**
Department of Health and Care Sciences, UiT The Arctic University of Norway, Tromsø, Norway

**Gyrd Thrane, Margit Alt Murphy and Katharina Stibrant Sunnerhagen**
Institute of Neuroscience and Physiology, Rehabilitation Medicine, Sahlgrenska Academy, University of Gothenburg, Gothenburg, Sweden

**Katharina Stibrant Sunnerhagen**
Center for Person-Centered Care (GPCC), University of Gothenburg, Gothenburg, Sweden

**Andrew Sexton and Chris A. McGibbon**
Institute of Biomedical Engineering, University of New Brunswick, Fredericton, NB E3B 5A3, Canada

**Alyssa Whelan, Colleen O'Connell and Chris A. McGibbon**
Faculty of Kinesiology, University of New Brunswick, Fredericton, NB, Canada

**Melony Jones and Colleen O'Connell**
Stan Cassidy Centre for Rehabilitation, Fredericton, NB, Canada

**Cristina Bayón, Óscar Ramírez and Eduardo Rocon**
Centre for Automation and Robotics (CAR), CSIC-UPM, Ctra Campo Real km 0.2 - La Poveda-Arganda del Rey, 28500 Madrid, Spain

**Teresa Martín-Lorenzo, Beatriz Moral-Saiz and Álvaro Pérez-Somarriba**
Hospital Infantil Universitario Niño Jesús, Madrid, Spain

**Teresa Martín-Lorenzo**
Universidad Rey Juan Carlos, Madrid, Spain

**Beatriz Moral-Saiz and Sergio Lerma-Lara**
Universidad de La Salle, Madrid, Spain

**Eduardo Rocon**
Universidade Federal do Espírito Santo Vitória, Brazil

**Michael D. Wood, Leif E. R. Simmatis, J. Gordon Boyd and Stephen H. Scott**
Centre for Neuroscience Studies, Queen's University, Botterell Hall, 18 Stuart St, Kingston, ON, Canada

**J. Gordon Boyd**
Department of Critical Care Medicine, Queen's University, Kingston, ON, Canada
Department of Medicine, Queen's University, Kingston, ON, Canada

**Jill A. Jacobson**
Department of Psychology, Queen's vUniversity, 318 Craine Hall, 62 Arch Street, Kingston, ON K7L 3N6, Canada

**Angelos Karatsidis, Jason M. Konrath, H. Martin Schepers and Giovanni Bellusci**
Xsens Technologies B.V, Pantheon 6, 7521 PR Enschede, The Netherlands

**Angelos Karatsidis and Peter H. Veltink**
Department of Biomedical Signals and Systems (BSS), Technical Medical Centre, University of Twente, Enschede, The Netherlands

**Rosie E. Richards, Josien C. van den Noort and Jaap Harlaar**
Department of Rehabilitation Medicine, Amsterdam Movement Sciences, VU University Medical Center, Amsterdam, The Netherlands

**Josien C. van den Noort**
Academic Medical Center, Musculoskeletal Imaging Quantification Center (MIQC), Department of Radiology and Nuclear Medicine, Amsterdam Movement Sciences, Amsterdam, The Netherlands

**Jaap Harlaar**
Department of Biomechanical Engineering, Delft University of Technology, Delft, The Netherlands

**Max K. Shepherds**
Shirley Ryan AbilityLab, Room 11-1414, 355 E Erie St, Chicago, IL 60611, USA
Department of Biomedical Engineering, Northwestern University, 663 Clark St, Evanston, IL 60208, USA

**Max K. Shepherd, Alejandro F. Azocar and Elliott J. Rouse**
Neurobionics Lab, University of Michigan, Ann Arbor, MI 48109, USA

**Alejandro F. Azocar, Alejandro F. Azocar and Elliott J. Rouse**
Department of Mechanical Engineering, University of Michigan, Ann Arbor, MI 48109, USA
Robotics Institute, University of Michigan, Ann Arbor, MI 48109, USA

**Matthew J. Major**
Northwestern University Feinberg School of Medicine, Department of Physical Medicine and Rehabilitation, 710 North Lake Shore Drive, #1022, Chicago, IL 60611, USA

**Adam Trepczynski, Ines Kutzner, Verena Schwachmeyer and Georg N. Duda**
Julius Wolff Institute, Charité - Universitätsmedizin Berlin, corporate member of Freie Universität Berlin, Humboldt-Universität zu Berlin, and Berlin Institute of Health, Augusten burger Platz 1, 13353 Berlin, Germany

**Markus O. Heller**
Bioengineering Sciences Research Group, Faculty of Engineering and Physical Sciences, University of Southampton, Southampton, UK

**Tilman Pfitzner**
Center for Musculoskeletal Surgery, Charité - Universitätsmedizin Berlin, corporate member of Freie Universität Berlin, Humboldt-Universität zu Berlin, and Berlin Institute of Health, Berlin, Germany
Clinic for Adult Hip and Knee Reconstruction, Vivantes Spandau Hospital, Berlin, Germany

**Christopher Schmidt Easthope, Luca Renato Traini, Lea Awai, Martina Franz, Armin Curt and Marc Bolliger**
Spinal Cord Injury Center, Balgrist University Hospital, Forchstrasse 340, CH-8008 Zürich, Switzerland

**Lea Awai**
Sobell Department of Motor Neuroscience and Movement Disorders, University College London, London, UK

**Georg Rauter**
BIROMED-Lab, Department of Biomedical Engineering, University Basel, Basel, Switzerland

**Néstor J. Jarque-Bou, Margarita Vergara, Joaquín L. Sancho-Bru, Alba Roda-Sales and Verónica Gracia-Ibáñez**
Department of Mechanical Engineering and Construction, Universitat Jaume I, Avinguda Vicent Sos Baynat, s/n., 12071 Castellón, Spain

**Jacob Rammer, Susan Riedel and Gerald Harris**
Orthopaedic and Rehabilitation Engineering Center (OREC), Marquette University, Olin Engineering Suite 323, Milwaukee, WI 53201-1881, USA
Department of Biomedical Engineering, Marquette University, Olin Engineering Suite 323, Milwaukee, WI 53201-1881, USA
Department of Orthopaedic Surgery, Medical College of Wisconsin, Milwaukee, WI 53201-1881, USA

**Joseph Krzak and Gerald Harris**
Shriners Hospitals for Children, Chicago, IL, USA

**Joseph Krzak**
Midwestern University, Physical Therapy Program, 555 31st St., Alumni Hall 340C, Downers Grove, IL 60515, USA

**Brooke Slavens**
University of Wisconsin-Milwaukee, 2400 E Hartford Ave, Rm. 983, Milwaukee, WI 53211, USA

**Jack Winters**
Marquette University, Biomedical Engineering, Milwaukee, WI 53201-1881, USA

**Guangtao Chen, Le Ma and Rong Song**
Key Laboratory of Sensing Technology and Biomedical Instrument of Guangdong Province, School of Engineering, Sun Yat-sen University, Guangzhou, Guang Dong, China

**Le Li**
Department of Rehabilitation Medicine, Sun Yat-sen First Affiliated Hospital, Sun Yat-sen University, Guangzhou, China

**Xiaoyun Wang**
Guangdong Work Injury Rehabilitation Center, Guangzhou, China

**Kaiyu Tong**
Department of Biomedical Engineering, the Chinese University of Hong Kong, Hong Kong, China

**Louise Thomas and Carolyn J. Sparrey**
School of Mechatronic Systems Engineering, Simon Fraser University, SFU Surrey Campus, 250-13450 102 Ave, Surrey, BC, Canada

**Jaimie Borisoff**
British Columbia Institute of Technology, BCIT Centre for Applied Research and Innovation, 4355 Mathissi Pl, Burnaby, BC, Canada

**Louise Thomas, Jaimie Borisoff and Carolyn J. Sparrey**
International Collaboration on Repair Discoveries (ICORD), Blusson Spinal Cord Centre, 818 West 10th Avenue, Vancouver, BC V5Z 1M9, Canada

**K. K. A. Bakhti, I. Laffont, M. Muthalib, J. Froger and D. Mottet**
Euromov, University of Montpellier, Montpellier, France

**K. K. A. Bakhti and I. Laffont**
Physical Medicine and Rehabilitation, Montpellier University Hospital, Montpellier, France

**M. Muthalib**
Silverline Research, Brisbane, Australia

**J. Froger**
Physical Medicine and Rehabilitation, Nîmes University Hospital, Le Grau du Roi, France

**K. K. A. Bakhti, I. Laffont, J. Froger and D. Mottet**
Federative Institute for Research on Handicap, Paris, France

**Amber R. Emmens, Edwin H. F. van Asseldonk and Herman van der Kooij**
Department of Biomechanical Engineering, University of Twente, Drienerlolaan 5, 7522 Enschede, NB, the Netherlands

**Mark Brinton**
Department of Bioengineering, University of Utah, 20 S. 2030 E., Salt Lake City, UT 84112, USA

**Yossi Mandel**
Faculty of Life Sciences, Bar Ilan University, 5290002 Ramat-Gan, Israel

**Ira Schachar and Daniel Palanker**
Department of Ophthalmology, Stanford University, 2452 Watson Court Palo Alto, Stanford, CA 94303, USA

**Daniel Palanker**
Hansen Experimental Physics Laboratory, Stanford University, 452 Lomita Mall, Stanford, CA 94305, USA

# Index


A
Adenosine Triphosphate, 209, 212, 216
Adolescent Idiopathic Scoliosis, 12, 21-22
Adrenergic Pathways, 209
Angular Velocity, 2-3, 5, 26, 28, 32, 37-38, 40-43, 52, 155,
Ankle-foot Orthosis, 10-11, 196-197
Anteroposterior, 196, 199, 201

B
Body Mass Index, 86

C
Carpometacarpal, 144, 147
Center of Pressure, 12-13, 136, 199
Central Nervous System, 18, 21, 58, 70, 139, 192
Cerebral Palsy, 46, 50-51, 56, 58, 70-72, 195

D
Discrete Wavelet Transform, 12-13, 16
Dorsiflexion Stiffness, 108-109, 114
Dysvascular, 109, 112-113

E
Elbow Extension, 25-26, 31, 36-37, 41, 43, 187, 193, 195
Electrical Vasoconstriction, 209, 216-217
Electroceuticals, 209
Electrode, 48, 99-100, 102-103, 138, 140, 148, 162, 171, 211-212, 215-216
Electromyography, 23, 35, 70, 125, 138, 147-148, 161-162, 170-171
Exoskeleton, 1-6, 8-11, 58-61, 63, 65, 71, 83, 196-197, 203, 207-208

G
Gait Pattern, 10, 59, 61, 63, 85, 93-94, 126-127, 135
Gait Training Therapy, 58
Gastrocnemii, 117, 120-121, 123
Grasping Intention, 23-26, 28, 31, 34
Guanethidine, 209-210, 212, 214-216

H
Hemiplegia, 56, 97, 101, 138, 195
Hemorrhage, 106, 216-218

J
Joint Contact Forces, 95, 117, 123, 125

K
Kinematics, 1, 3, 6, 8, 34, 36, 41, 43, 45, 64, 67, 69, 71-72, 86-88, 111-115, 118-119, 125-127, 132, 135-137, 146, 149-152, 154, 164, 173, 175, 186, 190, 193-197, 202-203, 207
Kinesiophobia, 64-65, 68
Knee Flexor, 46-47, 52
Knee Osteoarthritis, 85, 94-95, 117, 125

L
Limb Spasticity, 56-57
Locomotion, 10-11, 107-108, 124-127, 134-137, 195, 204

M
Manual Wheelchair, 149, 151, 155-160, 172-173, 177-178, 181-183
Markerless Motion Capture, 149-150, 160, 194-195
Metacarpophalangeal, 144, 147
Mobility, 1, 58, 61, 68, 74, 107, 110-112, 115, 117, 124, 149-150, 154-155, 158, 172-173, 177-178, 181-183, 207
Motor Impairment, 36, 38, 106, 138, 147, 195
Muscle Co-contraction, 117-118, 123-125
Muscle Forces, 117-118, 120-121, 123-125, 171
Muscle Spasticity, 46-47, 50-52, 56, 106
Musculoskeletal Loading Conditions, 117
Musculoskeletal Models, 117-118, 149, 151
Musculotendon Excursion, 149, 154-156
Myoelectric Control, 23, 34-35, 139, 148
Myoelectric Prosthesis, 25, 34

N
Neural Pathways, 209, 212, 216
Neurological Rehabilitation, 97, 160, 194
Neuron Injury, 46
Norepinephrine, 209, 212, 216, 218

O
Orthosis, 1, 10-11, 170, 196-197, 207

P
Pediatric Rehabilitation, 149
Pendulum Test, 46-57
Phantom Limb, 25, 34

Phenoxybenzamine, 209-210, 212, 214
Phenylephrine, 209-210, 212, 214, 216
Post-stroke Pain, 98, 105
Postural Stability, 12-13, 15, 18, 20-22, 207
Proximal, 11, 28, 33, 35, 74, 106, 119, 141, 144, 146-147, 151, 184-187, 189, 191-194, 210
Purinergic Pathway, 209, 216

R
Relaxation Index, 46, 49, 52-53

S
Scoliosis, 12, 21-22
Sensory Impairments, 37, 98, 220
Sensory System, 12-16, 18-21
Somatosensory System, 13, 18-20
Spasticity, 46-57, 62, 99, 101, 106, 194
Spinal Cord Injury, 22, 46-47, 50-51, 56-57, 126, 136-137, 147, 150, 158-159, 162, 170, 206-207
Stroke, 2, 9-11, 22, 36-46, 50-51, 56-57, 66, 73, 82-84, 96-106, 132, 134, 136-138, 147, 150-151, 159, 161-165, 167-171, 184-185, 187, 189, 191-195, 207
Support Vector Machines, 23

T
Telerehabilitation, 97-98, 102-105, 158
Transcranial Direct Current Stimulation, 97, 105-106

U
Unilateral Dominant, 4-6
Upper Extremity, 36-37, 40-41, 43-44, 99, 102, 106, 139, 147, 149-152, 154, 157-160, 185, 194-195
Upper Limb Prosthesis, 23, 34-35

V
Vasoconstriction, 209-210, 212, 214-218
Vestibular, 12-22, 112
Vestibular System, 12-13, 20-21, 112

W
Waveform, 26, 97, 101, 128
Wheelchair Biomechanics, 149, 156

Z
Zero Impedance, 196, 198, 203

CPSIA information can be obtained
at www.ICGtesting.com
Printed in the USA
LVHW061657270323
742706LV00003B/269

9 781632 429414